Proceedings in Life Sciences

Plant Growth Substances 1985

Proceedings of the 12th International Conference on Plant Growth Substances, Held at Heidelberg, August 26-31, 1985

Edited by Martin Bopp

With 135 Figures

Springer-Verlag
Berlin Heidelberg New York Tokyo

Professor Dr. MARTIN BOPP
Botanisches Institut der Universität
Im Neuenheimer Feld 360
6900 Heidelberg, FRG

ISBN-13: 978-3-642-71020-9 e-ISBN-13: 978-3-642-71018-6
DOI: 10.1007/978-3-642-71018-6

Library of Congress Cataloging-in-Publication Data. International Conference on Plant Growth Substances (12th:1985:Heidelberg, Germany). Plant growth substances 1985. (Proceedings in life sciences). Bibliography: p. Includes index. 1. Plant hormones–Congresses. 2. Plant regulators–Congresses. I. Bopp, Martin, 1923– . II. Title. III. Series. QK731.I55 1985 581.3 86-10219.

Softcover reprint of the hardcover 1st edition 1986

Offsetprinting and bookbinding: Brühlsche Universitätsdruckerei, Giessen
2131/3130-543210

Preface

The 12th International Conference on Plant Growth Substances was held from 26th to 31st August 1985 in Heidelberg, F.R.G., under the auspices of the IPGSA (International Plant Growth Substances Association) and the University of Heidelberg in its 599th year.

As many as 750 participants from 40 countries all over the world attended the conference, including guests and staff members of the local organizers. Fine days provided an excellent background for a fruitful and pleasant meeting and all the activities accompanying the scientific programme. During the conference all current aspects concerning growth substances were treated. Altogether the participants presented 207 oral reports organized in four parallel sessions and about 300 posters, for which 2 hours' poster sessions were reserved each day. The conference gained in perspective from the arrangement of five workshops in which special aspects and the most recent results could be presented by specialists in the particular fields. The topics of the workshop were: actual methods of hormone detection (organizer H. Kende), auxin transport (organizer R. Hertel), growth substances and tumour formation (organizer J. Schröder), evolution of the hormone system (organizer W. Jacobs) and problems of application (organizer J. Jung). The abstracts of all presentations were collected in a Book of Abstracts available during the conference, giving a rough survey of the whole field of plant growth substances in its present state.

The present volume of Proceedings in the Life Sciences series follows two volumes from earlier meetings (1976, 1979), and should be used as a guideline to the development of research in growth substances. It contains all invited lectures, including the reports from the workshops. To cover the many different aspects, the contributions of the meeting were divided into 23 sections, for each of which one main speaker was invited. Therefore we hope that no important aspect has been totally neglected, even if certain chapters may be missing. Questions of hormone receptor substances are, for example, not the subject of active research at the moment, and we have omitted this aspect. Also the mechanisms of hormone effects – how the hormones do their job – is with few exceptions almost completely unkown and even some of these exceptions, like auxin effects on elongation, have recently been questioned. The organization of the volume does not follow the time schedule of the conference, but is composed so that related topics are treated in one chapter. The lectures of the workshops are included in the relevant chapters. It is, of course, impossible to arrange all contributions in a completely homogeneous way. Nevertheless, we hope that the book may serve as a source of information in the field of plant hormones.

It is a great pleasure for me to express my thanks to all the speakers who have prepared their manuscripts for this volume; to all members of the local organization committee for their help before and during the conference: Dr. B. Knoop, Dr. W. Rademacher and Dr. B. Zbell; Prof. I. Essigmann-Capesius, who organized the entertainment; Dr. G. Bopp-Hassenkamp and Mrs. H. Knoop, responsible for the Ladies' Programme. Many members of the staff of the Botanical Institute worked hard to prepare the conference and its smooth organization, above all, my secretary Mrs. B. Nickel.

The conference would not have been such a success without the financial support of the Deutsche Forschungsgemeinschaft, the Land Baden-Württemberg, the P.F. Saunders Memorial Fund and the following organizations: Amersham Buchler GmbH & Co. KG, Braunschweig; ARCO Solar, Inc., California; BASF, Ludwigshafen; Bayer AG, Leverkusen; Boehringer Mannheim GmbH, Mannheim; Celamerck GmbH & Co. KG, Ingelheim/Rhein; Ciba-Geigy AG, Division AGRO, Basel; FESPP (Federation of European Societies of Plant Physiology); Hoechst AG, Frankfurt; Kali und Salz AG, Kassel; Schering AG, Berlin; F. Springer Verlag, Heidelberg; Union Carbide Agricultural Products Company, Inc., Research Triangle Park, N.C. 27709, USA.

Finally I wish to express my gratitude to the members of Springer-Verlag for their support during the preparation of this volume.

Heidelberg, June 1986 MARTIN BOPP

Contents

IV. Hormones: Effects

Contributors

You will find the addresses at the beginning of the respective contributions

Plant Cell Responses and the Role of Growth Substances*

P.F. WAREING[1, 2]

1 Introduction

It is commonplace that there is a bewildering variety of effects of applied plant growth regulators on plant tissues, and there is a mass of observational and descriptive data on plant cell responses, but very little understanding of the underlying events and processes, and almost no unifying concepts. In spite of the wide variety of responses, we tend to generalize from observations made on a single type of response, but I shall argue that we need to be much more discriminating between different types. I propose, therefore, to examine some of the characteristics of plant cell responses especially in relation to the role of growth substances.

The early studies on the role of auxin in phototropism and geotropism, especially in the *Avena* coleoptile, led to the Cholodny-Went hypothesis, according to which these tropic growth curvatures could be attributed to the differential distribution of auxins on opposite sides of coleoptiles and roots. These early studies in which an apparently good correlation could be established between endogenous growth substance levels and growth responses, led to a more general concept of hormonal control of growth and development, which embraced the later-discovered classes of regulators [18].

However, it has to be admitted that many subsequent attempts to demonstrate a clear correlation between endogenous growth substance levels and developmental processes have proved indecisive. One possible explanation of the lack of correlation appeared to be attributable to the crudity of the techniques available for determining endogenous growth substance levels in plant tissues, but even with enormous improvements in analytical techniques, such as gas chromatography-mass spectrometry and immunoassay methods, which provide a high degree of accuracy in the quantitation of specific endogenous growth substance, correlations between endogenous levels and developmental processes still remains ambiguous in many cases. Moreover, even where variations in endogenous growth substance levels have been found [22], it has been argued that the magnitude of the changes is not sufficient to account for the associated growth responses (p. 4).

This situation has led to a high degree of scepticism in certain quarters, and to the assertion that modulation of endogenous growth substance levels is not important

* Introductory lecture of the president of IPGSA

1 Department of Botany and Microbiology, University College of Wales, Aberystwyth, Dyfed, Great Britain

Plant Growth Substances 1985
(ed. by M. Bopp)

in regulating growth and development, and it has been suggested that it is the modulation of *sensitivity* to growth substances, which is the primary controlling factor in growth and development [2, 20, 21].

2 Sensitivity

It is obvious that a tissue must be sensitive to applied plant growth regulators before we can obtain a response of any interest. As is well-known, tissues vary in their sensitivity during the successive phases of cell division, vacuolation and maturation [29].

The concept of sensitivity has been recognized for at least 50 years, and was clearly stated by Went and Thimann [28]. Nevertheless, even though the importance of sensitivity is not in doubt, until now it has not been clear how we could investigate it further, since the necessary techniques have not been available. Until we know the molecular basis of cell maturation, we will not be able to study and understand what processes underlie variation in sensitivity.

It is frequently assumed that variation in sensitivity depends upon the levels of growth substance receptors. The recent findings of Sanders et al. [14] that there is a close correlation between sensitivity and ethylene binding in pea epicotyls is consistent with the hypothesis. However, other possible bases for variation in sensitivity can be envisaged. For example, variation in the sensitivity of a coleoptile to IAA depends upon changes in the plastic properties of the cell wall during cell vacuolation and maturation, but there is no evidence that these changes are necessarily controlled by variation in the levels of IAA receptor, and it is not difficult to envisage other mechanisms.

Variation in sensitivity is undoubtedly an important factor in plant responses to growth substances, but to abandon the idea of any possible regulatory role for growth substances may be "throwing out the baby with the bath water" and I wish to examine briefly possible alternative explanations for the lack of correlation between observed endogenous growth substance levels and developtmental events.

3 The Diversity of Responses to Growth Substances

Firstly, there is a very wide diversity of responses to growth substances, and I suggest that we need to examine the characteristics of the various types and to distinguish between them, instead of lumping them all under one head. There are at least five contrasting categories of response, which can be recognized, as follows:

1. quantitative v^{s} qualitative,
2. all-or-nothing v^{s} graded,
3. reversible v^{s} irreversible,
4. inductive v^{s} non-inductive,
5. differential growth v^{s} turgor changes.

These differences occur in various combinations, such as irreversible, graded growth responses (e.g. in coleoptiles) and reversible, turgor-mediated graded responses (e.g. in stomatal movements), so that the overall diversity of plant responses is clearly highly complex. It would be tedious to discuss the various combinations in detail and at this point I simply wish to stress that it is necessary to take account of this diversity in considering the role of growth substances since, as I shall show, the pattern of modulation of endogenous growth substances is likely to vary in different types of response. I shall confine myself to a discussion of the implications for hormonal control of all-or-nothing as compared with graded responses.

3.1 All-or-Nothing Responses

Examples of all-or-nothing responses include the induction and breaking of dormancy of seeds and buds, fruit-setting, abscission and senescence. In each of these responses we are dealing with a quantitative, irreversible change which is not graded – a seed has either germinated or it has not, and there is no half-way stage between an intact and an abscised leaf.

If an all-or-nothing response is initiated by modulation of endogenous growth substance levels, it is probably dependent upon the attainment of a certain *threshold concentration,* by either an increase or a decrease in the level of growth substance. We may anticipate that in some cases the variation in growth substance would be large, whereas the crossing of a threshold may involve only small changes in growth substance concentration. The onset of senescence in flowers and leaves and of ripening in fruits is accompanied by marked increases in ethylene production [4, 13], but whether the rises in ethylene levels initiate these processes is still not clear. There is considerable evidence that the onset of abscission is correlated with high ethylene levels [15].

In considering all-or-nothing responses, we also have to distinguish between inductive and non-inductive situations. Inductive responses are those in which certain conditions (internal or external) are necessary to bring about a change in development, but which do not necessarily need to continue to operate after the switch has occurred. Examples of this situation are seen in artificial parthenocarpy, stimulation of germination by gibberellic acid and the stimulation of flower senescence by ethylene, in which a single application of the growth regulator initiates a long-term effect. Thus, there may need to be only a transitory increase or decrease in growth substance levels to initiate the response. In such cases we may speak of "trigger" effects of a growth substance. There is some evidence that such transitory variations in endogenous growth substance may be involved in normal development – as, for example, in the stimulus to fruit setting provided by the auxin present in pollen. Such transient peaks in endogenous growth substance may be important as "primers" of a biosynthetic pathway for specific growth substance, as in the autocatalytic production of ethylene in ripening fruits [31].

Where there are inductive effects, possibly involving only transient changes in an endogenous growth substance, a correlation between the growth substance and developmental events may well prove highly elusive, but the absence of correlation does not exclude the possibility of control by modulation of growth substance levels.

There are several good examples of all-or-nothing responses in which changes in sensitivity are important. Thus, the corollas of morning glory are only sensitive to applied ethylene after the flowers have opened, and ripening of fruits of tomato in response to ethylene does not occur until a certain stage in development is reached [4]. A clear example of variation in sensitivity to ABA is provided by the work of Smart and Trewavas [17] on the production of turions in the duckweed, *Spirodela polyrrhiza,* in which it was shown that there is only a relatively brief "window" during the development of fronds when exogenous ABA stimulates turion formation. Changes in sensitivity of cambial tissue to applied IAA appear to be important in seasonal variation in cambial activity in Sitka spruce (*Picea sitchensis*), in which endogenous IAA levels remain high in the fall, but the cambium becomes unresponsive to applied IAA at this time [7].

3.2 Graded Responses

The most obvious graded responses are those involving modulation of stem extension, but including also special cases of extension growth, such as those seen in the coleoptile and the epicotyl hook of *Phaseolus.* Other graded responses include apical dominance, epinasty and variation in stomatal aperture, but it would seem that there are fewer types of graded than of all-or-nothing responses.

By contrast with all-or-nothing responses, in which it was suggested that levels of growth substance or some other factor need to cross a certain threshold, in graded responses there is, by definition, no specific threshold, but a continuous gradation of response to increasing growth substance concentration.

The concept of hormonal control of plant growth and development arose from the classical studies of Went and others on phototropism. Doubts about the validity of the Cholodny-Went hypothesis have arisen from the fact that in experiments, in which coleoptile segments are immersed in IAA solution, there is a linear relationship between the *log* concentration of applied IAA and coleoptile extension growth, and on this basis it has been suggested that the differential growth rates on the two sides of a coleoptile in response to a difference in IAA concentration of 65:35 would not be sufficient to account for observed curvatures [3].

The "log-linear" relation between growth substance concentration and the magnitude of a response has been found to extend over a very wide range of concentrations – over as much as four to five orders of magnitude in some cases, as with gibberellins. It has been claimed that in this respect the situation in relation to plant growth regulators is very different from that occurring with animal hormones, where the linear phase is said to cover only two to three orders of magnitude of hormone concentration [6]. Much significance has been attached to this supposed difference between plant growth regulators and animal hormones and it has been argued [20] that it indicates that plant growth regulators cannot, in fact, *regulate,* since observed variations in endogenous growth substance concentrations are too small to evoke a significant response. However, this argument would appear to be invalid on several grounds.

Firstly, it appears that there is not a fundamental difference between plant growth substances and animal hormones in the range over which there is a log-linear relation between concentration and response. R.Firn has carried out an extensive survey of the growth substance concentration/response data from a large number of publications on plant and animal hormones, and the data indicate that the situation is very similar in plants and animals (pers. commun.). Thus, in the majority of these studies a log-linear relation was obtained over a range of two to three orders of magnitude in both plants and animals, and the small minority of instances in which a wider concentration range was reported occurred with equal frequency in both plants and animals. Moreover, steroid hormones of animals show a log-linear relation over a wide range of hormone concentrations, yet these are the animal hormones for which the receptors have been best characterized and show typical binding properties, and their regulatory function is not in doubt. Hence when we meet a situation in plants in which there is a log-linear relation between growth substance concentration and response over a wide range of concentrations, we are not justified in drawing the conclusion that plant growth substances do not have a regulatory function.

There are several possible reasons for a log-linear relation extending over a wide range of growth substance concentrations. Growth substance concentration/response curves are not derived from studies on a single enzyme-controlled process in vitro, but from experiments with a complex systems, such as a barley grain from which the embryo has been removed, and when this is treated with a solution of GA_3 there is undoubtedly a complex series of processes between the initial application of the GA_3 and the final response as measured by starch hydrolysis, involving possibly carrier-mediated uptake of the GA, followed by transcription, m-RNA processing, translation, secretion of α-amylase by the aleurone layer, to final starch hydrolysis. If these successive processes have differing kinetic parameters, then it is not difficult to see that the system as a whole will show a response over a wider range of GA concentrations than would a single process carried out in vitro. In some systems, such as the barley aleurone, there may well be more than one receptor for a given growth substance, and here also the overall response can be expected to reflect the complexity of the system.

The auxin concentration/growth relationship has recently been rigorously examined by Shinkle and Briggs [16], who have shown that the response curve is biphasic. The curve is log-linear over the range 0.003–0.4 μM, but above this there is a plateau, followed by a further growth increase from 4 to 10 μM. The authors suggest that there are two modes of uptake of IAA into the cell: carrier-mediated uptake over the lower concentration range, and uptake via chemiosmotic diffusion at the higher concentrations. They point out that the auxin dose-response curve for growth is a composite of the response of the auxin-uptake system and a separate growth transduction system. Thus, the wide range of the response curve does not imply the existence of an auxin receptor showing binding over the same concentration range.

There is evidence that conjugation and metabolism of applied growth substance may result in a much lower concentration in the cell than in the external solution. Indeed, Bogers et al. [1] have reported that when IAA was supplied to tobacco-pith explants, the bulk of it was rapidly metabolized to indoleacetyl aspartate (IAAsp) and an unknown substance, so that at equilibrium only 2% of the IAA taken up by the cell was present as free IAA. Thus, the initial free IAA concentration was much lower than that applied, a fact which clearly must be taken into account in attempting to interpret growth substance/response relations. The only relevant growth substance concentration is that at the site of action.

Bearing in mind these various complexities, it is perhaps surprising that log-linear response curves have, in fact, been found in so many cases.

There remains the question of whether, if the magnitude of a graded response is related to the logarithm of the growth substance concentration, the degree of variation of endogenous growth substance levels is sufficient to result in a significant growth response. This question cannot be answered unequivocally at present and the problem will only be resolved by further detailed studies on specific responses, in which accurate determinations of modulation of endogenous growth substance concentrations at the site of action are related to accurate response curves.

Clearly, there are still a number of important problems to be resolved concerning the relation between variation in endogenous growth substance concentrations and growth responses, but it would appear that the arguments which have been advanced against the "hormonal control" theory are themselves untenable. On the other hand, some arguments, which have been used to support the hormonal control theory, are also invalid. Thus, the demonstration that the presence of a specific growth substance is essential for a given growth process is not, of itself, evidence in support of the hormonal control theory, which postulates that growth and development are regulated by *modulation* of endogenous growth substance concentrations.

The antithesis between variation in sensitivity and modulation of growth substance concentrations is a false one, since it is likely that both methods of control occur.

4 Competence

As with sensitivity, the concept of competence is central to thought about responses to plant growth regulators. The concept originated from studies of animal development, but in plant development the term is generally used to refer to the ability of different cells or tissues to respond in a characteristic and specific manner to the same hormonal signal. Thus, the wide spectrum of effects of a growth substance, such as IAA, in different tissues of a plant can only be attributed to differences in the competence or "programming" of the various tissues.

Again, we have no idea how the specific competence of a given cell is controlled. It is often assumed that the differentiation of a particular cell reflects its complement of receptors. However, to account for the specificity of a given cell's competence we would have to postulate that either there are specific receptors for the same growth regulator in different target cells, or, alternatively, that there is a single type of receptor (for a given growth regulator) and that the specificity lies in the subsequent events initiated by the growth substance-receptor complex. At present, it is not possible to decide between these two alternative hypotheses, but the recent work of Hornberg and Weiler [5] suggests that stomatal guard cells contain an ABA-receptor, which is absent or present in only low levels in other epidermal cells.

Although it is, perhaps, natural to think of competence differences in term of receptors, other possibilities appear entirely feasible. Thus, the increasing evidence that oligosaccharides may be involved in cell-recognition processes and may be carriers of highly specific signals [19] indicates that it would be wise to keep an open mind on the possible molecular basis of competence.

A special case of competence is constituted by the "target cells" concept, studied by Osborne and co-workers in relation to be effects of ethylene on abscission in *Phaseolus vulgaris* [30] and *Sambucus nigra* [11]. During shoot development a special layer of cells is formed, which have the competence to respond to ethylene produced during senescence of the leaf, by producing cellulase and other hydrolytic enzymes, which bring about the degradation of the adjacent cell walls, so leading to separation and abscission. In *S. nigra* the cells of the abscission zone contain specific proteins and unique antigenic determinants, which distinguish them from cells in the neighbouring tissues [12].

5 Determination

The concept of determination, like competence, arose from studies on animal development. In the developing animal various regions become successively committed to specific pathways of development, so that the cells become increasingly specialized and show progressive restriction of developmental potential. This progressive commitment to particular pathways is referred to as *determination.* The determined state is remarkably stable and can be transmitted without reversal through many cell generations.

In plants the progressive commitment observed during animal development is less conspicuous in the plant as a whole because of its "open" pattern of growth by apical meristems. Nevertheless, the occurrence of stable cell differences, which are transmitted through cell division, has been well-known in normal plant development for many years. In vernalization, certain normally irreversible changes occur in the shoot apices of embryos of rye (*Secale cereale*) during the chilling period, which lead to flower initiation at a later stage. Thus, the apical meristem cells of the embryo become determined and the vernalized state is transmitted undiluted through many cell generations. Another example of determination in shoot apices is seen in the phenomenon of phase change in woody plants [23].

There is good reason to think that, in plants, differences in the state of determination do not involve irreversible genetic changes, but rather relatively stable differences in *gene expression,* which are transmitted through cell division, but which are reversible under certain conditions. Such differences are said to be *epigenetic.*

Meins [8–10] has pointed out that the phenomenon of habituation in tissue cultures, involving differences in the requirement for auxin and/or cytokinin is an example of cell determination and in a series of elegant experiments he has demonstrated that the changes involved in habituation are epigenetic in nature.

It is commonly assumed that the cells of the shoot and root apical meristems are totally uncommitted, but I have suggested from a priori considerations, that this may not be so, and that the cells of the shoot and root meristems may be committed to the "shoot" and "root" pathways of development respectively [24, 25]. We have recently obtained experimental evidence which is consistent with this hypothesis.

Callus cultures were established from shoot and root apical meristems of *Euphorbia heterophylla*, consisting of only the apical dome and two leaf primordia from shoots, and of the apical 0.5 mm of the root tips [27]. On a medium containing naphthalene acetic acid (NAA) and benzylaminopurine (BAP), the "shoot" callus was green, whereas the "root" callus was non-green. After transfer to liquid cell suspension culture, cell clones were established, from which further callus cultures of shoot and root cells were obtained. After three passages on a medium containing NAA and BAP, the differences in the appearance of the two types of callus were still visible. On transfer to a medium containing only cytokinin, the shoot callus produced buds, and the root callus did likewise after becoming green.

These preliminary results seem to indicate that cells of apical meristems are not necessarily totally uncommitted, but may show stable differences which can be eliminated by appropriate growth substance regimes, so that these differences appear to be epigenetic.

6 Conclusions

The concepts of sensitivity and competence have been recognized in plants for many years, but plant scientists have, I think, tended to underestimate the importance of determination. Sensitivity, competence and determination are clearly of great importance for an understanding of the role of growth substances in the control of plant growth and development, but they are also of great practical importance in tissue culture. It is important for modern techniques of plant breeding and genetic engineering to be able to regenerate whole plants from cultured cells, but this has not yet been achieved with many species of practical importance. If we had a better understanding of competence and determination at the molecular level, progress on this problem could be achieved. Until now, it has not been possible to study the molecular basis of the phenomena, but the techniques are now available to make a start and I suspect that exciting advances will be made in the next few years.

References

1. Bogers RJ, Kulescha Z, Quint a, Van Vliet TB, Libbenga KR (1980) Plant Sci Lett 19:311
2. Cleland R (1983) Trends Biochem Sci 8:354
3. Digby J, Firn RD (1976) Curr Adv Plant Sci 8:953
4. Grierson D, Slater A, Maunders M, Crookes P, Tucker GA, Schuch W, Edwards K (1984) In: Roberts JA, Tucker GA (eds) Ethylene and plant development. Butterworths, London, p 147
5. Hornberg C, Weiler EW (1984) Nature 310:321
6. Kende H, Gardner G (1976) Annu Rev Plant Physiol 27:267
7. Little CHA, Wareing PF (1981) Can J Bot 59:1480

8. Meins F (1982) In: Earle ED, Demarley Y (eds) Variability in plants regenerated from tissue culture. Praeger, New York, p 202
9. Mein F (1983) Annu Rev Plant Physiol 34:327
10. Meins F (1986) This volume, p 333
11. Osborne DJ, Sargent JA (1976) Planta (Berl) 130:203
12. Osborne DJ, McManus MT, Webb J (1984) In: Roberts JA, Tucker GA (eds) Ethylene and plant development. Butterworths, London, p 197
13. Roberts JA, Tucker GA, Maunders M (1984) In: Roberts JA, Tucker GA (eds) Ethylene and plant development. Butterworths, London, p 267
14. Sanders IO, Smith AR, Hall MA (1986) Physiol Plant (in press)
15. Sexton R, Lewis LN, Trewavas AJ, Kelly P (1984) In: Roberts JA, Tucker GA (eds) Ethylene and plant development. Butterworths, London, p 173
16. Shinkle JR, Briggs WR (1984) Plant Physiol (Bethesda) 74:335
Smart CC, Trewavas A (1983) Plant Cell Environ 6:507
17. Smart CC, Trewavas A (1983) Plant Cell Environ 6:507
18. Thimann KV, Bonner J (1933) Proc R Soc Lond B Biol Sci 113:126
19. Tran Thanh Van K, Toubart P, Cousson A, Darvill AG, Gollin DJ, Chelf P, Albersheim P (1985) Nature 314:615
20. Trewavas A (1981) Plant Cell Environ 4:203
21. Trewavas AJ (1983) Trends Biochem Sci 8:354
22. Van Staden J, Webb DP, Wareing PF (1972) Planta (Berl) 104:110
23. Wareing PF (1978) Bot Mag Tokyo Spec Issue 1:3
24. Wareing PF (1979) BPGRG Monograph No 3. Control of development, p 1
25. Wareing PF (1982) In: Smith H, Grierson D (eds) The molecular biology of plant development. Blackwell Scientific, Oxford, p 517
26. Wareing PF, Al-Chalabi (1976) Biologia Plantarum, Prague (in press)
27. Al-Ani, Wareing PF (unpubl.)
28. Went FW, Thimann KV (1937) Phytohormones. MacMillan, New York
29. Wright STC (1961) J Exp Bot 12:303
30. Wright M, Osborne DJ (1974) Planta (Berl) 120:163
31. Yang SF, Hoffman NE (1984) Annu Rev Plant Physiol 35:155

I. Methods

The Use of Immunological Techniques in Plant Hormone Analysis

A. CROZIER[1], G. SANDBERG[2], A.M. MONTEIRO[1,3], and B. SUNDBERG[2]

1 Introduction

Immunoassays were first used by Berson and Yalow [4] to measure insulin in human blood and the technique was subsequently adapted for analysis of a variety of compounds involved in human and animal metabolism [1]. Antisera raised against indole-3-acetic acid (IAA) and gibberellin A_3 were initially used in an assay based on inhibition of activation of modified bacteriophage T_4 [6, 8, 9]. However, these reports received only limited attention and it was not until the development of radioimmunoassays (RIA) [2, 10, 14, 18–21, 25] and enzyme-linked immunoabsorbent assays (ELISA [3, 5, 22, 24] for plant hormones that the potential value of immunological techniques for growth regulator research began to be more generally appreciated.

2 Immunoassays

The appeal of immunoassays in plant hormone analysis is based on their low limits of detection, the fact that they require little sophisticated equipment and their potential for processing large numbers of samples. There is also a belief that the selectivity of immunoassays is such that they can be used to analyse trace quantities of hormones in plant extracts in circumstances where it would be impossible to use alternative procedures without extensive sample purification [23]. This view is open to question and requires close scrutiny.

1 Department of Botany, The University of Glasgow, Glasgow, G12, 8QQ, Great Britain
2 Department of Forest Genetics and Plant Physiology, The Swedish University of Agricultural Sciences, S-901 83 Umea, Sweden
3 Permanent Adress: Department of Plant Physiology, UNICAMP, 13100, Campinas, SP, Brazil

Plant Growth Substances 1985
(ed. by M. Bopp)

2.1 Verification of Accuracy

2.1.1 Internal Standardisation

The main approach that has been taken to verify the accuracy of immunoassay-based quantitative estimates of the hormone content of plant extracts has been to assay replicate samples in the presence of increasing amounts of standard [14]. The apparent endogenous hormone content obtained by comparing values of percent immunotracer bound with a standard curve is plotted against the amount of exogenous hormone added. This procedure makes it possible to detect inactivation of the antibody or binding of compounds other than the hormone. In principle, the exogenous hormone curve should be a straight line with unit slope and a *y*-intercept equal to the amount of hormone originally present in the sample. The presence of interfering compounds that adversely affect accuracy will be indicated by a change in the slope of the curve. Typical data that are obtained with this procedure are presented in Fig. 1, which illustrates curves obtained when RIA was used to estimate the IAA content of an impure and a purified extract from *Pinus sylvestris* seedlings [16]. Interference, which, if undetected would result in an overestimate of the amount of IAA present, is evident in the impure sample.

The internal standardisation procedure described above is not foolproof and it can fail to detect certain compounds which compete with the hormone for antibody-binding sites. This point was clearly demonstrated by Pengelly and Meins [14] when they analysed known amounts of IAA by RIA in the presence and absence of either 2 μg 2,4-dichlorophenoxyacetic acid (2,4-D) or 20 ng α-naphthaleneacetic-acid (α-NAA). The cross-reactivities of these compounds with the anti-IAA antiserum were 0.4% and 25% respectively. The data obtained are illustrated in Fig. 2. The α-NAA curve is linear with a slope of one but is displaced upwards on the *y*-axis giving a false estimate of the IAA content. In contrast, the 2,4-D curve differs significantly in slope from the expected value. Thus, the presence of compounds,

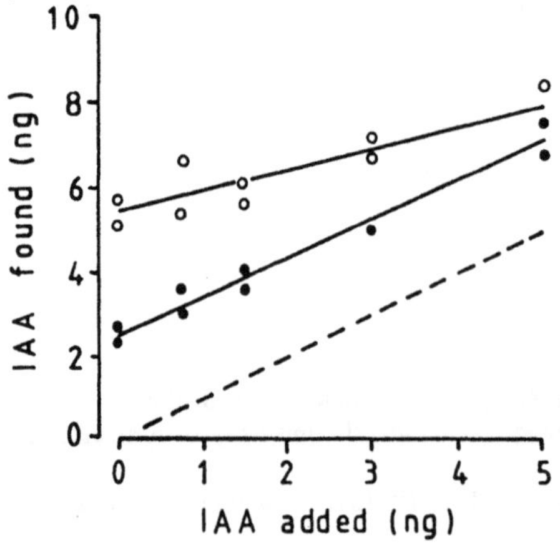

Fig. 1. Internal standard IAA-RIA curves obtained with impure (○) and purified (●) extracts from *Pinus sylvestris* seedlings; IAA standards alone (-----) [16]

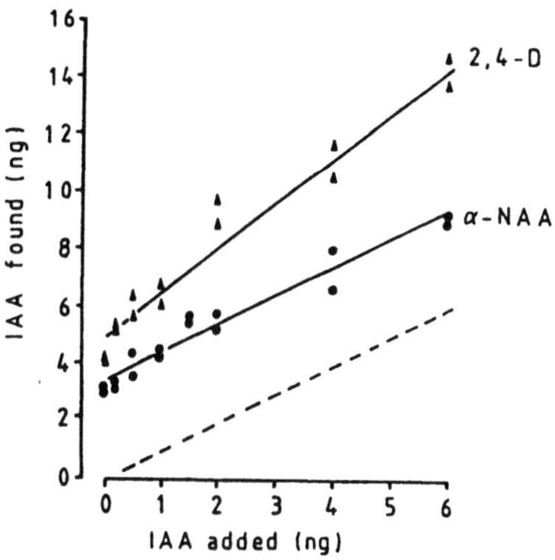

Fig. 2. Internal standard IAA-RIA curves obtained with IAA standards alone (-----) and IAA standards assayed in the presence of either 2 μg 2,4-D (▲) or 20 ng α-NAA (●) [14]

which bind to the antibody considerably less effectively than IAA, are detected readily, while components with similar affinities are not.

Internal standardisation can also be carried out using a range of sample dilutions and parallel curves are an indication of the absence of interference. This test will detect high affinity impurities which compete for antibody-binding sites, provided they have a different dose: response curve to the compound of interest.

2.1.2 Successive Approximation

A general view of the relationship between sample purity and the accuracy of immunoassay data can be achieved through the application of a "successive approximation" [15]. When given a sample purporting to contain a given quantity of hormone (E_1), a successive approximation consists of purifying the sample and re-estimating the hormone content (E_2). An internal standard is used so sample losses encountered during purification can be assessed. If E_1 is accurate, E_2, taking into account the precision of the method, should not be significantly different. If a difference is found, E_1 must be rejected as inaccurate and E_2 tested by further purification and analysis. This process is continued until an estimate is obtained that does not change on purification. There are then no grounds for believing the estimate to be inaccurate.

This procedure was used by Sandberg et al. [16] to determine the purification steps that are required in order to accurately analyse IAA in extracts from *Pinus sylvestris* seedlings by RIA. For comparative purposes, samples were also analysed by high performance liquid chromatography (HPLC) with a fluorimetric detector. An outline of the successive approximation procedure employed is presented in Fig. 3. Seven replicate extracts were analysed and IAA was quantified at the points indicated. The data obtained are presented in Table 1. Internal standardisation, as described in the previous section, detected the presence of impurities which affected the accuracy of RIA estimates of endogenous IAA at quantification points 1 and 2. Interference

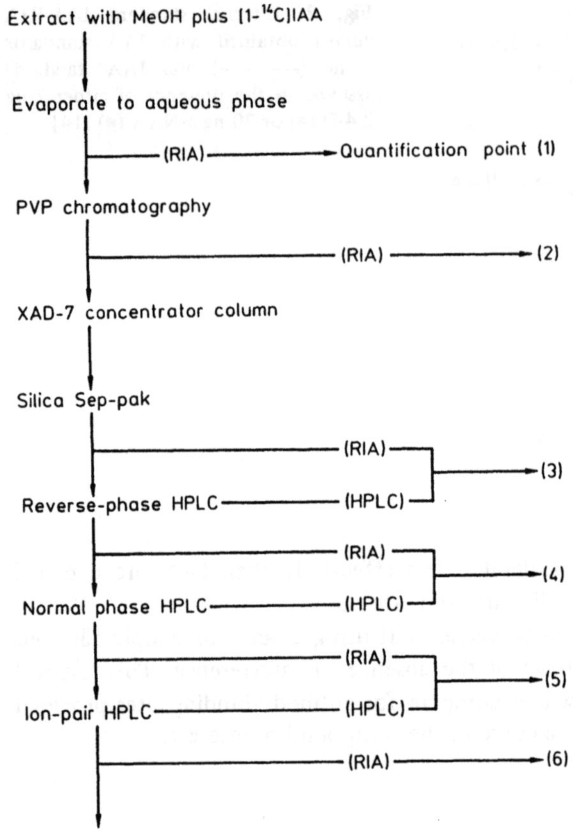

Fig. 3. Successive approximation procedures used when analysing IAA in extracts from *Pinus sylvestris* shoots by RIA and HPLC [16]

Table 1. Successive approximation of estimates of the endogenous IAA content of *Pinus sylvestris* extracts obtained with RIA and HPLC with a fluorimeter [16][a]

Quantification point	RIA	HPLC
1	87.6 ± 36.1 (41.2%)	–
2	62.3 ± 21.5 (34.6%)	–
3	51.5 ± 11.0 (21.4%)	43.5 ± 7.6 (17.4%)
4	54.0 ± 8.8 (16.3%)	40.6 ± 7.0 (17.0%)
5	50.2 ± 8.7 (17.4%)	44.5 ± 6.4 (14.0%)
6	53.4 ± 10.3 (19.2%)	–

[a] Data expressed as ng IAA g^{-1} fresh wt. ± standard deviation (percent precision)

was not apparent at quantification point 3 and beyond. The precision of the RIA estimates improved with sample purification up to point 3. It is, thus, apparent that unless samples are purified adequately, contaminants have detrimental effects on both the precision and the accuracy of RIA estimates. The HPLC estimates provide an example of the purity required for the analysis of IAA by a selective, physicochemical method. Estimates obtained at quantification point 3 did not change significantly with further purification. It was not possible to make realistic measurements prior to quantification point 3, because samples contained large amounts of fluorescent contaminants. Although the estimates were consistently lower than those obtained with RIA, the figures were not significantly different. The data in Table 1 thus indicate that HPLC with a fluorimetric detector provides an accurate estimate of the IAA content of *Pinus sylvestris* seedling extracts at the same stage as immunoassay.

There is a general belief that the great advantage of immunoassays over other techniques in plant hormone analysis is their ability to assay relatively impure samples. This is clearly not the case, at least as far as the IAA content of extracts of Scots pine is concerned. The degree of purification that is required will vary with samples from different sources and has to be assessed by successive approximation with typical extracts before routine quantification of an endogenous hormone in a particular species, tissue or organ is undertaken. Once this is achieved RIA and ELISA are valuable techniques as the large number of samples that can be analysed on a daily basis makes it possible to include internal standardisation checks with every quantitative estimate. Combined gas chromatography-mass spectrometry operating in the multiple-ion monitoring mode also provides a ready check of accuracy through the m/z response ratio. Although the speed at which samples can be analysed does not approach that of RIA or ELISA, this is not always a problem. For instance, when measuring IAA in *Pinus sylvestris,* the limiting step in the analytical sequence is the speed at which samples can be purified rather than the rate at which they are analysed.

3 Immunoaffinity Chromatography

In addition to being utilised in immunoassays, antibodies raised against growth regulators can also be used for immunoaffinity chromatography. In this role they offer the possibility of rapid and effective purification of extracts prior to quantitative analysis. Immunoaffinity supports for the purification of gibberellin A_3 [7], trans-zeatin riboside and isopentenyl adenine [11, 13] and the separation of (+) and (–) enantiomers of (^{3}H)abscisic acid [12] have been described. During the course of our own studies, we have developed an immunoaffinity chromatography system for IAA [17] that has provided extensive purification of samples from a variety of plant tissues.

3.1 Preparation of an Immunoaffinity Support for IAA

The antibodies used most frequently for IAA-immunoassays have a higher affinity for IAA methyl ester than for underivatised IAA, having been raised against IAA linked to bovine serum albumin (BSA) via C′-1 in the side-chain carboxyl group [21]. The anti-IAA antibodies for immunoaffinity chromatography were raised against BSA coupled to IAA via the nitrogen of the indole ring [14]. This leaves the side chain at the critical 3′-position free and, as a consequence, more specific antibodies with a higher affinity for IAA as a free acid are produced. When these antibodies are used for immunoaffinity chromatography the inconvenience of methylating impure samples is avoided.

The polyclonal rabbit immunoglobulin fraction containing anti-IAA antibodies was separated from the serum and covalently attached to a glutardialdehyde-activated silica support (Boehringer). Immunoaffinity chromatography of IAA was then carried out using a column with a 35x10 mm i.d. bed (ca. 1.7 g support). Samples were applied to the column dissolved in 2.5 ml 50 mM, pH 7.5 sodium phosphate buffer and impurities removed by elution with 50 ml of the buffer. IAA was then eluted from the column in 30 ml 0.1 M, pH 2.7 citrate buffer. Before being re-used, the column was eluted with 200 ml 1.0 M propionic acid and re-equilibrated in 50 mM, pH 7.5 phosphate buffer. The acidic citrate buffer eluate, containing IAA, was partitioned against diethyl ether, which was reduced to dryness, and the IAA content determined by ion-suppression reverse-phase HPLC with a fluorimetric detector. Experiments with ($1'$-^{14}C) IAA showed that the immunoaffinity column had a sample capacity of 80 ng IAA with a recovery of 90%.

3.2 Immunoaffinity Chromatography of IAA in Plant Extracts

3.2.1 Pinus sylvestris Shoots

Scots pine shoots were extracted with methanol in the presence of the antioxidant sodium diethyldithiocarbamate and a ($2'$-^{14}C) IAA internal standard. The extract was reduced to dryness in vacuo, dissolved in 10 ml 50 mM, pH 7.5 sodium phosphate buffer and split into four equal-sized aliquots. Three of these were subjected to immunoaffinity chromatography, the IAA fractions combined and analysed by ion-suppression reverse-phase HPLC. An acidic diethyl ether fraction was obtained directly from the fourth aliquot and similarly analysed so that the purification achieved by immunoaffinity chromatography could be assessed. The HPLC traces obtained are illustrated in Fig. 4 and demonstrate that immunoaffinity chromatography provided an extremely effective purification. Quantitative estimates indicated that the *Pinus* shoot extract contained 56 ± 3 ng IAA g^{-1} fresh wt. The accuracy of this figure was checked by collecting the putative IAA peak in the immunoaffinity-purified sample and analysing it by normal phase HPLC after which the fluorescent IAA-like component was again collected and re-analysed by ion-pair reverse-phase HPLC. The quantitative estimates of IAA content were not significantly different from that obtained by ion-suppression reverse-phase HPLC. The successive

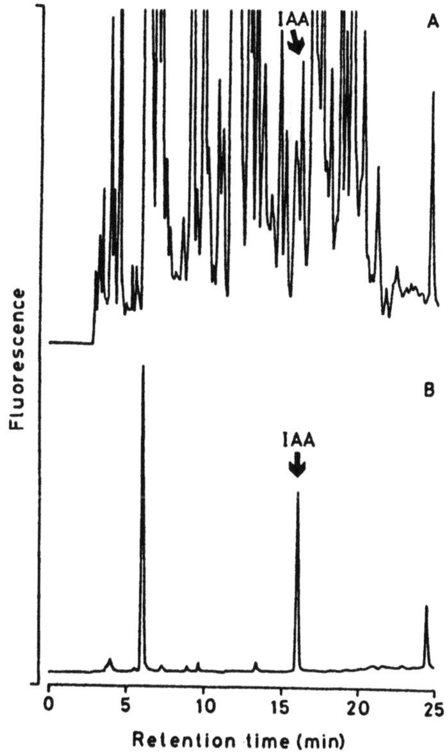

Fig. 4A, B. Effect of immunoaffinity chromatography on HPLC analysis of IAA in extracts from *Pinus sylvestris* shoots. *Sample:* **A** Acidic diethyl ether extract; **B** as **A**, but extract subjected to immunoaffinity chromatography. Column: 250x5 mm i.d. 5 μm ODS Hypersil. Mobile phase: 25 min gradient, 25-75% methanol in 1% aqueous acetic acid. Flow rate: 1 ml min^{-1}. *Detector:* Fluorimeter, excitation 280 nm, emission 350 nm

approximation thus indicates that the initial HPLC step, immediately after immunoaffinity chromatography, provided an accurate assessment of the endogenous IAA content of the *Pinus* shoot extract. Similar data have been obtained with extracts from seed and cambial tissue of *Pinus sylvestris* and germinating seed of *Dalbergia dolichopetala.*

3.2.2 Dwarf-1 Zea mays Shoots

Extracts from the shoots of dwarf-1 maize seedlings were analysed in the same manner as the *Pinus* shoot extracts described above. Ion-suppression reverse-phase HPLC indicated that once again extensive purification was achieved and that after immunoaffinity chromatography, the extract contained a large Gaussian-shaped fluorescent peak, which co-chromatographed with IAA (Fig. 5A, B). However, when this peak was collected and re-analysed by ion-pair reverse-phase HPLC, it became evident that IAA was only a minor component and that the major constituent was a contaminant (Fig. 5C). Subsequent analysis by normal phase HPLC confirmed the identity and the homogeneity of the IAA peak obtained from the ion-pair system.

In view of the data obtained when extracts from other tissues were purified by immunoaffinity chromatography and analysed by ion-suppression reverse-phase HPLC, it would have been very tempting to assume that the application of identical

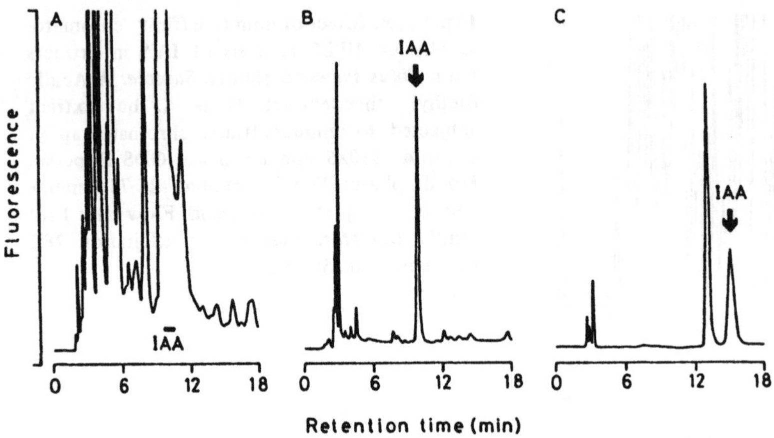

Fig. 5 A–C. Effect of immunoaffinity chromatography on HPLC analysis of IAA in extracts from the shoots of dwarf-1 *Zea mays* seedlings. *Sample:* **A** Acidic, diethyl ether extract; **B** as **A**, but subjected to immunoaffinity chromatography. *Column:* 250x5 mm i.d. 5 μm ODS Hypersil. Mobile phase: 45% methanol in 1% aqueous acetic acid. *Flow rate:* 1 ml min^{-1}. *Detector:* Fluorimeter, excitation 280 nm, emission 350 nm. *Sample:* **C** IAA-like peak from **B**. Mobile phase: 35% methanol in 50 mM phosphate buffer and 20 mM tetrabutylammonium hydrogen sulphate at pH 6.5. *Column, flow rate and detector:* as **A** and **B**

procedures would also have provided acceptable estimates of the IAA content of *Zea* shoot extracts. The data in Fig. 5B, C show that such an assumption would have been incorrect, because IAA in the purified *Zea* extracts co-chromatographed with a fluorescent contaminant when analysed by ion-suppression reverse-phase HPLC. Accurate estimates of IAA can, however, be obtained when the samples are analysed by ion-pair rather than ion-suppression reverse-phase HPLC.

4 Conclusions

Each tissue contains not only varying amounts of hormones, but also, in far greater quantities, its own characteristic assortment of impurities. As a consequence, standardised analytical procedures cannot be applied "cookbook style" to samples from diverse sources. In the circumstances, when investigating endogenous hormones in a particular plant material, it is essential to evaluate purification and analytical procedures, immunological or otherwise, by methods such as a successive approximation, before routine quantitative analyses are undertaken.

Acknowledgements. G.S. and A.C. would like to thank the British Council for U.K.-Sweden travel grants. Studies on immunoaffinity chromatography of IAA were supported by a grant from the Swedish Council for Forestry and Agricultural Research (P 425) to G.S.; A.M.M. was in receipt of a fellowship from CNPq (Brazil). Dwarf-1 *Zea mays* seed was kindly supplied by Prof. B.O. Phinney, University of California, Los Angeles, U.S.A.

References

1. Abraham GE (1977) Handbook of radioimmunoassay. Dekker, New York
2. Atzhorn R. Weiler EW (1938a) Planta (Berl) 159:1
3. Atzorn R, Weiler EW (1983b) Planta (Berl) 159:7
4. Bersen SA, Yalow RS (1959) Ann NY Acad Sci 82:338
5. Daie J, Wyse R (1982) Anal Biochem 119:365
6. Fuchs S, Fuchs Y (1969) Biochim Biophys Acta 192:528
7. Fuchs Y, Gertman E (1974) Plant Cell Physiol 15:629
8. Fuchs S, Haimovich J, Fuchs Y (1971) Eur J Biochim 18:384
9. Fuchs Y, Mayak S, Fuchs S (1972) Planta (Berl) 103:117
10. Khan SA, Humayun MZ, Jacob TM (1977) Anal Biochem 83:632
11. MacDonald EMS, Morris RO (1985) Methods Enzymol 110:347
12. Mertens R, Stüning M, Weiler EW (1982) Naturwissenschaften 69:595
13. Morris RO, Akiyoshi DE, MacDonald EMS, Morris JW, Regier DA, Zaerr JD (1982) In Wareing PF (ed) Plant growth substances 1982. Academic Press, London, p 175
14. Pengelly WL, Meins F Jr (1977) Planta (Berl) 136:173
15. Reeve DR, Crozier A (1980) (ed) Encyclopedia of plant physiology, new series, vol 9. Hormonal regulation of development I. Molecular aspects of plant hormones. Springer, Berlin Heidelberg New York, p 203
16. Sandberg G, Ljung K, Alm P (1985) Phytochemistry (Oxf) 24:1439
17. Sundberg B, Sandberg G, Crozier A (1986) Pytochemistry (Oxf) 25:295
18. Walton D, Dashek W, Galson E (1979) Planta (Berl) 146:139
19. Weiler EW (1979) Planta (Berl) 144:255
20. Weiler EW (1980) Planta (Berl) 149:155
21. Weiler EW (1981) Planta (Berl) 153:319
22. Weiler EW (1982) Physiol Plant 54:510
23. Weiler EW (1984) Annu Rev Plant Physiol 35:85
24. Weiler EW, Jourdan PS, Conrad W (1981) Planta (Berl) 153:561
25. Weiler EW, Wieczorek U (1981) Planta (Berl) 152:159

Immunoassays for the Quantitation of Plant Growth Regulators Using Monoclonal Antibodies*

E.W. WEILER, J. EBERLE, and R. MERTENS[1]

1 Introduction

Plant growth regulators (PGR) can be quantitated by bioassay, physiochemical assay and immunoassay. Because of several distinct advantages, immunoassay currently attracts much attention. It is the purpose of this workshop to critically evaluate this novel technique in relation to others. To do so, this contribution will provide a brief discussion of the basic technology involved, while the following articles will deal more specifically with assay validation, comparisons of immunoassay and other approaches to PGR analysis and potential uses of PGR antibodies in other areas, such as immunoaffinity chromatography.

Historically radioimmunoassays (RIA) comprised the first generation of immunoassays which were sensitive enough to cope with PGR at physiological levels. These assays made use of polyclonal antisera raised in rabbits. A review of these techniques can be found in Weiler [9–11]. Such assays are available for indole-3-acetic acid (IAA), abscisic acid (ABA), xanthoxin, the gibberellins A1, A3, A4, A5, A7, A9, A19, A20, A24 and for cytokinins of the trans-zeatin und isopentenyladenine-type. They employ tritium or, less frequently, iodine-125-labeled PGR or PGR derivatives and allow to quantitate as little as 20-100 pg of the respective PGR. It has been found more difficult to establish assays based on enzyme-labeled PGR in conjunction with antisera, although several have been described [1].

With the introduction of the monoclonal antibody technology [3] the applications of immunology have entered a new dimension. It is the purpose of this contribution to briefly discuss this potential and point out routes for its realization.

2 Monoclonal Antibodies – Generation and Properties

The fusion of antibody-secreting B-lymphocytes with suitable myeloma cells results in fusion products (hybridomas) which will grow in culture (B-lymphocytes will not) and may continue to secrete antibody. Since any given B-lymphocyte produces only a single – the monoclonal – antibody provided it originated from the fusion

* The following contributions (pp. 22–51) are part of the workshop *Radioimmunoassay*

1 Universität Osnabrück, Fachbereich Biologie/Chemie (Pflanzenphysiologie), Postfach 4469, D-4500 Osnabrück, FRG

Plant Growth Substances 1985
(ed. by M. Bopp)

of one lymphocyte with a myeloma cell which in itself does not produce antibody light or heavy chains. After a tedious subcloning process, stable hybridomas can usually be selected which continue to grow rapidly in vivo or in vitro and secrete large amounts of the desired monoclonal antibody (MAB). A detailed account of the technical aspects of hybridoma generation is beyond the scope of this article and can be found in Goding [2] or Campbell [1].

Despite the considerable investments in time, manpower and capital required to run a hybridoma laboratory, the effort is worthwhile because of a number of major advantages of this system over the production of sera (Table 1).

In practical terms, it is of major importance whether or not these theoretical advantages can be materialized in the form of an improved generation of immunoassays. Evidence now available strongly suggests that it can. Monoclonal antibodies contribute substantially to increased assay specificity (see Table 2, for an example), allow a further simplification of assay protocols, are easily compatible with enzyme-labeled PGR, and are more precise. The selection of high-affinity MAB by RIA techniques has furthermore allowed to achieve assay sensitivities which are comparable to those of antiserum-based assays. Table 3 provides somes characteristics of MAB-based enzyme immunoassay (MAB-EIA) as developed in the author's laboratory in comparison with conventional immunoassays. A detailed protocol for MAB-EIA of PGR can be found in Weiler [11].

Table 1. Comparison of properties of polyclonal and monoclonal antibody systems

Polyclonal	Monoclonal
Supply limited and variable	Unlimited availability
Variable properties of individual sera frequent	Constant properties of MAB
Antigen binding to multiple sites of unknown individual properties	Only a single site involved
Antigen must be pure for immunization	Impure antigens or antigen mixtures can be used
Cross-reactions occur as a result of contaminant, low-affinity antibodies	No cross-reactions due to low affinity interactions
Sera may perform unpredictably if used with labeled-PGR derivatives of altered structure (e.g. in EIA)	Can be selected to perform equally well in RIA, EIA and other types of assays
Individual assay protocols required	Identical protocols for all assays

Table 2. Comparisons of cross-reactions of anti-IAA antisera and MAB[a]

Compound	MAB 17-II-AI	Polyclonal antibody
Indole-3-acetic acid	100	100
Indole-3-acetone	5.2	35
Indole-3-propionic acid	0.5	2.5
Indole-3-acetaldehyde	0.05	6.3
Indole-3-acetamide	1.0	8.6
Tryptamine	0.01	0.2
5-Hydroxyindole-3-acetic acid	0.02	0.3
IAA-aspartate	0.5	6.5

[a] Values (%) are given on molar basis. All compounds subjected to diazomethane treatment prior to analysis

Table 3. Some properties of MAB against PGR as compared to polyclonal sera (tZR=trans-zeatinriboside, R: RIA, E:EIA)

PGR	MONOCLONAL			POLYCLONAL	
	Hybridoma code	Affinity (l/mol)	Measuring range (pmol)	Affinity (l/mol)	Measuring range (pmol)
ABA	15-I-C5	2.6×10^8	0,02-2 (E)	1.8×10^9	0.1-300 (E)
IAA	17-II-AI	1.0×10^8	0.2-20 (E)	1.7×10^8	0.02-50 (E)
tZR	J3-I-B3	9.1×10^7	0.02-5 (E)	4.2×10^{10}	0.06-30 (R)

3 Immunological Hormone Analysis

Being based on immunological reagents, which can be produced in unlimited amount, MAB-based PGR immunoassays bear considerable potential as easily accessible and standardized reference methods for PGR analysis. With the use of such assays in an increasing number of laboratories and the subsequently accumulating experience with their performance under varying experimental conditions, a much more detailed assessment of the potential of immunoassay will be possible. Available evidence [5, 6, 8] suggests that immunoassay is both accurate and precise, if appropriately applied. Checks for assay performance should always include the following steps before proceeding to the routine analysis of the plant material to ensure the validity of the determinants (a more detailed treatment of this important aspect is to be found in Pengelly [6] and Weiler et al. [11].

1. Indispensable for any plant material for which experience with immunoassay ist not yet available: careful analysis of immunoreactivity in separated extracts (immunohistograms, preferably in conjunction with HPLC).
2. Analysis at various levels of purification (successive approximation [7]).
3. Dilution analysis with internal standardization (this only reveals gross interference).
4. Whenever possible: comparison of results with those obtained by other techniques for which accurate data are already at hand.

4 Conclusions

Whenever comparisons with different techniques have been made (for some references, see above), it has been found that immunoassays can be applied to the analysis of plant growth regulators containing fractions of lower purity than required for other techniques. In exceptional cases, crude fractions [e.g. 4, 8], may be analyzed, but usually, extract clean-up will involve the removal of lipophilic material (e.g. by silgel-C18 passage), phenolic compounds (e.g. by PVP), followed by solvent extraction and TLC or HPLC. The degree of clean-up required is largely dependent upon the PGR to be analyzed (ABA requiring little, if no, pre-purification, while GAs usually require an HPLC step) and upon the actual analytical problems to be solved. Established techniques for PGR extraction and separation may not be as appropriate for immunoassay as they are for physicochemical assay which proceeds largely in organic phase. Consequently, the full potential of immunoassay may only be realized in conjunction with an optimized pre-assay processing of the plant material. This could rely on rapid and reproducible column chromatographic steps to remove interference (lipids, pigments, phenols) and selectively remove and at the same time concentrate the PGR of interest (immunoaffinity matrices). Preliminary results in our laboratory indicate that a fast and high-yield preparation of PGR shall be possible this way.

References

1. Campbell A (1984) Monoclonal antibody technology. Elsevier, Amsterdam, p 265
2. Goding JW (1983) Academic Press, London, p 276
3. Köhler G, Milstein C (1975) Nature 256:495
4. Mertens R, Deus-Neumann B, Weiler EW (1983) FEBS Lett 160:269
5. Oden PC, Weiler EW, Schwenen L, Graebe IE (1985) Physiol Plant 64:21A
6. Pengelly WL, Bandurski RS, Schulze A (1981) Plant Physiol (Bethesda) 68:96
7. Reeve DR, Crozier A (1980) In: MacMillan J (ed) Hormonal regulation of development I. Springer, Berlin Heidelberg New York, p 203 (Encyclopedia of plant physiology, vol 9)
8. Wang TL, Futers TS, McGreary F, Cove DJ (1984) In: Crozier A, Hillman JR (eds) The biosynthesis and metabolism of plant hormones. Cambridge University Press, Cambridge, p 135
9. Weiler EW (1982) Physiol Plant 54:230
10. Weiler EW (1984) Annu Rev Plant Physiol 35:85
11. Weiler EW, Eberle J, Mertens R, Atzorn R, Feyerabend M, Jourdan PS, Arnscheidt A, Wieczorek U (1986) In: Wang T (ed) Immunology in plant science. Cambridge University Press, Cambridge (in press)

Immunoassays for Plant Growth Regulators – A Help or a Hindrance?

T.L. WANG, P. GRIGGS, and S. COOK[1]

1 Introduction

The improvements in our ability to detect and quantify endogenous plant growth regulators have reflected advances in analytical techniques to a certain extent. The literature reveals that detection and quantification of plant hormones initially relied entirely on biological assays (bioassays), later on a combination of simple purification schemes followed by bioassays, then on more complex purification systems followed by physicochemical detection; the main detection system at present being the mass spectrometer. It is clear that immunological methods will add another facet to research on plant growth regulators though whether they will supersede bioassays as has been suggested [10] or other physicochemical methods remains to be seen. It is important to note, however, that although technical details may change (e.g. thin-layer chromatography replaced by high-pressure liquid chromatography), very few methods become extinct because each has a slightly different function. Hence, one cannot totally replace the bioassay by an immunoassay simply because the bioassay, by definition, detects biological activity, whereas the immunoassay does not. Thus, the antibody is an important and *different* tool which provides us with novel methods for PGR research.

2 Uses of Antibodies in PGR Research

For simplicity, the uses of antibodies can be divided into two groups – direct and indirect. Direct methods employ an antibody raised directly against the plant hormone as a hapten, while the indirect methods use, for example, an antibody raised against a molecule with which the hormone interacts, e.g. a metabolic enzyme. Indeed, it is likely that the more exciting advances in our understanding of plant hormones and their modes of action will stem from these indirect methods. Examples of these uses are shown in Table 1.

1 John Innes Institute, Colney Lane, Norwich NR4 7UH, Great Britain

Plant Growth Substances 1985
(ed. by M. Bopp)

Table 1. The use of antibodies in PGR research

Direct	Indirect
Identification	Enzymology
Quantification	Receptor technology
Detection	Transport carriers
Immunopurification	New PGRs
Immunolocalisation	

2.1 Direct Uses of Antibodies

To date, the main use of antibodies raised against plant hormones have been in immunoassays, whether they be radioimmunoassays or enzyme-linked immuno-assays. Much of this work has been reviewed elsewhere in the literature [e.g. 19, 20] and in this volume (see Weiler, this volume) and has been concerned with the quantification of plant hormones. Immunoassays could additionally be used to identify a hormone in a plant extract (though considering the number of other compounds in a plant extract that may interfere with the antigen-antibody binding, this is unlikely without supplementary information from other techniques) or simply to detect the hormone without providing an accurate estimate of its level.

2.1.1 Quantification

Although one of the advantages of immunoassays for plant hormone measurement was envisaged to be their potential for use with crude plant extracts [18], many problems can be encountered with this application. It appears in many instances that interference in the antigen-antibody binding is prevalent when crude plant extracts are assayed. If no account is taken of the possibility that interference exists then the results obtained cannot be assessed for their accuracy. One problem associated with immunoassays is that the quantification relies on "competition" of endogenous plant hormone with the tracer for the antibody. Thus, the assay can produce false positives due to non-specific interference or cross-reactions from the plant extract. There are, nevertheless, several ways in which one can assess possible interference from a plant extract and determine if the results are likely to be accurate. The theoretical considerations of antigen-antibody binding and the binding interference are considered by Pengelly (this volume) and we will confine ourselves to ways in which one can check for interference and examples of the methods.

2.1.2 Checks for Interference

The simplest technique is to perform a dilution series of the extracts and to show additivity, or parallelism to the standard curve (plotted for convenience as a logit-log plot). Figure 1 shows a dilution series of *Cercospora rosicola* culture medium measured using a monoclonal antibody (McAb) to abscisic acid (ABA); in this instance, a straight line is obtained, but not with additivity – twice the quantity of culture medium assayed does not produce twice the value of ABA. Hence, interference is present. Figure 2 shows an example of logit plots; in this instance, from several tissues of *Phaseolus vulgaris* using an antiserum against dihydrozeatin riboside (DHZR). Such plots are not very sensitive to interference; the root extracts show interference using other checks. It should also be noted that logit transformations

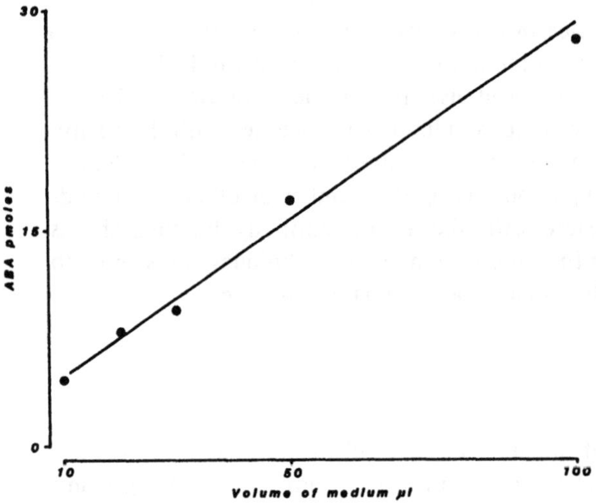

Fig. 1. ABA in the culture medium of *Cercospora rosicola.* Immunoreactive material was measured by radioimmunoassay using AFRC MAC 62 anti-(+)-ABA McAb in aliquots of 100-fold diluted culture filtrate (a gift from S.J. Neill, Aberystwyth)

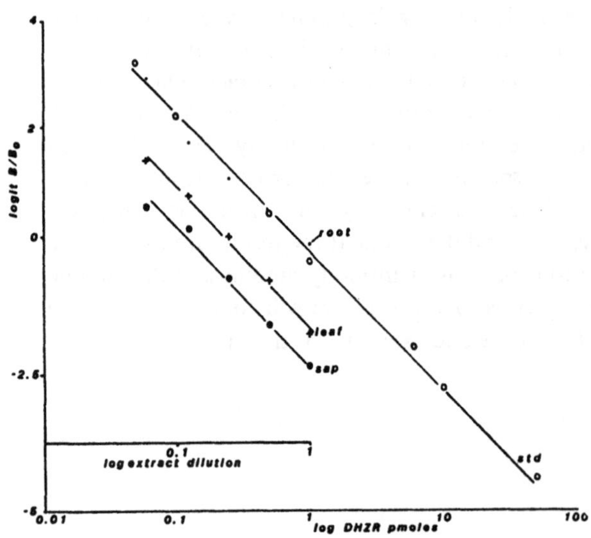

Fig. 2. Logit transformations for a series of tissues from *Phaseolus vulgaris* L. Leaves, roots and sap from *P. vulgaris* were assayed by radioimmunoassay using a rabbit anti-DHZR antiserum

for linearising standard curves (the most frequently employed method for immunoassays), though convenient for assessing parallelism, show larger variances at either extreme of the line than in the central part (heteroscedasticity; [1, 11]) and are thus not very precise in these regions.

Pengelly and Meins [8] adopted the method of internal standardisation to check their samples for interference in an immunoassay for indoleacetic acid (IAA). In this method increasing amounts of the plant hormone are added to the sample und the values obtained in the assay plotted against the amount of hormone added. If no interference is present, one will obtain a line parallel to the standard line and its intercept with the y-axis corresponds to the amount of endogenous compound present. In such comparisons, it is also an advantage to use more than one dilution of extract [11]. Figure 3 shows an example of a series of parallel lines obtained using a McAb against ABA, where no interference is apparent, while Fig. 4a illustrates a series where interference apparently increases the binding of the antigen to the antibody. In this latter case, however, HPLC analysis indicates only a single immunoreactive peak co-chromatographing with ABA (Fig. 4b). Tests for parallelism will not detect interference if the interfering compound has similar binding characteristics to the compound of interest ([8], Pengelly, this volume). For example, Fig. 5a shows an internal standard graph for DHZR in sap from *Phaseolus vulgaris*, which is parallel to the standard line (the regressions are not significantly different) though this sap is known also to contain a significant amount of DHZ from HPLC analysis (Fig. 5b) and mass spectrometry. More often binding is decreased by interfering substances and the slope of the sample line has a value greater than unity [e.g. 12].

If interference is present then it must be accounted for either by reference to a standard curve [12] or, as is preferred, removed by purification. After purification one can again test the extract using internal standardisation for parallelism to the standard curve. It is better, however, to further purify the sample and adopt the method of successive approximation [10, 13] since this allows one to determine a suitable purification procedure and assess the accuracy of the estimate.

Finally, one can check the immunoassay against a physicochemical technique (such as mass spectrometry) and there are now several examples of such comparisons

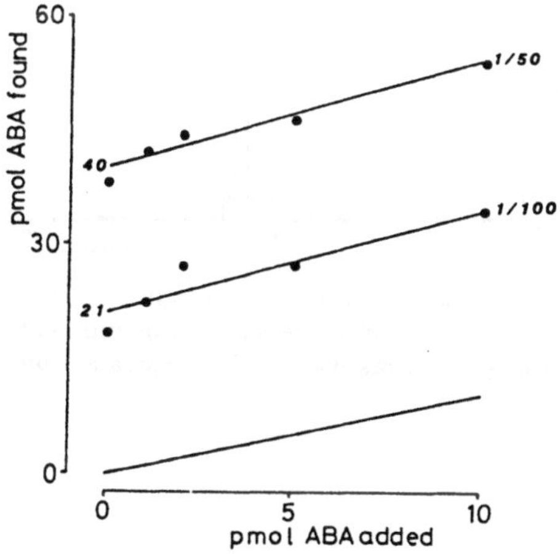

Fig. 3. Internal standardisation of plant extracts. Immunoreactive material in two dilutions of buffer-extracted material from a pea embryo was determined (as in Fig. 1) after the addition of different amounts of (+)–ABA

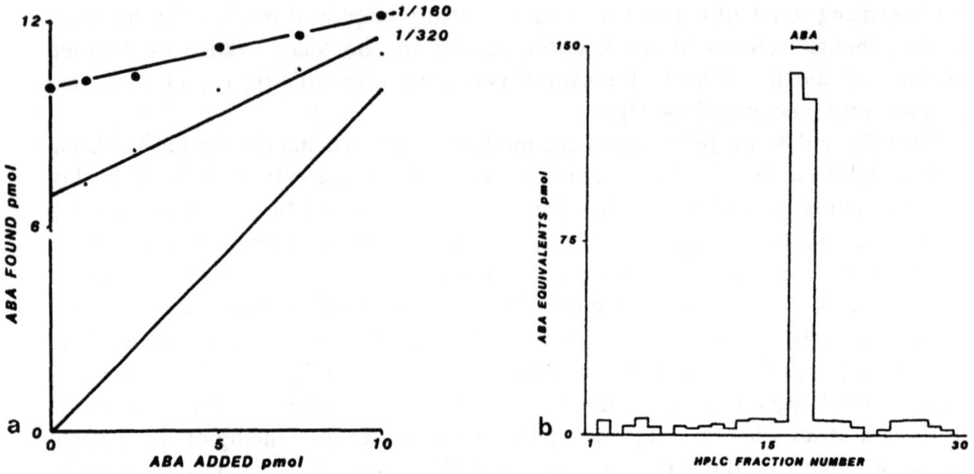

Fig. 4a, b. Assay interference from plant extracts. **a** Immunoreactive material in two dilutions of methanol-extracted material from a pea embryo was determined (as in Fig. 1) after the addition of different amounts of (+)–ABA. **b** Radioimmunoassay of a methanol-extract of pea embryos following reverse-phase HPLC separation on Hypersil ODS

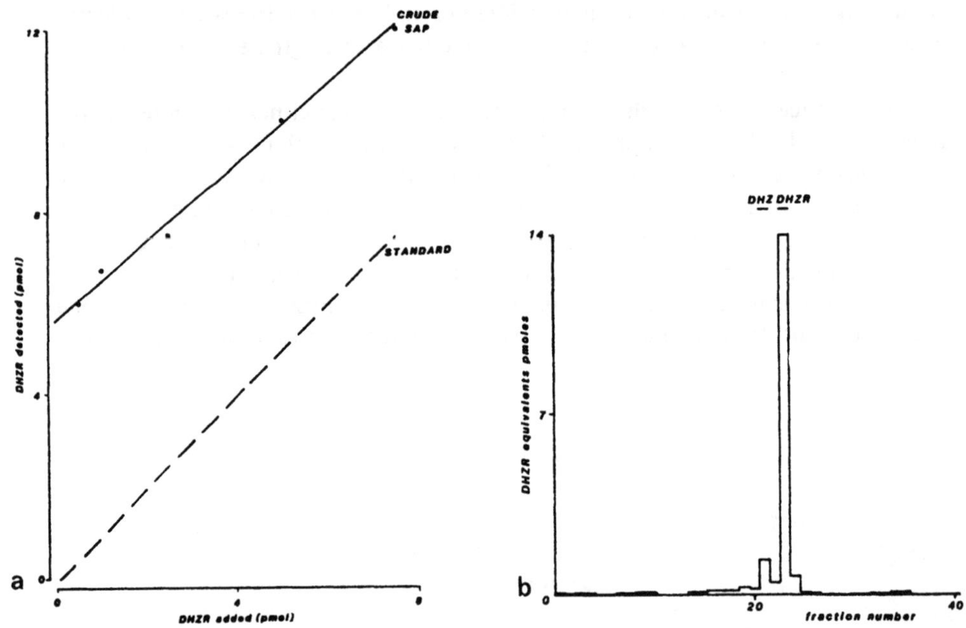

Fig. 5a, b. Internal standardisation of sap from *Phaseolus vulgaris* L. **a** Immunoreactive material in ten-fold diluted sap from *P. vulgaris* (assayed as in Fig. 2) after the addition of different amounts of DHZR. **b** Radioimmunoassay of sap following reverse-phase HPLC separation on Hypersil ODS

[4, 9, 12, 13, 17]. In some of these comparisons [13], validation was only possible upon purification of the extracts. Unfortunately, many of the advantages of the immunoassay as a method for quantifying plant hormones are lost if extensive purification is required. As Sandberg [13] has pointed out, perhaps the main advantage of the immunoassay resides in the rapidity of the technique.

2.2 Detection

The rapidity of the technique should not be ignored, however, since it allows the immunoassay to be used very effectively as a detection system alone. If one considers the purification scheme in Fig. 6, which is of the type used to purify a cytokinin prior to mass spectral identification, each step requires a detection system to locate the elution position of the plant hormone. In this instance, the soybean bioassay was used, which meant that each step required a 3 week delay before one could proceed to the next step. Other analyses, which provided information on the nature of the compound isolated using this scheme (at the bottom of the figure), also involved using a bioassay as a detector. Although this is an extreme example and many assays are much quicker, an immunological assay would speed up the process considerably. The repeated use of the assay in such a scheme would also lend its application to the successive approximation method. An advantage, in this case, may be an assay which used a wide-range antibody or group-selective antibody initially rather than the highly specific ones mainly in use at present. This would avoid serial assays, especially at early stages, for each species of antigen (e.g. Fig. 6; [17]. Such antibodies may be obtained during screening after immunisation with a single compound, if McAb are being sought [15]. Relative cross-reactivity can also be

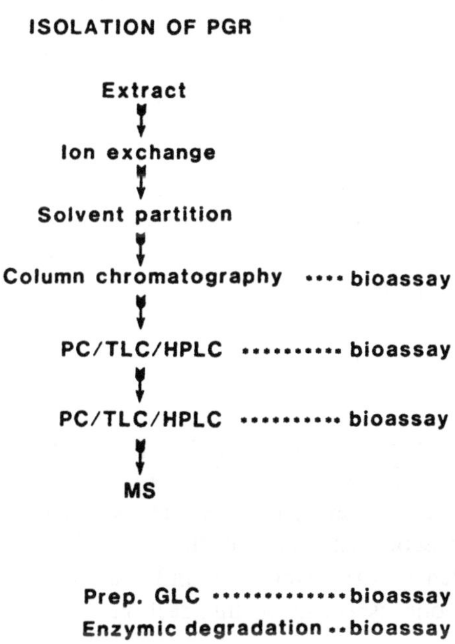

Fig. 6. Flow chart for the isolation of cytokinins from plant tissues. The chart indicates the steps involved in the isolation and purification of a cytokinin for mass spectrometry. The points at which it is necessary to detect the biologically active material by bioassay are indicated

Table 2. Cross-reactivity of cytokinins (%) with anti-DHZR antiserum using different tracers

	Tracer		
	[^{3}H]-DHZR	[^{3}H]-DHZ	[^{3}H]-DHZR dialcohol
Competitor			
DHZR	100.0	100.0	100.0
DHZ	23.3	60.0	40.0
ZR	7.8	18.8	5.8
Z	0.4	1.4	1.0

altered by using different tracers (Table 2) with the possibility of increasing the range of the antibody. Nevertheless, it should be reiterated that the immunoassay cannot be used as a substitute bioassay for detecting biological activity. Thus, with a complete unknown, biological activity must be assessed.

2.3 Immunopurification

If interference remains a problem and access is available to other physicochemical techniques, it may be better to use the antibody to shorten the conventional purification techniques by producing affinity columns. The preparation of columns for immunoaffinity purification was dealt with in detail by Davis in this workshop. With the development of reliable HPLC-MS techniques based on thermospray ionisation [2, 16], one can envisage the purification procedure being reduced to a single passage through an immunoaffinity column followed by HPLC-MS, thus considerably reducing the possibility of losses through purification. Since resolution of different plant hormones within a given group can be achieved using HPLC, it may be better to employ antibodies that show group selectivity as mentioned above to ensure all types of plant hormone within a class are retained. An alternative to this would be to produce mixed-bed affinity columns, where the columns contain antibodies, say, for the cytokinins, to zeatin-like *and* isopentenyladenine-like cytokinins or group selective antibodies, e.g. all N^6-substituted adenines. Such columns could then be followed with others of increased selectivity if desirable.

2.4 Immunolocalisation

Antibodies have been used extensively in plant science to localise antigens in particular cells in the plant and within these cells (see [6] for a review). There are several problems with the localisation of plant hormones, however, notably the fact that they are soluble in most solvents used in microscopy and may readily be leached from the tissue and the fact that their abundance is relatively low and thus, the likehood of labelling them with antibodies is poor, especially in thin sections. The attempts to localise plant hormones are few – one using low temperature techniques for cytokinins [21] and one using chemical coupling to fix ABA in situ to proteins in the cell and thus prevent its loss during processing [14]. There is no doubt, however, that all these techniques may produce artifacts and, therefore, it is essential that biological controls are used in addition to the normal chemical and immunological ones [6]. Hence, for example, one should have two similar tissues to compare, in which it is known that the hormone is present at markedly different levels. In the work on *Chenopodium,* this was achieved using water stress to increase the ABA content of the tissues. It is highly desirable also to know whether the plant hormone of interest can actually be obtained from the plant or organ in question.

3 Indirect Uses of Antibodies

As mentioned, the indirect uses of antibodies in PGR research may well be more beneficial than the immunoassay. Antibodies against metabolic enzymes allow the possibility of localising the sites of biosynthesis and catabolism much more easily and accurately than any attempts to locate a pool of hormone in a tissue. If enzymes are used, many of the problems associated with the localisation of the free hormone itself by microscopy are negated. Furthermore, they provide more concrete evidence for the sites of metabolism of the plant hormones, since pools in any event may merely indicate the site of accumulation, and they open the way to molecular biological approaches [3] for studying hormone metabolism. Antibodies against receptors and other molecules with which the plant hormone may interact (e.g. transport carriers, [5]) provide information on its mode of action. In addition, the use of McAb can circumvent the need to obtain such antigens in pure form in many instances. Thus, anti-receptor antibodies may be obtained from the same serum as anti-hormone antibodies or may be obtained by re-injecting the anti-hormone antibody, making use of the anti-idiotypic response [7]. An antibody against an auxin transport carrier has been obtained by screening a series of McAb raised against a general pea membrane preparation for binding to a specific auxin transport inhibitor [5]. Finally, it may be possible to obtain indirectly new PGRs by isolating unknown immunoreactive fractions in plant extracts which have similar conformational properties to the plant hormones (and, hence, would bind to antibodies raised against them) and, therefore, possibly similar biological properties.

4 Conclusions

In plant growth regulation research, there is little doubt that immunoassays will be used to an increasing extent to quantify the levels of plant hormones. The ease of use and the rapidity of the assays almost ensures this. In this respect, they have features in common with many bioassays which, it has been suggested, made the latter so popular [10]. Nevertheless, if we are to make certain that they do not fall into the disrepute that befell bioassays, we must be aware of their limitations. Thus, each tissue should be assessed for the possibility that it contains substances that will interfere in the immunoassay and extract purification applied when necessary. This could mean that each sample in a series, let alone from a single tissue, may need checking if the interference varies. Immunoassays, therefore, may become less attractive for quantifying plant hormones. In a recent article, Rosher et al. [12] stated that 24 assays per sample were required to obtain sufficient precision in their ABA immunoassay.

The use of antibodies is increasing rapidly in all branches of the plant sciences. Even if the attractiveness of the immunoassay for quantification is reduced, antibodies clearly have a gamut of other uses in PGR research. As long as we ensure that their limitations are recognised and they are integrated with the other techniques available, progress in our understanding of plant hormones must surely be helped and not hindered.

References

1. Chard T (1982) In: Work TS, Work E (eds) Laboratory techniques in biochemistry and molecular biology, vol 6(2). Elsevier/North Holland, Amsterdam
2. Chapman JR, Pratt JAE (1985) Lab Pract 34:7
3. Cuming AC, Williams RS, Cullimore JV (1986) In: Wang TL (ed) Immunology in plant science. Cambridge University Press, Cambridge, p 137 (Society for experimental biology series, 29)
4. Ernst D, Schafer W, Oesterhelt D (1983) Planta (Berl) 159:216
5. Jacobs M, Gilbert SG (1983) Science 220:1297
6. Knox RB (1982) In: Bullock GR, Petrusz P (eds) Techniques in immunocytochemistry, vol 1. Academic Press, London, p 205
7. Marx JL (1985) Science 228:162
8. Pengelly W, Meins F Jr (1977) Planta (Berl) 136:173
9. Pengelly W. Bandurski RS, Schulze A (1981) Planta (Berl) 68:96
10. Reeve DR, Crozier A (1980) In: MacMillan J (ed) Molecular aspects of plant hormones. Springer, Berlin Heidelberg New York, p 203 (Encyclopedia of plant physiology new series, vol 9)
11. Robard D (1974) Clin Chem 20:1255
12. Rosher PH, Jones HG, Hedden P (1985) Planta (Berl) 165:91
13. Sandberg G, Ljung K, Alm P (1985) Phytochemistry (Oxf) 24:1439
14. Sotta B, Sossountzov L, Maldiney R, Sabbagh I, Tachon P, Migniac E (1985) J Histochem Cytochem 33:201
15. Trione EJ, Krygier BB, Banowetz GM, Kathrein JM (1985) J Plant Growth Regul 4:101
16. Vestal ML (1984) Anal Chem 56:2590
17. Wang TL, Futers TS, McGeary F, Cove DJ (1984) In: Crozier A, Hillman JR (eds) The biosynthesis and metabolism of plant hormones. Cambridge University Press, Cambridge, p 135 (Society for experimental biology seminar series 23)
18. Weiler EW (1982) Physiol Plant 54:230
19. Weiler EW (1984) Annu Rev Plant Physiol 35:85
20. Weiler EW, Eberle J, Mertens RM, Atzorn R, Feyerabend M, Jourdan PS, Arnscheid A, Wieczorek U (1986) In: Wang TL (ed) Immunology in plant science. Cambridge University Press, Cambridge, p 27 (Society for experimental biology seminar series, 29)
21. Zavala ME, Brandon DL (1983) J Cell Biol 97:1235–1239

Validation of Immunoassays

W.L. PENGELLY[1]

1 Introduction

Immunoassays provide rapid and sensitive methods for the quantitative analysis of compounds in complex mixtures. The ability to analyse crude tissue extracts can be attributed to the high specificity of antibody-antigen binding. A high degree of binding specificity does not guarantee a valid assay, however, and substances with weak affinities for the antibody will interfere with antigen binding if present at sufficiently high concentrations. Interference may also arise from nonspecific inhibition, in which case the interfering compound need not show structural similarity to the antigen of interest. Since it is not possible to predict the nature or occurrence of interfering substances, it is important that immunoassays be validated for accuracy.

The most rigorous validation of immunoassays is by direct comparison with analytical methods which provide unequivocal evidence of chemical purity, e.g., mass spectrometry [3]. Since routine comparisons with other methods would abrogate the advantages of immunoassay, validation by direct comparison is usually limited to a few examples. This may be useful for many clinical applications where it may be assumed that the examples chosen are representative. In most studies of plant growth substances, however, the investigator is interested in comparing species or tissues in different physiological states or stages of development, where the composition of tissue extracts may vary markedly from sample to sample. Thus, validation based on a few examples may fail to uncover serious errors resulting from interference with the assay. In this paper I briefly describe the principles and limitations of simple procedures for routine validation of immunoassays. These procedures depend on changes in antibody-antigen binding behaviour which occur in the presence of interfering substances.

1 Department of Chemical, Biological and Environmental Sciences, Oregon Graduate Center, 19600 N.W. Von Neumann Drive, Beaverton, OR 97006-1999, USA

Plant Growth Substances 1985
(ed. by M. Bopp)

2 Antibody-Antigen Binding

All immunoassays provide a direct or indirect measure of antibody-antigen binding at equilibrium. The dissociation constant, K_d, for the binding reaction can be described by the equation:

$$K_d = \frac{[Ab] \cdot [L]}{[AbL]}, \tag{1}$$

where L is the concentration of the free ligand or antigen, Ab is the concentration of free antibody-binding sites, and AbL is the concentration of bound antigen.

A useful derivation of the binding equation relates the fraction of antigen bound (B) to the total antigen concentration:

$$B = \frac{[AbL]}{[AbL] + [L]} = \frac{[Ab] \cdot [L]}{K_d\,([AbL] + [L])}. \tag{2}$$

A semi-logarithmic plot of B as a function of total antigen concentration yields a standard curve with which unknown samples are compared (Fig. 1). The fraction of antigen bound can be measured directly using radiolabeled antigen as in radioimmunoassay (RIA), or indirectly using antigen-enzyme conjugates as in enzyme-linked immunosorbent assay (ELISA) [5]. Various linear and nonlinear curve-fitting procedures have been used to describe the standard curve for precise comparison with samples [4].

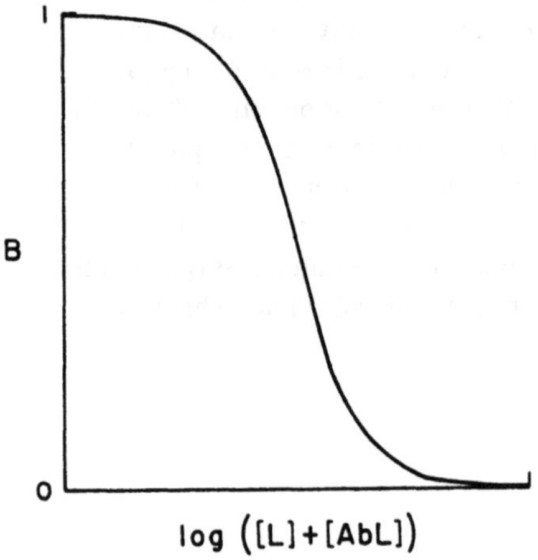

Fig. 1. The immunoassay standard curve as described by Eq. (2)

2.1 The Effect of Antibody Concentration on the Binding Curve

From Eq. (1) it is evident that the concentrations of both free and bound antibody are important factors determining the binding equilibrium. The influence of antibody concentration on the binding curve can be illustrated when half of the antigen is bound (i.e., B = 0.5). Since the concentrations of free and bound antigen are equal, then:

$$K_d = [Ab] \tag{3}$$

Since [Ab] is constant, it follows that [AbL] will increase with increasing antibody concentration. Since free and bound antigen are at equal concentrations, the point on the binding curve where half the antigen is bound will be displaced to a higher total antigen concentration. Thus, increasing the antibody concentration will shift the binding curve shown in Fig. 1 to the right, making assays less sensitive.

Conversely, assays can be made more sensitive by lowering the antibody concentration, with the limit of sensitivity determined by K_d. For example, it can be seen that the relationship described in Eq. (3) will only be true if the total antibody concentration is greater than the dissociation constant. Thus, when the total antibody concentration is less than or equal to K_d, half of the antigen can no longer be bound. In other words, the total antibody concentration determines the theoretical limit for B.

A general expression which relates K_d to [Ab] at any fraction bound can be derived from Eqs. (1) and (2):

$$K_d = \frac{(1 - B)}{B} \cdot [Ab]. \tag{4}$$

This relationship, shown graphically in Fig. 2, describes the theoretical limit for B at any antibody concentration. It can be seen that in order to bind nearly all of the

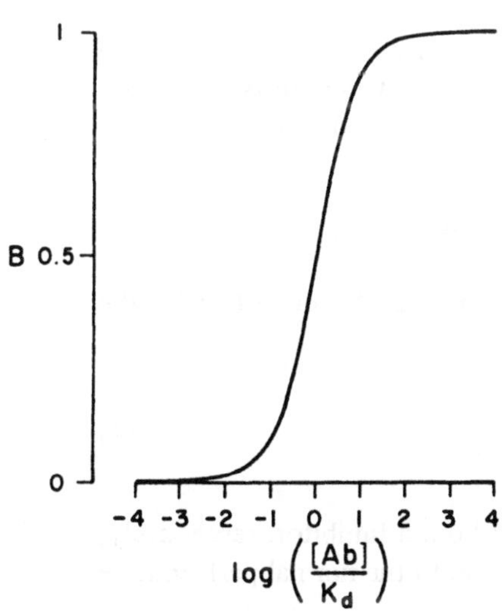

Fig. 2. The relationship between the fraction of antigen bound (**B**) and the free antibody concentration. (See text for details)

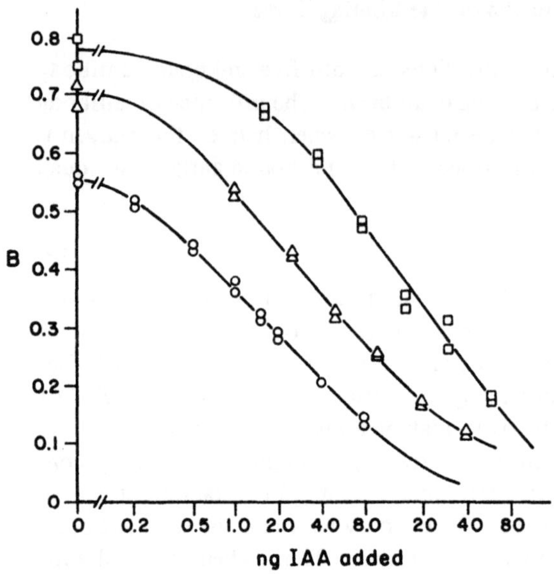

Fig. 3. The effect of antibody titer on the RIA standard curve. Antiserum dilutions used were 1:10 (□), 1:20 (△), and 1:40 (○). (Data from Pengelly and Meins [2])

antigen, the total antibody concentration must exceed the binding constant by at least two orders of magnitude. On the other hand, if the antibody is diluted to a level well below K_d, only a small fraction of the antigen can be bound.

When the total antibody concentration is near K_d, small changes in antibody concentration cause relatively large changes in B. Figure 3 shows RIA-binding curves obtained at three different dilutions of antibody. With dilution of the antibody, the binding curves are displaced to the left, making the assay more sensitive. However, antibody dilution also reduces the maximum level of binding, decreases the slope of the binding curve, and thereby lowers the resolution of the assay. Adjusting antibody concentration so that the maximum value for B is about 0.5 provides a good compromise between maximizing sensitivity and minimizing the loss of resolution.

From the above discussion it is clear that a change in antibody concentration leads to a displacement of the binding curve. It follows that inhibitors, which act by inactivating antibodies, will have the same effect on the antibody-antigen binding.

2.2 The Effect of Competitive Inhibitors on Antigen Binding

The binding of inhibitors by antibodies at equilibrium can be described by the equation:

$$K_i = \frac{[Ab] \cdot [I]}{[AbI]}, \tag{5}$$

where I and AbI are the concentrations of free and bound inhibitor, respectively, and K_i is the dissociation constant for inhibitor binding. In the normal performance of immunoassays, it is assumed that antibody not bound to antigen is in the free form. With competitive inhibition, however, the concentration of antibody not bound to antigen is the sum of [Ab] and [AbI]. Thus, the apparent binding constant of the antigen, K_{app}, can be defined as follows:

$$K_{app} = \frac{([Ab] + [AbI]) \cdot [L]}{[AbL]} . \tag{6}$$

By subsitution for AbI using Eq. (5) an expression is obtained which relates K_{app} and K_d:

$$K_{app} = K_d \cdot \left(1 + \frac{[I]}{K_i}\right). \tag{7}$$

This shows that the apparent binding constant differs from the true binding constant by a factor related to both the concentration and affinity of the inhibitor. It is also evident from Eq. (7) that inhibitors with different antibody affinities can bring about the same K_{app}, if present at the appropriate concentrations.

A special form of Eq. (7) occurs when the inhibitor has an affinity constant which is equal to that of the antigen:

$$K_{app} = K_d + [I] . \tag{8}$$

In this case, K_{app} is determined only by free-inhibitor concentration. Since an analogous relationship would occur if the immunoassay were contaminated with an unknown amount of antigen, it is clear that the binding of inhibitor and antigen cannot be distinguished when they have the same antibody affinity.

The value for K_{app} is dependent on both antigen and inhibitor concentrations, since a change in either would affect [I]. It is this change in K_{app}, with the addition of antigen or inhibitor, which provides the basis for assay validation.

3 Detection of Inhibitors

By decreasing the availability of antibody-binding sites, inhibitors make immunoassays more sensitive than they should be, giving rise to spurious values which are too large. One way to detect interference is by the addition of internal standards of antigen to the sample, making sure that the total assay volume is not changed. In the following examples, calculations were based on a total antibody concentration equal to K_d, giving a standard curve with a maximum value for B of 0.5.

3.1 Nonspecific Inhibition

The loss of antibody-binding sites due to nonspecific inhibition results in a change in the binding curve as shown in Fig. 3. Figure 4 shows what happens when binding curves altered in this way are compared with the less sensitive standard curve. In the

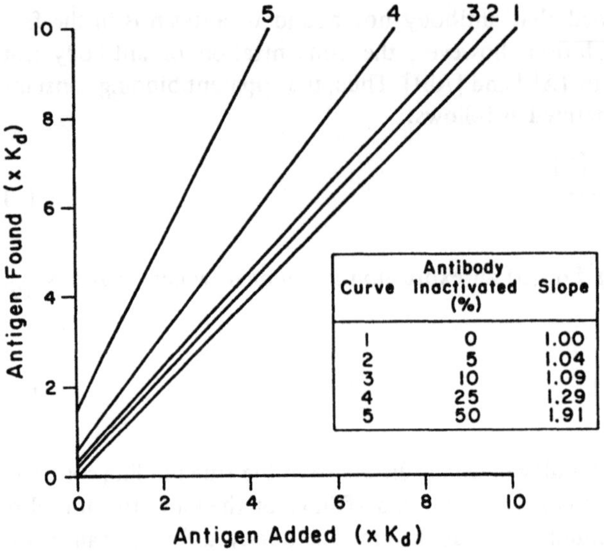

Fig. 4. The effect of nonspecific inhibitors on antigen binding. *Inset* shows the slope of each curve and the fraction of antibody destroyed. The initial antibody concentration was equal to K_d

absence of inhibitor there is no loss of antibody, and a plot of antigen found as a function of antigen added yields a straight line with a slope of unity that passes through the origin. With the progressive loss of antibody sites, there is an increase in the *y*-intercept due to the decline in the maximum value for B, and an increase in the slope of the line due to the increased sensitivity of the assay. Therefore, when the assay is inhibited nonspecifically, the addition of antigen to the assay will cause a greater-than-expected increase in the amount of antigen found, and the deviation of the answer from the expected value will be proportional to the degree of interference.

3.2 Competitive Inhibition

Competitive inhibitors also reduce the availability of binding sites, but in contrast to nonspecific inhibition, these binding sites are not irreversibly lost and may be occupied by antigen under appropriate conditions. The availability of inhibitor-occupied sites for antigen binding depends on inhibitor affinity (K_i).

Figure 5 shows the effect of K_i on the measurement of antigen. In the absence of inhibitor, a plot of antigen found as a function of antigen added again yields a straight line passing through the origin with a slope of unity. The figure also shows five lines that would be obtained in the presence of inhibitors with different antibody affinities. In each case, the amount of inhibitor was calculated to give the same amount of inhibitor-antibody complex in the absence of antigen, and hence, each of the lines has the same *y*-intercept and would provide the same wrong answer. The slopes of these lines differ markedly depending on the value of K_i. When K_i equals K_d, then the slope of the line is unity, showing that the inhibitor cannot be distinguished from the antigen. When K_i differs from K_d, then the slope of the line deviates from unity, giving values less than 1.0 when K_i is greater than K_d, values greater than 1.0 when K_i is less than K_d.

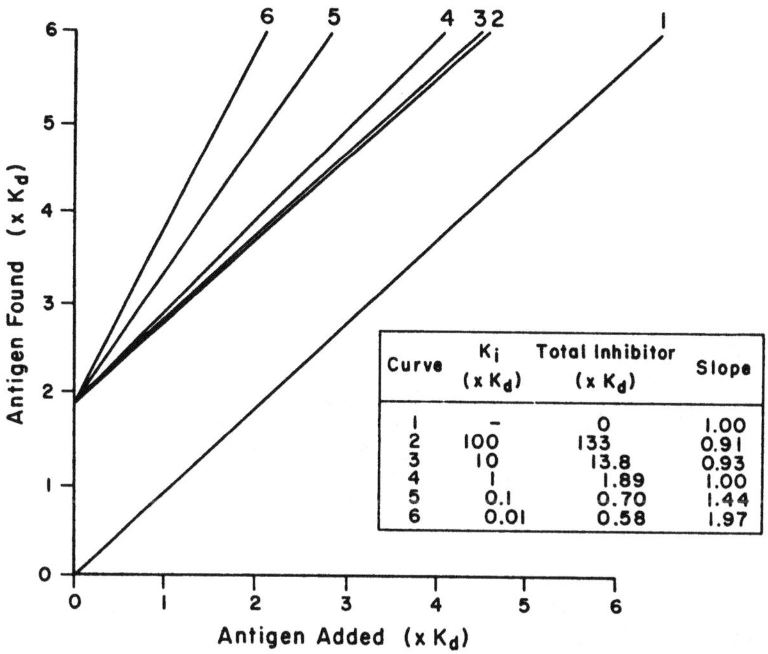

Curve	K_i (x K_d)	Total Inhibitor (x K_d)	Slope
1	-	0	1.00
2	100	133	0.91
3	10	13.8	0.93
4	1	1.89	1.00
5	0.1	0.70	1.44
6	0.01	0.58	1.97

Fig. 5. The effect of competitive inhibitors on antigen binding. *Inset* shows the slopes for curves obtained at different values of K_i and the amount of each inhibitor present in the assay. The total antibody concentration was equal to K_d

It is also evident from Fig. 5 that high-affinity inhibitors deflect the slope of the internal standard curve much more effectively than low-affinity inhibitors. This asymmetry can be attributed to the amount of compound required to inhibit the assay (see inset, Fig. 5). When the inhibitor has a low affinity, then the inhibitor concentration must exceed that of the antibody (K_d) to occupy a sufficient fraction of the binding sites. Under these conditions most of the inhibitor is in the free form (cf. Fig. 2) and changes in [I] will lead to relatively small changes in K_{app} [Eq. (7)]. In contrast, high-affinity inhibitors must occur at concentrations near to that of the antibody to occupy a substantial portion of the binding sites. Under these conditions, the inhibitor concentration exceeds K_i, most of the inhibitor occurs in the bound form, and even a small change in [I] can result in a substantial change in K_{app}. Thus, high-affinity inhibitors are easier to detect than low-affinity inhibitors.

Another method of assay validation is to simply assay different volumes of the sample, again keeping the total assay volume constant. With this method, both antigen and inhibitor concentrations are changed proportionately. A plot of antigen found as a function of assay volume will give a straight line which shows additivity if there is no interference or if K_i equals K_d (Fig. 6). If K_i differs from K_d, then this plot fails to show additivity and is nonlinear. When K_i is less than K_d, then the line curves markedly upward. Conversely, the line curves downward when K_i is greater than K_d, although curvature is so small that the plot appears to be linear.

Alternatively, one can test for parallelism by plotting results with the same method used to display the standard curve, assuming a linear, curve-fitting procedure

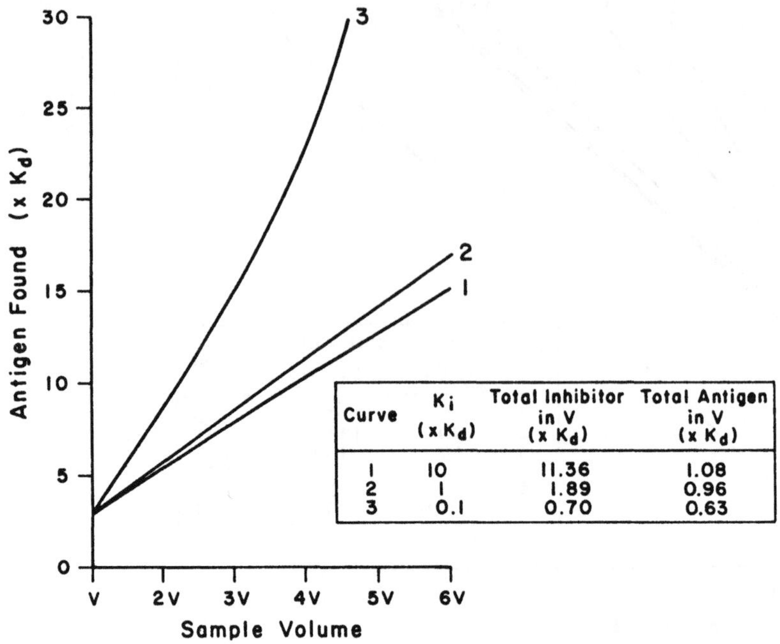

Fig. 6. Detection of competitive inhibitors by analysis of different volumes of sample. *Curve 2* is linear and shows additivity. *Curves 1* and *3* are nonlinear and do not show additivity

is used. In this case, the logarithm of sample volume is substituted for the logarithm of antigen concentration on the abscissa. This plot will yield a line parallel to the standard curve if there is no interference, whereas the line will be nonparallel with an inhibited assay [4].

4 Conclusions

Interference in immunoassays reduces the availability of antibody-binding sites and changes K_{app} for the antigen. In theory, all interference can be detected except for that caused by a competitive inhibitor with the same antibody affinity as the antigen. In practice, however, methods for detecting inhibition are not very sensitive to competitive inhibitors showing dissociation constants greater than K_d [1], and compounds with binding constants close to that of the antigen may be difficult to detect [2]. It is important, therefore, to establish statistical quality control for routine evaluation of binding curves [4]. It is also clear that analysis of antigen-binding behavior cannot provide an absolute test for accuracy, and direct comparison of immunoassays with other methods remains an important part of assay validation.

Acknowledgments. Support to the author from the National Science Foundation, DMB-8417087, and a travel award from the National Science Foundation to the American Society of Plant Physiologists, DMB-8504608, are gratefully acknowledged.

References

1. Cekan Z (1975) J Steroid Biochem 6:271
2. Pengelly WL, Meins F Jr (1977) Planta (Berl) 136:173
3. Pengelly WL, Bandurski RS, Schulze A (1981) Plant Physiol (Bethesda) 68:96
4. Rodbard D (1974) Clin Chem 20:1255
5. Weiler EW (1984) Annu Rev Plant Physiol 35:85

Immunoaffinity Columns for the Isolation and Analysis of Plant Hormones

G.C. DAVIS, M.B. HEIN, D.A. CHAPMAN, B.C. NEELY, C.R. SHARP, R.C. DURLEY, D.K. BIEST, B.R. HEYDE, and M.G. CARNES[1]

1 Introduction

One of the most promising developments to emerge in recent years has been the application of immunochemical methods for plant hormone analysis. Antibodies to the major plant hormones are being used as a means of quantitiation and assays have been developed with detection limits down into the picomole range. In the near future antibodies may well provide the ability to establish the site(s) of hormone synthesis and storage via immunocytochemical techniques. Antibodies can be extremely selective chemical reagents whose potiential in plant hormone research is only now beginning to be realized.

In the development of a radio- or enzyme-immunoassay, a certain amount of attention is directed toward the antibody selectivity. This is due to the fact that the selectivity translates directly into the speed of analysis (the more selective the antibody, the less sample purification prior to final analysis). In terms of analyzing plant hormones, one has to question how desirable or prudent it is to expend a lot of time and effort in raising highly selective antisera. If one considers the number of identified gibberellins (over 70), it soon becomes apparent that to raise antibodies unique to each structure would be a Herculean task (if not impossible). A more viable approach would be to prepare antibodies of defined cross-reactivity that would be applicable to the majority of gibberellins that are found in higher plants. These antibodies could then be used in combination with a simple HPLC clean-up/purification step for qualitative or quantitative analysis by subjecting fractions of the HPLC eluent to immunoassay. This approach has been used by Morris and his colleagues for the analysis of cytokinins [3].

Alternatively, one could employ immunoadsorption columns of immobilized, cross-reactive antisera prior to a physicochemical method of analysis. These columns would act as selective chemical filters, passing all of the unwanted biochemical "noise", while retaining the compound of interest. Immunoaffinity columns have been used successfully for a number of years for large molecule isolation, but their use in small molecule isolation is only beginning. The method is, in fact, being pioneered by groups working with plant hormones [1, 2].

1 Monsanto Agricultural Products Company, 700 Chesterfield Village Parkway, Chesterfield, MO 63198, USA

Plant Growth Substances 1985
(ed. by M. Bopp)

The preparation and use of immunosorbents for the isolation and purification of plant hormones is discussed in this paper. Key measures of immunoaffinity column performance are capture efficiency, column capacity, specificity of adsorption, and speed of separation. A number of chemical and physical parameters can be manipulated to alter column performance. Of these, the influence of flow rate, linkage chemistry, and chromatographic support in optimizing the immunosorbent are examined in this report.

2 Material and Methods

Antibodies were raised in rabbits against plant hormones by immunizing with a hormone-protein conjugate. The hormones were linked to the surface of the protein carrier (bovine serum albumin) through the epsilon amino group of lysine. The covalent attachment of cytokinin ribosides and (+) abscisic acid to the protein carrier has been described elsewhere [5, 6].

Each rabbit was injected subcutaneously and intramuscularly with 1 mg of conjugate (multiple sites), every 3 weeks. Ten to 14 days following the third immunization, the serum was checked for titer. Rabbits displaying a measureable titer were boosted with 200 μg of conjugate approximately 2 weeks later. Boosts were given on a 3 to 4 week basis and the titer was checked approximately 10 days after each boost. In many instances, a serum titer for the hormones was in the 1:12 000 range with a few rabbits yielding titers of 1:40 000. The rabbit serum against individual hormones was pooled and then retitered. The pooled serum was used for column preparation.

The immunoglobulin G fraction of the antisera was isolated by passing the serum over a Protein A column. The IgG fraction was then dialyzed and concentrated prior to immobilization on a number of chromatographic supports. The immobilization of the antibodies to cyanogen bromide-activated Sepharose 4B (Pharmacia, Uppsala, Sweden) Affi-gel 10 (Bio Rad, Richmond, California), and Ultraffinity EP (Beckman Instruments, Inc. Berkely, California) were performed according to product guidelines. The attachment of the antibodies to glycophase silica (Diagnostic Specialities, Metuchen, New Jersey) and Fast Flow Sepharose 4B was performed after the protocol of Roy et al. [4].

3 Results and Discussion

For the initial evaluation of plant hormone immunosorbents, cyanogen bromide-activated Sepharose 4B was selected due to its availability and simple linking procedure. The antibody forms an isourea linkage with the support in carbonate buffer (pH 8.3). Typical capacities for Sepharose 4B columns prepared with the described antibodies are in the 1 to 2 μg range (4 ml bed volume). The factors which affect the capacity of the resultant column are the original serum titer and the amount of protein linked to the support. Since only a fraction (~5%) of the Protein A purified immunoglobulin represents antibodies against the solute of interest, the capacity could be increased by affinity purifying the antibody prior to attachment. However, this step is usually unnecessary for trace analysis as most samples have only nanogram quantities of hormone present. Thus, the preparation of these columns with antibodies of sufficient titer can be quite simple. Affinity purification may be a necessity if the titer of the serum is very low. A column prepared from serum with a titer of below 1:5000 would undoubtably benefit from this step.

Several Sepharose 4B columns have been prepared in our laboratory with Protein A purified antibody (anti-zeatin riboside, anti-dihydrozeatin riboside, anti-isopentenyl adenosine and anti–(+)abscisic acid). In some cases, mixed-antibody supports have been formulated in order to create a mixed-bed immunosorbent. For example, a soybean seed extract was passed through an immunosorbent column of antibodies against zeatin-, dihydrozeatin-, and isopentenyl-type cytokinins. The cytokinins were then eluted from the immunoaffinity column and injected onto a reverse-phase HPLC column. The cytokinins in the sample were easily separated and identified (Fig. 1). Cytokinins in the crude extract would have been impossible to quantitate if the extract had been injected directly onto the HPLC.

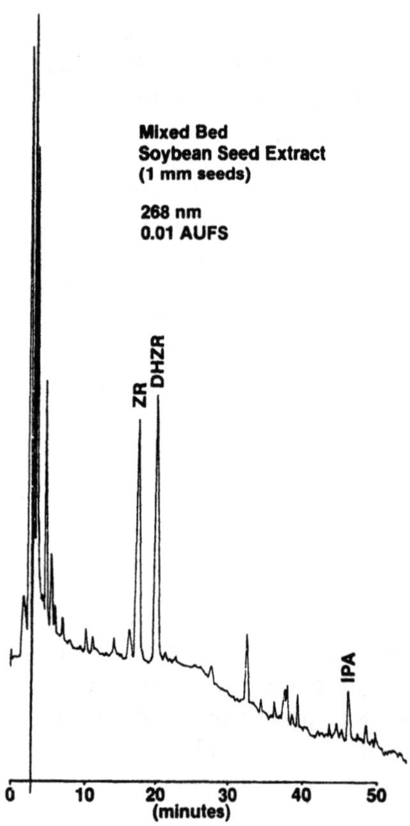

Fig. 1. HPLC separation of soybean seed extract (40 mg fresh weight) after affinity purification on a mixed-antibody immunosorbent column for the isolation of zeatin- and isopentenyl-type cytokinins. Column, Beckman Ultrasphere ODS; solvent A, 40 mM acetic acid adjusted to pH 3.35 with triethylamine; solvent B, acetonitrile; gradient conditions, 11% B for 18 min followed by a linear gradient to 18% B in 5 min followed by a linear gradient to 25% in 20 min; flow rate, 1.5 ml min^{-1}; temperature, ambient. *ZR* Zeatin riboside, *DHZR* dihydrozeatin riboside, *IPA* isopentenyl adenosine

The eluent used to remove the hormone from the column was 100% methanol. Chaotropic agents, high salt, and changes in pH were all ineffective in eluting the hormones from the column. Initially it was thought that such rigorous elution conditions as 100% methanol would irreversibly denature, or at best reduce the lifetime of the columns. This has not been the case as these columns are still in use after dozens of separations. Undoubtably, there has been some loss in capacity; however, since the column capacity far exceeds the sample content of hormone, these losses are never noticed under routine use. Another benefit of using methanol is that it is easily removed from samples by vacuum or evaporation under nitrogen. This facilitates the sample processing following the immunosorbent.

The flow rate through the Sepharose anti-cytokinin immunoaffinity column was 1 ml min^{-1} as controlled by a peristaltic pump at the column outlet. Under the conditions in which these columns were configured, a flow rate of 1 ml min^{-1} was near the maximum one could reasonably employ. When using Sepharose based supports one must be concerned with the column back pressure as the supports tend to fracture at pressures generated under high flow rates. An investigation of flow rate and capture efficiency demonstrated that even at the maximum flow, the columns were capturing 100% of the material being applied. This indicated that the antibodies were not limited in capture efficiency at low flow rates and that higher flow rates might be possible. Since this would speed sample throughput and lend the whole process to automation, other supports, which could tolerate higher back pressures and flow rates, were evaluated.

Glycophase silica, Fast Flow Sepharose 4B, and Ultraffinity EP were all examined as supports for plant hormone immunosorbents. Tables 1–3 list the results from flow studies on column capture efficiency. At all flow rates examined, the capture

Table 1. Effect of flow rate on capture efficiency of anti-zeatin riboside Fast Flow Sepharose

Flow rate (ml min^{-1})	Linear velocity (cm min^{-1})	Capture efficiency (%)
1.0	1.6	100
1.5	2.4	97
2.0	3.2	100
2.5	4.1	96
3.0	4.9	97
Zeatin riboside applied:	0.58 μg	
Bed height:	3.3 cm	
Bed volume:	2.0 ml	
Column diameter:	0.9 cm	
Column capacity	2.0 μg zeatin riboside	

Table 2. Effect of flow rate on capture efficiency of anti-dihydrozeatin riboside (glycophase-) silica

Flow rate ($ml min^{-1}$)	Linear velocity ($cm min^{-1}$)	Capture efficiency (%)
1.0	1.6	100
1.5	2.4	100
2.0	3.2	98
2.5	4.1	98
3.0	4.9	98
Dihydrozeatin riboside applied:	0.5 μg	
Bed height:	2.1 cm	
Bed volume:	1.3 ml	
Column diameter:	0.9 cm	
Column capacity:	1.0 μg dihydrozeatin riboside	
Silica:	500 A pore size, 200-400 mesh	

Table 3. Effect of flow rate on capture efficiency of anti-dihydrozeatin riboside Ultraffinity EP

Flow rate ($ml min^{-1}$)	Linear velocity ($cm min^{-1}$)	Capture efficiency (%)
1.0	5.9	100
2.0	11.8	99
3.0	17.6	98
4.0	23.3	97
Dihydrozeatin riboside applied:	0.5 μg	
Bed height:	5.0 cm	
Bed volume:	0.83 ml	
Column diameter:	0.46 cm	
Column capacity:	1.0 μg dihydrozeatin riboside	

efficiency was essentially 100%. This illustrated that there was a great amount of latitude in selecting the eluent flow rate and that flow rate limitations had not been approached. This makes sense from a theoretical point of view if one considers the binding of the antibody to hormone as being instantaneous. Instantaneous binding is a reasonable assumption as antibody Ka's of 10^{10} to 10^{13} have been measured.

Assuming instantaneous binding, the limiting factor for solute retention is the lateral diffusion of molecules in the stationary mobile phase within the support. Lateral diffusion to antibody-binding sites in the support is opposed by the longitudinal movement (i.e., linear velocity) of the solute molecules in the column. Given sufficient antibody density in the gel, flow rate is not a problem for small molecule diffusion into the gel. For large molecule separations (e.g., proteins) lateral diffusion is a limiting factor and low flow rates are recommended to maximize protein-antibody binding. In addition, it is recommended that the ligand (antibody) density be kept low to avoid steric problems with multiple site attachment of proteins. In the case of small molecule immunosorption, it is advisable to have the antibody density as high as possible. High density immunosorbents permit the use of smaller columns. This reduces the volume of eluent needed to elute the compounds from the column and also minimizes the amount of nonspecific background adsorption that can occur. Limiting the column volume also allows higher flow rates if column cross-sectional area is kept constant, since back pressure is a function of column-bed height.

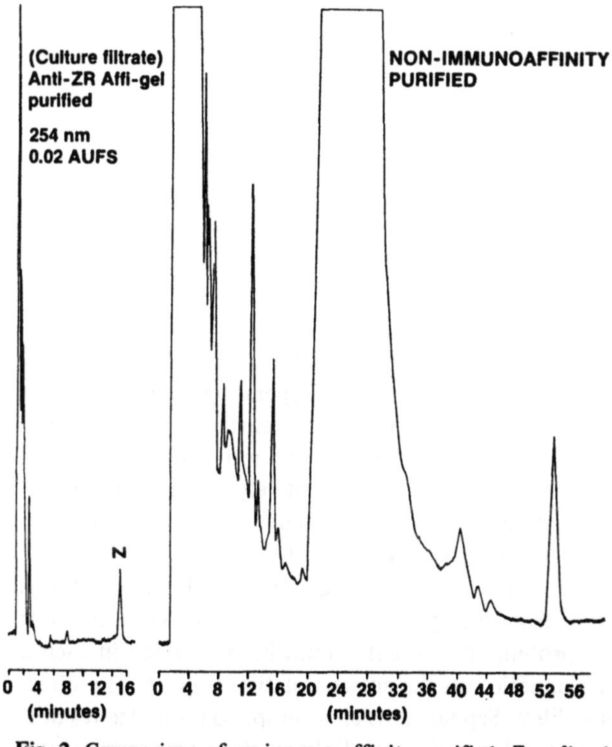

Fig. 2. Comparison of an immunoaffinity purified. *E. coli* culture filtrate to nonpurified filtrate by reverse-phase HPLC. The filtrate was from a culture of transformed *E. coli* (LE392) grown in minimal glucose medium. *E. coli* was transformed with pMON7036, which contains a clone (pKC7) of the vir region of an agrobacterium Ti plasmid. Column, Beckman Ultrasphere ODS; solvent, 40 mM acetic acid adjusted to pH 3.35 with triethylamine in 11% acetonitrile; flow rate, 1.5 ml min^{-1}; column temperature 40°C. One ml of culture filtrate was used for each sample. *Z* zeatin

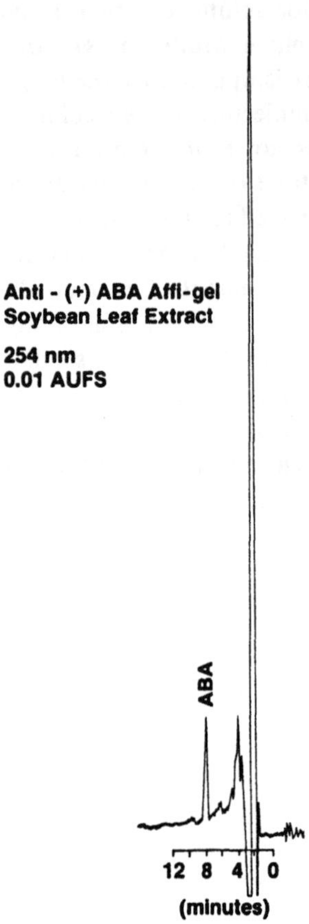

Fig. 3. Examination of a soybean leaf extract (200 mg) by HPLC following purification on an anti-(+)-ABA immunoaffinity column. Column, Bioanalytical Systems, Inc. Biophase C18; solvent, 25% acetonitrile in water adjusted to pH 3.0 with phosphoric acid; flow rate, 1.5 ml min^{-1}; temperature, ambient. *ABA* abscisic acid

Nonspecific background adsorption remains a critical hurdle in making the running of these columns a one-step purification procedure. Depending upon the linkage chemistry, a multitude of charged sites can be left on the surface of the gel, thus inviting ionic interactions of charged species with the surface. In the case of cyanogen bromide-activated Sepharose 4B the isourea linkage is positively charged at the pH's normally ecountered with affinity separations. This phenomenon accounts for much of the background observed as small peaks in Fig. 1. Traditionally, this has been a problem with the isourea linkage. In the case of Affi-gel 10, nonblocked carboxyl groups on the support surface arise from the hydrolysis of the N-hydroxysuccinimide ester from the spacer arm. The protein attachment should be performed in such a way as to avoid leaving charged moieties on the surface of the support. The direct coupling of the antibody to Fast Flow Sepharose is an attempt to minimize residual charged groups in the column bed. This and other linkage chemistry is currently being evaluated in our laboratory to overcome the problems of nonspecific adsorption.

Affi-gel 10 is a support that meets several criteria of an ideal immunosorbent. It has desirable flow properties under gravity conditions (2 ml min^{-1}), thus eliminating the need for a pump. Several columns could be run in parallel thereby increasing the sample through-put without expensive equipment. With this support, up to 35 mg of protein per ml of bed can be immobilized compared to the 5 to 10 mg recommended for cyanogen bromide-activated Sepharose 4B. This relatively high antibody immobilization helps limit the amount of support necessary in the column bed which in turn allows faster flow rates, reduced nonspecific adsorption, and reduced amount of eluent required for sample elution. Figures 2 and 3 demonstrate the use of Affi-gel as a support for use in purifying plant extracts prior to HPLC.

4 Conclusions

Immunochemical techniques offer tremendous potential for isolation and analysis of small molecules. Immobilized antibodies have been used for a number of years in the isolation of macromolecules, but are only beginning to be exploited for the isolation of small organic molecules. Data presented here clearly demonstrate that a number of solid supports can be used effectively for the immunoaffinity isolation of plant hormones. There has long been a need for such an economical and reliable method of isolating these molecules from plant samples. Immunoaffinity columns provide an ideal preparative separation prior to immunological, chromatographic, or mass spectral methods of determination. As such they provide an important advance in the investigation of the function of plant hormones.

References

1. Davis GC, Hein MB, Neely BC, Sharp CR, Carnes MG (1985) Anal Chem 57:638
2. MacDonald EMS, Morris RO (1985) Methods Enzymol 110:347
3. MacDonald EMS, Akiyoshi DE, Morris RO (1981) J Chromatogr 214:101
4. Roy SK, Weber DV, McGregor WC (1984) J Chromatogr 303:225
5. Weiler EW (1980a) Planta (Berl) 149:155
6. Weiler EW (1980b) Planta (Berl) 148:262

II. Hormones: Synthesis, Metabolism and Biochemistry

Dwarf Mutants in Maize – The Gibberellin Biosynthetic Pathway and its Molecular Future*

B.O. PHINNEY[1], M. FREELING[2], D.S. ROBERTSON[3], C.R. SPRAY[1], and J. SILVERTHORNE[1]

1 Introduction

Single gene dwarf mutants have been used to unravel the GA biosynthetic pathway in shoots of *Zea mays* (maize) [18, 19]. These GA mutants have provided critical evidence leading to the conclusion that GA_1 is the main endogenous gibberellin active per se in the control of shoot growth in maize [17]. The mutants also provide definitive evidence that shoot elongation is dependent on the level of endogenous GA_1, not on the total amount of GAs in the plant [19].

This paper reviews and brings up to date the genetic and chemical evidence for the unitary control of shoot elongation by GA_1. In addition, information will be presented on a new class of GA-responding mutants in maize, Mutator (*Mu*) dwarfs. These dwarf mutants are valuable tools to study GA-controlled growth, since *Mu*, an active transposon in mutator lines, can be used to clone the specific genes for GA biosynthesis. The approach can lead to information on the molecular control of hormonal level. The cloned genes can be used eventually to obtain mutant and normal enzymes in amounts that can be studied biochemically.

2 The Maize Gibberellins and the Early 13-Hydroxylation Pathway

The nine gibberellins, GA_{53}, GA_{44}, GA_{19}, GA_{17}, GA_{20}, GA_{29}, GA_{29}-catabolite, GA_1 and GA_8 have been identified by gas chromatography-mass spectrometry (GC-MS) in extracts from maize tassels and shoots ([9, 10] and Spray et al., unpubl. data) (Fig. 1 and 2). These GAs fit into the early 13-hydroxylation pathway, which originates from GA_{12}-aldehyde. GA_{12}-aldehyde in turn comes from mevalonic acid (MVA) via cyclization of geranylgeranylpyrophosphate (GGPP) to copalylpyrophosphate (CPP) to *ent*-kaurene. *ent*-Kaurene is then oxidized and rearranged to give GA_{12}-aldehyde. Feeds and re-feeds of labeled substrates to maize have demon-

* This talk was also given by BOP (Stephen Hales Lecture) at the ASPP Annual Meetings, Brown University, Providence, RI, USA, June 23–28, 1985

1 Department of Biology, University of California, Los Angeles, CA 90024, USA
2 Department of Genetics, University of California, Berkeley, CA 94720, USA
3 Department of Genetics, Iowa State University, Ames, IA 50010, USA

Plant Growth Substances 1985
(ed. by M. Bopp)

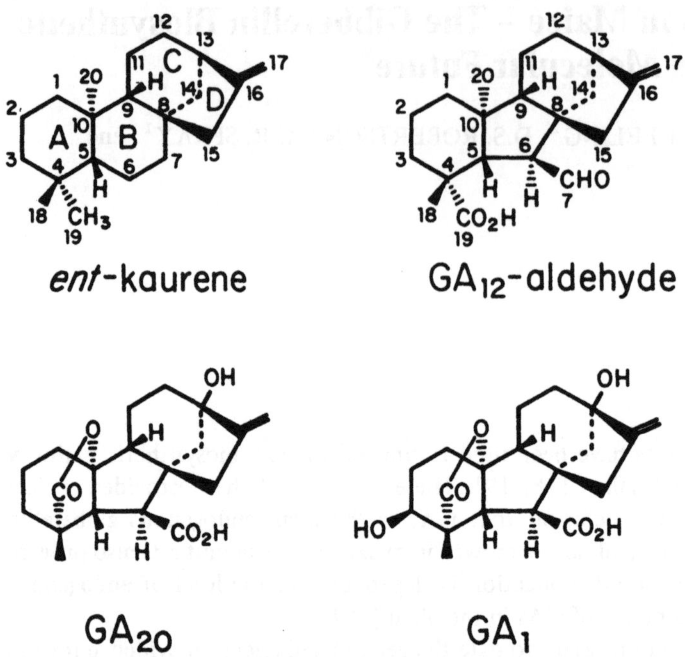

Fig. 1. Structures and numbering system for selected gibberellins and their precursors

strated the occurrence of the following steps in this pathway: CPP to *ent*-kaurene [8], *ent*-kaurenol to *ent*-kaurenal to *ent*-kaurenoic acid [28], GA_{53} to GA_{44} [10], GA_{20} to GA_{29} [27], GA_{20} to GA_1 [24, 27], and GA_1 to GA_8 ([4, 5] and Spray and Phinney, unpubl. data). Thus, there is good evidence for the presence of the early 13-hydroxylation pathway in the shoots of maize. Knowledge of the GAs endogenous to the maize shoot, together with definitive information on the position of these GAs in the pathway, have provided a critical background of information for the genetic studies reviewed in this paper.

3 Spontaneous Origin Dwarf Mutants

Dwarf-1, *dwarf*-2, *dwarf*-3, *dwarf*-5, and *anther ear*-1 mutants were isolated by maize geneticists more than 50 years ago as random mutations that appeared in field grown stocks (Fig. 2 [6]). The mutants were originally used to locate genes on chromosomes. The discovery that these dwarfs were gibberellin (GA)-requiring mutants [15] led to the current studies on the GA control of shoot growth in maize [18, 19].

The five mutants are especially useful as tools for experimental studies on the control of shoot growth for the following reasons: (1) each mutant is a simple recessive, implying a simple unitary control of growth at the biochemical level; (2) the mutants are nonallelic to each other, implying a different primary control for each mutant gene; (3) the dwarf phenotypes are expressed in the dark as well as

Fig. 2. The five GA-responding mutants of maize, *dwarf*-1, *dwarf*-2, *dwarf*-3, *dwarf*-5 and *anther ear*-1. The position of the metabolic block for each mutant is indicated below each plant. Note that the precise step controlled by *an*1 is not known; bioassay data place it prior to *ent*-kaurene

in the light ([26] and Phinney, unpubl. data), which may well eliminate light as a factor in the mechanism responsible for the dwarf phenotype; (4) the dwarf phenotypes are expressed in the stem from the early seedling stage to maturity, thus allowing the use of plant material over a wide range of growth stages; and (5) the phenotypes are highly uniform in their growth habit, especially when segregating from either inbred or hybrid lines.

The five dwarfing genes control shoot elongation by the blockage of specific steps in the metabolic pathway leading to the highly active gibberellin, GA_1 [18, 19] (Fig. 2). The evidence for this control is based on three lines of evidence: (1) growth responses of each mutant to specific maize GAs and their precursors. The patterns of responses strongly suggest the position of each mutant in blocking a specific step in the pathway; (2) radiolabeled feeding studies, where the metabolic fate of the label is followed after incubation of substrate with a mutant. The rationale is that substrates before the block will not be metabolized to product(s) after the block; on the other hand, substrates subsequent to the block will be further metabolized; and (3) comparison of the endogenous gibberellins present in normals and mutants. Depending on the position of the genetic block, specific endogenous GAs present in the normal, should be absent or present at very low levels in the mutants.

(There is also the possibility that substrates before the block may accumulate to levels higher than those found in the normals.)

The location of the genetic block for each of the five mutants (Fig. 2) was based initially on the relative bioactivities of the maize GAs and their precursors. The data suggested that the *dwarf*-1 mutant blocks the reaction, GA_{20} to GA_1; the *dwarf*-2 mutant, GA_{53}-aldehyde to GA_{53}; the *dwarf*-3 mutant, GA_{12}-aldehyde to GA_{53}-aldehyde; and the *dwarf*-5 and the *anther ear*-1 mutant, a step prior to *ent*-kaurene. [It should be noted that bioassay data locate the blocking position of the *dwarf*-5 and *anther ear*-1 mutants only in respect to *ent*-kaurene, since all precursors to *ent*-kaurene are biologically inactive as growth promotors in maize (Phinney, unpubl. data)]. There are now three lines of evidence that locate the steps blocked by the *dwarf*-5 and *dwarf*-1 mutants, as follows:

3.1 The *Dwarf*-5 Mutant

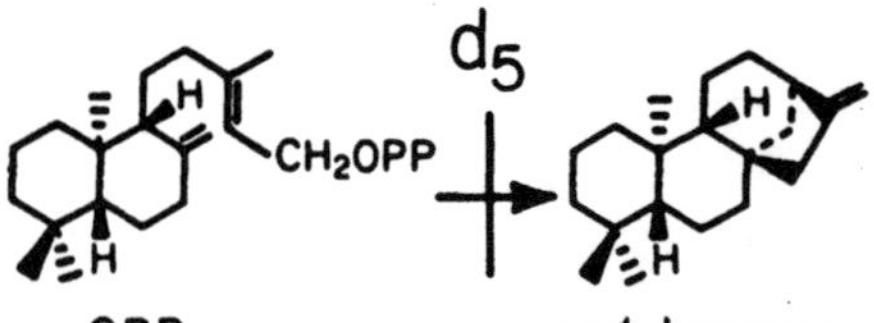

1. Growth Responses. The *dwarf*-5 mutant gives a normal-type growth response to a wide array of gibberellins [3, 16]. Among these, GA_1, GA_{20}, and GA_{53} (GAs endogenous to maize) are bioactive when assayed on *dwarf*-5 seedlings [18]. In addition, the GA-precursors, GA_{53}-aldehyde, GA_{12}-aldehyde, *ent*-kaurenoic acid, *ent*-kaurenol and *ent*-kaurene are also active [12, 18]. These bioassay data suggest that the genetic block is prior to *ent*-kaurene.

2. Feeding Studies. *Dwarf*-5 mutants synthesize *ent*-kaurene from MVA, GGPP, and CPP at levels one-fifth that of the normals. This information comes from feeds of [2-^{14}C]MVA, [^{14}C]GGPP, and [^{3}H]CPP to cell-free preparations of seedling shoots. In these feeds the label was recovered in *ent*-kaurene [8]. The marked reduction in the synthesis of *ent*-kaurene by the *dwarf*-5 mutant places the genetic block between CPP and *ent*-kaurene, a step controlled by *ent*-kaurene B synthetase [25].

3. Endogenous Gibberellins. (Spray et al., unpubl. data). We have now identified by GC-MS GA_{44}, GA_{19}, GA_{20}, and GA_1 from extracts of normal seedling shoots of maize. Similar extracts from *dwarf*-5 seedlings show no evidence for the presence of GA_{44}, GA_{19}, and GA_1, and only trace amounts of GA_{20}. These low levels or absence of endogenous GAs and their precursors would be expected if the *dwarf*-5 mutant is controlling a step prior to *ent*-kaurene.

3.2 The *Dwarf*-1 Mutant

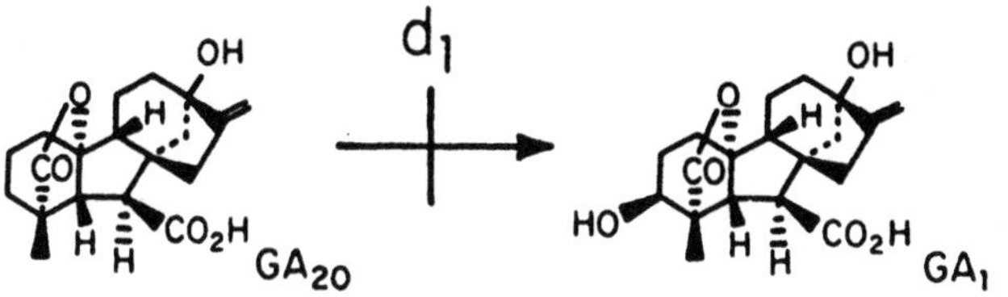

1. Growth Responses. In *dwarf*-1 assays, GA_1 is highly active whereas GA_{20} and GA_{53} are less than 1% as active as GA_1 when assayed on *dwarf*-1 seedlings; earlier precursors in the pathway are inactive when assayed on this mutant. These response data suggest that the *dwarf*-1 mutant is blocking the conversion of GA_{20} to GA_1 [18].

2. Feeding Studies. While normal and *dwarf*-5 seedlings have been shown to metabolize [^{13}C, ^{3}H]GA_{20} to [^{13}C]GA_1, no evidence has been obtained for this conversion in similar feeds to *dwarf*-1 seedlings [27]. The feeding studies support the conclusion that the *dwarf*-1 mutant is blocking the conversion of GA_{20} to GA_1.

3. Endogenous Gibberellins. (Spray et al., unpubl. data). GC-MS analyses of extracts from *dwarf*-1 seedlings demonstrate the presence of GA_{44}, GA_{19}, and GA_{20}. No evidence was found for the presence of GA_1. The apparent absence of GA_1 in *dwarf*-1 seedlings and its presence in normal seedlings also supports the conclusion that the *dwarf*-1 mutant is blocking the conversion of GA_{20} to GA_1.

3.3 Interpretations and Significance

Two conclusions come from the genetic and biochemical studies on the control of GA-dependent shoot elongation in maize. Firstly, it is only GA_1 that is active per se in the control of shoot elongation. Other GAs in the biosynthetic pathway are bioactive through their metabolism to GA_1. Interestingly, if other members of the pathway showed bioactivities similar to GA_1 there would be no dwarf mutants that show a differential growth response to exogenous GAs, no mutants that accumulate GA intermediates, and no mutants that show differential patterns of GA metabolism. Secondly, it is the amount of endogenous GA_1 that controls shoot elongation, not the total quantity of GAs in the shoot. It follows that studies correlating total amounts of endogenous GAs with growth have limited meaning in the control of GA-dependent shoot elongation.

The reader should be aware that the fate of GA_1 in terms of mechanism of action is a black box. It is not known whether GA_1 binds directly to a protein, or is metabolized to a product which in turn binds to the protein. One candidate for further metabolism of GA_1 is GA_3. However, it appears that metabolism of GA_1 to GA_3 is not necessary for the observed bioactivity of GA_1 [13]. This approach involves the synthesis of 1- and 2-methyl derivatives of GA_1. These derivatives cannot be metabolized to GA_3 because of the chemical blockage of carbons-1 and -2. An example of this is 2,2-dimethyl-GA_1 [2] which was found to be highly active in the *dwarf*-1 assay. Thus, the bioactivity of GA_1 is *not* dependent on its metabolism to GA_3.

Finally, recent evidence in the literature suggests that the early 13-hydroxylation pathway may be widespread in the shoots of higher plants. Single gene dwarf mutants are now known to block specific steps in the early 13-hydroxylation pathway for *Pisum sativum* (pea) [11, 20]; *Oryza sativa* (rice) [14], *Lycopersicum esculentum* (tomato), and *Arabidopsis thaliana* (wall cress) (Zeevaart, unpubl. data).

4 Transposable Element Dwarf Mutants

The general subject of transposable elements in plants has been reviewed by Freeling [7]. Transposable elements are small pieces of mobile DNA that are present in specific strains of plants and animals. Their random insertion into the genome can result in the production of visible mutants. The phenotype expressed by a particular mutant depends on the position of insertion of the transposable element into the genome. These mutants may be stable or unstable. (The basis for suspecting the presence of such elements in a particular strain is an observed high mutation frequency in that strain.) Robertson's Mutator (*Mu*) was discovered as a result of the genetic analysis of strains of maize that showed such elevated mutation rates [21–23]. Since then *Mu 1* has been isolated, cloned and sequenced [1]. *Mu 1* is a 1.367 Kb fragment of DNA. It has been [^{32}P]-labeled and, by the Southern blot technique, used as a probe to identify particular restriction fragments in nuclear DNA. Thus, the restriction fragment that is genetically linked to a particular phenotype may also be identified, isolated and cloned. Such an approach has been adopted in our investigation of *Mu* derived of dwarf mutant of maize. These studies should provide information on the control of GA biosynthesis at the molecular level. They offer the potential for the production of relatively large amounts of enzymes that catalyze biologically significant steps in the GA biosynthetic pathway, e.g., the 3β-hydroxylase enzyme (GA_{20} to GA_1), and the *ent*-kaurene synthetase B enzyme (CPP to *ent*-kaurene).

We have recently isolated a series of GA-responding dwarf mutants, each of which is presumably the result of the insertion into the genome of the transportable element, *Mu*. These five mutants, (*Mu*) *d*A, *d*B, *d*C, *d*D, and *d*E originated among F_2 progeny coming from selfed lines carrying multiple copies of *Mu*. The phenotypes of these five mutants will be described; preliminary studies on their molecular biology is presented as well as preliminary data on the location of the specific block for each mutant.

In the early seedling stage (Fig. 3) the dwarf phenotype of each mutant is indistinguishable from the typical GA-responding dwarf mutants of maize. Coleoptiles and mesocotyls are short; the first leaf blade is also short, broad, and deep green. At maturity the number of internodes of the mutants is similar to normals. Thus, the reduced stature is due primarily to length of internode, not number of internodes. All five mutants have anthers in the ears, i.e., they are andromonoecious,

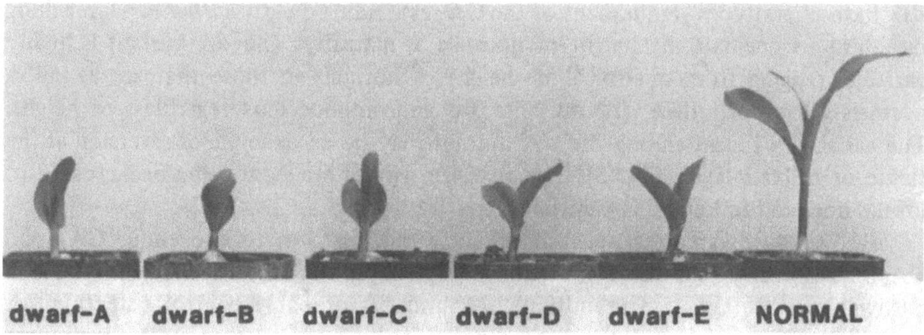

Fig. 3. Ten day old seedlings showing the dwarf phenotypes [Robertson's mutator (*Mu*)] (*left*) compared to a normal (*right*). Segregating mutat seedlings were uniform and clearly distinguishable from normals in the early coleoptile stage of growth

a property typical of the known GA mutants of maize. At maturity, the mutator dwarfs, *d*B and *d*E are very short, being one-fourth to one-fifth the height of normals (Fig. 4). Mature leaves are deep green, short, and crinkly, having phenotypes very similar to the GA-responding mutants, *dwarf*-1 and *dwarf*-5. The dwarfs, *d*C and

Fig. 4. *Mu d*B phenotype (*left*); normal (*right*). At maturity mutants are one-fifth the height of normals

*d*D have phenotypes reminiscent of the GA-responding dwarf, *anther ear*-1, reaching one-third to one-half the height of normals at maturity. The *d*A mutant is highly variable, ranging from one-third the height of normals to those that are as tall as normals. However, these *d*A mutants are andromonoecious regardless of height. The variability found among the *d*A mutant may be an example of reversion at the tissue or cellular level (Fig. 5). The five *Mu* dwarfs are now being backcrossed to inbred lines and to known GA mutants.

All five *Mu* dwarf mutants respond by normal growth to exogenous GA_1 and GA_{20}, suggesting that these mutants are blocking steps in the GA biosynthetic pathway prior to GA_{20}. The mutants, *d*A and *d*D probably control a step(s) prior to *ent*-kaurene since they give growth responses to both *ent*-kaurene and *ent*-kaurenoic acid. In contrast, the lack of response of the mutants, *d*B, and *d*E to *ent*-kaurene and *ent*-kaurenoic acid suggests that they block step(s) subsequent to *ent*-kaurenoic acid. Thus, we have a series of mutator dwarfs which probably control specific and different steps in the maize gibberellin biosynthetic pathway.

We have isolated nuclear DNA from the leaves of the *Mu* dwarfs, *d*A, *d*B, *d*E, and from non-*Mu* normals. Yields range from 200 μg to 1.0 mg DNA per plant. The DNA from each plant has been digested with the restriction enzymes, Bam HI and Hind III,

Fig. 5. Three *Mu d*A phenotypes (*right*); one normal plant (*left*). While the mutants were uniform in the seedling stage, variability soon appeared, with some mutants reaching the height of normals at maturity

and the fragments resolved on a 0.75% agarose gel, and transferred to nitrocellulose filters. The DNA from the shoots was then probed with the plasmid pMJ9. This plasmid had been previously labeled with [^{32}P] by nick translation. (pMJ9 contains the *Mu 1* sequence cloned into the Bam HI/Hind III site of pBR322). Using Southern blots, we have shown the presence of not more than 20 copies of *Mu 1* related sequences in any one dwarf plant. By comparing the banding patterns in sister dwarf plants, we have narrowed the number of unique copies to four. We are now in a position to clone the bonds of interest from the agarose gels. This approach can lead to the isolation and sequencing of the dwarfing gene(s) (and normal genes); the approach can provide the gene products (enzymes) that can be analyzed at the chemical/biochemical level.

5 Conclusions

1. GA_1 is the main gibberellin active per se in the control of elongation in the vegetative shoot of maize.
2. Other gibberellins in the shoot of maize are active through their metabolism to GA_1.
3. The amount of shoot elongation is correlated with the amount of endogenous GA_1, not the amount of all GAs in the shoot.
4. The unitary control of shoot elongation by GA_1 is probably widespread among flowering plants (Angiospermae).
5. Dwarf mutants induced by transposable elements provide powerful tools for the studies of the molecular basis for the control of hormone levels in plants.

Acknowledgments. Portions of the studies reported here were supported by a grant from the National Science Foundation (USA) (PCM 83-14388).

References

1. Barker RF, Thompson DV, Talbot DR, Swanson J, Bennetzen J (1984) Nucleic Acids Res 12:5955
2. Beale MH, MacMillan J, Spray CR, Taylor DA, Phinney BO (1984) J Chem Soc Perkin Trans I:541
3. Crozier A, Kuo CC, Durley RC, Pharis RP (1970) Can J Bot 48:867
4. Davies LJ, Rappaport L (1975) Plant Physiol (Bethesda) 55:620
5. Davies LJ, Rappaport L (1975) Plant Physiol(Bethesda) 56:60
6. Emerson RA, Beadle GW, Fraser AC (1935) Cornell University Agricultural Experiment Station Memoir, 180:1
7. Freeling M (1984) Annu Rev Plant Physiol 35:277
8. Hedden P, Phinney BO (1979) Phytochemistry (Oxf) 18:1475
9. Hedden P, Phinney BO, Heupel R, Fujii D, Cohen H, Gaskin P, MacMillan J, Graebe JE (1982) Phytochemistry (Oxf) 21:391
10. Heupel RC, Phinney BO, Spray CR, Gaskin P, MacMillan J, Hedden P, Graebe JE (1985) Phytochemistry (Oxf) 24:47

11. Ingram TJ, Reid JB, Murfet IC, Gaskin P, Willis CL, MacMillan J (1984) Planta (Berl) 160:455
12. Katsumi M, Phinney BO, Jefferies PR, Henrick CA (1964) Science 144:849
13. MacMillan J (1985) In: Randall DD et al. (eds) Current topics in plant biochemistry and physiology, Vol 4, University of Missouri, USA, p 53
14. Murakami Y (1972) In: Carr DJ (ed) Plant growth substances. Springer, Berlin Heidelberg New York, p 166
15. Phinney BO (1956) Proc Natl Acad Sci, USA 42:185
16. Phinney BO (1961) In: Klein RM (ed) Plant growth regulation. Iowa State University Press p 489
17. Phinney BO (1984) In: Crozier A, Hillman JR (eds) The biosynthesis and metabolism of plant hormones. Cambridge University Press, Cambridge, p 17
18. Phinney BO, Spray CR (1982) In: Wareing PF (ed) Plant growth substances 1982. Academic Press, London, p 101
19. Phinney BO, Spray CR (1985) In: Randall DD (eds) Current topics in plant biochemistry and physiology, vol 4, University of Missouri, USA, p 67
20. Reid JB, Murfet IC, Potts WC (1983) J Exp Bot 34:349
21. Robertson DS (1978) Mutat Res 51:21
22. Robertson DS (1980) Genetics 94:969
23. Robertson DS (1981) Mol Gen Gent 183:51
24. Rood SP, Koshioka M, Douglas TJ, Pharis RP (1982) Plant Physiol (Bethesda) 70:1614
25. Shechter I, West CA (1969) J Biol Chem 244:3200
26. Sembdner G, Schreiber K (1965) Flora oder Allge Bot Zeitung Abt A 156:359
27. Spray CR, Phinney BO, Gaskin P, Gilmour SJ, MacMillan J (1984) Planta (Berl) 160:464
28. Wurtele ES, Hedden P, Phinney BO (1982) J Plant Growth Regul 1:15

Gibberellins: Structure-Activity Relationships and the Design of Molecular Probes

M.H. BEALE[1], R. HOOLEY[2], and J. MacMILLAN[1]

1 Introduction

In general, two different approaches have been used in the study of gibberellin-mediated growth and development of plants. The first is concerned with the effects of exogenous application of GAs, usually GA_3, to plant systems. These studies range from gross morphological effects (including commercial uses) to the mechanism of action at the molecular levels. The other approach comprises the study of the natural hormonal regulation of plant growth by the endogenous GA-hormone(s). It proceeds from the identification of the native GAs in a plant species to the elucidation of the metabolic pathways of the native GAs. It then seeks to distinguish the true hormonal GA(s) from the large number (72 at present) of the natural GAs and then to determine the metabolic regulation and mode of action of the native GA-hormone(s).

Thus, the two approaches have the same objective of defining the mode of action of GAs at the molecular level. Each approach has made important advances recently. For example, in the extensively studied cereal aleurone system, gene transcription studies using nuclei isolated from aleurone protoplasts of *Hordeum vulgare* cv Himalaya [10] and of *Avena fatua* [18] have provided evidence that GA_3 and GA_4 appear to regulate the transcription of the genes for α-amylase and rRNA. Also, through the use of GA-responding single gene mutants of *Pisum sativum* [9] and of *Zea mays* [13, 16], GA_1 has been attributed the role of the GA-hormone for stem elongation in these, and probably other, plants [12]. Thus, molecular probes for GA-receptors in aleurone protoplasts, and for GA_1-receptors in shoot apices of *P. sativum* and *Z. mays*, are now appropriate. For aleurone protoplasts a direct approach, such as photo-affinity labelling, can be envisaged. For shoot apices it may first be necessary to identify the cellular and sub-cellular localisation of GA_1 or, failing that, of the 3β- and 2β-hydroxylases which catalyse respectively the formation of GA_1 from GA_{20} and the deactivation of GA_1 to GA_8 and GA_{20} to GA_{29}. For such localisation studies, monoclonal antibodies for GA_{20}, GA_1, GA_{29} and GA_8, and for the interconnecting enzymes, are obvious tools.

1 AFRC Research Group, School of Chemistry, The University, Bristol BS8 1TS, Great Britain
2 Molecular Biology, Long Ashton Research Station, University of Bristol, Long Ashton, Bristol B518 9AF, Great Britain

Plant Growth Substances 1985
(ed. by M. Bopp)

In designing photo-affinity labels, GA-antigens and affinity columns for enzyme purification, we have adopted a unified approach, based upon the use of bioactive GA-derivatives. A necessary preliminary has been the reinvestigation of structure-activity relationships.

2 Structure-Activity Relationships

These relationships have been the subject of many studies (for reviews see [2, 4, 14]. Previous reports relate to a wide range of GA-bioassays. For the reasons indicated in Section 1, we have concentrated on two types of assay only. (1) GA-induced synthesis of α-amylase in aleurone protoplasts of *A. fatua* and (2) GA-induced stem elongation in Tan-ginbozu rice and d-5 maize. We have also concentrated on GA-derivatives which have substituents at the 1-, 2 and 17-carbons since previous work had shown that chemical modification (except hydroxylation) at these positions do not greatly reduce biological activity and may even enhance it. Approximately 40 such derivatives have been prepared, based mainly on GA_1 and GA_4. Only a selection of these compounds are discussed; their structures are shown in Fig. 1–4 and are referred to in the text by numerals, given in these figures.

2.1 Stem Elongation

The 2-sbustituted derivatives of GA_4 (I) and GA_1 (XI) (Fig. 1) are considered first. Hoad et al. [5, 6] reported that 2,2-dimethyl GA_4 (II) was much more active than either GA_4 (I) or 2β-methyl GA_4 (IV) in the d-5 maize and Tan-ginbozu bioassays; in sharp contrast, 2,2-diethyl GA_4 (III) was inactive in these assays. The same authors also reported that in the GA_1 series, 2,2-dimethyl GA_1 (XII) was much more active

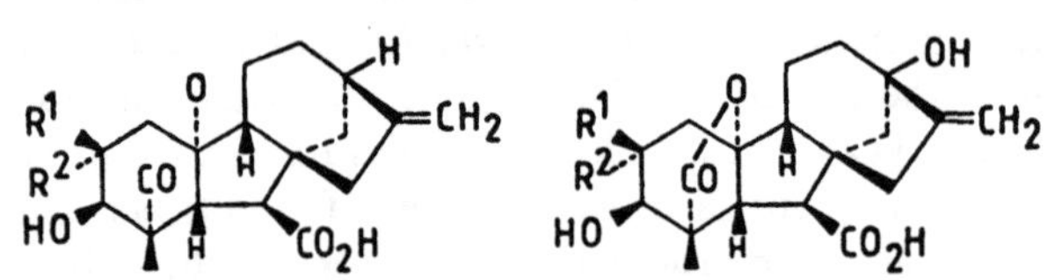

I: $R^1=R^2=H$
II: $R^1=R^2=CH_3$
III: $R^1=R^2=CH_2CH_3$
IV: $R^1=CH_3$, $R^2=H$
V: $R^1=CH_2CH_3$, $R^2=H$
VI: $R^1=(CH_3)_3C$, $R^2=H$
VII: $R^1=(CH_3)_2CH$, $R^2=H$
VIII: R^1, $R^2=(CH_3)_2C=$
IX: $R^1=H$, $R^2=CH_3$
X: $R^1=H$, $R^2=CH_2CH_3$

XI: $R^1=R^2=H$
XII: $R^1=R^2=CH_3$
XIII: $R^1=CH_3$, $R^2=H$
XIV: $R^1=CH_2CH_3$, $R^2=H$
XV: $R^1=CH_2CH_2CH_3$, $R^2=H$
XVI: $R^1=CH_2CH_2CH_2CH_3$, $R^2=H$
XVII: $R^1=H$, $R^2=CH_3$
XVIII: $R^1=H$, $R^2=CH_2CH_3$

Fig. 1. Structures of the 2-substituted GA_4 and GA_1

than GA_1 (XI) and 2β-methyl GA_1 (XIII) in the d-5 maize assay and that 2β-ethyl GA_1 (XIV), 2β-n-propyl GA_1 (XV) and 2β-n-butyl GA_1 (XVI) where less active in both the d-5 maize and Tan-ginbozu rice assays.

Recent results of A. Fowles [1] have shown that 2β-t-butyl GA_4 (VI), 2β-isopropyl GA_4 (VII) and 2β-isopropylidene GA_4 (VIII) are inactive in the Tan-ginbozu rice assay. Thus, steric bulk at carbon-2 is not the factor which confers high activity on 2,2-dimethyl GA_4 (II). Neither are two substituents at carbon-2 required for high activity as shown by the following results. In the d-5 maize and Tan-ginbozu rice assays, 2α-methyl GA_4 (IX) and 2α-methyl GA_1 (XVII) are as active as 2,2-dimethyl GA_4 (II) and 2,2-dimethyl GA_1 (XII). Furthermore, the 2α-ethyl derivatives (X) and (XVIII) are even more active than 2,2-dimethyl GA_4 (II). It appears, therefore, that it is a 2α-alkyl substituent which confers high biological activity on GA_4 and GA_1 for stem elongation in d-5 maize and Tan-ginbozu seedlings.

For 1-methyl derivatives of GA_4 and GA_1 (Fig. 2), the reverse stereochemistry confers high activity ([11], Willis, unpubl. data). Thus, in the d-5 maize and Tan-ginbozu rice assays, 1α-methyl GA_4 (XIX) has comparable activity to GA_4 (I), but the 1β-methyl derivative (XXI) has much higher activity than the parent GA_4 (I). The same relationship holds for the 1α-methyl GA_1 (XX) and 1β-methyl GA_1 (XXII).

Of the 17-substituted GAs (Fig. 3), only 17-carboxy GA_4 (XXIII) has so far been tested. In the Tan-ginbozu rice assay it has about 10% of the activity of GA_4 (I).

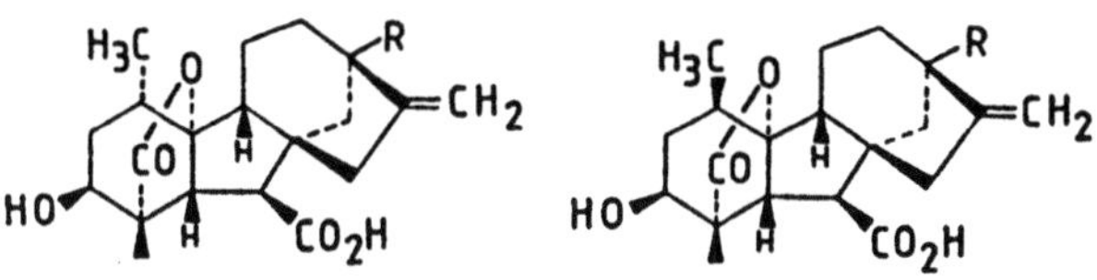

XIX: R=H XXI: R=H
XX: R=OH XXII: R=OH

Fig. 2. Structures of 1-substituted GA_4 and GA_1

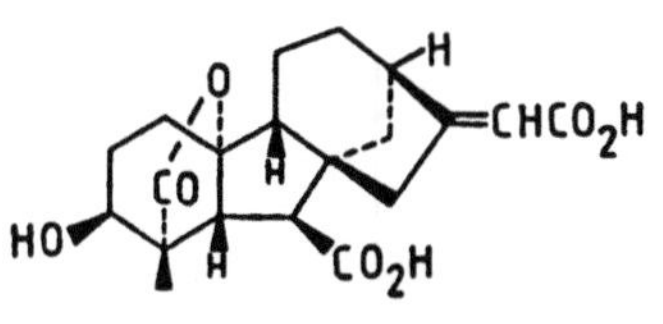

XXIII

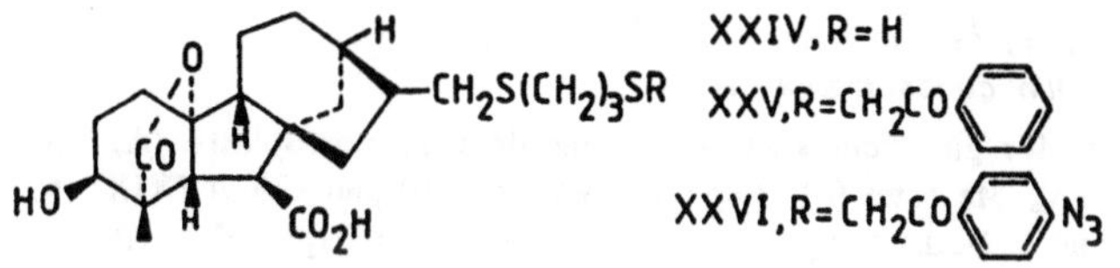

Fig. 3. Structures of 17-substituted GA_4

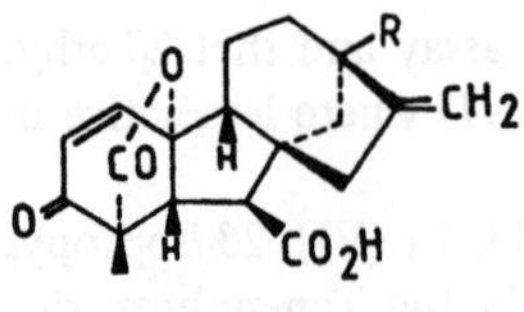

XXVII, R = H
XXVIII, R=OH

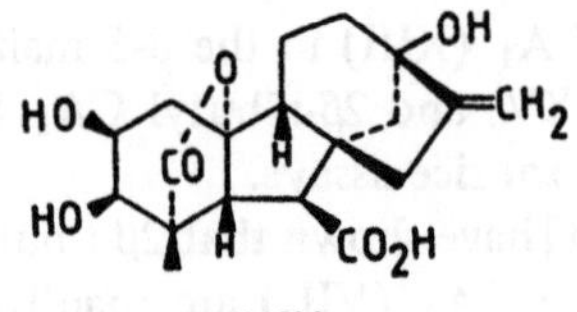

XXIX

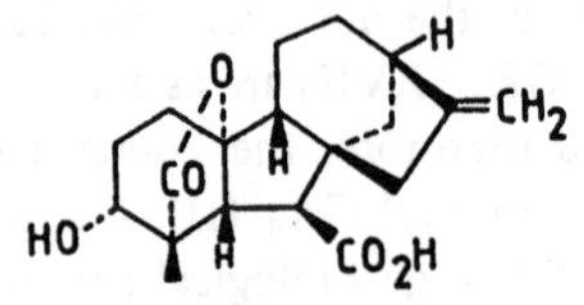

XXX

Fig. 4. Structures of miscellaneous gibberellins

2.2 α-Amylase Assay Using Aleurone Protoplasts

Hooley [7] has shown that protoplasts, prepared from aleurone cells of *A. fatua*, respond to GA_3 by α-amylase synthesis and secretion. He obtained a significant response at 1.6×10^{-13} M GA_3 and a maximum response at 1.6×10^{-7} M GA_3. We now report that GA_3, GA_1 and GA_4 have the same activities in this assay system. Gibberellin A_4 (I) was therefore used as the reference compound in the comparisons with the GA_4-derivatives which have been assayed. The dose-response curve for GA_4 (I) is shown in Fig. 5A.

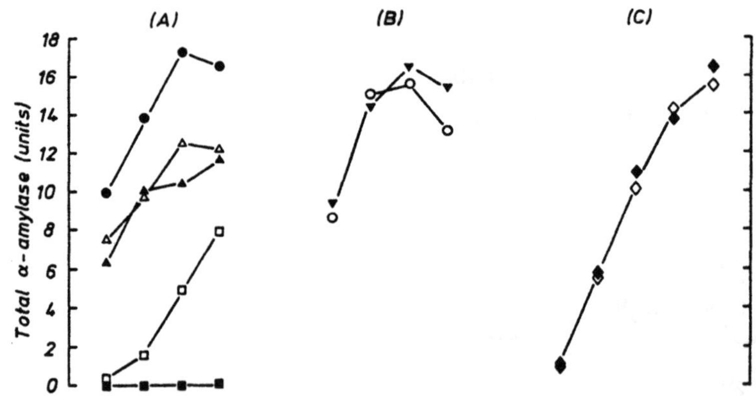

Fig. 5 A–C. Dose response data for gibberellins assayed using aleurone protoplasts. (A) GA_4 (I) (●), 3-epiGA_4 (XXX) (▲), GA_4 Me ester (△), 3-epiGA_4 Me ester (□) and GA_8 (XXIX) (■). (B) 2,2-diMe GA_4 (II) (▼) and 2,2-diEt GA_4 (III) (○). (C) GA_4-17-$S(CH_2)_3SH$ (XXV) (◆) and GA_4-17-$S(CH_2)_3SCH_2COPh$ (XXV) (◇)

Although the number of 2-substituted derivatives of GA_4 (I) so far examined is small, the results are surprising in view of the results, described in the previous section for stem elongation assays. For example, 2,2-dimethyl GA_4 (II) and 2,2-diethyl GA_4 (III) were of similar activity to GA_4 (I) in the protoplast assay (Fig. 5B), although the former derivative (II) was much more active than GA_4 (I), and the latter derivative (III) was very much less active than GA_4 (I), in the d-5 maize and Tan-ginbozu rice assays. Also t-butyl GA_4 (VI), which was inactive in the stem extension assays, showed moderate activity in the protoplast assay (Fig. 6A). A comparison of the activities of GA_4 (I), 2,2-diethyl GA_4 (III) and 2-t-butyl GA_4 (VI) in protoplasts (Fig. 6A) and in aleurone cells (Fig. 6B) was made to determine if the cell wall presented a barrier to the 2-substituted GAs. The data show that the activities of all three compounds were less in the aleurone cell assay than in the protoplast system. The activities of GA_4 (I) and 2,2-diethyl GA_4 (III) were the same in both systems, but the reduction in activity from protoplasts to aleurone cells was greater for the bulky 2-t-butyl derivative (VI) (Fig. 6C).

Three of the 17-substituted derivatives of GA_4, shown in Fig. 3, were also assayed in the protoplast system. The two dithio-derivatives (XXIV) and (XXV) showed high activity (Fig. 5C) and encouraged the exploration of photo-affinity labelling based upon structure (XXVI) (see later). Equally encouraging in this regard were 17-carboxy GA_4 (XXIII), which showed moderate activity, and the two conjugated ketones (XXVII and XXVIII, Fig. 4), which were as active as GA_4 (I) (data not shown).

Inactive GA-derivatives are also necessary to provide controls for photo-affinity labelling. Surprisingly, GA_4 methyl ester, 3-epiGA_4 (XXX, Fig. 4) and 3-epiGA_4 methyl ester showed some activity (Fig. 5a). However, GA_8 (XXIX, Fig. 4) was inactive (Fig. 5A) and so were the methyl esters of the two conjugated ketones (XXVII) and (XXVIII), shown in Fig. 4.

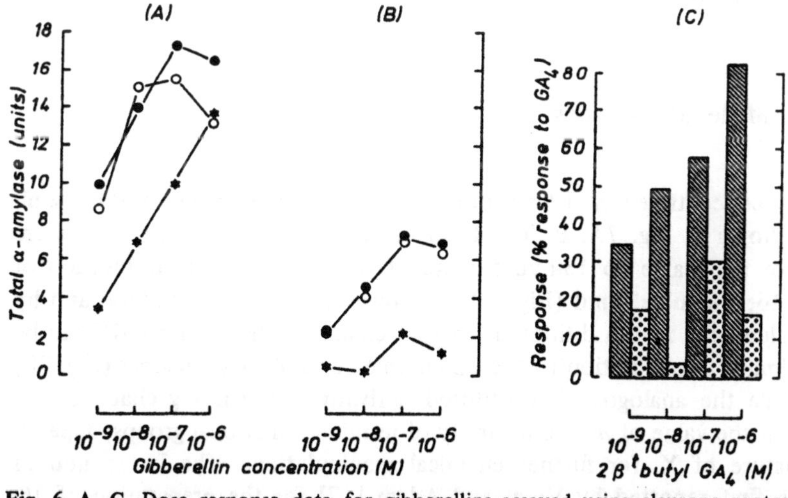

Fig. 6 A–C. Dose response data for gibberellins assayed using aleurone protoplasts (A) and aleurone layers (B). GA_4 (I) (•), 2,2-diEt GA_4 (III) (○) and 2β-t-butyl GA_4 (VI) (*). In (C) the response of protoplasts (▨) and intact cells (▩) to 2β-t-butyl GA_4 (VI) are, for each concentration, expressed as the percentage of the response to the same concentration of GA_4 (I)

2.3 Comparison of the Results

The bioassay results, presented in the two preceding sections, confirm that 1-, 2- and 17-substituted GAs are suitable derivatives on which to base the design of molecular probes. Some of the results, however, deserve further comment.

It is remarkable that the protoplast system responds to such a range of structural modifications to the GA_4-molecule. The activities of the methyl esters of GA_4 (I) and 3-epiGA_4 and of 3-epiGA_4 (XXX) were unexpected, although the difficulty in removing traces of highly active impurities must be stressed. The observed in-activity of GA_8 (XXIX) is therefore a relief. The similar activities of GA_4 (I), 2,2-dimethyl GA_4 (II) and 2,2-diethyl GA_4 (III) in the protoplast system and the dis-similar activities of these compounds in the d-5 maize and Tan-ginbozu assays is notable.

It is not surprising that the whole plant assays are more discriminating. In these assays, there are several factors that can intervene between the site of application and site of action. Metabolism of the applied compound may occur to less or more active metabolites. The original idea [5] that 2β-substituents might prevent deactivation by 2β-hydroxylation and lead to enhanced or prolonged bio-activity seems less likely in view of the high activity of the 2α-methyl and 2α-ethyl derivatives of GA_4 and GA_1. However, the high activity of these 2α-methyl and 2α-ethyl derivatives of GA_4 and GA_1, and of 1β-methyl GA_4, compared to the parent GAs, is more pronounced at high dosages and the difference becomes greater, the longer the assay is continued. This fact suggests a greater persistence of these compounds in the plant. However, another discriminating factor, which may operate in higher plants, is the transport of the compound from the site of application to the site of action. It may be that the increased lipophilicity of the 1β- and 2α-alkylated GAs aids both their transport to, and their binding at the active site. These aspects are currently being explored.

3 Design of Molecular Probes

So far we have concentrated on 17-substituted derivatives, making use of the chemical reactions shown in Fig. 7A, B. The Peterson reaction, shown in Fig. 7A is unfortunately not applicalbe to 13-hydroxylated GAs. The free radical addition of thiols to the 16,17-double bond (Fig. 7B) is, however, a versatile method and by varying the value of n in the dithiol, a range of chain lengths at carbon-17 can be prepared. A similar thiol addition to the 1,2-double bond of ring A enones (Fig. 7C) can also provide the analogous 1-substituted derivatives of varying chain length, depending upon the value of n, and a variety of terminal functional groups, depending on the nature of X, for further chemical manipulations. The latter method (Fig. 7C) was first reported by Voigt and Adam [17] for the preparation of 1β-derivatives; we find that solid phase acid-catalysis provides the 1α-derivatives stereoselectively.

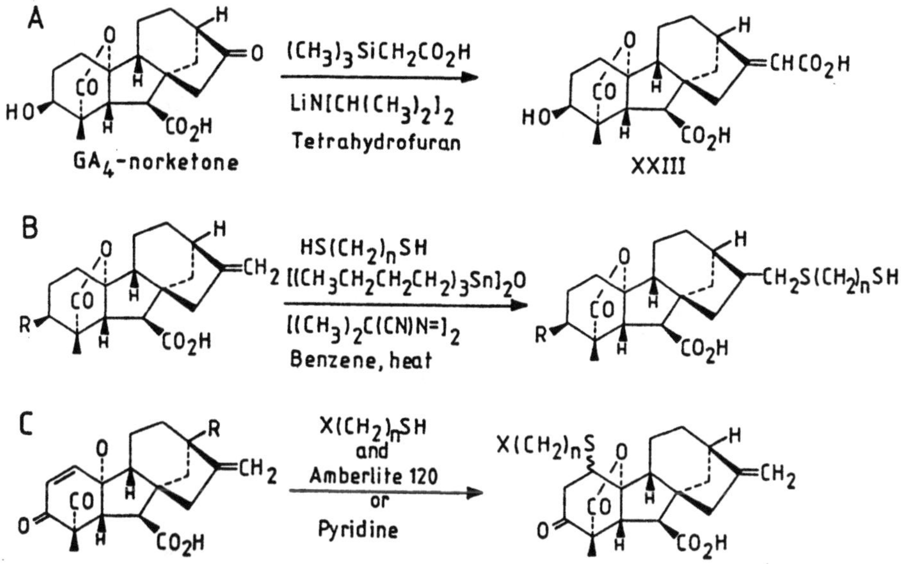

Fig. 7 A–C. Chemical procedures for 1- and 17-substituted GAs

3.1 Photo-Affinity Labels

The two enones (XXVII) and (XXVIII), shown in Fig. 4, are possible candidates for photo-affinity labelling experiments in aleurone protoplasts. They are active in this systems and their methyl esters would provide bio-inactive controls. Such enones have been used in steroid-receptor labelling [3] and in ABA-guard cell protoplast labelling [8].

The dithio compound (XXV, Fig. 4), prepared from the thiol (XXIV, Fig. 4) and phenacyl bromide, is bio-active in the protoplast system. The azide (XXVI), prepared in an analogous manner from the thiol (XXIV, Fig. 4) and p-azidophenacyl bromide, is another promising candidate for photo-affinity labelling in aleurone protoplasts. It has λ_{max} 287 (ϵ, 15,000) and is stable when stored as a solid at $-10^{o}C$ or in solution in the dark. The corresponding compound with the azido-group attached at carbon-1 provides another potential photo-affinity label. The attraction of these thiol-based affinity labels is the facility of preparing them at two positions which are known not to affect biological activity and the possibility of varying the chain lenght by varying the value of *n* (Fig. 7B, C).

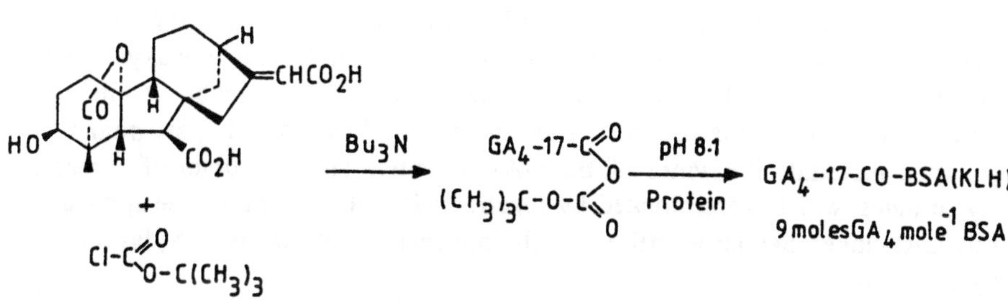

Fig. 8. Gibberellin A_4-protein conjugation at carbon-17 by the mixed anhydride method

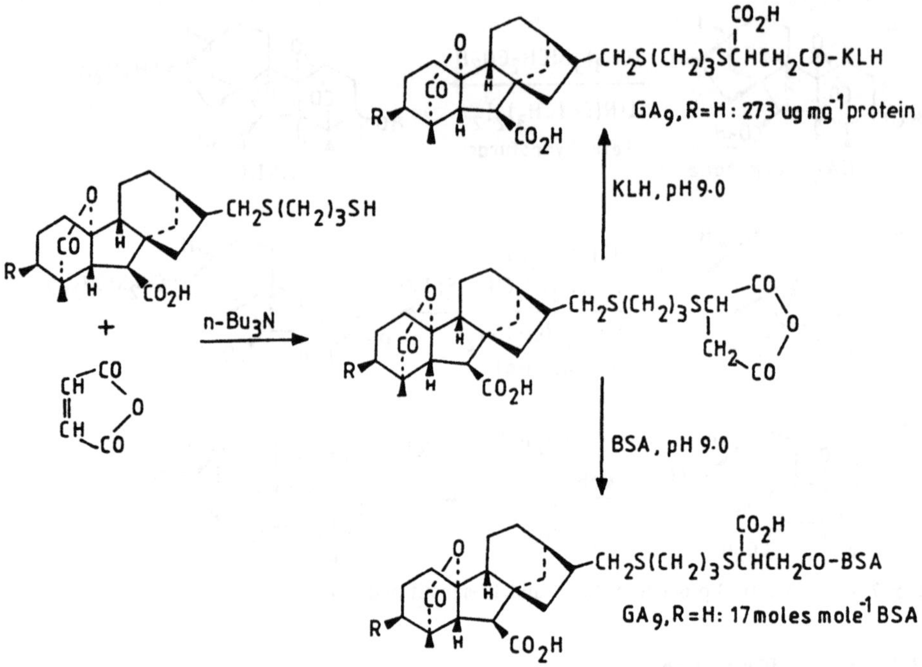

Fig. 9. Conjugation of GA_4 and GA_9 to BSA and KLH through carbon-17 and maleic anhydride

3.2 Gibberellin Antigens

Gibberellin A_4 (I) has been linked through the 17-carbonyl derivative to BSA and KLH as shown in Fig. 8. The BSA-conjugate contains 9 molecules GA_4 per molecule protein. A similar incorporation in the KLH conjugate is assumed. Gibberellins A_9 and A_4 have each been linked via the 17-thiol (XXIV, Fig. 3), as shown in Fig. 9. The incorporations of the GA_9-derivative, shown in Fig. 9, were determined using radio-active GA_9. Similar incorporations, using the cold GA_4-derivative, are assumed. Since all of these conjugates are linked through the 17-position of GA_4, we have also prepared a GA_1-KLH conjugate linked through the 3-hemisuccinate. Thus, a range of GA-antigens with different GA-epitomes are available for raising monoclonal antibodies.

3.3 Affinity Columns for GA-Hydroxylases

A thio-derivative of GA_4 has been prepared by the method shown in Fig. 7C, where R=OH and n=6. This thiol, which completely inhibits the GA-2β-hydroxylase from *Phaseolus vulgaris* seeds [15] at 10 μM concentration (Weir, unpubl. results), was reacted with epoxy-Sepharose 6B. The resultant gel contained, per gramme of freeze-dried material, 126 μmol of the GA_4-derivative and 382 μmol of unreacted epoxy residues, which were blocked by reaction with ethanolamine. The application of this GA_4-linked Sepharose 6B gel to the purification of GA-hydroxylases is currently being assessed.

In future, such columns may be useful for the purification of GA-receptor molecues.

4 Conclusion

This paper presents a progress report on the design of molecular probes, based upon bio-active GA-derivatives. Through a unified approach, a versatile method of preparing 17-thiols of varying chain length is described. The further elaboration of these thiols to promising photo-affinity labels, GA-antigens and affinity columns is described.

References

1. Fowles AM (1985) Ph D Thesis, Bristol University, Bristol, UK
2. Graebe JE, Ropers H-J (1978) In: Phytohormones and related compounds – a comprehensive treatise. The biochemistry of phytohormones and related compounds, vol 2, p 107
3. Gronemeyer H (1985) Trends in biological science, p 264
4. Hoad GV (1983) In: Crozier A (ed) The biochemistry and physiology of gibberellins, vol 2. Praeger, New York, p 57
5. Hoad GV, Phinney BO, Sponsel VM, MacMillan J (1981) Phytochemistry (Oxf) 20:703
6. Hoad GV, MacMillan J, Smith VA, Sponsel VM, Taylor DA (1982) In: Wareing PF (ed) Plant growth substances. Academic Press, London, p 91
7. Hooley R (1982) Planta (Berl) 154:29
8. Hornberg C, Weiler EW (1984) Nature 310:321
9. Ingram TJ, Reid JB, Murfet IC, Gaskin P, Willis CL, MacMillan J (1984) Planta (Berl) 160:455
10. Jacobsen JV, Beach LR (1985) Nature 316:275
11. MacMillan J, Willis CL (1985) J Chem Soc Perkin I, 2177
12. Phinney BO (1984) In: Crozier A, Hillman JR (eds) The biosynthesis and metabolism of plant hormones. Cambridge University Press, Cambridge, p 17
13. Phinney et al., this volume
14. Reeve DR, Crozier A (1974) J Exp Bot 25:431
15. Smith VA, MacMillan J (1984) J Plant Growth Regul 2:251
16. Spray C, Phinney BO, Gaskin P, Gilmour SJ, MacMillan J (1984) Planta (Berl) 160:464
17. Voigt B, Adam G (1979) Pharmazie 34:362
18. Zwar JA, Hooley R (1986) Plant Physiol (Bethesda) 80:459

Gibberellin Biosynthesis from Gibberellin A_{12}-Aldehyde

J.E. GRAEBE[1]

1 Introduction

The first part of this lecture is a general introduction to the portion of gibberellin (GA) biosynthesis that lies after GA_{12}-aldehyde in the pathway. In the second part I will present some new results concerning the presence of endogenous GA_{12}-aldehyde in germinating pea seeds.

Figure 1 shows a general outline of GA biosynthesis to locate the position of GA_{12}-aldehyde in the pathway. Mevalonic acid is converted by soluble enzymes in nine steps to the hydrocarbon *ent*-kaurene. Since *ent*-kaurene and its immediate metabolites are extremely apolar, it is only logical that the subsequent steps, by which GA_{12}-aldehyde is formed, are associated with the endoplasmic reticulum [11]. These steps were recently reviewed by Graebe [8] and Hedden [12]. Gibberellin

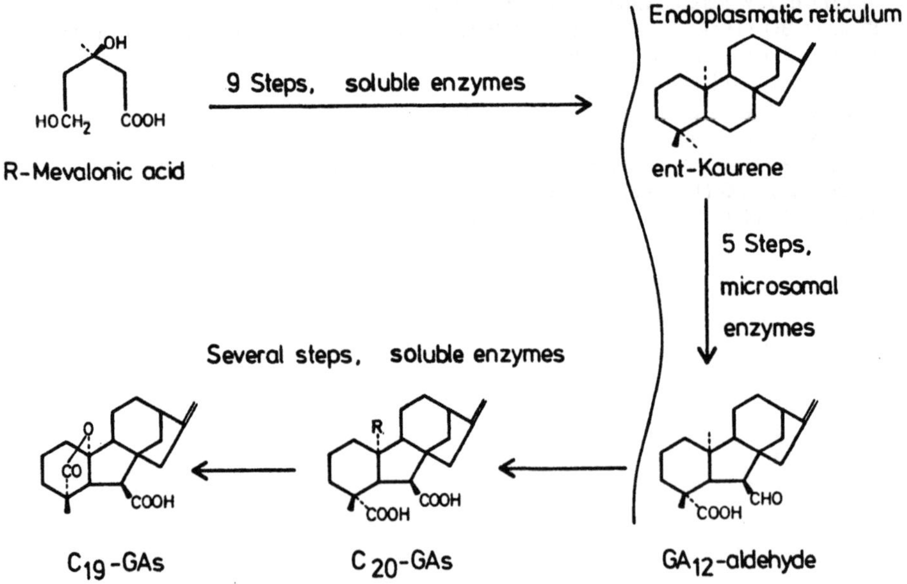

Fig. 1. General scheme of gibberellin biosynthesis

1 Pflanzenphysiologisches Institut, Untere Karspüle 2, 3400 Göttingen, FRG

Plant Growth Substances 1985
(ed. by M. Bopp)

A_{12}-aldehyde is converted to C_{20}- and C_{19}-GAs by soluble enzymes requiring α-ketoglutarate and Fe^{2+} for activity [13, 18].

Much of the information on the GA pathway described in this lecture was gained by the use of isotopically labelled GA_{12}-aldehyde obtained by incubation of labelled mevalonate, preferably R-mevalonate, with a cell-free system from *Cucurbita maxima* endosperm [9] (see [28] for the most recent procedure).

2 The General Pathway After GA_{12}-Aldehyde

Figure 2 shows the general conversion of GA_{12}-aldehyde in a scheme that emerges if species- and tissue-specific hydroxylations are disregarded. In the first reaction, the C-7 aldehyde group is oxidized to a carboxyl group. In the second reaction, the C-20 methyl group becomes hydroxylated. The resulting hydroxy-acid is oxidized

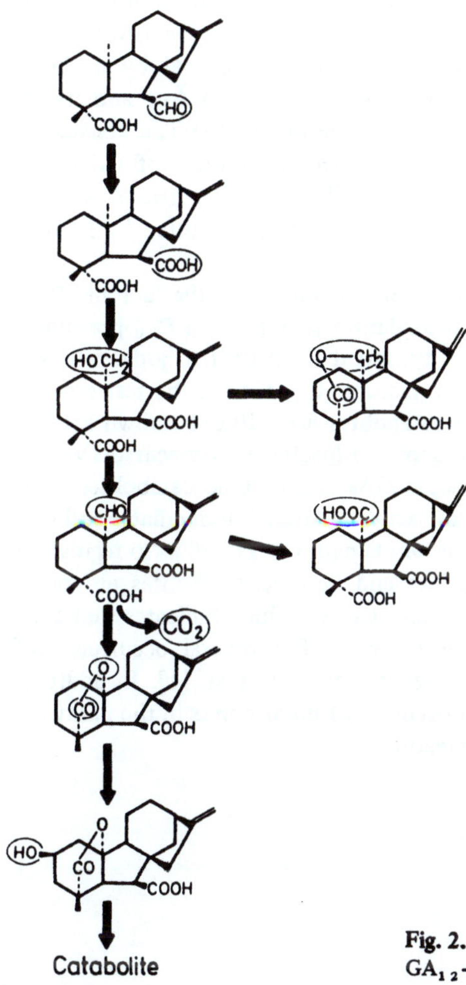

Fig. 2. Gibberellin biosynthesis. General pathway after GA_{12}-aldehyde

further to a C-20 aldehyde and the aldehyde is either oxidized to a tricarboxylic acid, or C-20 is lost as CO_2, which results in a lactonic C_{19}-GA. Finally, the C_{19}-GA becomes 2β-hydroxylated and the 2β-hydroxylated GA may be oxidized further to an α, β-unsaturated ketonic catabolite, which no longer prosseses a γ-lactone. The structure of this type of catabolite will be shown later (Fig. 3).

Some interesting points in the scheme (Fig. 2) deserve consideration. (1) The second intermediate after GA_{12}-aldehyde is a hydroxy-acid. The GA intermediates of this type easily form lactones as indicated, especially at pH-values below 7. Therefore, the lactone form is obtained by the usual type of acid extraction and it is also the form that has been assigned a GA-number (e.g. GA_{15} and GA_{44}). However, as has been shown for GA_{15} in a cell-free system from *C. maxima* [13] and for GA_{15} and GA_{44} in a cell-free system from *Pisum sativum* [18], the lactone must be hydrolyzed before this type of GA is accepted by the enzyme as a substrate for further conversion to the C-20 aldehyde. (2) The conversion from C_{20}- to C_{19}-GAs occurs when C-20 is at the oxidation stage of an aldehyde. This has been shown in the cell-free systems from *C. maxima* and *P. sativum* for the conversion of GA_{36} to GA_4 and for the conversion of GA_{24} to GA_9, respectively [10, 18]. In both cases, the corresponding tricarboxylic acids were not converted to C_{19}-GAs. The tricarboxylic acids have no known physiological role and are usually regarded as by-products, although they may be major endogenous constituents as, e.g. GA_{13} and its 2β-hydroxylated derivative, GA_{43}, in *C. maxima* endosperm [3]. (3) The conversion of C_{20}-GAs to C_{19}-GAs has been previously studied in cultures of *Gibberella fujikuroi* [2, 4, 20]. A lecture on this reaction in a cell-free system from *P. sativum* and the identification of the leaving fragment as CO_2 will be presented by Kamiya et al. (this Conference, Abstract No. 0203).

The physiological activity is believed to reside mainly with the lactonic C_{19}-GAs and is strongly dependent on the hydroxylation pattern. The C_{20}-precursors in many cases do have some biological activity, but it is difficult to judge whether this is due to activity per se or conversion to lactonic C_{19}-GAs in the test objects. The 2β-hydroxylated GAs and the ketonic catabolites are without known activity [26]. Thus, the flow of activity is from less active or inactive C_{20}-precursors via the active C_{19}-GAs to inactivated 2β-hydroxylated GAs and ketonic catabolites. It is important to remember that the physiological active GAs are intermediates and that their levels are sensitive to fluctuations in the whole pathway as well as to regulation of individual steps. The 2β-hydroxylated GAs and ketonic catabolites are often found in larger amounts than other GAs in plant tissues, which indicates that they are end-point metabolites or converted very slowly. The general occurrence of ketonic metabolites has recently been discussed by Albone et al. [1]. In addition to the reactions mentioned, conjugation can occur at different points in the pathway, but conjugation will not be discussed in this lecture.

3 The 13-Hydroxylation Pathway of Peas

Figure 3 shows the biosynthesis of GAs in immature pea seeds as an actual example of conversion of GA_{12}-aldehyde. On the left-hand side, the pathway is identical with the general pathway shown in Fig. 2. In addition, the pea seeds feature 13-hydroxylation, resulting in 13-hydroxylated C_{20}-GAs. These are converted further by the same reactions as the non-hydroxylated GAs, so that two pathways result. The 13-hydroxylation is associated with the endoplasmic reticulum [8, 11] and it takes place early in the pathway in conjugation with the microsomal steps by which GA_{12}-aldehyde is formed. The right leg of the pathway has become known as the "early 13-hydroxylation" pathway, and it plays an important role in many of the plants investigated to date.

The reactions of both pathways, from GA_{12}-aldehyde to GA_{51} and GA_{29}, have been shown in a cell-free system from developing embryos of *P. sativum* [18]. Overlapping, the reactions from GA_9 and GA_{20} to the end of the pathway have

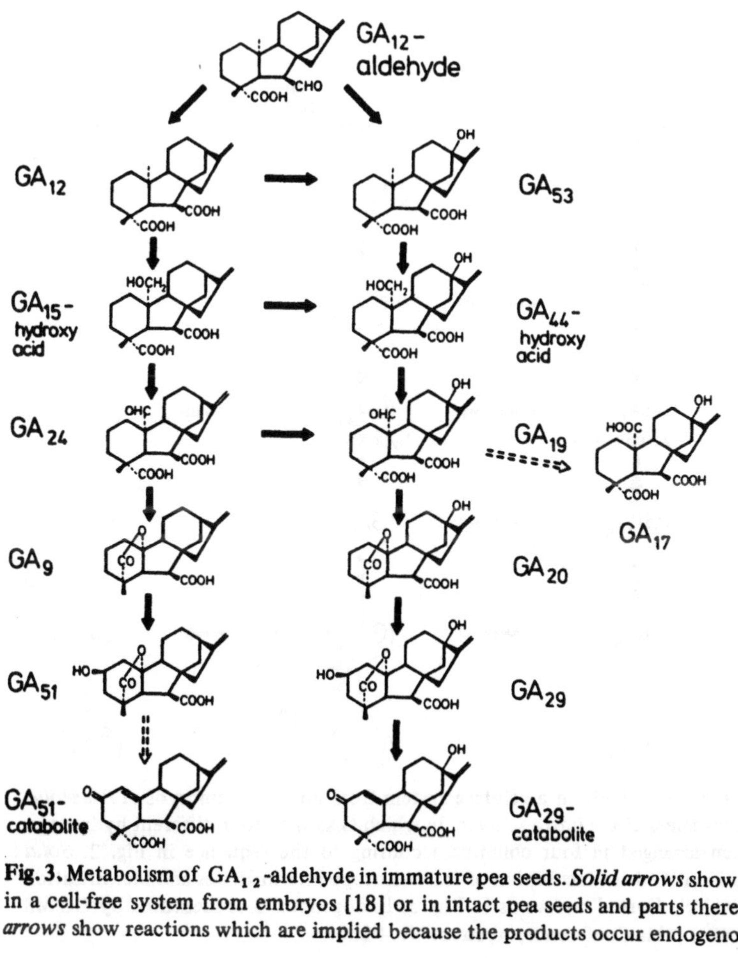

Fig. 3. Metabolism of GA_{12}-aldehyde in immature pea seeds. *Solid arrows* show reactions identified in a cell-free system from embryos [18] or in intact pea seeds and parts thereof [25, 26]. *Dotted arrows* show reactions which are implied because the products occur endogenously

been shown by feeding labelled GAs to intact peas in many experiments [25, 26]. Sponsel [26] has shown that 2β-hydroxylation mainly occurs in the cotyledons of developing pea seeds, whereas the conversion of GA_{29} to the ketonic catabolite takes place in the testa.

Gibberellins A_9, A_{17}, A_{19}, A_{20}, A_{29}, A_{44}, A_{51}, A_{29}-catabolite and A_{51}-catabolite have been identified as endogenous GAs in immature pea seeds [6, 26]. A comparison of these GAs with the reactions in Fig. 3 shows that only GA_{17} and GA_{51}-catabolite have not been obtained biosynthetically. On the other hand, GA_{12}, GA_{15}, GA_{24} and GA_{53} have not been identified as endogenous constituents, although they have been obtained biosynthetically in vitro. Obviously, these intermediates remain at low levels in vivo, thus escaping detection.

4 Some Further Hydroxylation Patterns

Figure 4 shows the reactions after GA_{12}-aldehyde in a cell-free system prepared from the embryos of immature *Phaseolus vulgaris* seeds. The reactions have been

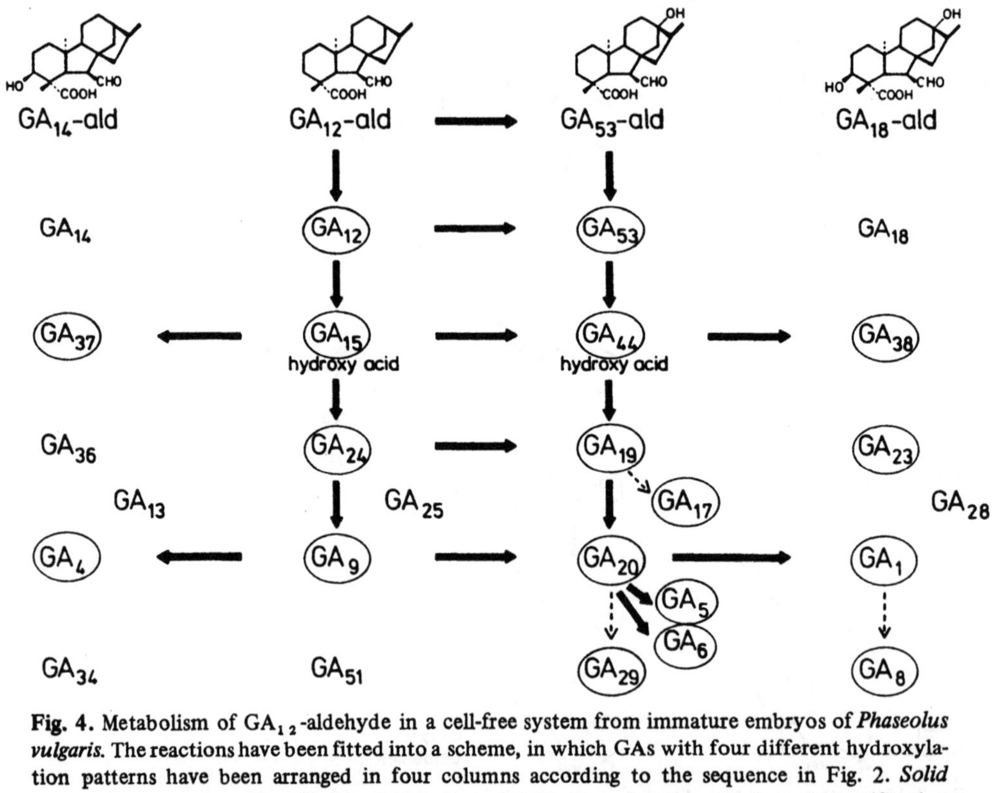

Fig. 4. Metabolism of GA_{12}-aldehyde in a cell-free system from immature embryos of *Phaseolus vulgaris*. The reactions have been fitted into a scheme, in which GAs with four different hydroxylation patterns have been arranged in four columns according to the sequence in Fig. 2. *Solid arrows* show reactions identified by incubation of stable-isotope intermediates and identification of the products by GC-MS. *Dotted arrows* as in Fig. 3. Endogenous GAs identified by GC-MS are *encircled* (Kamiya et al., this Conference, Abstract No. 0203)

placed in a scheme, in which series of 3β-hydroxylated, non-hydroxylated, 13-hydroxylated and 3β,13-dihydroxylated GAs have been arranged in the same sequence as presented for non-hydroxylated GAs in Fig. 2. These series are not supposed to represent pathways; it is a systematic listing of the GAs with the hydroxylation patterns mentioned. Pathways and endogenous GAs can be marked in such a scheme for easy comparison of results obtained with different objects. Figure 4 shows that the same two pathways as in peas (Fig. 3) operate in *P. vulgaris* as well. In addition, there is 3β-hydroxylation of both non-hydroxylated GAs (the conversions of GA_{15} to GA_{37} and of GA_9 to GA_4) and of 13-hydroxylated GAs (the conversions of GA_{44} to GA_{38} and of GA_{20} to GA_1). In contrast to 13-hydroxylation, 3β-hydroxylation is catalyzed by soluble enzymes, at least in cell-free systems from *P. vulgaris* [19] and *C. maxima* [13]. This may explain why 3β-hydroxylation tends to take place at intermediate or late steps rather than at the beginning of the pathway (Fig. 4). In addition to the pattern of 3β-hydroxylation, the bean system differs from the pea system by the formation of GA_5 and GA_6. A three-dimensional variety of the scheme (Fig. 4) is shown in the abstract by Kamiya et al. (this Conference, Abstract No. 0203).

A comparison of the reactions obtained in the *P. vulgaris* cell-free system with endogenous GAs identified in the same species (encircled in Fig. 4) shows that all of the endogenous GAs, except GA_{17}, which is a minor component, and the 2β-hydroxylated GAs, GA_{29} and GA_8, have been obtained in vitro. It is unknown why no 2β-hydroxylation was obtained, since it has been shown to occur in intact seeds of *P. vulgaris* [29]. Furthermore, the 2β-hydroxylation reactions have been studied with purified enzymes from germinating seeds of *P. vulgaris* [24].

An early 13-hydroxylation pathway combined with "late 3β-hydroxylation" has a special function in the seedlings of at least two species. Thus, in *Zea mays*, there is strong evidence for the 13-hydroxylation pathway from GA_{12}-aldehyde to GA_{20} and unequivocal evidence for the further conversion of GA_{20} to GA_{29} and GA_1 [27]. As has been shown by the work of Phinney ([23] and this vol.), the conversion of GA_{20} to GA_1 is blocked in plants that are homozygous for the recessive gene *dwarf-1*. Since GA_1 appears to be essential for normal internode elongation in this species, its absence results in a dwarf growth habit. The same is the case for pea seedlings, where the conversion of GA_{20} to GA_1 has also been unequivocally shown [17]. In this case, the reaction is controlled by the *Le*-locus, plants of the *le*-genotype having dwarf habit, because they do not produce GA_1 [16, 17]. Some aspects of this regulation are not completely clear at the time of writing, since GA_1 has been identified in seedlings of the pea cultivar Progress No. 9, which is of the *le*-genotype [6]. In spinach, photoperiodic control of defined steps in the 13-hydroxylation pathway forms the basis for internode elongation under long-day conditions ([7, 21, 22] Gilmour et al., this Conference, Abstract No. 0214).

5 Occurrence and Metabolism of GA_{12}-Aldehyde in Germinating Pea Seeds

Immature pea seeds contain substantial amounts of GAs, including GA_{20} as the main physiologically active component [6, 25, 26]. In the cultivar Progress No. 9, the biosynthesis of GA_{20} apparently ceases as the seeds mature, whereas further metabolism continues, so that only GA_{29} and GA_{29}-catabolite are present in the mature seeds [26] (cf. Fig. 3). Since GA_{29} and GA_{29}-catabolite are without known physiological activity, this raises the question whether pea seeds can germinate without physiologically active GAs or whether such GAs have to be synthesized before germination can occur. Sponsel [26], therefore, studied the effect of a growth retardant added to mature pea seeds at the time of imbibition. The growth retardant used blocks GA biosynthesis between mevalonate and *ent*-kaurene in *Gibberella fujikuroi* and presumably in higher plants as well (cf. Fig. 1). Sponsel found that the growth retardant had no effect on seed germination, but an inhibition of epicotyl growth, which could be counteracted by simultaneous application of GA_3, was apparent after 4 days. She concluded that de novo GA biosynthesis is not essential for pea seed germination, but is essential for the maintenance of normal seedling growth soon after germination.

We have done similar experiments with the tall pea cultivar "Grosshülsiger Schnabel mit gedrücktem Korn" and the dwarf pea cultivar "Kleine Rheinländerin" [5]. The plant growth retardant used, RSW 0411 (Bayer A.-G., Leverkusen), inhibits the three steps following *ent*-kaurene in the GA pathway [15]. The results were identical to those of Sponsel [26]; the germination of the pea seeds was not influenced, and differences between treated and untreated seeds became apparent after 4–5 days only. However, we also discovered that mature seeds of our cultivars contain endogenous GA_{12}-aldehyde (Schwenen et al., this Conference, Abstract No. 0251).

Neutral ethyl acetate extracts were made from mature, dry seeds of the tall and dwarf cultivars, purified by TLC and reverse-phase HPLC and methylated. Methyl GA_{12}-aldehyde was identified by full mass spectra in the appropriate fractions of both extracts. Quantitative measurements of GA_{12}-aldehyde at different times during imbibition and germination were made by the isotope dilution methd, using $[^{14}C]GA_{12}$-aldehyde as internal standard. The tall cultivar contained 17.6 ng g^{-1} fresh weight (2.8 ng $seed^{-1}$) and the dwarf cultivar 16.5 ng g^{-1} fresh weight (4.0 ng $seed^{-1}$) of GA_{12}-aldehyde before water was added. In both cultivars, the amounts decreased to almost zero in the first 24–30 h of imbibition (Fig. 5).

The discovery that GA_{12}-aldehyde is present in mature seeds and disappears during the very first stages of germination raised the question whether it was converted to physiologically active C_{19}-GAs. To study this, $[^{14}C]GA_{12}$-aldehyde in acetone was added to dry mature seeds, which were then imbibed with water and allowed to germinate for different times. Analysis of the $[^{14}C]$-products by extraction, TLC and GC-MS showed that $[^{14}C]GA_{12}$-aldehyde was indeed converted to products behaving like C_{19}-GAs on TLC already after 3 h of imbibition. No clear mass spectra were obtained by GC-MS at this time, but well after 9 h and at subsequent times as shown in Table 1. Although it was clear from the relative intensities of the $[M^+ + 8]$-ions that the $[^{14}C]$-label had been diluted by endogenous material, the exact specific

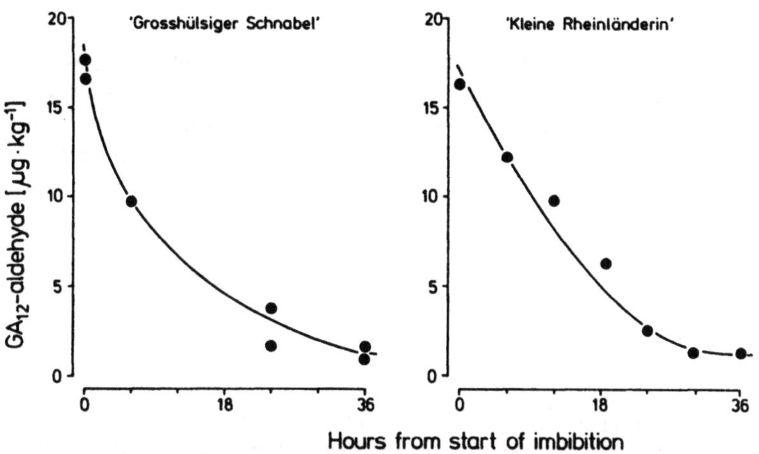

Fig. 5. Changes in endogenous GA_{12}-aldehyde in germinating pea seeds of a tall (G.S.) and a dwarf (K.R.) cultivar [5]. Quantitative estimation by isotope dilution using [^{14}C]GA_{12}-aldehyde (6.36 GBq $mmol^{-1}$) and GC-MS, scanning the molecular ion cluster. Imbibition and germination as in Table 1

Table 1. Incorporation of [^{14}C]GA_{12}-aldehyde into gibberellins in germinating pea seeds [14]

Hours after begin of imbibition [a]	[^{14}C]-Products identified by GC-MS
9	GA_{19}
15	GA_{19}, GA_{20}, GA_{29}, GA_{29}-catabolite
40	GA_{20}-conjugate

[a] Imbibition was completed after 10–12 h. Radicle protrusion was 59% after 29h, 97% after 48 h

activites could not be determined because of impurities. The data in Table 1 and the information in the footnote show that GA_{12}-aldehyde is converted to C_{19}-GAs already during imbibition, several hours before most of the seeds have germinated as indicated by radicle protrusion. Since this conversion would bypass the inhibitors of GA biosynthesis used by us and Sponsel [26], the question whether C_{19}-GAs play a role in germination of pea seeds may be raised again.

6 Conclusion

The GA biosynthetic pathway after GA_{12}-aldehyde basically consists of a defined series of reactions catalyzed by soluble enzymes (Fig. 2). Depending on the species or the tissue, hydroxylations may occur at different points in the pathway. Hydroxylation at carbon-13 tends to occur early in the pathway; it is catalyzed by microsomal enzymes in at least two species. 3β-Hydroxylation tends to occur later in the pathway and is catalyzed by soluble enzymes. Other hydroxylation patterns are also known.

Endogenous GA_{12}-aldehyde has been identified in dry, mature seeds of two pea cultivars, where it becomes converted to GAs during the earliest stages of germination. This conversion would not be inhibited by known growth retardants, since these act at steps lying before GA_{12}-aldehyde in the GA pathway.

References

1. Albone KS, Gaskin P, MacMillan J, Sponsel VM (1984) Planta (Berl) 162:560
2. Bearder JR, MacMillan J, Phinney BO, Hanson JR, Rivett DEA, Willis CL (1982) Phytochemistry (Oxf) 21:2225
3. Blechschmidt S, Castel U, Gaskin P, Hedden P, Graebe JE, MacMillan J (1984) Phytochemistry (Oxf) 23:553
4. Dockerill B, Hanson JR (1978) Phytochemistry (Oxf) 17:701
5. Gabriel K (1984) Diplom-Thesis, Göttingen University, FRG
6. Gaskin P, Gilmour SJ, MacMillan J, Sponsel VM (1985) Planta (Berl) 163:283
7. Gianfagna T, Zeevaart JAD, Lusk WJ (1983) Plant Physiol (Bethesda) 72:86
8. Graebe JE (1982) In: Wareing PF (ed) Plant growth substances 1982. Academic Press, London, p 71
9. Graebe JE, Bowen DH, MacMillan J (1972) Planta (Berl) 102:261
10. Graebe JE, Hedden P, Rademacher W (1980) In: Lenton JR (ed) Gibberellins – chemistry, physiology and use. British Plant Growth Regulator Group, Wantage, p 31 (Monograph No 5)
11. Hafemann C (1985) Thesis, Göttingen University, FRG
12. Hedden P (1983) In: Crozier A (ed) The biochemistry and physiology of gibberellins, vol 1. Praeger, New York, p 99
13. Hedden P, Graebe JE (1982) J Plant Growth Regul 1:105
14. Herbst G (1983) Diplom-Thesis, Göttingen University, FRG
15. Hildebrandt E (1982) Diplom-Thesis, Göttingen University, FRG
16. Ingram TJ, Reid JB, Potts WC, Murfet IC (1983) Physiol Plant 59:607
17. Ingram TJ, Reid JB, Murfet IC, Gaskin P, Willis CL, MacMillan J (1984) Planta (Berl) 160:455
18. Kamiya Y, Graebe JE (1983) Phytochemistry (Oxf) 22:681
19. Kamiya Y, Takahashi M, Takahashi N, Graebe JE (1984) Planta (Berl) 162:154
20. Lewer P, MacMillan J (1984) Phytochemistry (Oxf) 23:2803
21. Metzger JD, Zeevaart JAD (1980) Plant Physiol (Bethesda) 66:844
22. Metzger JD, Zeevaart JAD (1982) Plant Physiol (Bethesda) 69:287
23. Phinney BO, Spray C (1982) In: Wareing PF (ed) Plant growth substances 1982. Academic Press, London, p 101
24. Smith VA, MacMillan J (1984) J Plant Growth Regul 2:251
25. Sponsel VM (1983a) In: Crozier A (ed) The biochemistry and physiology of gibberellins, vol 1. Praeger, New York, p 151
26. Sponsel VM (1983b) Planta (Berl) 159:454
27. Spray C, Phinney BO, Gaskin P, Gilmour SJ, MacMillan J (1984) Planta (Berl) 160:464
28. Turnbull CGN, Crozier A, Schwenen L, Graebe JE (1985) Planta (Berl) 165:108
29. Yamane H, Murofushi N, Takahasi N (1975) Pytochemistry (Oxf) 14:1195

Biosynthetic and Metabolic Aspects of Auxins

R.S. BANDURSKI, A. SCHULZE, and D.M. REINECKE[1]

1 Introduction

We will concentrate on the reactions and processes which we believe to be determinants of the amount of indole-3-acetic acid (IAA) in the tissue, and particularly those in which the greatest recent progress has been made. Then, we will turn to a test of our knowledge – to wit – can we account for the assymmetric distribution of IAA during the tropic stimulus response? If our knowledge passes this test, then we are beginning to understand hormonal homeostasis in a plant system.

2 Determinants of the Amount of IAA in a Plant Tissue

2.1 Branch Points as Control Points

A summary of the reactions known to contribute to the IAA pool is shown in Fig. 1 [1–3, 11, 34]. The reactions, which are likely to be determinants of IAA amount, are the branch point reactions, such as reaction "f", the oxidation of IAA to oxindole-3-acetic acid (OxIAA), which represents the beginning of the oxidative pathway for IAA in *Zea mays*; reactions "b" and "d" representing conjugate hydrolysis and synthesis; reactions "e" and "c" representing transport to, or away from, a particular point; and, reaction "a" representing some form of de novo synthesis.

2.2 Relative Importance of Pathways

Attempts to evaluate the relative contributions of the various pathways to the IAA pool have been frustrated by the fact that these pathways occur simultaneously, even in seedling tissue. Initially, we had hoped that a seedling with the seed still attached would represent a closed system in which only conjugate hydrolysis and synthesis would be pool-size determinants. We find that even the seedling has aromatic, biosynthetic capabilities [16]. This should not have been surprising, in view of the

1 Department of Botany and Plant Pathology, Michigan State University, East Lansing, MI 48824–1312, USA

Plant Growth Substances 1985
(ed. by M. Bopp)

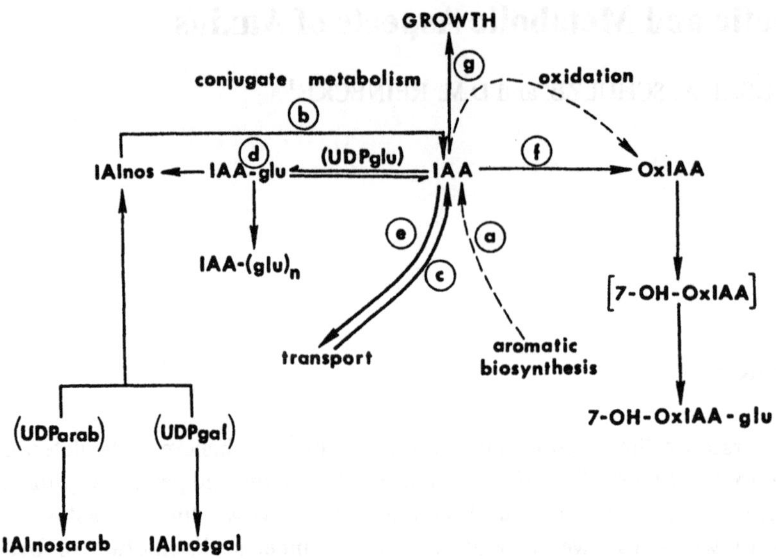

Fig. 1. A summary of the reactions which contribute to and take away from, the IAA pool of seedlings of *Zea mays.* The inputs are: *a* de novo synthesis; *b* conjugate hydrolysis; *c* transport. The outputs are: *d* conjugate synthesis; *e* transport; *f* catabolic oxidation; *g* IAA "use" during growth. (Adapted from P. Davies (ed) Biosynthetic and metabolic aspects of auxins. Martinus Nijhoff/Dr. W. Junk, The Hague)

capacity of an embryo in a chicken egg to synthesize phosphatides despite the presence of a large phosphatide pool in the yolk [17]. Nature, in her wish for protective redundancy endows the embryo with synthetic capacities – which may be feedback regulated – rather than trust a single source as input. A promising approach to understanding the importance of pathways is the use of mutants, as has been pioneered by Torti et al. [36]. This group selected endosperm-defective *Zea mays* and found strains which range from 8% to 81% of the total IAA as free IAA. Thus, the latter group is almost free of ester IAA. Such strains will be of value in determining the relative importance of the various pathways, and this is important, since not all workers agree on the importance of conjugate hydrolysis as a source of free IAA [38].

3 Reactions That Decrease the IAA Pool Size

3.1 Oxidation of IAA to Oxindole-3-Acetic Acid

Oxidative catabolism is an irreversible output and is likely to be an important determinant of the amount of IAA in the tissue [3]. Two catabolic routes have been studied: one pathway involves an early loss of the carboxyl carbon, whereas the other pathway, and the one we believe to be the primary pathway for *Zea mays,* involves oxidation of the IAA at the 2-position. The decarboxylation pathway has been reviewed elsewhere [31, 34] and will not be discussed here.

3.1.1 Occurrence of OxIAA

The pathway shown in Fig. 2 (pathway 1) is the major oxidative pathway for IAA in *Zea mays*. Both OxIAA and 7-hydroxy-OxIAA occur naturally in maize [24, 28, 29]. Interestingly, the pool size of IAA, OxIAA, and 7-hydroxy-OxIAA in shoots are similar, indicating that 7-hydroxy-OxIAA is not accumulating as an end product [24]. This suggests that the pathway may proceed to an end point metabolite, such as anthranilic acid. Labeled OxIAA and dioxindole-3-acetic acid (DiOxIAA) are produced from labeled IAA in maize [28] and broad bean [37] respectively, and D. Reinecke has shown that as much labeled OxIAA, as was expected, was obtained from the feeding of labeled IAA [28] using the turnover data for IAA earlier adduced by Epstein et al. [12]. This finding tends to validate the rapid rate of IAA turnover observed by Epstein et al., and also indicates that the OxIAA pathway is a major one.

Both OxIAA and DiOxIAA occur naturally in rice [20] and DiOxIAA has been observed to be produced from IAA in broad bean [37]. It is not known whether the DiOxIAA pathway is separate from the OxIAA pathway as shown in pathway 2 of Fig. 2 or whether OxIAA can be oxidized to DiOxIAA.

In maize the second oxidation is at the 7-position to form 7-hydroxy-oxindole-3-acetic acid-glucoside (7-hydroxy-OxIAA-glu) [24, 25]. 7-Hydroxy-oxindole-3-acetic acid has now been synthesized by two routes ([25] Paul Lewer, pers. comm.). This compound probably accounts for the zeanic acid isolated earlier [23, 32] from corn

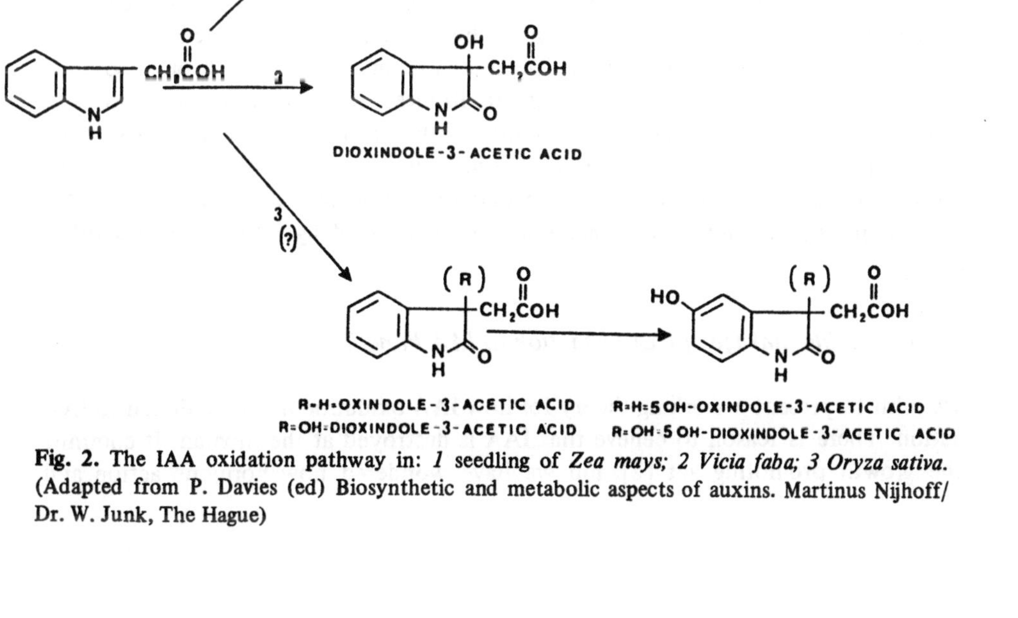

Fig. 2. The IAA oxidation pathway in: *1* seedling of *Zea mays; 2 Vicia faba; 3 Oryza sativa.* (Adapted from P. Davies (ed) Biosynthetic and metabolic aspects of auxins. Martinus Nijhoff/ Dr. W. Junk, The Hague)

steep liquor, since the strong acid and oxidizing conditions used in corn steep liquor production would oxidize, remove the glucose, and cyclize 7-hydroxy-OxIAA-glu to zeanic acid. The occurrence of beta acid in rice hulls [32] may indicate that in rice, the second and third hydroxylations occur at the 3- and 5-positions. Thus, there may be species differences as we move along in the oxidative pathway.

3.1.2 Turnover of OxIAA

H. Nonhebel measured the turnover of OxIAA in the seedling (personal communication). We had previously calculated that IAA in the shoot is being, "used up", at a rate of between 1 to 10 pmol shoot $^{-1}$ h^{-1}. If all the IAA is being oxidized to OxIAA, then the rate of turnover of OxIAA should also be of the order of 1 to 10 pmol shoot $^{-1}$ h^{-1}. If it is less than this, there must be an IAA-destroying system in addition to the IAA to OxIAA pathway. If it is more than this, there must be some source of IAA other than the 2 to 6 pmol shoot^{-1} h^{-1} that we had estimated to be supplied by IAA-*myo*-inositol [26]. The rate of oxidation of IAA to OxIAA in the shoot was found to be about 1 pmol shoot^{-1} h^{-1}. An assumption of constant pool size was made in calculating these data. Nonetheless, we have order of magnitude agreement, indicating we are dealing with major pathways of IAA input and output.

3.1.3 Enzymatic Synthesis of OxIAA

D. Reinecke obtained an enzyme preparation from kernels and vegetative shoot tissue of seedlings of *Zea mays* that catalyzes the oxidation of IAA to oxindole-3-acetic acid [30]. The enzymatic activity is destroyed by heating; it remains in the supernatant fluid following centrifugation at 100,000 xg; and is stable to freezing, ammonium sulfate fractionation, and dialysis. Enzymatic activity is dependent upon the presence of molecular oxygen and enzyme activity is not stimulated by the usual cofactors of a monooxygenase or dioxygenase, such as pyridine nucleotide or alpha-ketoglutarate but, there is a marked increase in activity upon the addition of a boiled triton extract of the enzyme. The "cofactor" was retained by a DEAE column, could be methylated with diazomethane, and its HPLC chromatographic properties were similar to those of an unsaturated fatty acid. Reinecke tried to replace his, "cofactor", material with various fatty acids and found a small stimulation with linoleic acid, but a three fold stimulation with linolenic acid. We do not know whether the endogenous, cofactor material is linolenic acid, but the fact that the activity of the endogenous material can be replaced by linolenic acid is interesting.

3.1.4 Possible Linkage of IAA Oxidation to IAA Function

We wish to present a working theory for how IAA oxidation might be linked to IAA action. There is reason to believe that IAA is destroyed at the moment it commits the growth promoting act [2] and we have wondered how hormone action and hormone destruction could be linked. Reinecke's finding of an IAA oxidase linked to a reaction using an unsaturated fatty acid presents the elements of a working theory. Whether right or wrong, the theory provides a set of postulates which suggest future experimentation. The postulates of the theory are: (1) IAA attaches to its binding site; (2) an oxygenase uses molecular oxygen to oxidize a membrane-localized unsaturated fatty acid to form a hydroperoxy fatty acid; (3) an "IAA oxidase" uses the hydroperoxy fatty acid to oxidize IAA; (4) the IAA is released from its binding site as OxIAA; and (5) the oxidation of the membrane-localized unsaturated fatty acid to a hydroxy fatty acid causes changes in membrane permeability leading ultimately to growth.

This is a working theory and we use it to visualize how growth promotion and IAA destruction might be linked. It does not preclude other effects of IAA since, by means of second messengers, the IAA could also promote the synthesis of new RNAs as has recently been so elegantly demonstrated [14, 35].

3.2 Conjugate Synthesis

There have been no new studies on the mechanism of the enzymatic synthesis of IAA-conjugates in plant systems since the original publications dealing with the synthesis of IAA-glucose, IAA-*myo*-inositol, IAA-*myo*-inositol galactose, and IAA-*myo*-inositol arabinose [e.g., 1, 2].

4 Reactions That Increase the Amount of IAA

4.1 De Novo Synthesis

Despite the large number of studies dealing with the conversion of tryptophan to IAA, and the certainty that tryptophan can be converted to IAA [34], there is still no evidence demonstrating that this is the major biosynthetic route. An encouraging development is the finding by Law and Hamilton [21] that L-tryptophan plus gibberellin A_3 or D-tryptophan alone are as effective as IAA in promoting the growth of pea stem sections. They believe that D-tryptophan is the immediate precursor of IAA and that the racemization of L to D is gibberellic acid-mediated. Owing to the disparity in pool size of L- and D-tryptophan, these experiments are difficult to quantitate, but the studies could lead to an understanding of how a plant can control the synthesis of micromolar amounts of IAA from millimolar amounts of tryptophan.

A further development has been the use of plants grown on D_2O to study the incorporation of deuterium into stable, nonexchangeable positions on the indole ring. Incorporation should be a measure of de novo aromatic biosynthesis, whereas failure to incorporate deuterium indicates a preformed source of IAA – such as IAA conjugates. The answer, obtained by two laboratories, is that there is some incorporation of deuterium into IAA [27, 33]. For the Lausanne group [33], there are indications of the incorporation of only one deuterium into the aromatic ring.

This may indicate that young corn seedlings use some preformed aromatic compound, for example, anthranilic acid to condense with a 4-carbon keto-enol tautomeric compound, thus leading to the incorporation of only one deuterium in the indole ring.

4.2 Conjugate Hydrolysis

J. Chisnell found that IAA-*myo*-inositol occurs in both the kernel and vegetative tissue of *Zea mays* [10] and confirmed a transport rate from kernel to shoot for IAA-*myo*-inositol of about 2 pmol plant^{-1} h^{-1} [9]. M. Komoszynski prepared [^{3}H]-indole-3-acetyl-*myo*-inositol-[^{14}C]-galactose and obtained data indicating that this compound is hydrolyzed in the scutellum to yield [^{3}H]-IAA-*myo*-inositol, which is then transported.

Two enzymes capable of hydrolyzing IAA conjugates have been extracted: (1) an enzyme from *Zea mays* kernels and shoots, which will hydrolyze IAA-*myo*-inositol [15], and an enzyme from *Phaseolus vulgaris,* which will hydrolyze IAA-alanine [6]. The maize enzyme was prepared from an acetone powder and the *Phaseolus* enzyme-required Triton-X-100. This need for lipid removal, or for a detergent, suggests membrane localization for the enzymes, which hydrolyze IAA conjugates.

5 Transport and the Geotropic Stimulus as a Pool Size Determinant

5.1 The Time Course for Geotropically-Induced IAA Asymmetry

There are two remarkable aspects to the time course for the development of an IAA asymmetry following a geotropic stimulus: the first is that the asymmetry, with greater IAA on the lower side, is established within 3 min after changing the plant from a vertical to a horizontal orientation [4], and the second is that *both* free and free plus ester IAA are asymmetric within the same time period [5]. We had earlier postulated that the asymmetry would be due to enhanced ester hydrolysis and Y. Momonoki has observed enhanced ester hydrolysis on the lower side of a geotropically-stimulated corn mesocotyl [5]. But there is a second aspect, since enhanced conjugate hydrolysis could not result in more free *and* conjugated IAA in the lower half of the mesocotyl cortex. With the data now at hand, we believe that more total IAA moved out of the vascular stele and into the cortex. Thus, this experiment indicates that the tropic stimulus induced more "leakage" of IAA from the bottom side of the stele into the bottom cortical cells and, thus, that transfer of IAA from stele to cortex is under metabolic control.

5.2 The Naphthylphthalamic Acid Effect

Many workers, including our own laboratory [8], had observed that N-1-naphthylphthalamic acid inhibited tropic curvature more than it inhibited growth. Inhibitors of this type have been extensively studied by Katekar [13] and have been observed to *inhibit* the downward transport of IAA, and recently, Jacobs and Gilbert have localized the naphthylphthalamic binding site [19]. We found that N-1-naphthylphthalamic acid *increased* by three to four fold the upward movement of radioactivity from [^{14}C]-IAA from kernel to the seedling and resulted in more radioactivity in both stele and cortex. The fact that low concentrations of an inhibitor can *increase* the amount of IAA moving up in the stele and out into the cortex is indicative of metabolic control.

5.3 Analogy Between Plasmodesmata and Gap Junctions

Lowenstein in 1968 [18], reviewed by Carr [7], indicated that plasmodesmata in plants might be analogous to the gap junction cells of animals and could permit the transfer of hormonal signals "directly from one cell interior to another". There is a similarity between plasmodesmata and gap junctions. Plants, owing to their cell wall and sedentary habitat, would require a plasmalemma linkage through the cell wall pore to permit cell to cell communication. Animal cells, lacking a thick wall, and with greater cell motility, and greater movement would find a plasmalemma linkage impossible. Thus, animals evolved a gap junction between cells in which polypeptide hexamers abut against each other to form low resistance junctions [18]. If this analogy is at all correct, then we may suspect that there is potential gating across the plasmodesmatal junctions of plant cells just as there is potential gating across the gap junction cells of animals [18]. This leads us to a working hypothesis for the geotropically-induced IAA asymmetry.

5.4 Potential-Gating Theory

To accomodate the disparate observations of: asymmetry of both free and ester IAA; the rapidity with which the asymmetry develops; the naphthylphthalamic effect; the gap junction analogy; and the known perturbations of the plant's bioelectric field following tropic stimulation [22], we developed the following postulates into a working theory [5]: (1) Gravity causes the settling of dense cell organelles; (2) The organelles carry a fixed charge and, in settling, cut and deform the cell's bioelectric field; (3) this deformation results in transient potential changes across the plasmodesmata such that: (4) the plasmodesmata open, and or close; resulting in (5) an altered movement of IAA from the vascular stele into the surrounding cortical cells. If plasmodesmatal gating also results in heightened ion gradients from one side of the tissue to the other, a device would also be at hand for amplifying the orignal weak signal generated by cutting the cell's bioelectric field.

6 Conclusions

With the possible exception of the route for the de novo synthesis of IAA, the "inputs to" and "outputs from" the IAA pool are known. Particular progress has been made in understanding the mechanism of IAA catabolism in *Zea mays.*

Attention must now be paid to the quantitative evaluation of these routes, using chemical and genetic methods, and to purifying the appropriate enzymes so that the mechanisms for their control might be understood. We have presented two working theories, one, for a possible linkage between IAA action and IAA catabolism and, two, a mechanism for the geotropic induction of the asymmetric distribution of IAA.

Acknowledgments. We gratefully acknowledge the support of the Metabolic Biology Section of the U.S. National Science Foundation, PCM-8204017, and the Space Biology Section of the National Aeronautic and Space Administration, NAGW-97. Manuscript preparation was aided by Ms. Wendy Withford and we acknowledge the use of unpublished data of A. Schulze, W. Domagalski, Y. Momonoki, H.M. Nonhebel, M. Komoszynski, and P. Lewer.

References

1. Bandurski RS (1980) In: Skoog F (ed) Plant growth substances 1979. Springer, Berlin Heidelberg New York, p 37
2. Bandurski RS (1982) In: Wareing PF (ed) Plant growth substances 1982. Academic Press, London, p 3
3. Bandurski RS (1984) In: Crozier A, Hillman JR (eds) The Biosynthesis and metabolism of plant hormones. Cambridge Press, Cambridge, p 183
4. Bandurski RS, Schulze A, Dayanandan P, Kaufman PB (1984a) Plant Physiol (Bethesda) 74:284
5. Bandurski RS, Schulze A, Momonoki Y (1984b) Physiologist 27:(S)123
6. Bialek K, Cohen JD (1984) Plant Physiol (Bethesda) 75:(S)108
7. Carr DJ (1976) In: Gunning BES, Robards AW (eds) Inter-cellular communication in plants: studies of plasmodesmata. Springer, Berlin Heidelberg New York, p 243
8. Ching TMT, Hamilton RH, Bandurski RS (1956) Physiol Plant 9:546
9. Chisnell JR (1984a) Plant Physiol (Bethesda) 74:278
10. Chisnell JR (1984b) Ph D Thesis, Michigan State University, East Lansing
11. Cohen JD, Bialek K (1984) In: Crozier A, Hillman JR (eds) The biosynthesis and metabolism of plant hormones. Cambridge Press, Cambridge, p 165
12. Epstein E, Cohen JD, Bandurski RS (1980) Plant Physiol (Bethesda) 65:415
13. Geissler AE, Katekar GF (1982) J Exp Bot 33:952
14. Hagen G, Guilfoyle TJ (1985) Mol Cell Biol 5:1197
15. Hall PJ, Bandurski RS (1986) Plant Physiol (Bethesda) 80:374
16. Hankins C, Largen MT, Mills SE (1976) Plant Physiol (Bethesda) 57:101
17. Hevesy GC, Levi HB, Rebbe OH (1938) Biochem J 32:2147
18. Hooper ML, Subak-Sharpe JH (1981) Int Rev Cytol 69:45
19. Jacobs M, Gilbert SF (1983) Science 220:1297
20. Kinashi H, Suzuki Y, Takeuchi S (1976) Agric Biol Chem 40:2465
21. Law PM, Hamilton RH (1985) Plant Physiol (Bethesda) 77:(S)2
22. Lund EJ, Rosene HF (1947) Bioelectric fields and growth. University of Texas Press, Austin
23. Matsushima H, Fukumi H, Arima K (1973) Agric Biol Chem 37:1865
24. Nonhebel HM, Bandurski RS (1984) Plant Physiol (Bethesda) 76:979
25. Nonhebel HM, Kruse LI, Bandurski RS (1985) J Biol Chem 260:12685
26. Nowacki J, Bandurski RS (1980) Plant Physiol (Bethesda) 65:422
27. Pengelly WL, Bandurski RS (1983) Plant Physiol (Bethesda) 73:445
28. Reinecke DM, Bandurski RS (1981) Biochem Biophys Res Commun 103:429
29. Reinecke DM, Bandurski RS (1983) Plant Physiol (Bethesda) 71:211
30. Reinecke DM, Bandurski RS (1985) Plant Physiol (Bethesda) 77(S):3

31. Reinecke DM, Bandurski RS (in press) In: Davies P (ed) Plant hormones and their role in plant growth. Martinus Nijhoff/Dr. W. Junk, Netherlands
32. Sahashi Y (1925) Biochem Z 159:221
33. Scharer S, Ribant JM (1983) Thesis, Lausanne
34. Sembdner G, Gross D, Liebisch HW, Schneider G (1980) In: MacMillan J (ed) Encyclopedia of plant physiology, vol 9. Springer, Berlin Heidelberg New York, p 281
35. Theologis A, Huyuh TV, Davis RW (1985) J Mol Biol 183:53
36. Torti G, Lombardi L, Manzocchi LA (1984) Maydica, p 335
37. Tsurumi S, Wada S (1980) Plant Cell Physiol 21:1515
38. Weiler EW, Wischnewski S (1984) Planta (Berl) 162:30

Cytokinin Biosynthesis and Metabolism

R. HORGAN[1]

1 Introduction

This article attempts to take a retrospectic look at the progress made in the area of cytokinin biochemistry since the subject was last reviewed at the 11th IPGSA conference at Aberystwyth in 1982. At that meeting six research papers were presented which gave an extensive view of cytokinin biosynthesis and metabolism. In this instance the subject is being covered by this single report which is therefore highly selective and somewhat subjective in its coverage of the subject.

In a sense the cytokinins are the "poor relations" amongst the known endogenous plant growth regulators. Inspite of the accumulation of extensive knowledge regarding their biochemistry, particularly the details of their metabolism in a wide variety of plants and plant tissues, their physiological role remains unclear. This is in marked contrast to the gibberellins and abscisic acid where the availability of suitable mutants has enabled links between their biochemistry and physiology to be forged. To date the only obvious cytokinin mutants are several cytokinin over-producer mutants (OVE mutants) of the moss *Physcomitrella patens.* Unfortunately, the biochemical basis of these mutations has not yet been elucidated. A comprehensive review of cytokinin metabolism in mutants of *P. patens* has been published [22].

From the point of view of attaching physiological relevance to investigations into cytokinin biosynthesis and metabolism, the most obvious systems to study are plant tissue cultures. Therefore, this review will focus in the main on plant tissue cultures, including crown gall tissues in which cytokinin biosynthesis is of particular relevance.

2 Cytokinin Biosynthesis

Undoubtedly, the most significant recent achievement in the area of cytokinin biochemistry has been the cloning and identification of the *tmr* gene from the T-DNA of a nopaline strain of *Agrobacterium tumefaciens.* Extensive work by several groups utilising transposon mutagenesis showed that tumour morphologies

1 Department of Botany and Microbiology, The University College of Wales, Aberystwyth SY23 3DA, Great Britain

Plant Growth Substances 1985
(ed. by M. Bopp)

were markedly influenced by the inactivation of several T-DNA genes [7]. The two most important of these were designated *tmr* and *tms*. Insertional inactivation of these genes gave rise to rooty and shooty tumors respectively. These genes were mapped to the T-DNA and were shown to be involved with the regulation of cytokinin and auxin levels in the tumors. Mutants in the *tms* locus were shown to contain elevated, total cytokinin levels, whilst those in the *tmr* locus contained considerably lowered cytokinin levels [1]. Thus, there was good circumstantial evidence that the *tmr* gene was involved in cytokinin biosynthesis. The *tmr* region of the nopaline Ti plasmid pTi T37 was cloned and the nucleotide sequence determined [6]. An investigation of this sequence suggested a protein product of 27 kD. A protein of this size had also been suggested as the product from an examination of the same region of an octopine Ti plasmid. A 27 kD protein had been shown to be expressed from the octopine *tmr* region in cell-free extracts of *E. coli* and *A. tumefaciens* and in *E. coli* minicells [18]. Barry et al. [3] have constructed a plasmid, designated pMON230, which contains the *tmr* sequence downstream from a *trp* promotor. This has enabled them to observe expression of the *tmr* gene both at the level of synthesis of [^{35}S]-labeled product in *E. coli* maxicells and the production of active enzyme by *E. coli* cultures. Cell-free extracts of *E. coli* carrying the pMON230 plasmid were shown to catalyse the reaction between 5'-AMP and Δ^2-isopentenyl-pyrophosphate to give a mixture of (Δ^2-isopentenyl)adenosine monophosphate ([9R-5'P]iP) and (Δ^2-isopentenyl)adenosine ([9R]iP). The latter compound presumably arose via phosphatase activity in the extract ([9R-5'P]iP) was identified by coelution on reversed-phase HPLC with a synthetic standard and by its conversion to [9R]iP with alkaline phosphatase followed by HPLC identification of the product. Cell-free extracts of *E. coli* not harbouring pMON230 showed no cytokinin synthetase activity. Thus, it has now been unequivocally established that the *tmr* gene of the T-DNA codes for the enzyme Δ^2-isopentenyl pyrophosphate: 5-AMP Δ^2-iospentenyl-transferase. This result has since been independently confirmed [2, 4]. Although of bacterial origin, this is the first plant growth substance biosynthetic gene to be cloned and expressed in vitro. The relationship between the bacterial enzyme and the previously described plant enzyme [5] is unclear. Substrate specificity appears to be the same, but the molecular weight of the plant enzyme is approximately twice that of the bacterial enzyme.

Although the function of the *tmr* gene has been established, several important questions still need to be answered regarding cytokinin biosynthesis in crown gall tissues and this system remains a valuable model for study of cytokinin biochemistry in plants.

It has previously been reported that gross cytokinin levels are influenced by the presence of functional *tms* genes [1]. It is now known that the *tms* locus contains two genes (gene 1 and gene 2) [8]. There is very strong evidence that these genes are involved in auxin biosynthesis via the pathway known to exist in *Pseudomonas savastanoi* [18]. Gene 1 appears to code for tryptophan monooxygenase and gene 2 for a non-specific amidohydrolase, which converts indole acetamide to IAA [21].

At Aberystwyth we have examined in detail the cytokinin levels in a series of *tms* mutants produced by Prof. Schilperoort's group at Leiden [16]. The results shown in Table 1 clearly demonstrate that inactivation of the *tms* locus markedly

Table 1. Cytokinin levels in Tobacco crown gall tissues carrying mutations in the *tms* and/or *tmr* genes. *Symbols in brackets* refer to inactivated genes. Measurements were made by isotope dilution mass spectrometry by the method of Scott and Horgan [19]

	Cytokinin levels (ngg^{-1} fresh wt.)			
Tissue	[9R-5'P]Z	[9R]Z	Z	[7G]Z
LBA 4001 (wt.)	25.1	17.5	22.3	10.7
LBA 4210 (*tmr*)	10.1	19.1	33.8	2.4
LBA 4060 (*tms*)	183.6	62.6	74.8	779.3
LBA 1501-1 (*tms*)	526.9	77.7	34.4	316.3
TSO 119 (*tmr, tms*)	10.9	26.6	35.1	8.0
SRI 4013-3	91.7	51.1	31.9	70.4
SRI (control)	10.9	16.3	7.2	4.3

influences the absolute and relative levels of cytokinins in these tissues. The biochemical nature of this effect is still not clear, although superficially it would appear that auxin levels in the tissue may influence cytokinin biosynthesis and/or metabolism. We are at present investigating these possibilities.

Although [9R-5'P]iP is the product of the gene 4 enzyme, its level in all tissues has proved to be too low to measure accurately on a routine basis. Thus, it would appear that the hydroxylation step, presumably involving the enzyme previously described by Chen [5], is very rapid in tobacco tissue and can take place at the nucleotide level.

In spite of the discoveries outlined above, investigations into the pathway and control of cytokinin biosynthesis in normal plant tissue is still very problematical. At Aberystwyth we have spent considerable effort in attempting to demonstrate cytokinin biosynthesis in decapitated *Phaseolus vulgaris* plants where cytokinin production is clearly occuring. In all cases rigorous purification of putative-labeled (diH OG)Z, the major cytokinin in this system, resulted in separation of the radioactivity and the endogenous compound. Similar results have been obtained with attempts to demonstrate cytokinin biosynthesis in *Zea mays* kernels (L.M.S. Plani, personal communication). However, confirmation of the results of [^{14}C]-adenine feeding experiments with crown gall tissues has recently been obtained by using stable isotopes and mass spectrometry to detect incorporation. [$^{15}N_5$]-adenine has been shown to be incorporated into [9R-5'P]Z, [9R]Z and Z with enrichments of the endogenous compounds of 10.2%, 1.5% and 1.0% respectively, in *Datura innoxia* crown gall tissue (L.M.S. Plani, personal communication). It is of interest to note that these results tend to confirm the picture of cytokinin metabolism in crown gall tissues suggested by Table 1 and discussed in detail in the next section.

3 Cytokinin Metabolism

An extremely comprehensive review of cytokinin metabolism was given by Letham at the 11th IPGSA Conference in 1982 [12] and two further comprehensive reviews have been published [11, 15]. The lenght of this article does not allow a reiteration of the numerous metabolic products which have been identified from both naturally occurring and synthetic cytokinins. However, with regard to the possible physiological significance of cytokinin metabolism, it is useful to classify the types of metabolic reactions which cytokinins have been shown to undergo. These may be briefy summarised as (1) interconversions of bases, nucleosides and nucleotides; (2) irreversible side-chain modification [e.g. Z to (diH)Z], (3) side-chain cleavage leading to loss of biological activity; and (4) conjugation with sugars or amino acids leading to retention or loss of biological activity. In addition, it may be said that aside from "phytochemical curiosity", studies on cytokinin metabolism are physiologically justifiable on two grounds. Firstly, the enable us to understand the metabolism of endogenous cytokinins and hence assess one component of the turnover, which presumably controls pool sizes in plant tissues. Secondly, they may help to understand structure/activity relationships of externally applied cytokinins both naturally occurring and synthetic. With regard to the first aspect of metabolic studies, it is probably true to say that very little has been achieved in the last 3 years. In spite of an impressive catalogue of cytokinin metabolites, the relevance of these to the turnover of endogenous cytokinins is not yet apparent. Indeed, it would seem that very often the fate of externally applied cytokinins does not mirror that of endogenous compounds. This is particularly apparent in tissues which exhibit high rates of side-chain cleavage. For example, the application of zeatin to cultures of *Vinca rosea* crown gall tissue does not produce the quantitative pattern of labeling of metabolites expected from the known levels of endogenous cytokinins [20]. This is understandable when one considers that this tissue contains high levels of cytokinin oxidase activity and that most of the applied zeatin is broken down to adenine and related compounds before further metabolism takes place. In tobacco tissues, however, there is a good correlation between the metabolism of externally applied cytokinins [9] and the spectrum of endogenous metabolites found in crown gall tissues.

Indeed, the results shown in Table 1 may be viewed from the stand point of cytokinin metabolism, in addition to points made in the biosynthesis section. The availability of different T-DNA mutants has enabled us to observe the relationship between endogenous cytokinin levels and the qualitative and quantitative spectrum of metabolites over a wide range of levels of the primary cytokinin that accumulates in the tissue, i.e. [9R-5′P]Z. They confirm the view that in tobacco tissues 7-glucosylation acts as an inactivation mechanism, which is sensitive to the concentration of applieds cytokinin, and they demonstrate that this applies to endogenous levels as well as to externally applied compounds. It is interesting to note that the level of Z in these tissues does not vary greatly, although it is undoubtedly the intermediate between [9R-5′P]Z and [7G]Z. It has also been suggested that the bases are the active forms of cytokinins [9]. These results may provide a crude picture of the relative turnover rates of endogenous cytokinins.

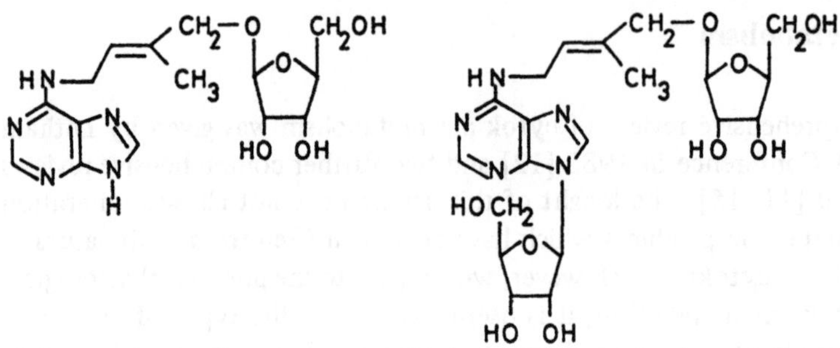

Fig. 1. Tentative chemical structures for metabolites of [8-^{14}C]-zeatin applied to embryos of *Phaseolus vulgaris* [10]

With regard to the modulation of the biological activity of cytokinins via metabolism, an interesting study has been made of the activity and metabolism of [9R]iP in callus cultures of *Phaseolus vulgaris* and *Phaseolus lunatus* [17]. In the former tissue, the lower activity exhibited by [9R]iP was correlated with its increased rate of breakdown to adenosine. In the *P. lunatus* cultures, [9R]iP had greater biological activity and its rate of side-chain cleavage was much reduced. These results confirm the importance of side-chain cleavage in regulating the observed activity of externally applied cytokinins. Undoubtedly, cytokinin oxidase is the key enzyme in this process [13, 14]. An important question that remains to be answered is the extent to which cytoinin oxidase plays a role in the turnover of endogenous cytokinins.

Two novel metabolites of zeatin have been tentatively identified by feeding [8-^{14}C]Z to embryos of *Phaseolus vulgaris* [10]. They were assigned the structures shown in Fig. 1 on the basis of their UV spectra and the mass spectra of their permethyl derivatives. The formation of O-ribosides represents a new pathway of

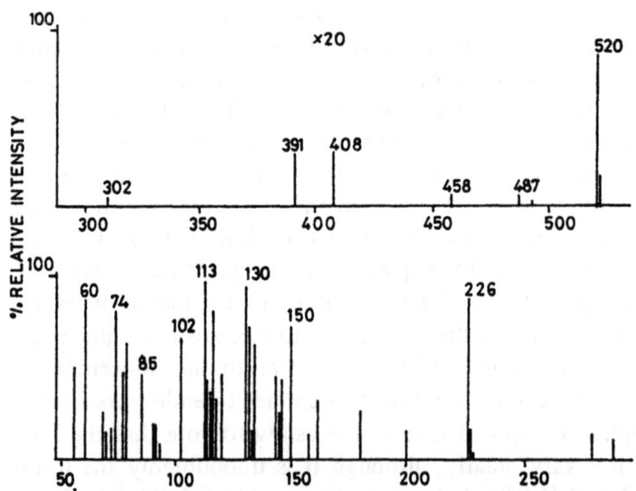

Fig. 2. Chemical ionisation mass spectrum of metabolites of [8-^{14}C]-benzyladenine applied to shoot cultures of *Gerbera* (Lenton and Horgan 1985, unpubl. results). The material was applied to a rhenium filament, which was heated at 20°C s. Ammonia was used as the reagent gas. The ion at m/z 520 represents the $(M+H)^+$ ion

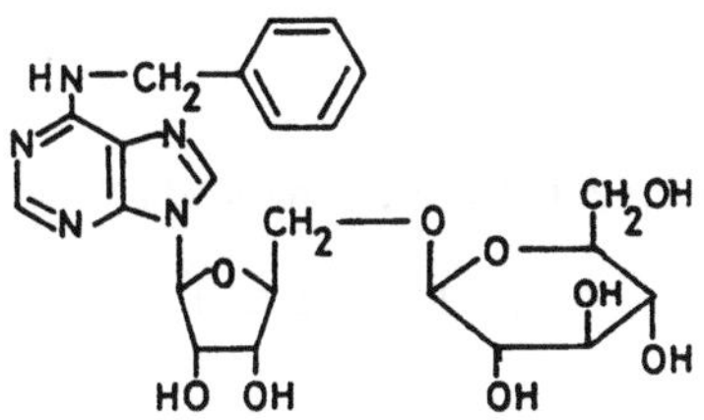

Fig. 3. Tentative chemical structure for the BA metabolite from *Gerbera* shoot cultures

cytokinin metabolism and as with the other forms of cytokinin conjugation, its biological significance is unknown. However, it is of interest to note that this form of metabolism was not detected in embryos of *Phaseolus lunatus.*

An investigation into the metabolism of [8-^{14}C]BA in *Gerbera* shoot cultures has resulted in the tentative identification of a novel metabolite of this synthetic cytokinin. β-Glucosidase treatment of a labeled, polar HPLC fraction gave [9R]BA. Attempts to obtain a direct probe MS of the purified unknown failed to produce a clear spectrum. However, a DCI probe spectrum using ammonia as the reagent gas produced the spectrum shown in a simplified form in Fig. 2. This indicated the sequential loss of a hexose and a pentose from the BA nucleus. The direct probe mass spectrum of the permethyl derivative further suggested the tentative structure shown in Fig. 3.

4 Conclusions

The subject of cytokinin biosynthesis and metabolism has not advanced as rapidly in the past 3 years as it did in the period prior to the previous IPGSA meeting. The questions of sites and control of cytokinin biosynthesis in plants remain to a large extent unanswered. Indeed, the pathway by which plants make cytokinins under normal conditions is still not clear. It has not yet proved possible to completely rule out the involvement of t-RNA in cytokinin biosynthesis. Since the demonstration of cytokinin biosynthesis in plants is such a technically difficult problem, new approaches are needed. One possibility may be to use antibodies raised against the *tmr* gene product, if it can be produced and purified in sufficient quantitites and if it is sufficiently similar to the plant enzyme, to locate sites of cytokinin biosynthesis in plant tissues at both the inter- and intracellular level. A complementary approach using antibodies to cytokinin oxidase might also prove valuable.

The area of cytokinin metabolism has proved to be considerably more lucid from the biochemical than from the physiological standpoint. The degradative metabolism of cytokinins is clearly important in influencing the biological activity of externally applied compounds. Unfortunately, the significance of the more complex forms of cytokinin metabolism remains obscure. However, it is to be hoped that as with the gibberellins and abscisic acid, suitable mutants may be found, which will enable some connections to be made between the biochemistry and physiology of the cytokinins, and that understanding of their mode of action will lead to a fuller appreciation of the physiological role of cytokinins in plants.

References

1. Akiyoshi DE, Morris RO, Hinz R, Mischke BS, Kosuge T, Garfinkel DJ, Gordon MP, Nester EW (1983) Proc Natl Acad Sci USA 80:407
2. Akiyoshi DE, Klee H, Amasino RM, Nester EW, Gordon MP (1984) Proc Natl Acad Sci USA 81:5994
3. Barry GF, Rogers SG, Fraley RT, Brand L (1984) Proc Natl Acad Sci USA 81:4776
4. Buchmann I, Marner FJ, Schroder G, Waffenschmidt S, Schroder J (1985) EMBO J 4:853
5. Chen C (1982) In: Wareing PF (ed) Plant growth substances. Academic Press, London, p 155
6. Goldberg SB, Flick JS, Rogers SG (1984) Nucleic Acids Res 12:4665
7. Hooykaas PJJ, Ooms G, Schilperoort RA (1982) In: Kahl G, Schell JS (eds) Molecular biology of plant tumors. Academic Press, London, p 373
8. Inze D, Follin A, Van Lijsebettens M, Simoens C, Genetello C, Van Montagu M, Schell J (1984) Mol Gen Gene 194:265
9. Laloue M, Pethe C (1982) In: Wareing PF (ed) Plant growth substances. Academic Press, London, p 185
10. Lee YH, Mok MC, Mok DWS, Griffin DA, Shaw G (1985) Plant Physiol (Bethesda) 77:635
11. Letham DS, Palni LMS (1983) Annu Rev Plant Physiol 34:163
12. Letham DS, Tago GQ, Parker CW (1982) In: Wareing PF (ed) Plant growth substances. Academic Press, London, p 144
13. McGaw BA, Horgan R (1983a) Planta (Berl) 159:30
14. McGaw BA, Horgan R (1983b) Phytochemistry (Oxf) 22:1103
15. McGaw BA, Scott IM, Horgan R (1984) In: Crozier A, Hillman JR (eds) The biosynthesis and metabolism of plant hormones. SEB Seminar series 23. Cambridge University Press, Cambridge
16. McGaw BA, Horgan R, Wullems G, Schilperoort RA (1985) Planta (Berl) (submitted)
17. Mok MC, Mok DWS, Dixon SC, Armstrong DJ, Shaw G (1982) Plant Physiol (Bethesda) 70:173
18. Schröder G, Waffenschmidt S, Weiler EW, Schröder J (1984) Eur J Biochem 138:387
19. Scott IM, Horgan R (1984) Planta (Berl) 161:345
20. Scott IM, McGaw BA, Horgan R, Williams PE (1982) In: Wareing PF (ed) Plant growth substances 1982. Academic Press, London
21. Van Onckelen H, Rudelsheim P, Inze D, Follin A, Messens E, Horemans S, Schell J, Van Montagu M, De Greef J (1985)
22. Wang TL, Futers TS, McGeary F, Cove DJ (1984) In: Crozier A, Hillman JR (eds) The biosynthesis and metabolism of plant hormones. SEB Seminar series 23. Cambridge University Press, Cambridge

The Nomenclature of Abscisic Acid and its Metabolites

G.L. BOYER[1], B.V. MILBORROW[2], P.F. WAREING[3], and J.A.D. ZEEVAART[1]

At the 1967 meeting of this Society in Ottawa, those involded with the characterisation of abscisic acid (ABA) agreed on its new name [1, 2]. Since then it has become evident that abscisic acid is the only active member of its class and so its nomenclature is far simpler than that of the gibberellins and even of the cytokinins. However, the continuing process of identifying metabolites, the awkward nomenclature that has sometimes been applied, and the changes to the rules of the International Union of Pure and Applied Chemistry (I.U.P.A.C.) require that the system of nomenclature applied to ABA and related compounds should be modified.

As noted in the Annual Review of Plant Physiology [5], *cis, trans* or 2*Z*, 4*E* is redundant: abscisic acid is defined as 2*Z*, 4*E*, only the 2-*trans* or 2*E* isomer needs to be specified. A serious deficiency is the frequent failure of authors to specifiy whether *S*-ABA or the racemate (*R, S*,) has been used in an experiment. (+) or (±) is equally acceptable.

We now propose a revision of the numbering system of abscisic acid with the methyl groups designated as in Scheme I. The 6′-*pro-S*-methyl group is designated 8′. A similar scheme appeared in a crystallographic analysis of ABA [7], but the authors omitted the primes on the ring methyl groups.

Hydroxylation of one of the 6′-Me groups produces "Metabolite C" or what is usually referred to as 6′-hydroxymethyl ABA. This compound from now on should be 8′-hydroxy ABA.

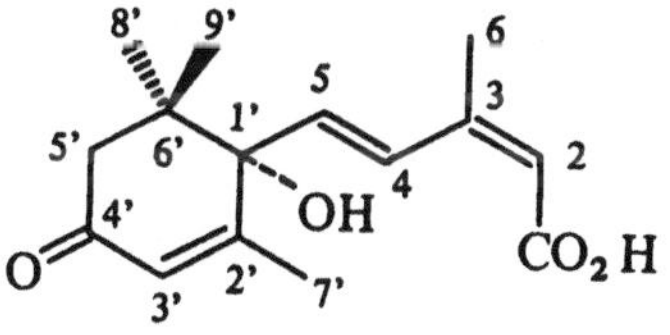

Scheme 1. (+)-*S*-Abscicic acid

Although the use of acronyms is discouraged by I.U.P.A.C., the frequency with which sugar conjugates of some phytohormones are referred to prompted the in-

1 MSU-DOE Plant Research Laboratory, Michigan State University, East Lansing, MI 48824, USA
2 School of Biochemistry, University of New South Wales, Kensington 2033, Australia
3 Department of Botany and Microbiology, University College of Wales, Aberystwyth SY23 3DA, Great Britain

Plant Growth Substances 1985
(ed. by M. Bopp)

troduction of GE for glucose ester and GS for glucoside [3]. Other conjugates and anomers will require specification as they are discovered. The full systematic name would require statement on its first usage and the relevant initials used thereafter. The other alternative for cumbersome terminology is to cite numerals in the text which refer to diagrams of the structures.

A major difficulty arises with phaseic acid (PA) because its ring system is numbered differently from that of ABA [4]. However, in metabolic work a clearer account can often be given if the numbering system of ABA is applied to PA, provided that it is specified.

It is sometimes necessary to refer to the faces of rings and the proposal of Rose et al. [6] has been used such that the face of the ring of ABA that is conventionally drawn uppermost (as in I) has its component atoms numbered in a clockwise sequence. This specifies the α-face and that from which the 1′-hydroxyl group projects is the β.

References

1. Addicot FT, Carns HR, Cornforth JW, Lyon JL, Milborrow BV, Ohkuma K, Ryback G, Smith OE, Thiessen WE, Wareing PF (1968) In: Wightman F, Setterfield G (eds) Biochemistry and Physiology of Plant Growth Substances. Runge, Ottawa, p 1527
2. Addicot FT, Carns HR, Cornforth JW, Lyon JL, Milborrow BV, Ohkuma K, Ryback G, Smith OE, Thiessen WE, Wareing PF (1968) Science 159:1493
3. Loveys BR, Milborrow BV (1981) Aust J Plant Physiol 8:571
4. Milborrow BV (1969) Chem Commun 966
5. Milborrow BV (1974) In: Briggs WR, Green PB, Jones RL (eds) Annu Rev Plant Physiol 25:259
6. Rose IA, Hansen KR, Wilkinson KD, Wimmer MJ (1980) Proc Natl Acad Sci USA 77:2439
7. Schmalle VHW, Klaska KH, Jarchow O (1977) Acta Crystalogr Sect 33:2218

Metabolism of Abscisic Acid

J.A.D. ZEEVAART, G.L. BOYER, K. CORNISH, and R.A. CREELMAN [1]

1 Introduction

The involvement of abscisic acid (ABA) in many physiological processes has been suggested [1, 36], but in most cases the role of ABA has not been established unequivocally. The main reason for this uncertainty is that it is not possible at present to grow plants devoid of ABA, since no ABA-less mutants are available and no specific inhibitors of ABA biosynthesis have been found. Nevertheless, it is clear that ABA plays a role in mediating responses to water stress. For example, application of ABA to turgid leaves causes rapid closure of stomata, and levels of ABA in leaves of mesophytic plants increase many times under conditions of water stress. In ABA-deficient mutants (wilty or droopy) applied ABA causes stomatal closure, but the ABA content of these mutants is not increased by water stress. Another phenomenon involving ABA is seed maturation. Observations made on ABA-deficient mutants of various species indicate that precocious germination of developing embryos (vivipary) is prevented by ABA [16, 18].

In order to understand how ABA controls various physiological processes, it is essential to know how the ABA level in a tissue is regulated. This requires knowledge of the biosynthesis and degradation of ABA, as well as of import and export of ABA into and from the tissue. These various aspects of ABA are briefly reviewed in this paper.

2 Biosynthesis of Abscisic Acid

The structure of ABA indicates that it is a sesquiterpenoid, which is assembled from three isoprene residues. It should be noticed, however, that the structure of ABA is unlike those of other sesquiterpenoids [30], but closely resembles the terminal rings of certain xanthophylls, such as violaxanthin and neoxanthin. Two pathways for ABA biosynthesis have been proposed: (1) the *direct* pathway, in which a C_{15} precursor, presumably farnesyl pyrophoshate, is cyclized and then converted to ABA; (2) the *indirect* pathway, in which ABA is synthesized by degradation of xanthophylls (C_{40}); cleavage of a terminal ring with six C's of the backbone would

1 MSU-DOE Plant Research Laboratory, Michigan State University, East Lansing, MI 48824, USA

Plant Growth Substances 1985
(ed. by M. Bopp)

yield a C_{15} compound, which would be further metabolized to ABA. It is possible that both these pathways operate in plants, depending on various environmental and internal physiological factors. In both pathways, mevalonic acid (MVA) would be the ultimate precursor. At present, no intermediates unique to ABA synthesis have been identified from higher plants in either pathway. Neill et al. [26] have recently reviewed biosynthesis of ABA.

2.1 Evidence in Favor of the Direct Pathway

Results from feeding labeled MVA to various plant tissues, in particular avocado mesocarp, indicate that three isoprene units are incorporated into ABA (see [22]). However, the yields were always very low. Possible reasons for this low incorporation are competition for precursor by other isoprenoid pathways, lack of penetration of the precursor to the site of synthesis, and low rates of ABA biosynthesis at the time of feeding. Ripening avocado fruit mesocarp gave the highest incorporation of MVA into ABA (1% to 2%), whereas before and after the critical time of incipient softening, the yield was only 0.001% [22].

A cell-free system containing plastids from avocado fruit incorporated labeled MVA into ABA (see [22]). On the other hand, no label from MVA was incorporated into ABA by chloroplasts isolated from spinach leaves, although incorporation was detected when protoplasts or cytoplasmic fractions were incubated with MVA [13]. Thus, the cellular location of ABA synthesis remains an open question.

In an attempt to differentiate between the direct and indirect pathway, phytoene labeled with ^{14}C was fed to avocado fruit along with MVA labeled with ^{3}H. ABA contained only ^{3}H, whereas both ^{3}H and ^{14}C were detected in carotenoids [22]. However, the results of this experiment are inconclusive, since it is not known if xanthophylls had become labeled. Moreover, a large (unlabeled) pool of xanthophylls was already in existence and conversion of only a very small fraction to ABA would be expected (see below). Thus, [^{14}C]-label in ABA would seem unlikely, even when the indirect pathway would be functioning.

2.2 Evidence in Favor of the Indirect Pathway

The idea that ABA is synthesized via the indirect pathway stems from the finding that xanthophylls, in particular violaxanthin, can either photochemically or enzymatically be degraded to the neutral growth inhibitor *cis,trans*-xanthoxin (Xan) [32]. Furthermore, radioactive Xan was converted to ABA and phaseic acid (PA) when fed to tomato and dwarf bean plants [33]. It is of interest to note in this context that the pheromone trisporic acid, which regulates sexual reproduction in the Mucorales, is a breakdown product of β-carotene [11].

If ABA is derived from xanthophylls, then blocking carotenogenesis with the use of inhibitors or with mutations in the carotenoid biosynthetic pathway should also prevent ABA accumulation. The herbicides fluridone and norflurazon block the desaturation of phytoene to phytofluene, the first noncolorless carotenoids [2]. Seedlings grown in the presence of these inhibitors in high-intensity light are white as a result of photobleaching of the chlorophyll and have greatly reduced ABA levels [7, 14, 24, 29]. For example, inhibitor-treated and control maize seedlings accumulated 15 and 307 ng ABA per g fresh wt, respectively, during an 8-h stress period. The corresponding carotenoid contents were 1.3 and 25.5 $\mu g\ g^{-1}$ fresh wt. [7]. However, these inhibitors are not specific; they cause changes in the lipid composition, and change the levels of carbohydrates and organic acids [7]. It could be further argued that a parallel pathway leading to ABA was inhibited as well. Thus, the results of these inhibitor studies with white seedlings are equivocal. When green leaves (which contained already a large pool of xanthophylls) were treated with norflurazon, ABA accumulation by stress was not inhibited [14]. However, results obtained with in situ labeled xanthophylls are more conclusive. In bean leaves labeled with $^{14}CO_2$ and then water-stressed, the specific activites of ABA and xanthophylls were similar on a carbon atom basis. When the leaves were treated with fluridone prior to labeling with $^{14}CO_2$, specific activities of ABA and xanthophylls were reduced to the same extent, but total ABA synthesis was not affected [19]. Clearly, this is strong evidence in favor of the indirect pathway.

Several mutants of maize are viviparous and carotenoidless. Embryos of these mutants contain much reduced (0 to 50% of wild-type) levels of ABA [4, 25]. It is possible that some of the ABA found in these embryos has a maternal origin. Interestingly, fluridone induced vivipary when applied to developing maize seeds 11 days after pollination [9].

Other green, ABA-deficient mutants are known in potato (droopy [28]), pea (wilty [37]), *Arabidopsis* (*aba* [18]), and tomato (*flacca, notabilis,* and *sitiens* [27]). In the case of *flacca* and *sitiens,* total carotenoids and the relative proportions of β-carotene, lutein, lutein epoxide, violaxanthin, and neoxanthin are very similar to those found in the wild-type "Rheinlands Ruhm" [27]. If ABA arises via the indirect pathway, then the lesions in these mutants must be between the xanthophylls and ABA.

In an attempt to determine the origin of the four oxygen atoms in ABA, stressed leaves of *Xanthium* and *Phaseolus* were incubated in an atmosphere containing $^{18}O_2$. Mass spectral analysis of the ABA isolated from these leaves indicated that one atom of ^{18}O was present in the carboxyl group [8]. This observation suggests that stress-induced ABA is derived from a stored precursor with oxygens already present in the 1′- and 4′-positions of ABA; it could be explained by an oxidative attack on violaxanthin similar to that which occurs when β-carotene is converted to retinaldehyde [12]. In a similar experiment, conducted with the fungus *Cercospora rosicola,* four atoms of ^{18}O were present in ABA (R. Horgan, D.C. Walton, pers. comm. [8]. This implies that the biosynthetic pathway of ABA in this organism is different from that for stress-induced ABA in higher plants. It would be interesting to conduct a similar $^{18}O_2$ experiment with a nonstressed tissue, such as avocado fruit, in which ABA is apparently synthesiszed via the direct pathway. In that case, one would also expect four ^{18}O atoms per ABA molecule. Unfortunately, such an experiment is difficult to perform, since the period during which ABA is synthesized at a high rate must be known precisely. Only in the case of stress can this be controlled at will.

Li and Walton [19] conducted experiments with *Phaseolus* leaves in which the epoxide oxygen of violaxanthin was labeled in situ with ^{18}O. When these leaves were stressed, the ABA contained ^{18}O in the 1′-hydroxyl group, although there was considerable dilution of the label. This result could be explained if ABA originated not only from violaxanthin, but from other xanthophylls, such as lutein, as well.

Thus, in their entirety the results discussed in this section support the indirect pathway for ABA biosynthesis. This raises the following questions: (1) What are the intermediates between xanthophylls and ABA in vivo? Is Xan the normal precursor as suggested by the work of Taylor and Burden [33]? (2) How is the conversion of xanthophylls to ABA regulated? For example, does the indirect pathway function only during water stress? If so, how does loss of turgor result in a greatly increased breakdown of xanthophylls? It is well-known that carotenoids are readily broken down in the presence of acids, light, oxygen, and in some cases also by base. For example, β-carotene in the presence of oxygen and light undergoes oxidative cleavage to β-ionone, dihydroactinidiolide, and desoxy-xanthoxin [15]. Violaxanthin is photo-oxidized to Xan, *trans,trans*-xanthoxin (*t*-Xan), butenone, and loliolide [32]. Since all four products are also found in plant extracts, the question arises: What is actually present in vivo and what is an artifact of extraction? The violaxanthin content of spinach leaves is 1.4 mg g^{-1} dry wt. [17], and the Xan and *t*-Xan content 240 and 720 ng g^{-1} dry wt., respectively [38]. So, breakdown of only a very small fraction of the violaxanthin would yield significant quantities of Xan and *t*-Xan. This observation raises some doubt concerning the role of Xan as an intermediate in ABA biosynthesis, since due to breakdown of xanthophylls, during extraction and purification procedures, the measured levels may be much higher than they actually are in vivo.

3 Catabolism of Abscisic Acid

ABA is rapidly metabolized in plants, giving rise to several metabolites and conjugates. This topic has been reviewed recently [20].

When stressed and subsequently rehydrated *Xanthium* leaves were kept in an atmosphere containing $^{18}O_2$, one atom of ^{18}O was incorporated into PA. Thus, conversion of ABA to PA requires molecular oxygen and the enzyme is an oxygenase [8]. Although there is ample evidence for high rates of ABA to PA conversion in vivo [39], no cell-free system for this step other than from *Echinocystis* liquid endosperm [10] has been obtained. ABA catabolites are generally thought to have little or no biological activity. An exception is the effect of PA in barley aleurone layers in which this metabolite, like ABA, inhibits gibberellin-induced α-amylase secretion. In this system, ABA appears to enhance its own catabolism to PA [34].

In our studies on ABA metabolism in *Xanthium* [40], a new metabolite was detected after feeding (±)-ABA to leaves. This metabolite has an optical rotary dispersion spectrum that is the mirror image of that of (+)-ABA (maximum at 246 nm, minimum at 289 nm), which led to the conclusion that this new metabolite is derived exclusively from (−)-ABA. We have proposed to expand the numbering system of the ABA skeleton to include the exocyclic methyl groups and have designated the new compound 7′-hydroxy (−)-*R*-ABA [3] (Fig. 1).

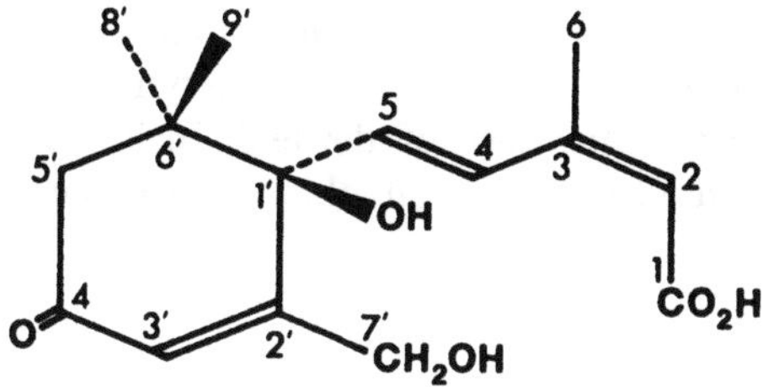

Fig. 1. Structure of 7′-hydroxy (−)-*R*-abscisic acid isolated from *Xanthium strumarium* leaves after feeding (*R*)-ABA

When radioactive (−)-ABA was fed to *Xanthium* leaves, 68% was conjugated to form the glucose ester, and 24% was metabolized to 7′-hydroxy (−)-*R*-ABA over a 24-h period. On the other hand, (+)-ABA was metabolized to PA (44%), ABA-glucose ester (12%), DPA (7%), and conjugates of PA and DPA (24%). Thus, the metabolism of the two enantiomers is distinctly different. Since it is now possible to resolve (+)- and (−)-ABA either by HPLC [35], or by immunoaffinity chromatography [21], future metabolic studies of ABA should be carried out with the natural enantiomer.

We have speculated that the oxygenase that forms 8′-hydroxy ABA from (+)-ABA in the PA pathway, also hydroxylates the 7′-methyl group in the unnatural (−)-isomer. The affinity of the enzyme for the latter substrate is presumably much less than that for (+)-ABA. Circumstantial evidence to support this idea comes from time course feeding studies with racemic ABA. In this case, PA is always formed rapidly, whereas 7′-OH *R*-ABA does not accumulate until the PA starts to level off. However, slower uptake of (−)-ABA than of (+)-ABA [23, 31] could be a factor as well. Whether a single enzyme hydroxylates the 8′-methyl of (+)-ABA and the 7′-methyl of (−)-ABA can ultimately only be established with a cell-free system.

4 Sites of Abscisic Acid Metabolism

It is well-known that ABA accumulates in seeds, fruits, buds, and leaves, but much less is known about ABA in roots. Recently, we have shown that excised roots of tomato and *Xanthium* can accumulate considerable amounts of ABA when stressed [6]. However, in contrast to leaves, maximum accumulation of ABA did not occur until the roots had lost 60% to 70% of their fresh weight. In both species, much greater relative increases in ABA were observed in roots than in leaves, although the absolute ABA content was always less than that of the corresponding leaves. Due to the water potential gradient that exists from the roots to the shoot, the former will undergo much greater dehydration than the latter. Therefore, during the early stages of drought the ABA content of the roots could increase considerably before the leaves start to wilt. Since ABA alters the hydraulic conductance of the roots, the stress-induced ABA could affect the plant's water economy before the leaves become stressed [6].

When mature *Xanthium* leaves are stressed, they rapidly accumulate ABA which upon rehydration is catabolized to PA [39]. In contrast, detached and stressed young leaves produce little ABA [5, 41]. On the other hand, ABA and PA levels in intact stressed plants were highest in the young leaves. These high levels are due to import from mature leaves and a low rate of catabolism [5]. Thus, the ability to synthesize and catabolize ABA depends on the developmental stage of the leaf. In half-expanded leaves ABA catabolism could be enhanced by stress, but this effect was not observed in young leaves [5].

5 Conclusions

Although earlier results indicated that ABA is synthesized directly from MVA, there are now several lines of evidence (inhibitors, mutants, ^{14}C-labeling, ^{18}O-incorporation) that favor the indirect pathway via carotenoids, at least for stress-induced ABA. It is conceivable that depending on various environmental and internal physiological conditions both pathways can function in plants.

Catabolism of (+)- and (−)-ABA is distinctly different. A novel metabolite of (−)-ABA produced in *Xanthium strumarium*, has been identified as 7′-hydroxy (−)-*R*-abscisic acid ABA. The numbering system of the ABA skeleton has been expanded to include the exocyclic methyl groups.

Rates of synthesis and degradation of ABA are determined by the developmental stage of the leaves, the youngest ones being least active in these processes.

In tomato and *Xanthium* roots, maximum accumulation of stress-induced ABA occurs at a much greater water loss than in leaves. It is conceivable that stress-induced ABA in the roots modifies the plant's water economy before the shoot exhibits stress symptoms.

Acknowledgments. Work supported by the United States Department of Energy under Contract DE-ACO2-76ERO-1338 and National Science Foundation grants PCM78-07653 and PCM83-14321.

References

1. Addicot FT (ed) (1983) Abscisic acid. Praeger, New York
2. Bartels PG, Watson CW (1978) Weed Sci 2:198
3. Boyer GL, Zeevaart JAD (1986) Phytochemistry (Oxf) 25
4. Brenner ML, Burr B, Burr F (1977) Plant Physiol (Bethesda) Suppl 59:76
5. Cornish K, Zeevaart JAD (1984) Plant Physiol (Bethesda) 76:1029
6. Cornish K, Zeevaart JAD (1985) Plant Physiol (Bethesda) 79:653
7. Creelman RA, Zeevaart JAD (1981) Plant Research '81. Annu Rep MSU-DOE Plant Res Lab East Lansing, USA, p 123
8. Creelman RA, Zeevaart JAD (1984) Plant Physiol (Bethesda) 75:166
9. Fong F, Smith JD, Koehler DE (1983) Plant Physiol (Bethesda) 73:899
10. Gillard DF, Walton DC (1976) Plant Physiol (Bethesda) 58:790
11. Gooday GW (1974) Annu Rev Biochem 43:35
12. Goodwin TW (1983) Trans Biochem Soc 11:473
13. Hartung W, Heilmann B, Gimmler H (1981) Plant Sci Lett 22:235
14. Henson IE (1984) Z Pflanzenphysiol 114:35
15. Isoe S, Hyeon SB, Katsumura S, Sakan T (1972) Tetrahedron Lett 2517

16. Karssen CM (1982) In: Wareing PF (ed) Plant growth substances 1982. Academic Press, London, p 623
17. Kirk JTO (1978) In: Kirk JTO, Tilney-Basset RAE (eds) The Plastids: their chemistry, structure, growth and inheritance. Elsevier, New York, p 1
18. Koornneef M (1985) In: King J (ed) Plant gene research. Springer, Berlin Heidelberg New York (in press)
19. Li Y, Walton D (1985) Plant Physiol (Bethesda) Suppl 77:23
20. Loveys BR, Milborrow BV (1984) In: Crozier A, Hillman JR (eds) The biosynthesis and metabolism of plant hormones. Cambridge University Press London, p 71 (Soc Exp Biol Sem Ser, vol 23)
21. Mertens R, Stüning M, Weiler EW (1982) Naturwissenschaften 69:595
22. Milborrow BV (1983) In: Addicott FT (ed) Abscisic acid. Praeger, New York, p 79
23. Milborrow BV, Rubery PH (1985) J Exp Bot 36:807
24. Moore R, Smith JD (1984) Planta (Berl) 162:342
25. Moore R, Smith JD (1985) Planta (Berl) 164:126
26. Neill SJ, Horgan R, Walton DC (1984) In: Crozier A, Hillman JR (eds) The biosynthesis and metabolism of plant hormones. Cambridge University Press, London, p 43 (Soc Exp Biol Sem ser, vol 23)
27. Nevo Y, Tal M (1973) Biochem Genet 10:79
28. Quarrie SA (1982) Plant Cell Environ 5:23
29. Quarrie SA, Lister PG (1984) Z Pflanzenphysiol 11:295
30. Robinson T (1980) The organic constituents of higher plants, 4th edn Cordus, North Amherst, Mass
31. Sondheimer E, Galson EC, Chang YP, Walton DC (1971) Science 174:829
32. Taylor HF, Burden RS (1972) Proc Soc Lond B Biol Sci 180:317
33. Taylor HF, Burden RS (1973) J Exp Bot 24:873
34. Uknes SJ, Ho TD (1984) Plant Physiol (Bethesda) 75:1126
35. Vaughan GT, Milborrow BV (1984) J Exp Bot 35:110
36. Walton DC (1980) Annu Rev Plant Physiol 31:454
37. Wang TL, Donkin ME, Martin ES (1984) J Exp Bot 35:1222
38. Zeevaart JAD (1974) Plant Physiol (Bethesda) 53:644
39. Zeevaart JAD (1980) Plant Physiol (Bethesda) 66:672
40. Zeevaart JAD, Boyer GL (1983) In: Wareing PF (ed) Plant growth substances 1982. Academic Press, London, p 335
41. Zeevaart JAD, Boyer GL (1984) Plant Physiol (Bethesda) 74:934

The Shapes of Abscisic Acid and the Active Site

B.V. MILBORROW[1]

1 Introduction

Interest in the influence that slight chemical modifications have on the interaction between the ABA molecule and active sites has revived recently. The supply of *S*- and *R*-ABA made available by an immunological resolution method [13] and milligram quantities by an HPLC method [31] now enables a more precise probing of metabolism and attachment to the active sites to be carried out.

2 Cyclization and Conformation

In 1981 Bennett, Norman and Maier [3] reported the [^{13}C]-NMR spectrum of *S*-ABA formed from [1, 2-$^{13}C_2$]acetate by the fungus *Cercospora rosicola*: one of the 6′-Me groups (24.3δ) showed ^{13}C-^{13}C coupling to C-6′, while the other (23.1δ) did not and so was derived from C-2 of mevalonate.

The downfield [^{1}H] 6′-Me signal of ABA (I) was irradiated using an I.N.E.P.T. sequence by M. Batley and J. Redman, which showed that this Me group produced the upfield 6′-Me signal in the ^{1}H spectrum. The next step was to relate its position to the known absolute configuration at C-1′ using the 300 MHz NMR spectra of the 1′,4′-*cis*- (II) and 1′,4′-*trans*-diols (III).

ABA exchanges (Fig. 1) first its C-5′ hydrogen atoms, then the C-3′ and the 2′-Me hydrogen atoms with the medium under either alkaline or acidic conditions. If ABA is carefully titrated in the NMR with NaOD in D_2O at about pH 11, most of the axial C-5′ hydrogen atoms can be replaced with deuterium. This ABA was then reduced to give the diols. These were separated and subjected to 300 MHz [^{1}H]-NMR to relate the configuration of the C-5′ hydrogen atoms to the hydrogen atom of the *S*-C-4′ of the *trans*-diol and the *R*-C-4′ of the *cis*-diol. From this the conformation of the rings and the absolute configuration of the C-6′-Me derived from C-2 of mevalonate was established [20]. The cyclization mechanism is now revised to be identical with that proposed for the β- or ϵ-rings of higher plant carotenoids [5]. The NMR analysis also showed that the preferred conformation of

1 School of Biochemistry, University of New South Wales, P.O. Box 1, Kensington, N.S.W. 2033, Australia

Plant Growth Substances 1985
(ed. by M. Bopp)

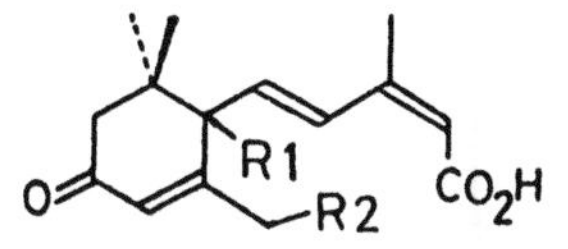

I

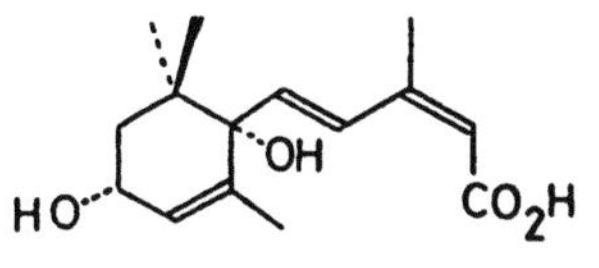

II

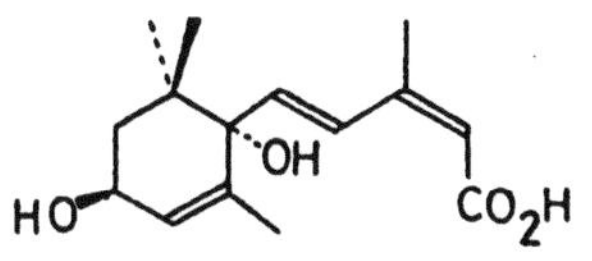

III

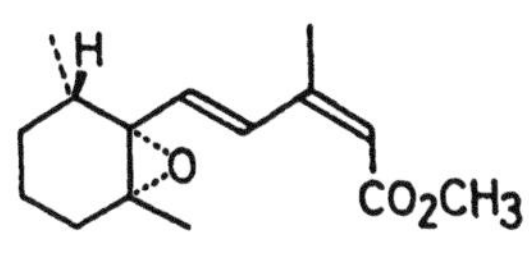

IV

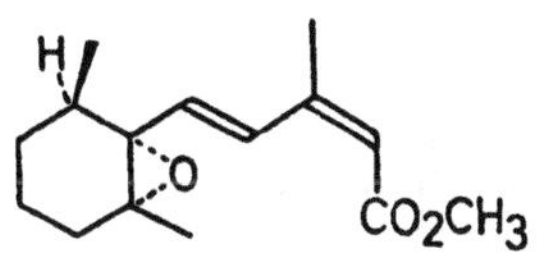

V

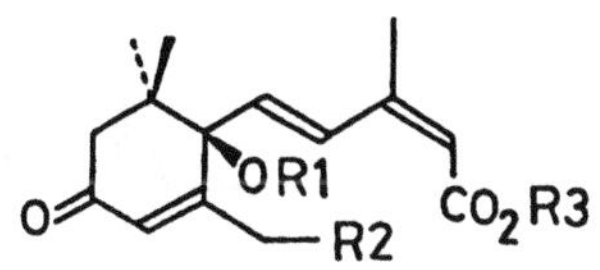

VI

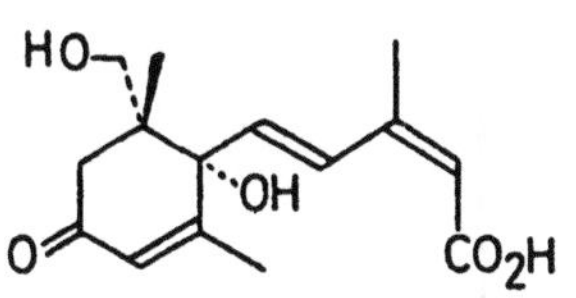

VII

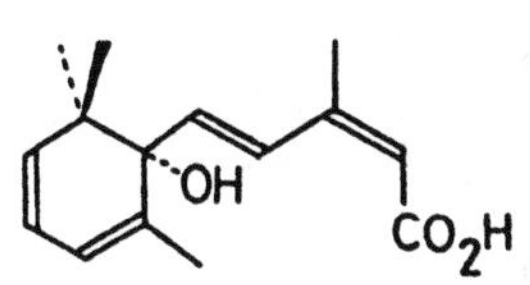

VIII

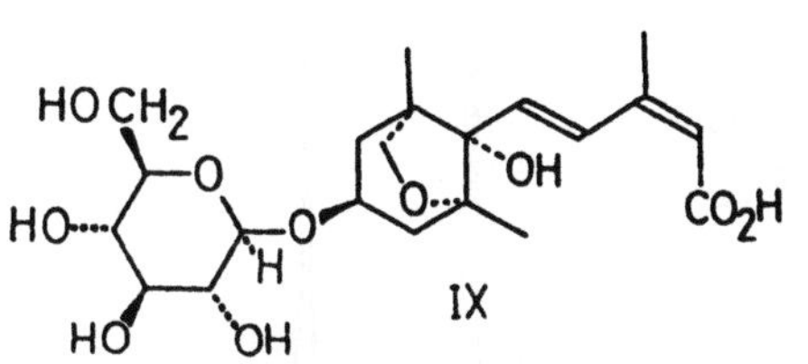

IX

Structure I. (+)-(*S*)-ABA R1 = - - -OH, R2 = H

Ia. deoxy ABA R1 = H, R2 = H

Structure II. 1′, 4′-*cis*-diol of ABA

Structure III. 1′,4′-*trans*-diol of ABA

Structure IV. Z 6′-demethyl epoxide

Structure V. E 6′-demethyl epoxide

Structure VI. 7′-hydroxy ABA R1, R2, R3 = H

VIa. R1 = H, R2 = glucosyl, R3 = CH_3

Structure VII. 8′-hydroxy ABA

Structure VIII. 4′-deoxy ABA

Structure IX. Dihydrophaseic acid-4′-O-β-D-glucopyranoside

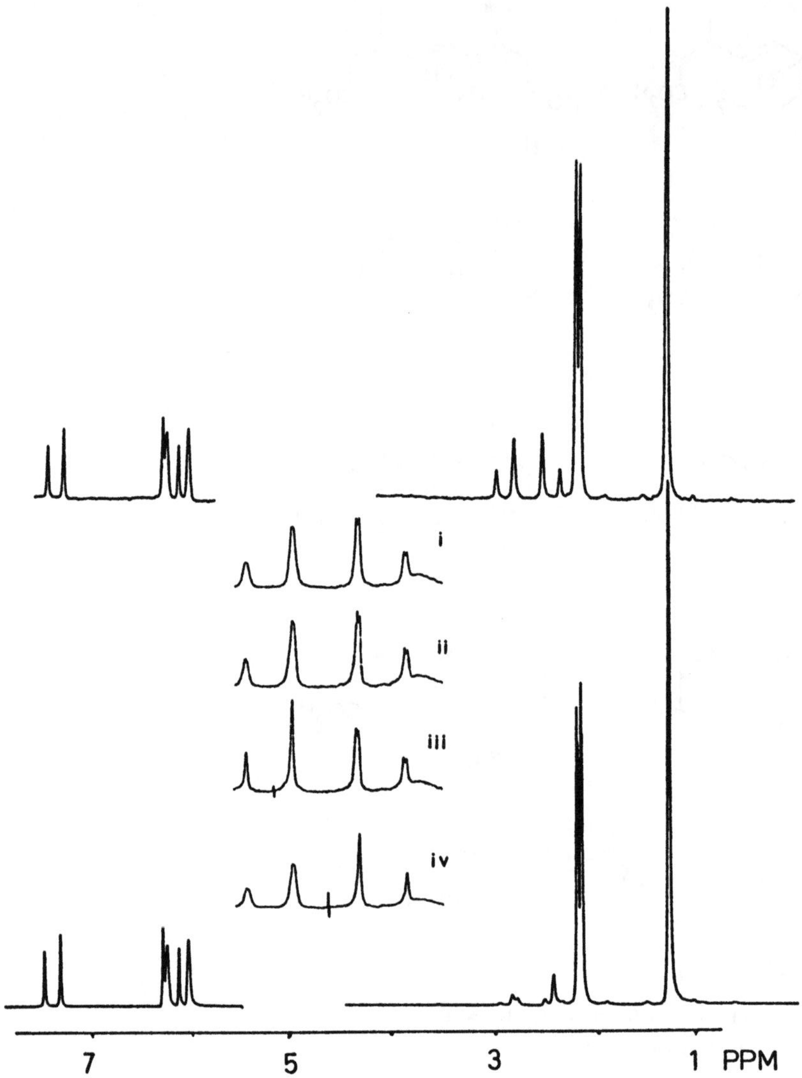

Fig. 1. ^{1}H 100 MHz NMR spectrum of ABA in D_2O + NaOD at pH 11.0 (*upper*). After 6 h (*lower*), most of the 5′-pro-*R*-hydrogen atoms (axial) have exchanged, but those of the C-2′-Me (2.14 ppm) and at C-3′ (2.25 ppm) have not

ABA in solution is with the ring in a pseudochair and with side chain axial, i.e. arched over the "upper" a-face of the ring (Fig. 2) [20]. It is the same as the conformation in the crystal [30].

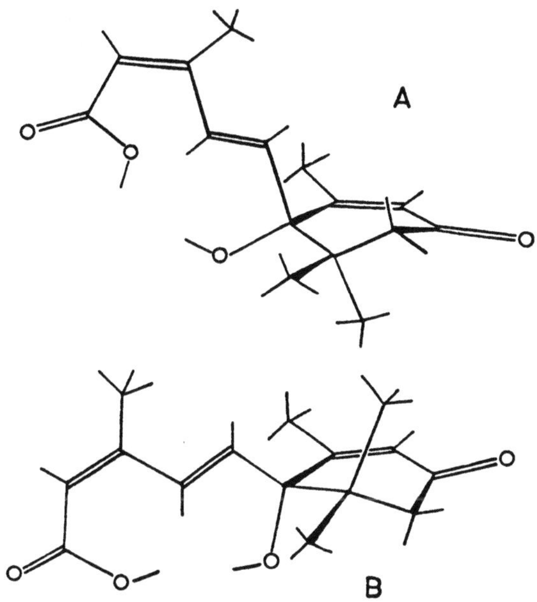

Fig. 2A, B. The two possible pseudochair conformations of ABA, A is preferred

3 Binding of ABA to the Uptake Carrier in Bean Roots

An investigation of the ABA uptake carrier [22] revealed that the *R,S*-1′,4-*cis*-diol competes for the uptake-carrier site with *RS*-$[^{14}C]$ABA at least as strongly as unlabeled *RS*-ABA (Ki 2.8 μM), while the *RS*-1′,4′-*trans*-diol was a weaker inhibitor (Ki about 16 μM).

If the 1′,4′-*cis*-diol binds to the carrier as strongly as ABA, then its new, 4′-hydroxyl group cannot interfere with the binding. On the other hand, the 1′,4′-*trans*-diol has its new 4′-hydroxyl group on the opposite face and this appears to render it much less effective (Fig. 3). The diols and ABA, therefore, are deduced to bind by their upper, α-faces [26]. In support of this 1′-deoxy ABA was also found to bind almost as strongly as ABA, this would not be expected if the 1′-hydroxyl group were involved in an interaction with the carrier.

The axial side chain (Fig. 2) makes a concave surface with the α-face of the ring so the active site of the carrier would be expected, on this basis, to be quite convex (Fig. 4). However, similar ring structures have been investigated Bernard and St. Jacques [4] and the energy to flex a dimethyl cyclohexenone ring through the unstable boat form is only about $\Delta G^{0\prime}$ 6.4 KCalmol^{-1} : a value which is within the range of thermal effects at normal temperatures.

The uptake carrier appears to accept *S*-, but not *R*-ABA, consequently the Km determined for *RS*-ABA (2.8 μM) [27] can probably be halved (1.4 μM).

If one makes what can be a slightly inexact assumption that $\Delta G^{0\prime} = -RT \ln Km$, then the free energy of binding to the upake carrier can be calculated (from 1.4 μM) as 7.85 KCalmol^{-1}.

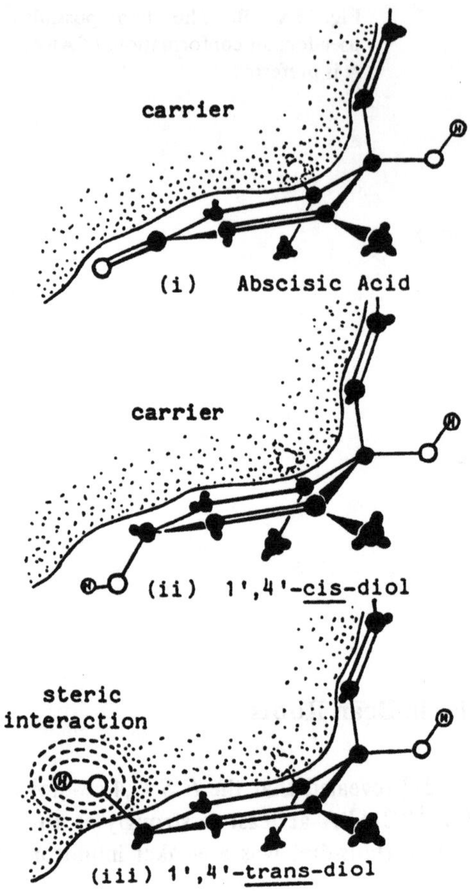

Fig. 3. (+)-*S*-ABA and the isomeric diols are shown as they are believed to bind to the uptake carrier

These figures show that ABA could easily bind to an active site in the less favoured, planar, pseudochair conformation.

The uptake carrier in root tips and the K^+ uptake carrier involved in stomatal closure react with *S*-ABA, but not or very weakly with *R* enantiomer [10, 18]. A dissociation constant (Ks) can be calculated from the data of Tucker and Mansfield [29] to be 0.3 μM, or 0.15 μM for *S*-ABA. This value is considerably smaller than those calculated for the growth inhibition site.

4 The Binding of ABA to the Growth Inhibition Receptor

The model proposed to account for the similar efficacy of *R*- and *S*-ABA postulated that the enantiomers bind to the growth inhibitory site by the same face of their rings, but that a 6′-Me group of unnatural *R*-ABA occupies the site, normally filled by the 7′-Me of *S*-ABA [16] (Fig. 5).

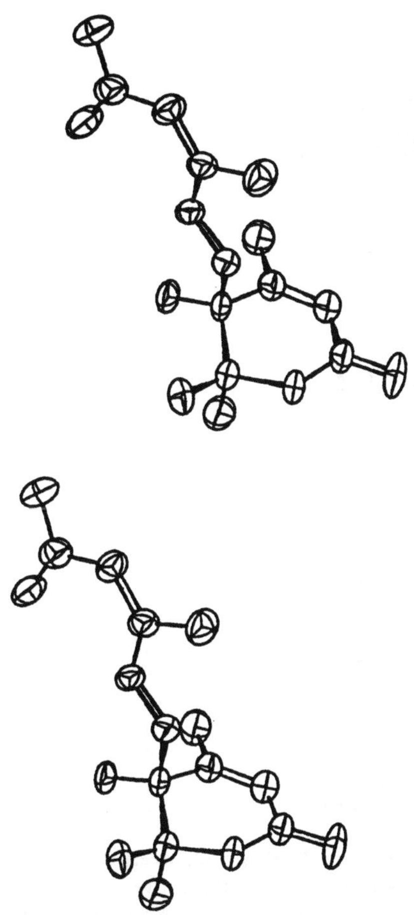

Fig. 4. Stereoscopic view of *S*-ABA. The 3-D effect can be obtained by viewing in a stereoscope or by focussing at infinity and holding the paper about 250 mm away. The side chain is axial and can be seen arched over the a-face of the ring. Carbon and oxygen atoms are represented as elipsoids of a size such that the vibrating atoms have a 50% probability of being found within them [30]

In intuitive consideration would suggest that the receptor site that can respond to the *R* as well as the *S* would need to bind ABA strongly in order to overcome the minor discomforts of fitting the *R* into the position that properly accommodates *S*-ABA and, conversely, that sites that bind the *S*-enantiomer only might be expected to bind less firmly.

Rothwell and Wain [27] measured wheat coleoptile growth in the presence of an array of concentrations of IAA and the Yellow Lupin inhibitor (ABA) and subjected the results to Lineweaver-Burke analysis. The lines of the reciprocal plots converged to a point on the X-axis indicating non-competitive inhibition of IAA promoted growth by ABA so effects of IAA can be ignored.

The Ks 1.3 μM is based on Ks = Ki = the ABA concentration, which halves the maximal IAA-stimulated growth rate. The inhibitor was later found to be almost half ABA [8]. The Ks, therefore, is about 0.6 μM.

The concentration of *RS*-ABA reducing growth of oat mesocotyl sections [14] by half gives a value of Ks=Ki 0.57 μM (Fig. 6). The measurement of true concentrations in experiments such as these, which last for 18–24 h, is fraught with dif-

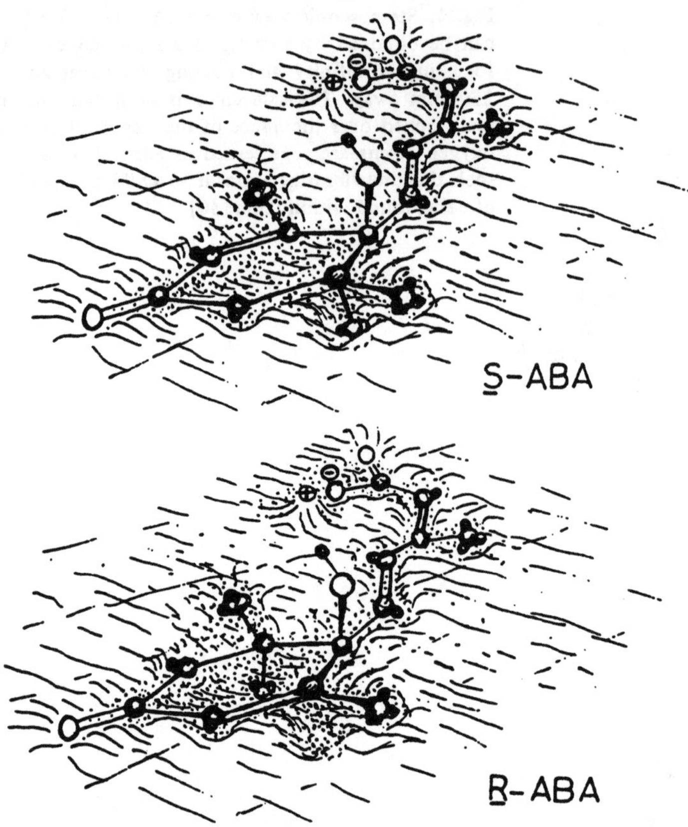

Fig. 5. Schematic representation to illustrate how *S*-ABA and *R*-ABA could exert their almost equal inhibitory activity on growth. The unnatural *R*-enantiomer fits the active site if its 2′-methyl group occupies the position normally taken by the 6′-Me group of *S*-ABA and vice versa

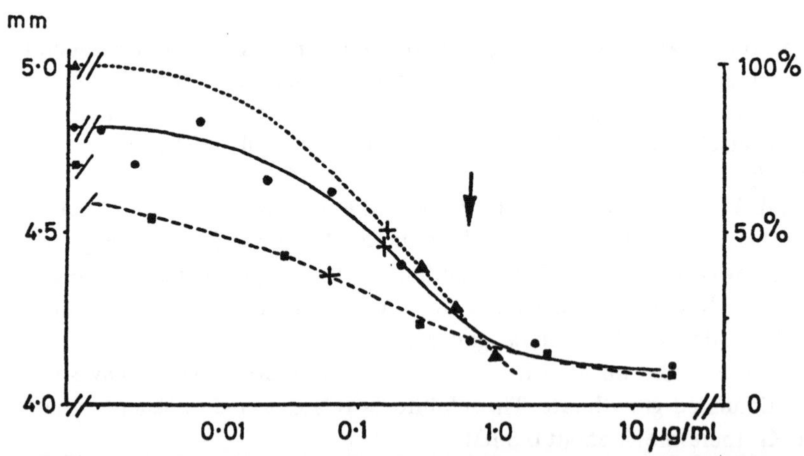

Fig. 6. Growth of oat mescotyl 4 mm sections (•) [14] inhibited by *RS*-ABA, and wheat coleoptile 10 mm sections (▲) [27] inhibited by an *S*-ABA fraction isolated from yellow lupin pods. Stomatal closure by *RS*-ABA (■, relative units [29]) as only *S*-ABA is active the Ks value should be halved. The *arrow* marks the Km value determined for the uptake carrier using *RS*-ABA. The 1/2 Vmax values obtained from the graphs are equated with Ks values. The true value for wheat may be about half the value shown (0.16 ppm = 0.61 μM) because the sample was later found to be approximately half *S*-ABA

ficulties, the intercepts of the lines of the double-reciprocal plots fall off with increasing concentrations of ABA, suggesting that some other influence is at play. The bathing solutions used by Rothwell and Wain [27] and Milborrow [14] were buffered at pH 5.0 so the cytoplasm of the cells, if near neutrality, could act as a trap for the ABA anions. On the other hand, metabolism would tend to lower the internal concentration of the ABA absorbed.

Harrison and Walton [9] reported that (*RS*-[^{14}C]) ABA fed to wilted and unwilted bean shoots turned over with a half-time of about 3 h and a similar turnover rate in wilted tomato shoots has recently been measured [14]. *S*-[^{14}C]-ABA showed a half-time of 14 h when fed to tomato shoots to give a concentration of 7% of the endogenous concentration. Thus, the presence of an excess of ABA over the normal, endogenous concentration probably stimulates rapid metabolism. Which of the competing effects: anion trap accumulation or catabolism, has a greater effect on the endogenous concentration within the tissues of wheat coleoptiles was investigated under conditions as close as possible to those used by the original authors, but the added ABA was labelled with [^{14}C]. The coleoptiles were sampled after 6, 12 and 18 h and their total, ethanolic extracts subjected to silica gel thin-layer chromatography. The majority of the [^{14}C]-ABA taken up by the coleoptile segments remained unmetabolized at the end of the experiment and there was no marked increase in the rate of metabolism with increasing concentration (Table 1). Even a specific decrease may be apparent at 2.0 ppm, which suggests that the catabolic rate of the tissue may be approaching saturation. A similar falling off has been seen with IAA [25] above 1.75 ppm. It must be emphasized that the coleoptile segment is not a uniform block of tissue and fast absorption of ABA by the cut ends could initiate rapid metabolism, while the same amount spread evenly might not do so. The amounts of [^{14}C]-ABA in the coleoptile tissues are slightly less than in the external solution at the end of the exposure period when total growth was measured. The relation between time and the integral of growth inhibition has not been analyzed, but if one takes the amount of [^{14}C]-ABA in the tissues half way through the experiment as an approximation of the effective amount of [^{14}C]-ABA

Table 1. Uptake and metabolism of *S*-[2-^{14}C]ABA by wheat coleoptile sections, ten per tube, 1 ml solution, 20°, other conditions as specified originally [27]

	2.00 ppm free [^{14}C]ABA in coleptiles (ng)	% dpm in metabolites	0.2 ppm free [^{14}C]ABA in coleoptiles (ng)	% dpm in metabolites	0.02 pm free [^{14}C]ABA in coleoptiles (ng)	% dpm in metabolites
6 H	75	18	9.5	25	1.21	24
12 H	103	20	10.8	34	1.57	36
18 H	155	38	15.7	45	1.98	41

[a] The 2.00 ppm solution contained 17,000 dpm, the 0.2 ppm 40,000 dpm and the 0.02 ppm 4,000 dpm. The percentage of [^{14}C] in metabolites was calculated from the total dpm present in the ethanolic extract of the ten sections. The coleoptile sections at H_0 contained 3.5 ng free ABA (i.e. 3.5 μg ABAkg^{-1})

in the coleoptiles, it is about half the nominal concentration of the test solution. Whether the exogenous ABA mixes with, and is inactivated at the same rate as, the endogenous ABA has not been determined. Furthermore, the distribution has been averaged over the whole tissue and the amounts in the cytoplasm, vacuole and apoplast have not been measured. However, the data do suggest that the overall concentration within the tissues may be lower than the nominal concentrations and, hence, the Ks for inhibition of growth may be even smaller than that calculated from the concentration of ABA in the external medium.

The strongest calculated interaction between ABA and an active site is that for the stomata. As *R*-ABA is almost without effect in closing stomata, the specificity of the receptors must be attributable to steric discrimination against *R*-ABA and not merely to weak binding.

Interestingly, a 6′-demethyl ABA analogue (V), synthesized by Adachi et al. [1] and which has a single 6′-Me on the *a*-face of the ring, opposite to the 1-epoxide group (1′,6′-*E*), is ten times a stronger growth inhibitor than the 1′,6′-*Z*-isomer (VI). This is a expected if the *a*-face of the ring lies against the growth inhibition receptor and provided that the rates of conversion of (IV) and (V) into their appropriate 6′-demethyl *S*-abscisic acid analogues and their rates of catabolism are similar. By analogy only the 1′-enantiomer corresponding to *S*-ABA would be expected to be converted into ABA [21].

5 Hydroxylation of *R*-ABA

Indirect evidence supporting the model for the interchangeability of the 6′- and 2′-Me groups of ABA at an active site (Fig. 5) comes from the work of Lehmann et al. [11]. They characterized a metabolite of *RS*-[^{14}C]-ABA as 7′-hydroxymethyl ABA (VI). Boyer and Zeevart [6] found that the same compound in *Xanthium* is formed from unnatural *R*-ABA only. More recently, two conjugates of the methyl ester of the same 7′-hydroxy compound have been isolated from sunflower leaves fed with *RS*-Me ABA and identified as 1′-*R*-7′-hydroxy ABA methyl ester 1′-O-β-D-glucopyranoside (Loveys and Milborrow, unpubl.).

This suggests that the enzyme that normally oxidizes a 6′-Me group of *S*-ABA to give 8′-hydroxylmethyl ABA (VII) can accept *R*-ABA provided that the C-7′-Me comes to occupy the site that normally oxidizes the C-8′-*pro-S*-Me.

An interesting feature of the 7′-Me hydroxylation of *R*-ABA can be deduced from an examination of Dreiding molecular models. The favoured conformation of *S*-ABA in solution places the 6′-*pro-S*-Me axial [20] and yet it is this methyl group that is hydroxylated and forms the oxymethylene bridge of phaseic acid [15, 17]. If the ring is flexed to the less favoured pseudochair (Fig. 2), i.e. the side chain becomes equatorial and planar with the ring, then the 6′-*pro-S*-Me becomes equatorial and thereby equivalent to the 2′-hydroxymethyl group of *R*-ABA. It is also noteworthy that equatorial methyl groups, not axial ones, of geminal pairs are hydroxylated, suggesting that the reaction occurs in a highly restricted hydrophobic pocket in the enzyme [7]. Thus, *S*-ABA is probably in the less favoured pseudochair conformer when bound to the oxygenase. Whether or not the 2′-Me of *R*-ABA and the 6′-*pro-S*-

Me of *S*-ABA are hydroxylated by the same enzyme could be established, if unlabeled *R*-ABA were shown to compete with small quantities of *S*-[^{14}C]-ABA, and vice versa, in a cell-free system.

6 The 1′,4′-*trans*-diol of ABA

That pea plants formed 4′-deoxy ABA (VIII) from *RS*-[^{14}C]-ABA [28] led to the discovery that the supposed 4′-deoxy metabolite was an artefact, produced by dehydration of the 1′,4′-*trans*-diol in the mass spectrometer [19], left unanswered the question of why the 1′,4′-*trans*-diol was formed from added ABA and it has now been found to occur naturally in pea plants (Vaughan and Milborrow, unpubl.).

Both enantiomers of ABA were reduced by pea seedlings and a ratio of 0.7:1.0 (1′-*R*:1′*S*) was found in the free acid, but in experiments in which *S,S*-[^{14}C]1′,4′-*trans*-diol was fed to the seedlings almost seven times more radioactivity was found in dihydrophaseic acid 4′-O-B-D-glucopyranoside (DPAGS) (IX) and other oxidized metabolites than was formed from the *R,R*-1′,4′-*trans*-diol. The latter was glucosylated at C-4′-0 or converted into ABA and its conjugates. There are a number of possible pathways by which DPAGS could be formed from the 1′,4′-*trans*-diol. The major question being whether the 4′-hydroxyl group was oxidized to a ketone (thereby allowing spontaneous cyclization) or whether the 4′-hydroxyl group of the *trans*-diol was retained as the 4′-hydroxyl group of DPAGS (if this occurred then an

Table 2. Relative intensities of the major parent ion fragmentations of the tetraacetyl Me ester of the 4′-O-B-D-glucopyranoside of endogenous dihydrophaseic acid and material formed from [4′-^{2}H]1′,4′-*trans*-diol (C.I.M.S., methane as ionizing gas) [a]

m/z	Unlabeled material Intensity (%)	Material derived from [4′-^{2}H]1′,4′-diol Intensity (%)
608	0	0
609	65.3	70.39
610	28.56	22.65
611	5.18	5.66
612	0.73	1.05
613	0.23	0.24
614	0	0

[a] The loss of 18 mass units from the M+1 molecular ion has been identified by [^{18}O]-labeling as the loss of the 1′-hydroxyl group [23]. The sample derived from the 1′,4′-*trans*-diol contained no deuterium

enzymically catalyzed cyclization would be required). The fate of the 4′-hydrogen atom of DPA atoms (numbered as in ABA) can provide the crucial test between these routes. If the 4′-hydroxyl group of the 1′,4′-*trans*-diol, labeled with deuterium at C-4′, were oxidized to a ketone by the plant, then the deuterium would be lost. Any retention of a deuterium atom at C-4′ of DPAGS, therefore, would establish that the *trans*-diol is hydroxylated and cyclized without oxidation of its 4′-hydroxyl group to a ketone.

The absence of a ketone group to activate the 2′-double bond of the 1′,4′-*trans*-diol makes it extremely unlikely that its 8′-hydroxymethyl derivative could cyclize spontaneously to give DPA under intracellular conditions [32]. When [U-^{3}H]-ABA and [2-^{14}C]1′,4′-*trans*-diol, with its 4′ protium atom replaced by deuterium, were fed to tomato plants, there was a rapid formation of DPA 4′-0-β-D-glucopyranoside. When this material was subjected to chemical ionization mass spectrometry, it was found that no deuterium had been retained (Table 2). Hence, the hydroxyl group had been oxidized to a ketone (Vaughan and Milborrow, unpubl.). We have not yet established whether the 1′,4′-*trans*-diol is hydroxylated to 4′-hydroxy, 8′-hydroxy ABA and then oxidized to 8′-hydroxy ABA or whether the 1′,4′-*trans*-diol is first oxidized to ABA before undergoing 8′-hydroxylation.

7 Conclusion

This work has established that the relative amounts of metabolites vary quite widely between different plant species [12] and compounds are formed from labeled, racemic abscisic acid, which do not occur naturally.

Acknowledgments. Thanks are due to Drs. B.R. Loveys and H.M. Nonhebel for permission to cite results which have not yet been published, to Drs. J. Redman, M. Batley and J. Saunders for some of the NMR spectra, Mr. G. Grossman for carrying out the [^{2}H]-exchange and to Dr. A.M. Duffield for the GC/MS CI spectra. The work was supported by the Australian Research Grants Scheme.

References

1. Adachi T, Oritani T, Yamashita K (1975) Agric Biol Chem 38:1681
2. Astle MC, Rubery PH (1980) Planta (Berl) 150:312
3. Bennet RD, Norman SM, Maier VP (1981) Phytochemistry (Oxf) 20:2343
4. Bernard M, St Jacques M (1973) Tetrahedron 29:2539
5. Britton G, Goodwin TW, Lockley WJS, Mundy AP, Patel NJ (1979) Chem Commun, p 27
6. Boyer GL, Zeevaart JAD (1985) Plant Physiol (Bethesda) 77:76
7. Cornforth JW (1983) In: Simpkins MA (ed) Int Symp Med Chem (Oxf) p 7
8. Cornforth JW, Milborrow BV, Ryback G, Rothwell K, Wain RL (1966) Nature 211:742
9. Harrison MA, Walton DC (1975) Plant Physiol (Bethesda) 56:250
10. Hornberg C, Weiler EW (1984) Nature 310:321
11. Lehmann H, Preiss A, Schmidt J (1983) Phytochemistry (Oxf) 22:1277
12. Loveys BR, Milborrow BV (1983) In: Crozier A, Hillman JR (eds) SEB Symposia, Cambridge, p 71

13. Mertens R, Stüning M, Weiler EW (1982) Naturwissenschaften 69:595
14. Milborrow BV (1966) Planta (Berl) 70:155
15. Milborrow BV (1969) Chem Chommun 966
16. Milborrow BV (1974) In: Runeckles VC, Sondheimer E, Walton DC (eds) Recent Advances in Phytochemistry, vol 7, Academic Press, New York, p 57
17. Milborrow BV (1975) Phytochemistry (Oxf) 14:1045
18. Milborrow BV (1980) Aust J Plant Physiol 7:749
19. Milborrow BV (1983) J Exp Bot 34:303
20. Milborrow BV (1984) Biochem J 220:325
21. Milborrow BV, Garmston M (1973) Phytochemistry (Oxf) 12:1597
22. Milborrow BV, Rubery PH (1985) J Exp Bot 36:807
23. Milborrow BV, Vaughan GT (1982) Aust J Plant Physiol 9:361
24. Nonhebel HM, Milborrow BV (1985) Proc Aust Soc Plant Physiol, p 21
25. Nonhebel HM, Hillman JR, Crozier A, Wilkins MB (1985) J Exp Bot 36:99
26. Rose IA, Hansen KR, Wilkonson KD, Wimmer MJ (1980) Proc Natl Acad Sci, USA, 77:2439
27. Rothwell K, Wain RL (1964) In: Nitsch JP (ed) Régulateurs naturels de la croissance végétale. Centre National de la Recherche Scientifique, Paris, p 363
28. Teitz D, Dörffling K, Whorle D, Erxleben I, Liemann F (1979) Planta (Berl) 147:168
29. Tucker DJ, Mansfield TA (1971) Planta (Berl) 98:157
30. Ueda H, Tanaka J (1977) Bull Chem Soc (Jpn) 50:1506
31. Vaughan G, Milborrow BV (1984) J Exp Bot 35:110
32. Walton DC, Dorn B, Fey J (1973) Planta (Berl) 112:87

Enzymes of Ethylene Biosynthesis

H. KENDE, A.B. BLEECKER, W.H. KENYON, and R.G. MAYNE[1]

1 Introduction

The biosynthesis of ethylene has been reviewed recently [23]. Our paper aims mainly at describing progress that has occurred since then with regard to the enzymes that catalyze the formation of ethylene. There are at least two enzymes in this pathway, one that converts S-adenosylmethionine (SAM) to 1-aminocyclopropane-1-carboxylic acid (ACC) and one (or perhaps more than one) that oxidizes ACC to enthylene. To fully understand ethylene biosynthesis, it will be necessary (1) to elucidate the enzymatic mechanisms of ACC and ethylene formation and (2) to purify the respective enzymes as a first step towards determining their mode of regulation at the molecular level.

2 ACC Synthase

Shortly after Adams and Yang [2] discovered that ACC is the immediate precursor of ethylene, Boller et al. [6] described an enzyme activity from tomato pericarp tissue that catalyzed the conversion of SAM to ACC. A report on properties of ACC synthase was also published by Yu et al. [25]. In most cases investigated thus far, ethylene biosynthesis in higher plants is regulated at the level of ACC synthase, i.e., the availability of ACC is the limiting factor. Hence, conditions and chemicals that induce ethylene formation, e.g., stress and auxin at high concentrations, often enhance the activity of ACC synthase (for a review see [23]).

Because the activity of ACC synthase often determines the developmental fate of plant tissues and because ACC synthase is rapidly induced by a variety of factors, there is considerable interest in understanding how this enzyme is regulated. Acaster and Kende [1] and Chappell et al. [7] used density labeling to determine whether ACC synthase was newly synthesized in wounded tomato pericarp tissue and elicitor-treated cells of parsley, respectively. In both instances incorporation of deuterium from deuterated water indicated that the enzyme was synthesized de novo rather than activated from a preexisting protein. However, these results do not permit the conclusion that enhancement of ACC-synthase activity is based on induction of

1 MSU-DOE Plant Research Laboratory, Michigan State University, East Lansing, MI 48824, USA

Plant Growth Substances 1985
(ed. by M. Bopp)

de novo ACC-synthase formation. Reduced breakdown of continuously synthesized ACC synthase would also lead to increased enzyme levels and isotope incorporation into the protein. It is not clear whether or not RNA synthesis is required for enhancement of ACC-synthase activity. In wounded tomato pericarp tissue, the increase in ACC-synthase activity was inhibited by cordycepin, an inhibitor of RNA synthesis [1]; in cultured parsley cells, it was not [7]. Antibodies and cDNA probes will be required to elucidate the regulation of ACC synthase in different cell types and under different conditions.

2.1 The Stereochemistry of ACC Formation

ACC is formed from SAM via a cyclization reaction in which the α-carbon of SAM attacks the γ-carbon in an S_N2-type displacement of the 5′-methylthio-5′-deoxyadenosine moiety. In all known pyridoxal phosphate (PLP)-mediated γ-elimination reactions of this kind, the reaction proceeds via a vinylglycine-type intermediate, where a hydrogen at the β-carbon is stereospecifically exchanged. Adams and Yang [2] and Yu et al. [25] suggested that the positive sulfonium function could allow the reaction to proceed by a direct γ-elimination without the involvement of the β-carbon. Recently, Ramalingam et al. [19] demonstrated that no exchange of β-hydrogens occurred and that the γ-elimination proceeded with inversion of configuration at the γ-carbon (Fig. 1).

2.2 Mode of Action of Aminoethoxyvinylglycine (AVG)

Lieberman et al. [13] discovered that rhizobitoxine and some of its analogs, such as AVG, inhibit ethylene biosynthesis in higher plants. Adams and Yang [2] proposed that AVG blocks ethylene synthesis by inhibiting the formation of ACC. Based on enzyme kinetics, Boller et al. [6] provided evidence that ACC synthase was inhibited by AVG and that this inhibition was competitive with regard to SAM. More recent studies (Kieliszewski and Bleecker, unpubl.) revealed a more complex mode of action of AVG. ACC synthase incubated in PLP-free buffer containing AVG and then chromatographed on a Sephadex column to remove AVG exhibited very low enzymatic activity. This activity could be largely restored by reconstituting the enzyme with PLP. In PLP-free medium lacking AVG, very little PLP was removed from ACC synthase. These experiments indicate that AVG interacts with PLP at the active site of the enzyme, probably through the formation of a Schiff base, resulting in removal of PLP from ACC synthase.

2.3 Purification of ACC Synthase

The level of ACC synthase in plants is very low, even in fully induced tissues. In the pericarp of ripening (light-red) tomato fruits, a relatively rich source of the enzyme, the level of ACC synthase was found to be 3×10^{-6} units per gram fresh weight or per miligram soluble protein (1 g tissue contains ca 1 mg soluble protein; 1 enzyme unit = 1 μmol min^{-1}) [10a]. How much protein does such enzyme activity represent?

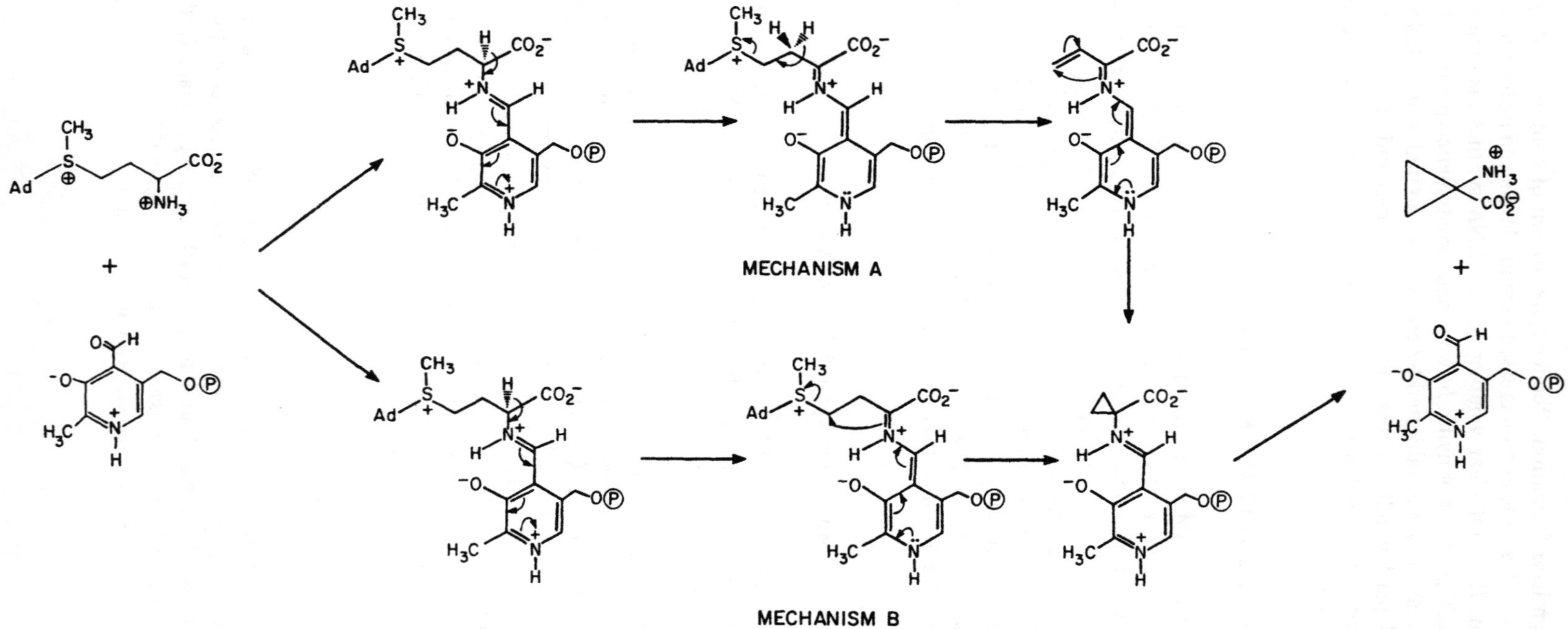

Fig. 1. Two possible mechanisms of pyridoxal phosphate-dependent conversion of SAM to ACC by ACC synthase. Evidence reported by Ramalingam et al. [19] suppports mechanism B

To calculate this, one would have to know the turnover number, k_3, of the enzyme ($V_{max} = k_3[E_T]$). The turnover number of most enzymes is in the range of 1 to 10^4 s^{-1} [20]. Using $k_3 = 1\ s^{-1}$ for our calculation, the level of ACC synthase (MW 50,000) is ca. 2 ng mg^{-1} soluble protein (= 0.0002%) in light-red tomato pericarp tissue. If $k_3 > 1\ s^{-1}$, as is the case with most enzymes, the amount of ACC synthase per milligram soluble protein may be up to several orders of magnitude lower.

Two procedures are currently used to increase ACC-synthase activity in tomato pericarp tissue. First, the fruits are cut in halves, and all the central tissues (including the septa, intralocular pulp and seeds) are removed. The resulting cavity is filled with a solution of 100 mM LiCl which enhances ACC-synthase activity [5]. After ca. 20 h, the LiCl solution is poured out, the pericarp tissue is cut finely to enhance further ACC-synthase activity, and the wounded tissue is incubated for 3 h before homogenization. Together, these two treatments raise ACC-synthase activity in tomato pericarp tissue up to 50-fold.

Using a combination of ammonium sulfate preciptation and hydrophobic interaction chromatography, Acaster and Kende [1] achieved a 60- to 70-fold purfication of ACC synthase. At this level of purification, ACC synthase could not be recognized as a distinct band on a Coomassie-blue-stained sodium dodecyl sulfate (SDS) polyacrylamide gel. The rapidity and high resolution of HPLC permitted further purification steps, each resulting in 10- to 20-fold purification of ACC synthase, to be carried out in relatively short time and without major losses of enzyme activity. By combining traditional column chromatography, using phenyl-Sepharose, hydroxylapatite, and affigel blue (the latter suggested by T. Boller, pers. commun.) with protein HPLC, using preparative and analytical anion exchange and analytical hydroxylapatite columns, an ACC-synthase activity of up to 0.5 units per milligram protein was obtained. This preparation was subjected to two-dimensional polyacrylamide gel electrophoresis (PAGE). A nondenaturing gel was used in the first dimension, and ACC synthase was localized on this gel by incubating gel slices in buffer containing pyridoxal phosphate and SAM. SDS-PAGE was used in the second dimension. Following Coomassie blue staining of the SDS polyacrylamide gel, only one protein band was found to correspond to the enzyme activity that had been detected in the first dimension. This protein had a molecular weight of ca. 49,000, which agreed well with molecular-weight determinations using HPLC gel filtration. If this band is indeed ACC synthase, we estimate from the amount of enzyme activity loaded on to the nondenaturing gel that the purified enzyme has a specific activity of ca. 3 units per milligram protein. The recovery of ACC-synthase activity following the last chromatographic step (hydroxylapatite HPLC) was approximately 5–10% of the level of activity originally present in the ammonium sulfate fraction.

3 Conversion of ACC to Ethylene

Most plant tissues are capable of converting applied ACC to ethylene. This indicates that the enzyme catalyzing this reaction is constitutive and usually not rate limiting [23]. Although ACC-dependent ethylene formation can be demonstrated readily in intact tissues, no enzyme has yet been isolated that displays all the properties of the in vivo activity. Inspite of this, important insights have been gained into the mechanism by which ACC is oxidized to ethylene. Progress has been made using the three following approaches.

3.1 The Stereochemistry of Ethylene Formation

To clarify the stereochemistry of ethylene formation, ACC was labeled with deuterium at positions 2 and 3, and the configuration of these deuterium atoms was determined in ethylene evolved from deuterated ACC. Adlington et al. [3] synthesized [2,2,3,3-$^{2}H_4$]-ACC and found that it was converted by apple slices into [$^{2}H_4$]ethylene without loss of deuterium. Next, Adlington et al. [4] synthesized the two isomers of *cis*-2,3-dideuterio-ACC and (±)-*trans*-2,3-dideuterio-ACC and found that both were converted to a 1:1 mixture of *cis*- and *trans*-1,2-dideuterioethylene by apple tissue (Fig. 2). When the same isomers were oxidized chemically to ethylene with NaClO, according to Lizada and Yang [14], the *cis* and *trans* configuration of the deuterium atoms was retained. The results of Adlington et al. [4] were fully confirmed by Pirrung [18], who fed *cis*-2,3-dideuterio-ACC to apple and cantaloupe disks and mungbean hypocotyls, all three of which produced a 1:1 mixture of *cis*- and *trans*-dideuterioethylene. This mode of oxidation could be reproduced by converting ACC to ethylene through electrolysis (Fig. 2). As in the biological systems, the resultant ethylene was stereochemically scrambled, i.e., *cis*-2,3-dideuterio-ACC was converted to a 1:1 mixture of *cis*- and *trans*-1,2-dideuterioethylene. Based on these results, Pirrung proposed that ACC was oxidized stepwise both in the biological and electrochemical systems. Accordingly, an initial single-electron oxidation would be following by ring opening and a second single-electron oxidation. The stereochemical analogies of the biological and electrochemical oxidation

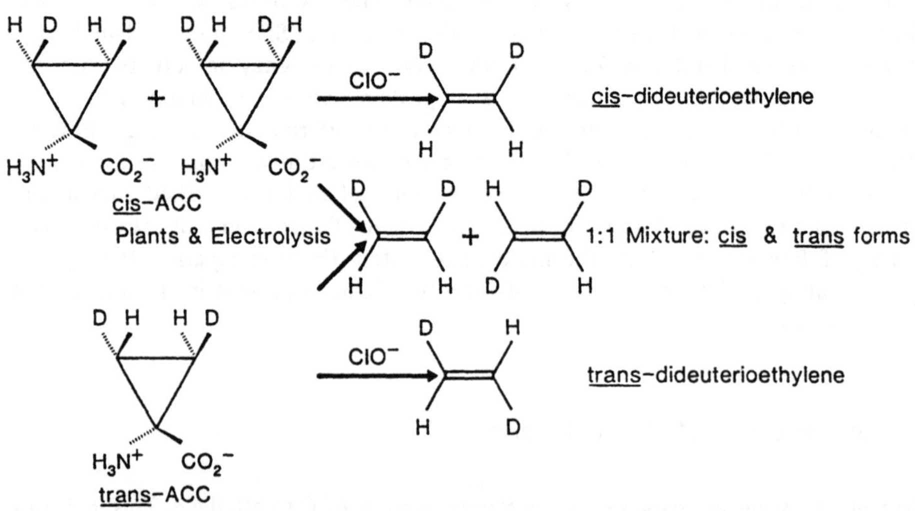

Fig. 2. The stereochemistry of ethylene production from ACC. Summary of results obtained by Adlington et al. [4] and Pirrung [18] using chemical model systems and plant tissues

of ACC led Pirrung to the suggestion that biogenesis of ethylene in plant tissues was accompanied by formation of cyanide from C-1 and the nitrogen atom of ACC as has been observed in the electrochemical reaction.

3.2 In Vivo Products of ACC Oxidation

Following the discovery by Lieberman et al. [12] that methionine is degraded to ethylene in a chemical system consisting of Cu^{2+}, ascorbate, and H_2O_2. Yang et al. [24] described another model system in which ethylene is formed from methionine in the presence of flavin mononucleotide and light. In this free-radical-mediated reaction, methionine is broken down to CO_2, formic acid, ethylene, methyl disulfide, and ammonia. These products arise from C-1, C-2, C-3, and C-4, the methylmercapto and the amino groups of methionine, respectively [24]. In 1974, Yang [22] suggested a scheme for ethylene biosynthesis in plants, according to which the products of methionine breakdown were the same as those in the flavin mononucleotide system except for the methylmercapto group, which was incorporated into S-methylcysteine and eventually recycled into methinonine. Recently, Peiser et al. [17] examined the fate of the carbon atoms and the amino group of ACC in mungbean hypocotyls and in the epicotyls of *Vicia sativa*. Carbon atoms 2 and 3 of ACC (corresponding to C-3 and C-4 of methionine) were converted to ethylene and the carboxyl group (corresponding to C-1 of methionine) to CO_2. In mungbean, C-1 of ACC (corresponding to C-2 of methionine) was incorporated into asparagine, in *V. sativa* into β-cyanoalanine and γ-glutamyl-β-cyanoalanine. This would be expected if C-1 and the amino group of ACC gave rise to cyanide. Since C-1 of ACC and [^{14}C]cyanide were metabolized in identical fashion in mungbeans and *V. sativa*, Peiser et al. [17] proposed that C-1 and the amino group of ACC are released as cyanide during ethylene formation.

3.3 Conversion of ACC to Ethylene in Isolated Vacuoles

Very soon after the discovery of ACC as the immediate precursor of ethylene, Konze and Kende [11] described a cell-free system from etiolated pea epicotyls that was capable of converting ACC to ethylene. This system had several properties of the in vivo activity, e.g., requirement for O_2 and sensitivity to inhibitors. On the other hand, it had some features that differed from the in vivo system, e.g., lack of saturability (or very high K_m) with respect to ACC. Because of this, it was proposed that conversion of ACC to etylene in pea homogenates was based on a chemical reaction between ACC and the product of an enzyme reaction. The discovery by Hoffman et al. [10] that the ACC-dependent ethylene-forming system is stereospecific provides an excellent tool to examine whether or not an in vitro ethylene-forming system conforms to the in vivo activity. It was found that the cell-free system from etiolated peas [11] was little or not at all stereoselective and, therefore, different from the in vivo ethylene-forming activity [9, 16, 21]. A number of other cell-free ethylene-forming preparations, consisting of microsomal or mitochondrial fractions, also lacked stereospecificity [21]. The microsomal ACC-dependent ethylene-forming system was recently characterized by Lynch et al. [15]. These workers provided evidence that in vitro conversion of ACC to ethylene by microsomal membranes was mediated by lipoxygenase-generated lipid hydroperoxides rather than by the ethylene-forming enzyme, which operates in vivo. These examples make it clear that all cell-free ACC-dependent ethylene-forming systems have to be critically examined, especially with regard to their stereospecificity, before claims are made concerning their physiological significance.

The first cell-free system converting ACC to ethylene in a manner similar to that of the in vivo activity was reported by Guy and Kende [9]. When [^{14}C]methionine

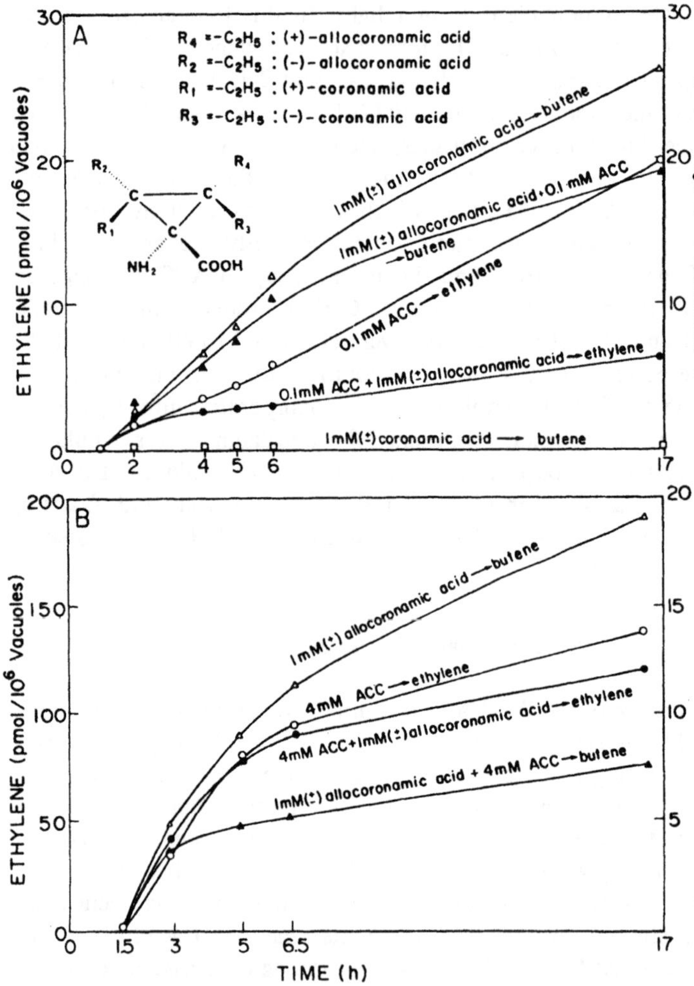

Fig. 3 A, B. Stereospecificity of the ethylene-forming system in pea vacuoles. Vacuoles were prepared from protoplast that had been treated with 200 μM AVG during their isolation. Ethylene production was monitored in the presence of ACC (○) and ACC + (±)-allocoronamic acid (●). 1-Butene formation was monitored in the presence of (±)-allocoronamic acid (△), (±)-coronamic acid (□), and (±)-allocoronamic acid + ACC (▲). The concentrations of the precursors were as indicated in the graph. *Inset:* The structure of the four stereoisomers. For each isomer one substituent is always an ethyl group (R = C_2H_5), the other three substituents being hydrogen atoms (R = H). From Guy and Kende [9]

was fed to protoplasts isolated from pea leaves, the specific radioactivity of C-2 and C-3 of ACC formed from methionine was much higher than that of ethylene originating from ACC [8]. This indicated that components of the ethylene-forming system were localized in different cellular compartments. The newly formed radioactive ACC obviously did not equilibrate rapidly with the pool of ACC from which ethylene was synthesized. Subsequently, it was shown that vacuoles isolated from pea leaf protoplasts contained more than 80% of the cellular ACC and that vacuoles themselves were capable of converting ACC to ethylene [9]. In fact, over 80% of the ethylene produced by protoplasts could be accounted for as originating from the vacuole. The properties of the vacuolar ethylene-forming system resembled those of the intact tissue in every respect tested. This included inhibitor sensitivity, saturation kinetics, and most importantly, stereospecificity (Fig. 3).

The properties of the ethylene-forming system in vacuoles isolated from *Vicia faba* leaves were examined further by Mayne and Kende (unpublished). It was confirmed that ethylene synthesis ceases following lysis of the vacuoles. Attempts to prevent loss of ethylene-forming activity or to reconstitute it from lysed vacuolar preparations failed. This included rupture of vacuoles in a minimum volume to reduce dilution of vacuolar contents and addition of protecting agents (e.g., *p*-chloromercuribenzoate, dithiothreitol), of nucleotides, and of cofactors. When the mannitol concentration in the incubation medium was lowered to lyse vacuoles, the rate of ethylene synthesis was positively correlated to the number of surviving vacuoles.

Two lines of evidence indicated that ethylene synthesis in isolated vacuoles may be dependent on a transmembrane ion gradient. First, the level of ethylene evolution depended on the pH of the incubation medium, decreasing as the pH was lowered from 7.5 to 5. Second, the protonophores, dinitrophenol and carbonyl cyanide *m*-chlorophenylhydrazone (CCCP) and the ionophores valinomycin, nigericin, and A23187 also inhibited ethylene formation in isolated vacuoles at concentrations where vacuolar survival was not significantly affected.

4 Conclusions

Research on the enzymology of ethylene biosynthesis has progressed on a number of fronts. The stereochemistry of both enzymatic reactions, the conversion of SAM to ACC and ACC to ethylene, has been clarified. A purification scheme for ACC synthase has been worked out, and the enzyme has been identified tentatively on a SDS polyarcylamide gel. A number of nonphysiological cell-free systems have been described that are capable of oxidizing ACC to ethylene. The stereospecificity of the ACC-dependent ethylene-forming enzyme offers an excellent tool to distinguish between "model systems" of ethylene formation and ethylene-forming activity that represents the in vivo enzyme. A cell-free system consisting of isolated vacuoles is capable of converting ACC to ethylene in a fashion similar to that of the in vivo enzyme. Ethylene formation in isolated vacuoles depends on integrity of the tonoplast, probably because of the requirement for a transmembrane ion gradient.

Acknowledgments. The work of the authors was supported by the U.S. Department of Energy under Contract No. DE-AC02-76ERO1338 and by the National Science Foundation through Grant No. DCB 8416492.

References

1. Acaster MA, Kende H (1983) Plant Physiol (Bethesda) 72:139
2. Adams DO, Yang SF (1979) Proc Natl Acad Sci USA 76:170
3. Adlington RM, Aplin RT, Baldwin JE, Rawlings BJ, Osborne D (1982) J Chem Soc Chem Commun 1086
4. Adlington RM, Baldwin JE, Rawlings BJ (1983) J Chem Soc Chem Commun 290
5. Boller T (1985) In: Fuchs Y, Chalutz E (eds) Ethylene: biochemical, physiological and applied aspects. Nijhoff/Junk, The Hague, p 87
6. Boller T, Herner RC, Kende H (1979) Planta (Berl) 145:293
7. Chappell J, Hahlbrock K, Boller T (1984) Planta (Berl) 161:475
8. Guy M, Kende H (1984a) Planta (Berl) 160:276
9. Guy M, Kende H (1984b) Planta (Berl) 160:281
10. Hoffman NE, Yang SF, Ichihara A, Sakamura S (1982) Plant Physiol (Bethesda) 70:195
10a. Kende H, Boller T (1981) Planta (Berl) 151:476
11. Konze JR, Kende H (1979) Planta (Berl) 146:293
12. Lieberman M, Kunishi AT, Mapson LW, Wardale DA (1965) Biochem J 97:449
13. Lieberman M, Kunishi AT, Owens LD (1974) In: Ulrich R (ed) Facteurs et règulation de la maturation des fruits. Editions du CNRS, Paris, p 161
14. Lizada MCC, Yang SF (1979) Anal Biochem 100:140
15. Lynch DV, Sridhara S, Thompson JE (1985) Planta (Berl) 164:121
16. McKeon TA, Yang SF (1984) Planta (Berl) 160:84
17. Peiser GD, Wang T-T, Hoffman NE, Yang SF, Liu H-W, Walsh CT (1984) Proc Natl Acad Sci USA 81:3059
18. Pirrung MC (1983) J Am Chem Soc 105:7207
19. Ramalingam K, Lee K-M, Woodard RW, Bleecker AB, Kende H (1985) Proc Natl Acad Sci USA 82:7820
20. Stryer L (1981) Biochemistry. Freeman, San Francisco
21. Venis MA (1984) Planta (Berl) 162:85
22. Yang SF (1974) In: Runeckles VC, Sondheimer E, Walton DC (eds) The chemistry and biochemistry of plant hormones. Academic Press, New York, p 131 (Recent advances in phytochemistry, vol 7)
23. Yang SF, Hoffman NE (1984) Annu Rev Plant Physiol 35:155
24. Yang SF, Ku HS, Pratt HK (1967) J Biol Chem 242:5274
25. Yu Y-B, Adams DO, Yang SF (1979) Arch Biochem Biophys 198:280

Chemistry, Physiology and Agricultural Application of Brassinolide and Related Steroids

T. YOKOTA and N. TAKAHASHI[1]

1 Introduction

Brassinolide [(*1*); see Fig. 1] is a steroidal growth substance which was first isolated from rape pollen by Grove et al. in 1979 [15]. Since then a number of steroids related to brassinolide have been isolated from various plant sources: these natural steroids, as well as synthetic ones, are often called brassinosteroids and, in this text, are abbreviated as BRs. A considerable body of knowledge concerning the biological activity and mode of action of this new class of growth substances has also accumulated in the last few years, resulting in the recognition of BRs as endogenous plant growth substances.

This paper reviews recent advances in chemistry and biology in this area, as well as some aspects of the agricultural application of BRs.

2 Natural Occurrence in Plants

2.1 Distribution in Plants

The bioassay used in the isolation of brassinolide (*1*) was the bean second-internode assay [26]. Following the isolation of brassinolide [15], the rice-lamina inclination assay [21], which is so sensitive as to detect 0.1 $\mu g\ l^{-1}$ brassinolide [44, 46] has been used for the purification of BRs. Thus, dolicholide (*2*), castasterone (*3*). dolichosterone (*4*), typhasterol (*7*), homodolicholide (*10*), homodolichosterone (*11*) were isolated in pure states and their structures were rigorously elucidated by [^{1}H]-NMR and GC/MS. The other steroids (*5, 6, 9, 12–15*) were characterized by GC/MS or SIM without isolation (Fig. 1). Synthesis of all these BRs has been reviewed by Mori [27]. The synthetic BRs have been used as standard samples in a survey for natural BRs in plants. Fifteen BRs have thus been characterized from fourteen plant species including eleven angiosperms (eight dicots and three monocots), two gymnosperms and one alga (Table 1). These findings, coupled with the bioassay data of Mandava and Mitchell [22] suggest that BRs may occur

1 Department of Agricultural Chemistry, The University of Tokyo, Bunkyo-ku, Tokyo 113, Japan

Plant Growth Substances 1985
(ed. by M. Bopp)

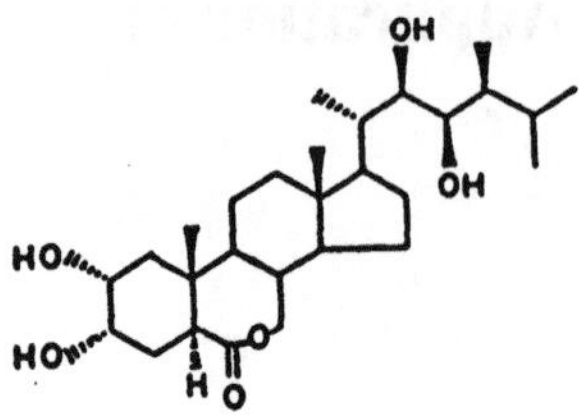

1 Brassinolide

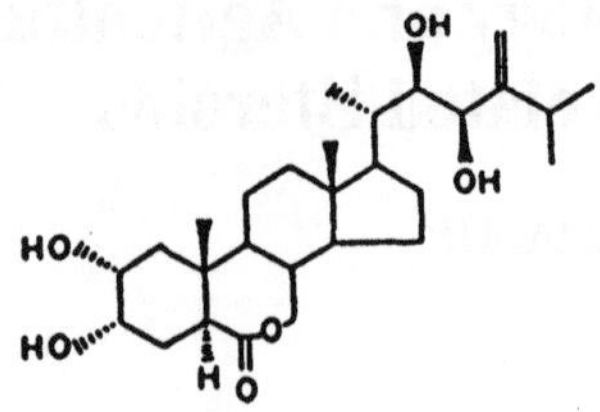

2 Dolicholide

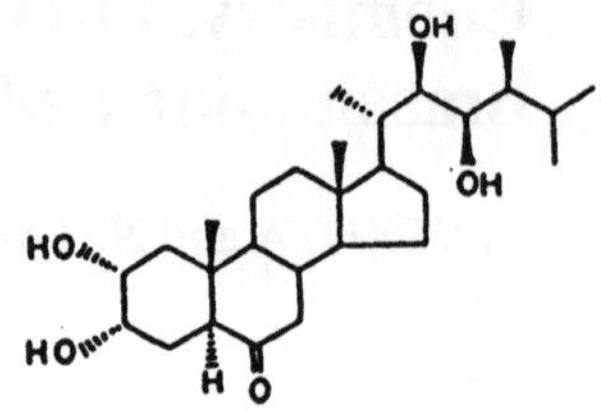

3 Castasterone

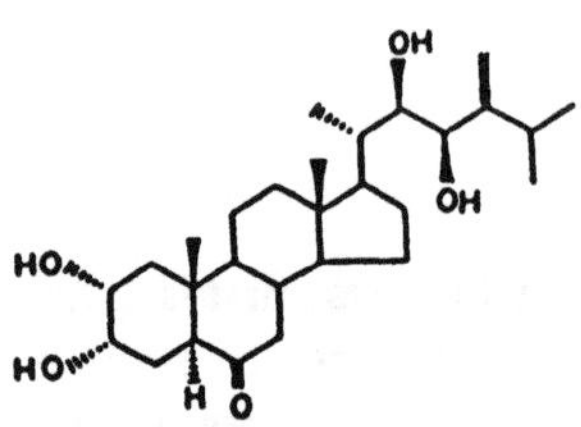

4 Dolichosterone

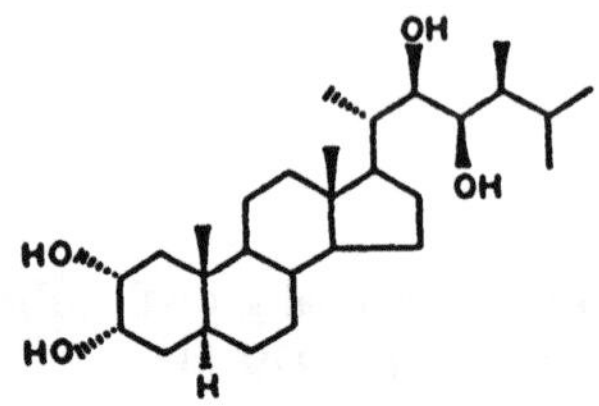

5 6-Deoxocastasterone

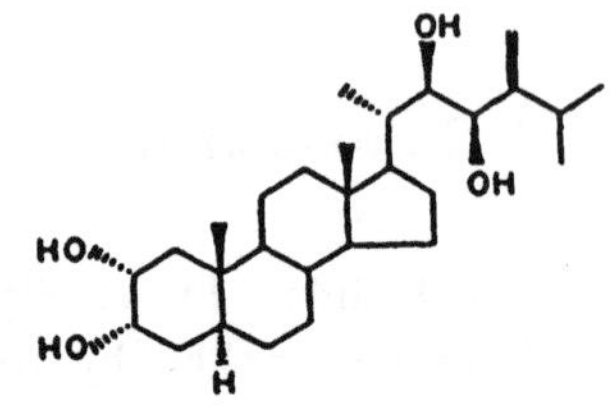

6 6-Deoxodolichosterone

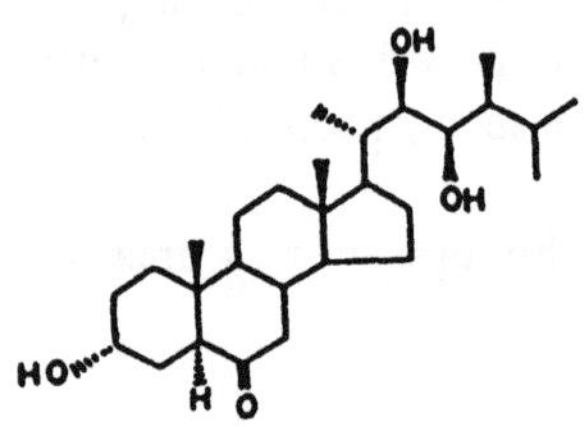

7 Typhasterol

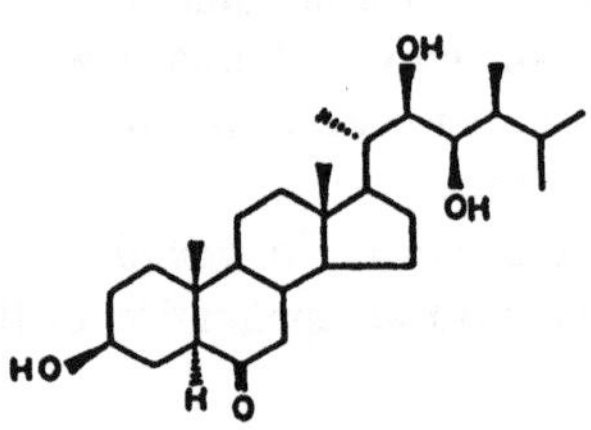

8 Teasterone

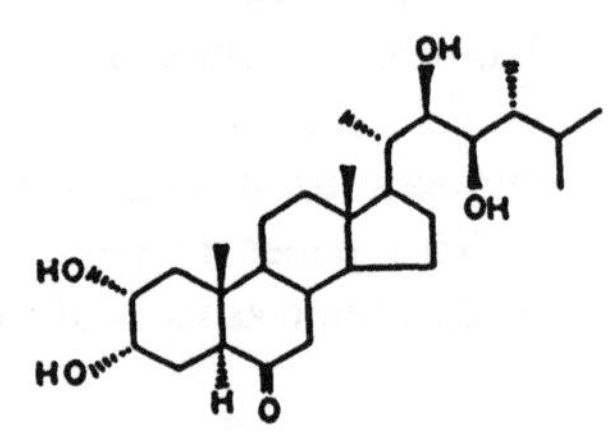

9 24-Epicastasterone

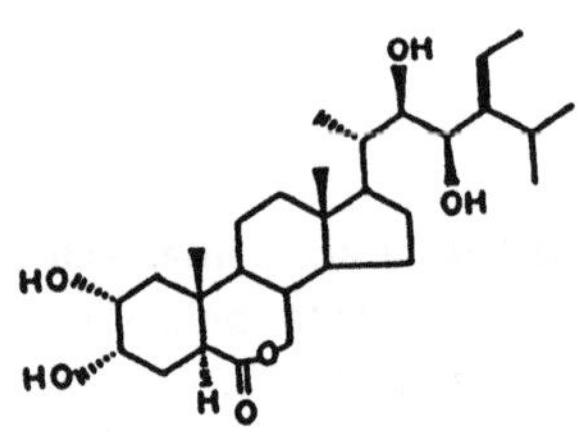

10 Homodolicholide

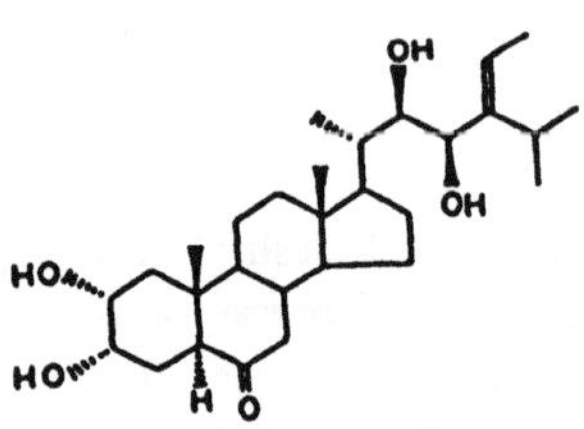

11 Homodolichosterone

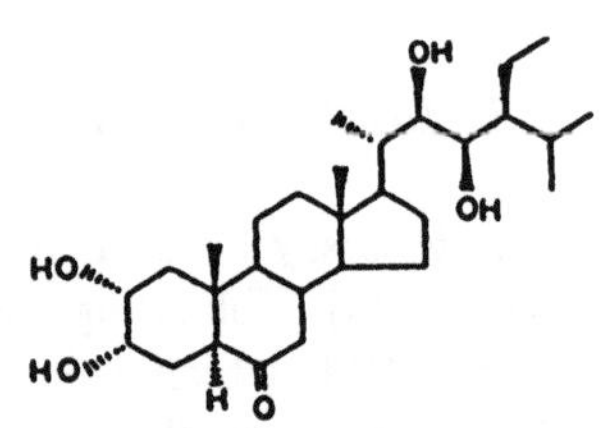

12 (24S)-24-Ethylbrassinone

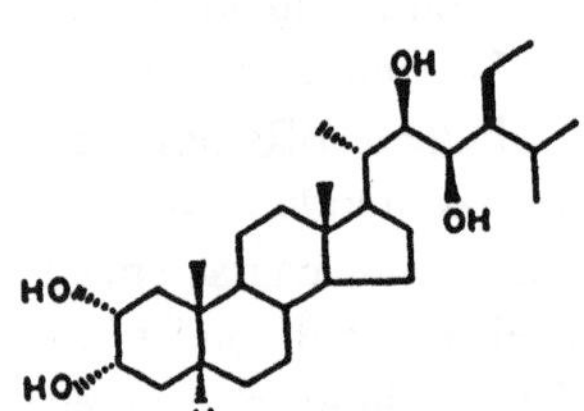

13 6-Deoxohomo-
dolichosterone

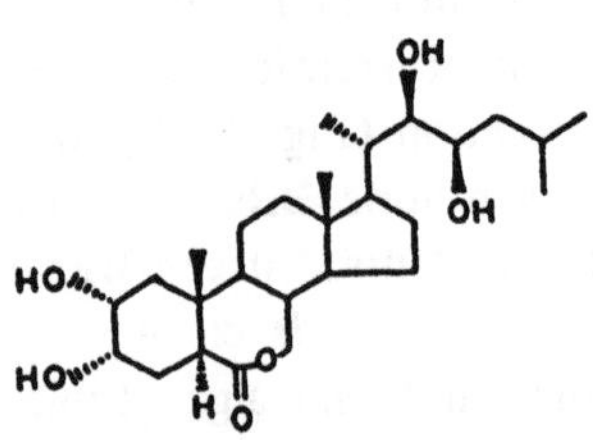

14 28-Norbrassinolide

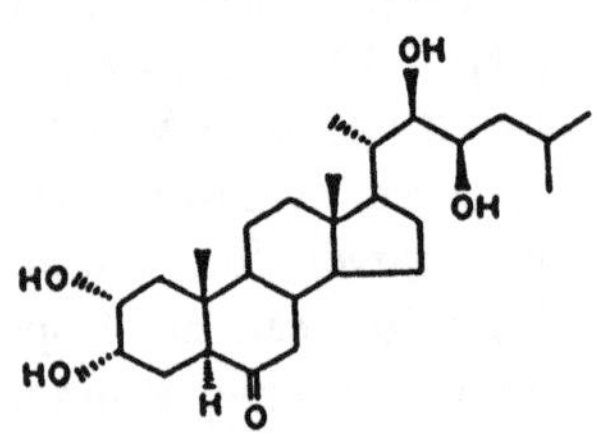

15 Brassinone

Fig. 1. Structures of natural BRs

Table 1. Distribution and contents of BRs in plants [a]

Species, part	C28 BR									C29 BR				C27 BR	
	1	*2*	*3*	*4*	*5*	*6*	*7*	*8*	*9*	*10*	*11*	*12*	*13*	*14*	*15*
Brassica napus, pollen [15]	7														
Brassica campestris, fruit [1, 2, 16]	2		2									1		2	1
Castanea crenata, gall [2, 5, 17, 48]	4		6		6	nd						nd		nd	6 [b]
stem, leaf, flower [5]			5		6										
Tea sinensis, leaf [2, 3, 16, 28]	2		4				3	3				1		nd	2
Distilium racemosum, gall [16]	nd		5											nd	5
leaf [16]	3		4											4	3
Dolichos lablab, seed [6, 49, 51, 53] [c]	5	7	5	6	5	5				5	1				
Phaseolus vulgaris, seed [52]	nq	nq	nq	nq	nq	nq							nq [d]		
Pharbitis purpurea, fruit [34]			5												4
Typha latifolia, pollen [33]							6								
Oryza sativa, shoot [4]			3	2											
Zea mays, pollen [e]			7				5	5							
Pinus Thunbergii, pollen [50]			5				6								
Picea sitchensis, shoot [54]			5				5								
Hydrodyction reticulatum, colony [d]									nq			6			

[a] Contents are roughly expressed as follows: 1. (< 1 ng kg^{-1}), 2. (1 ~10 ng kg^{-1}), 3. (10 ~10^2 ng kg^{-1}), 4. (10^2 ~10^3 ng kg^{-1}), 5. (10^3 ~10^4 ng kg^{-1}), 6. (10^4 ~10^5 ng kg^{-1}), 7. (10^5 ~10^6 ng kg^{-1}). nd: not detected; nq: identified, but not quantified

[b] Compound (*15*) was identified by Abe et al. [2], but could not be detected by us [5]

[c] Quantitative date are unpublished

[d] Yokota et al. 1985, unpubl.

[e] Suzuki et al. 1985, unpubl.

more widely than is presently established. The number of steroids belonging to the BR family will be likely to increase in future, as we have recently found that the immature seeds of *Phaseolus vulgaris* contain a number of unknown steroidal growth substances, together with some conjugates (Yokota et al. 1985, unpubl.). The BR contents in the plant tissues are described in Table 1, but these values are obviously underestimates due to the considerable losses encountered during purification.

BRs may be present in all parts of plants, although roots have not yet been examined. High levels of BRs have been found from various pollen (Table 1), indicating that pollen is one of the richest sources of BRs. Sasse [31] noted that the growth of pollen tubes is promoted by nM levels of the synthetic 24-epimer of brassinolide, suggesting that BR might be concerned with physiology of reproduction growth. Immature seed also contains high levels of BRs. Eight BRs have been characterized from seed of *Dolichos lablab* [53]. Among them, dolicholide is a major BR and its content reaches up to 200 $\mu g \, kg^{-1}$ fresh weight at a younger stage of the seed (Yokota et al. 1985, unpubl.), comparable to the content of brassinolide in rape pollen. Seeds of *Phaseolus vulgaris* (but not the pods) contain high BR activity, but the contents of the respective compounds are not high because they contain a variety of BRs other than those listed in Table 1 (Yokota 1985, unpubl.).

With respect to vegetative tissues the contents of BRs are rather low compared with those in pollen and immature seeds. In this respect it is worth noting that insect galls of *Castanea crenata* and *Distylium racemosum* contain higher amounts of BR than the normal tissues [5, 16], suggesting the possibility that BRs might be concerned with abnormal growth of plant tissues.

Brassinolide and catasterone are considered to be the most important BRs, not only because they are most widely distributed, but also because their biological activities are very high.

2.2 Structural Variations and Biosynthetic Consideration

BRs so far characterized from plants are all 5α-cholestane derivatives and can be classified into C_{27}, C_{28} and C_{29} steroid as is the case with typical phytosterols. These variations come from different substitutions at C-24, i.e. no substitution in C_{27} steroids, a methyl or an exomethylene in C_{28} steroids, and an ethyl or an (*E*)-ethylidene in C_{29} steroids. The alkyl (methyl or ethyl) substituents at C-24 are *S*-oriented in BRs of higher plants, but recently, a BR with a 24*R*-methyl (24-epi-castasterone) has been characterized from a green alga *Hydrodictyon reticulatum* (Yokota et al. 1985, unpubl.). It would be of interest to know the biosynthetic relationship between BR and the major phytosterols with respect to C-24 alkylation.

All BRs carry a vicinal glycol in the side chain at C-22*R* and C-23*R*. Thirteen BRs have another vicinal glycol system at C-2α and C-3α in the A-ring, but the remaining two BRs, typhasterol (*7*) and teasterone (*8*) carry instead a 3α-hydroxyl and 3β-hydroxyl group respectively. Therefore, it is suspected that the introduction of a vicinal glycol occurs first in the side chain and then in the A-ring. Typhasterol and teasterone might be considered as precursors of BRs [3, 33, 50].

With respect to B-ring oxidation. BRs can be grouped into either a 7-oxalactone, a 6-ketone or a non-oxidized type. It seems likely that ketonic BRs are converted to lactonic BRs by a Baeyer-Villiger-type oxidation, as is seen in the chemical synthesis of lactonic brassinosteroids [27]. BRs with non-oxidized B-rings have been considered to be precursors of ketonic BRs [52]. Compounds (*5*) and (*6*) with a non-oxidized B-ring, however, show very low activity in the rice-lamina inclination assay (Table 2), suggesting that they are not convertible into biologically active ketonic congeners.

3 Biological Effects

3.1 Physiology

BR has growth-promoting effects similar to auxin and gibberellin (GA). BR elicits pronounced stem elongation on dwarf pea hypocotyls, dwarf bean apical segments [23] and mung bean epicotyls [14], which are sensitive to GAs, but not to auxins. These authors, and more recently Katsumi [18], reported that the effects of BR and GA are additive. Ancymidol, a growth retardant, inhibits GA-mediated growth response, but has no influence on the effect of BR [14, 55], indicating that the biological effects of BR and GA are independent. Takeno and Pharis [41] have reported that BR is almost ineffective in the dwarf-rice growth test, which is known to be quite sensitive to GA. It is apparent that the actions of BR are different from those of GA.

Yopp et al. [56] showed the similarity between the effect of auxin and BR in selected auxin bioassays, e.g. elongation of dwarf pea hook segment, maize mesocotyl, azuki bean epicotyl and retardation of dwarf bean hypocotyl hook opening. They pointed out the strong synergism that exists between auxin and BR. This synergism, however, occurs only when tissues are treated first with BR and then with IAA; BR is ineffective when the order of treatment is reversed [11, 18]. More important is that the BR-elicited growth responses, such as hypocotyl elongation and rice-lamina bending, are nullified by the antiauxins, α-(*p*-chlorophenoxy)isobutyric acid (PCIB) or 2, 3, 5-triiodobenzoic acid (TIBA) [18, 41, 55]. A membrane-bound ATPase inhibitor, dicyclocarbodiimide also retarded the BR action [18]. These facts suggest that BR enhances the sensitivity of plant tissues to auxin, resulting in synergistic interaction with endogenous auxin, but does not itself have auxin-like functions. Cohen and Meudt [11] indicated that BR does not have a direct effect on IAA uptake, metabolism and transport.

In maize root segments, BR induces a significant stimulation of root growth, associated with an increase of acid secretion, but auxin inhibits such action [9, 10, 30]. Growth and acid extrusion in azuki bean epicotyls, however, are effected by both BR and auxin [8]. Based on these findings, it was claimed that in roots, the action of BR and auxin is elicited through different pathways. It should be noted that the effect of BR was not observed in the *Avena* curvature assay when BR was tested alone or in combination wit IAA (Yokota 1985, unpubl.).

Takematsu et al. [39, 40], using a number of plants, found that combinations of auxin and BR are more effective than those of auxin and benzyladenine with respect to the number of plant species whose callus growth was stimulated. BR shows cytokinin-like activity in the cucumber cotyledon expansion assay, but was ineffective in other cytokinin bioassays examined [23]. Recently, Wada et al. [47] found that BR is quite active in the wheat leaf unrolling test, in which cytokinins are moderately active, but auxins are inhibitory.

BR has been found to stimulate auxin-induced ethylene synthesis in mung bean hypocotyls between *S*-adenosylmethionine and 1-aminocyclopropane-1-carboxylic acid ([32] and references cited therein).

Sasse [31] examined the place of BR in the sequential response to growth substances in wheat coleoptile and dwarf pea segments and found that the peak of sensitivity lies between those of gibberellin (cytokinin) and auxin.

The importance of radiant energy and spectral quality in the BR-elicited growth response has been emphasized ([13, 20] and references cited therein). In soya bean seedlings and mung bean sections the growth-promoting effect of brassinolide occurs only in the light, but not in the dark, and that the most important region of light is red light (660 nm, 2.6 W cm^{-1}) (Kamuro 1985, pers. comm.).

In summary, BRs are quite distinct from any known plant growth substances with respect to physiological functions.

3.2 Agricultural Application

In 1972, the U.S. Department of Agriculture reported the biological effects of brassins (crude brassinolide extract of rape pollen) on young bean plants and young Siberian elm trees, and found that brassins caused overall growth [25]. In field and greenhouse trials synthetic brassinolide analogues significantly boosted the harvest of various vegetables and root crops, e.g. radish, lettuce, bean, pepper and potato [24], wheat and mustard [7] and barley [13]. These growth responses occur chiefly in the slow growing plants and, therefore, the phenotypic variability of the seedlings is reduced [13, 25].

In Japan, (22*S*, 23*S*)homobrassinolide and brassinolide itself have been tested for agricultural application in the last few years by several research groups and the following promising results were obtained [12, 39, 40]. (1) Growth acceleration of rice and tobacco results from treatment of seeds with homobrassinolide and this effect was prominent under low temperature conditions (Utsunomiya Univ.). (2) Yield increases of potato and sweet potato when they are treated with brassinolide at the planting stage (Zennoh). (3) Yield increases of fruits and grains are observed when flowers of wheat (Tokyo Univ.), maize (Zennoh), rice and tomato (Nissan) are treated with brassinolide and when flowers of rice, tomato, cucumber and egg plant are treated with homobrassinolide (Utsunomiya Univ.). This effect is mainly based on increase in the fruit setting of less viable flowers and subsequent growth promotion. In some of these cases, this effect was more pronounced under low temperature conditions. (4) Environmental stresses, such as injuries due to low temperature, herbicides, pathogenic microorganisms and salt, are protected or

Table 2. Relative activity of natural BRs and related steroids in the rice-lamina inclination test

Compound	Relative activity	
	cv Koshihikari [a]	cv Arborio Jl [b]
1	100	100
2	10	
3	20	50
4	4	
5	≪ 1	
6	~ 0	
7	1	
10	50	
11	10	
12		50
14		5
15	4	5
16		1
17		10
18	20	100
19		10

[a] Yokota et al. (unpubl.). As for assay procedure, see Arima et al. [5]
[b] Takatsuto et al. [36]

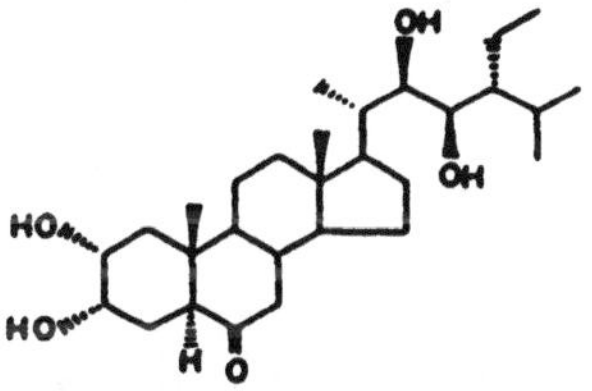

16 24-Epimer of 12

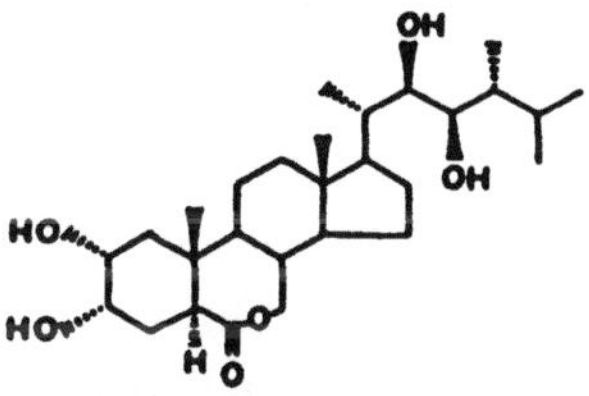

17 24-Epimer of 1

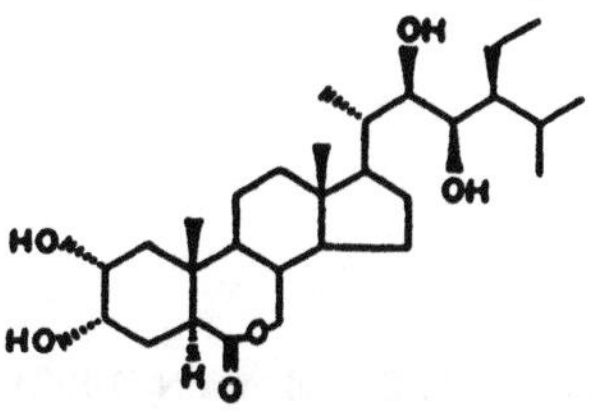

18 Homobrassinolide

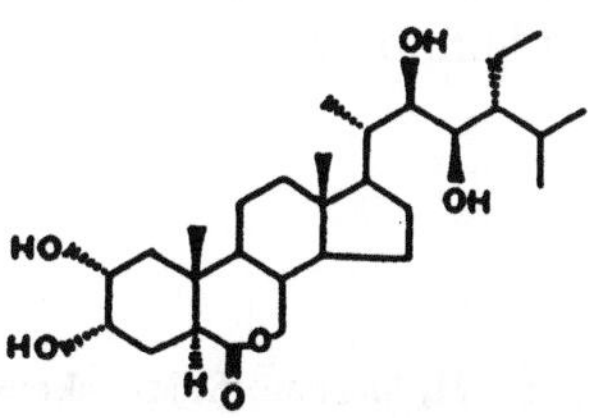

19 24-epimer of 18

alleviated by the treatment with brassinolide (Zennoh). In most of the above cases, the effective concentration range of brassinolide lies between 10^{-3} and 10^{-6} ppm, while that of homobrassinolide is much higher.

4 Structure-Activity Correlations

A number of BR analogues have been synthesized and examined for their structure-activity correlations by means of the bean second-internode assay and bean first-internode assay [42], and rice-lamina inclination assay [19, 29, 35, 36, 43, 45]. In these bioassays the following structural features were indicated as necessary for activity: (1) a *trans* A/B ring system; (2) a 7-oxalactone or a 6-ketone system in the B-ring (the former is preferable); (3) *cis* α-oriented hydroxyl groups at C-2 and C-3; (4) *cis* hydroxyl groups at C-22 and C-23 (22*R*, 23*R*-configuration is preferable); and (5) a methyl or an ethyl substituent at C-24 (24*S* configuration is preferable). In the rice test, however, the structural features are less sharply defined. Namely, activity differences between 7-oxalactones and 6-ketones are small (see also [46]) and a $2\alpha;3\alpha$-diol can be replaced by a $3\alpha;4\alpha$-diol or 3α-OH with little loss of activity [35]. We examined the relative activity of some of the natural BRs in the rice test according to Arima et al. [5] (Table 2). It is worth noting that replacement of the C-24 methyl group in brassinolide (*1*) by an exomethylene considerably reduces the biological activity as seen in dolicholide [49].

Takatsuto et al. [37] examined structure-activity correlations using elongation tests of cotyledon petioles and hypocotyls of intact radish and tomato seedlings, concluding that the structural requirements of BR in the two bioassay systems are similar to those described above, but are rather stringent. On the basis of the rice-lamina inclination test and radish test , Takatsuto et al. [38] found that the terminal methyl groups (C-26 and C-27) of brassinolide are not important for activity. BRs with 6-oxalactone, 6-azalactam and 6-azathiolactam have all been shown to be inactive [29, 35, 42].

5 Conclusion

From the data presented here it is clear that BRs are widely distributed in the plant kingdom and that they also have potent biological activities, which are distinct from any other known growth substances. BRs should be categorized as a new family of growth substances or plant hormones.

References

1. Abe H, Morishita T, Uchiyama M, Marumo S, Munakata K, Takatsuto S, Ikekawa N (1982) Agric Biol Chem 46:2609
2. Abe H, Morishita T, Uchiyama M, Takatsuto S, Ikekawa N, Ikeda M, Sassa T, Kitsuwa T, Marumo S (1983) Experientia (Basel) 39:351

3. Abe H, Morishita T, Uchiyama M, Takatsuto S, Ikekawa N (1984a) Agric Biol Chem 48:2171
4. Abe H, Nakamura K, Morishita T, Uchiyama T, Takatsuto S, Ikekawa N (1984b) Agric Biol Chem 48:1103
5. Arima M, Yokota T, Takahashi N (1984) Phytochemistry (Oxf) 23:1587
6. Baba J, Yokota T, Takahashi N (1983) Agric Biol Chem 47:659
7. Braun P, Wild A (1984) J Plant Physiol 116:189
8. Cerana R, Bonetti A, Marre MT, Romani G, Lado P, Marre E (1983a) Physiol Plant 59:23
9. Cerana R, Columbo R, Lado P (1983b) Physiol Veg 21:875
10. Cerana R, Lado P, Anastasia M, Ciuffreda P, Allevi P (1984) Z Pflanzenphysiol 114:221–225
11. Cohen JD, Meudt WJ (1983) Plant Physiol (Bethesda) 72:691
12. Fujita F (1985) Kagaku-to-seibutsu 23:717 (in Japanese)
13. Gregory LE (1981) Am J Bot 68:586
14. Gregory LE, Mandava NB (1982) Physiol Plant 54:239
15. Grove MD, Spencer GF, Rohwedder WK, Mandava N, Worley JF, Warthen JD Jr, Steffens GL, Flippen-Anderson JL, Cook JC Jr (1979) Nature 281:216
16. Ikekawa N, Takatsuto S, Kitsuwa T, Saito H, Morishita T, Abe H (1984) J Chromatogr 290:289
17. Ikeda M, Takatsuto S, Sassa T, Ikekawa N, Nukina M (1983) Agric Biol Chem 47:655
18. Katsumi M (1985) Plant Cell Physiol 26:615
19. Kondo M, Mori K (1983) Agric Biol Chem 47:97
20. Krizek DT, Mandava NB (1983) Physiol Plant 57:324
21. Maeda E (1965) Physiol Plant 18:813
22. Mandava NB, Mitchell JW (1971) Indian Agric 15:19
23. Mandava NB, Sasse JM, Yopp JH (1981) Physiol Plant 53:453
24. Maugh II TH (1981) Science 212:33
25. Mitchell JW, Gregory LE (1972) Nature 239:253
26. Mitchell JW, Livingston GA (1968) Agric Handb 336:26
27. Mori K (1985) J Synth Org Chem Jpn 43:849 (in Japanese)
28. Morishita T, Abe H, Uchiyama M, Marumo S, Takatsuto S, Ikekawa N (1983) Phytochemistry (Oxf) 22:1051
29. Okada K, Mori K (1983) Agric Biol Chem 47:89
30. Romani G, Marre MT, Bonetti A, Cerana R, Lado P, Marre E (1983) Physiol Plant 59:528
31. Sasse JM (1985) Physiol Plant 63:303
32. Schlagnhaufer C, Arteca RN, Yopp JH (1984) Physiol Plant 61:555
33. Schneider JA, Yoshihara K, Nakanishi K, Kato N (1983) Tetrahedron Lett 24:3859
34. Suzuki Y, Yamaguchi I, Takahashi N (1985) Agric Biol Chem 49:49
35. Takatsuto S, Ikekawa N (1984) J Chem Soc Perkin Trans I:439
36. Takatsuto S, Yazawa N, Ikekawa N, Morishita T, Abe H (1983a) Phytochemistry (Oxf) 22:1393
37. Takatsuto S, Yazawa N, Ikekawa N, Takematsu T, Takeuchi Y, Koguchi M (1983b) Phytochemistry (Oxf) 22:2437
38. Takatsuto S, Yazawa N, Ikekawa N (1984) Phytochemistry (Oxf) 23:525
39. Takematsu T, Takeuchi Y, Koguchi M (1983) Chem Regul Plants 18:38 (in Japanese)
40. Takematsu T, Takeuchi Y, Koguchi M (1985) Shokucho 18:2 (in Japanese)
41. Takeno K, Pharis RP (1982) Plant Cell Physiol 23:1275
42. Thompson MJ, Meudt WJ, Mandava NB, Dutky SR, Lusby WR, Spaulding DW (1982) Steroids 39:89
43. Wada K, Marumo S (1981) Agric Biol Chem 45:2579
44. Wada K, Marumo S, Ikekawa N, Morisaki M, Mori K (1981) Plant Cell Physiol 22:323
45. Wada K, Marumo S, Mori K, Takatsuto S, Morisaki M, Ikekawa N (1983) Agric Biol Chem 47:1139
46. Wada K, Marumo S, Abe H, Morishita T, Nakamura K, Uchiyama M, Mori K (1984) Agric Biol Chem 48:719
47. Wada K, Kondo H, Marumo S (1985) Agric Biol Chem 49:2249
48. Yokota T, Arima M, Takahashi N (1982a) Tetrahedron Lett 23:1275

49. Yokota T, Baba J, Takahashi N (1982b) Tetrahedron Lett 23:4965
50. Yokota T, Arima M, Takahashi N, Takatsuto S, Ikekawa N, Takematsu T (1983a) Agric Biol Chem 47:2419
51. Yokota T, Baba J, Takahashi N (1983b) Agric Biol Chem 47:1409
52. Yokota T, Morita M, Takahashi N (1983c) Agric Biol Chem 47:2149
53. Yokota T, Baba J, Koba S, Takahashi N (1984) Agric Biol Chem 48:2529
54. Yokota T, Arima M, Takahashi N, Crozier A (1985) Phytochemistry (Oxf) 24:1333
55. Yopp JH, Colclasure GC, Mandava NB (1979) Physiol Plant 46:247
56. Yopp JH, Mandava NB, Sasse JM (1981) Physiol Plant 53:445

Plant Growth Substances of Plant and Microbial Origin

G. SEMBDNER and D. GROSS[1]

1 Introduction

In addition to the classical groups of plant hormones, a high number of naturally occurring secondary metabolites are known, which possess different plant growth regulating properties, at least after exogenous application. Detection of plant growth substances of microbial and plant origin resulted from both extensive screening programmes and intensive plant hormone research. Some special groups, like polyamines and the hormonelike brassinosteroids, are studied extensively and will be reported separately. The present report demonstrates some plant growth substances recently isolated from higher plants as well as some microbial products selected according to their chemical structure and physiological activity. Among regulators of plant origin, jasmonic acid and related compounds will be discussed in more detail.

2 Microbial Products Which Affect Plant Growth

It is well-known that microorganisms can produce plant hormones, like auxins, ethylene, cytokinins, abscisic acid and gibberellins as so-called secondary metabolites. In addition, a high number of microbial, especially fungal substances have been published which affect plant growth and development. These compounds may be important for practical use and as templates for the chemical synthesis of analogues.

Some recently isolated substances, showing inhibitory activities in the wheat coleoptile bioassay, have been reviewed by Cutler [9]. Pergillin (I) and dihydropergillin were isolated from cultures of *Aspergillus ustus* and *A. variecolor;* the latter being more active. Hydroxyterphenyllin (II) was obtained from *Aspergillus candidus* and cladosporin (III) was re-discovered in cultures of *Aspergillus repens* and stereochemically elucidated.

Chaetoglobsin K (IV), obtained from *Diplodia macrospora,* is highly active giving significant inhibition at concentrations $\geq 10^{-7}$ M. Cytochalasin H (V) isolated

1 Akademie der Wirtschaftswissenschaften der DDR, Institut für Biochemie der Pflanzen, 4010 Halle (Saale), GDR

Plant Growth Substances 1985
(ed. by M. Bopp)

I
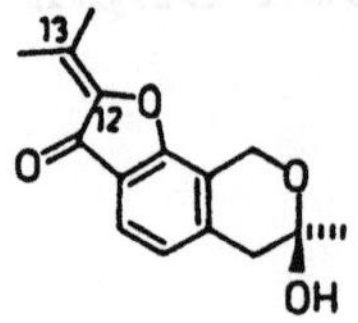
II
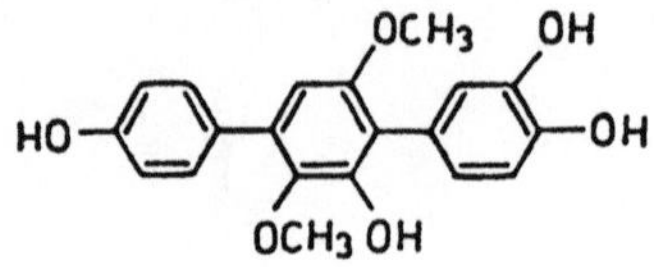
III
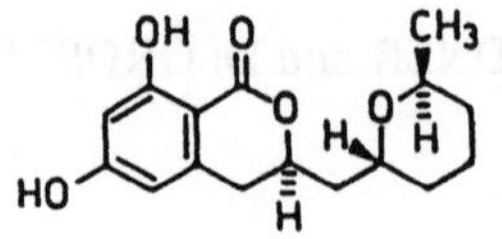

IV
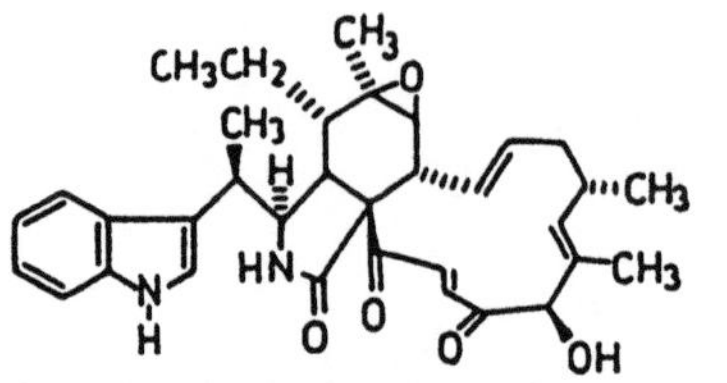
V
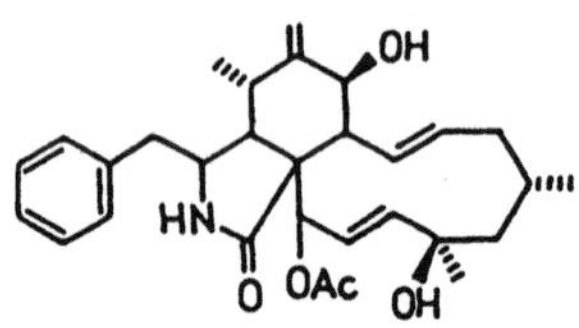
VI
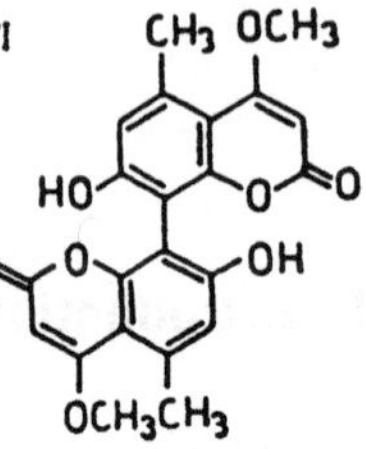

VII
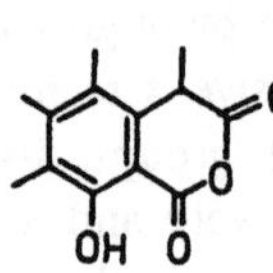
VIII
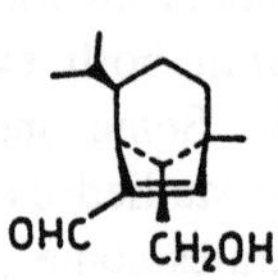
IX
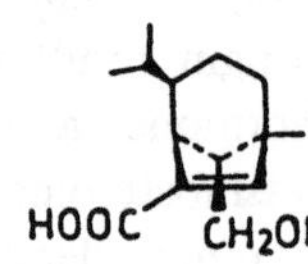
X
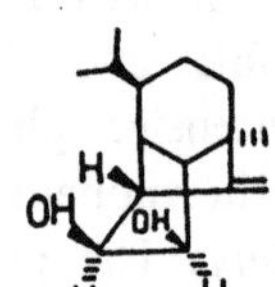
XI
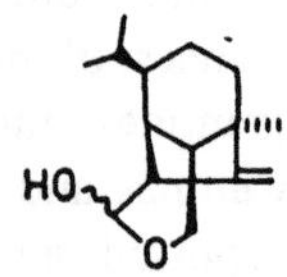

XII
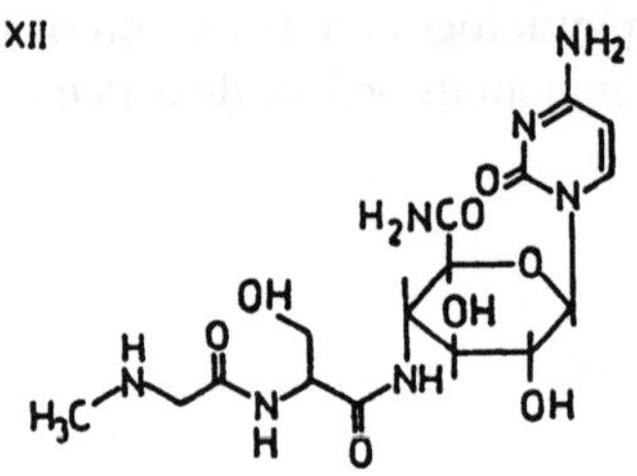
XIII
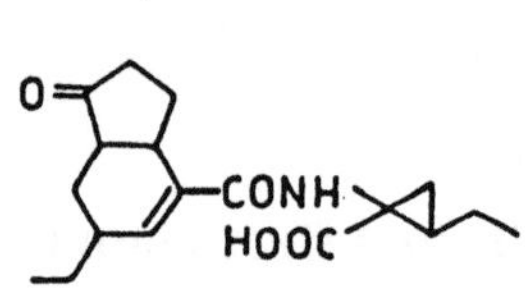
XIV
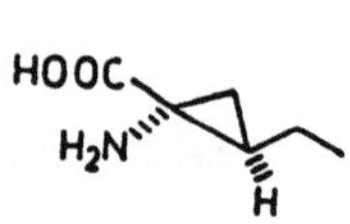

XV
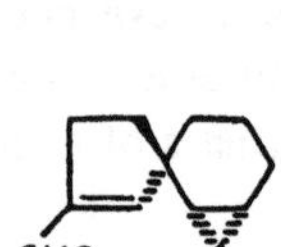
XVI
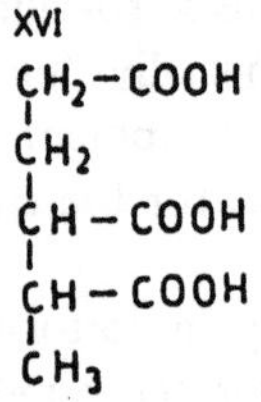
XVII
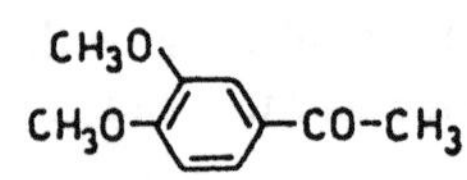

XVIII
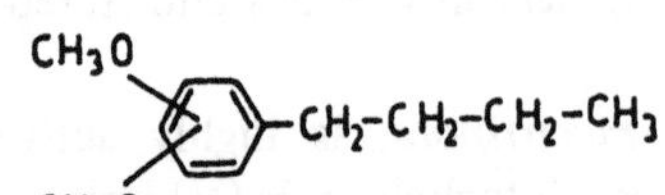
XIX
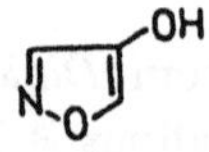
XX
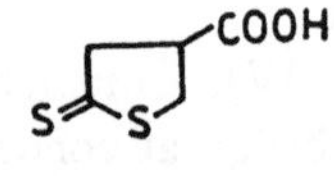
XXI
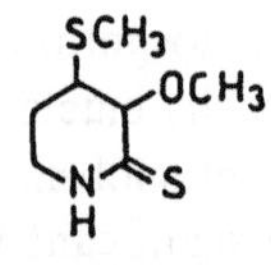

Schemes I–XXI

from *Phomopsis spp.* was somewhat less active, but induced additional curving and twisting of the coleoptile sections. The bicoumaryl compound orlandin (VI) isolated from *Aspergillus niger* inhibited coleoptile section growth at concentrations $\geqslant 10^{-6}$ M.

Sclerin (VII) and related compounds, like sclerotinin A and B, represent a long known group of fungal metabolites, the biosynthesis of which has been studied more recently [2]. Due to their physiological potencies, including stimulation of enzyme activities, promotion of root formation, dry weight increase and gibberellin-synergistic effects, they are termed plant growth hormones, which is incorrect in the author's view.

Some fungal macrolides are also known to possess plant growth regulating activities. Recently published examples are cladospolide A and B, isolated from *Cladosporium cladosporoides* [18]. In the lettuce seedling bioassay cladospolide A inhibited and B promoted root elongation. The sesquiterpenes helminthosporol (VIII), helminthosporic acid (IX) and cis-sativenediol (X) known from cultures of *Helminthosporium sativum* and *Colchiobolus setariae* possess interesting structural and biological relationships to gibberellins [19]. Prehelminthosporol (XI) was isolated from the plant pathogen *Dreschlera sorokiana* [9]. It was shown to stimulate wheat coleoptile section growth (at 10^{-5} M), but has not been tested in gibberellin-specific bioassays like the helminthosporins mentioned.

Gougerotin (XII), initially isolated from *Streptomyces gougerotii* as an antibiotic, has been found to inhibit both autonomous and gibberellin-stimulated growth of dwarf rice seedlings [35].

Coronatin (XIII) is produced by plant-pathogenic strains of *Pseudomonas syringae* pv. *coronafaciens* [11, 21, 39, 40]. The physiologically active substance causes several effects on plant cells and tissues, like necrosis in host plants, hypertrophy on potato tuber discs, root elongation in wheat and hypocotyl elongation in mung beans. Apparently its primary action is related to changes in the plasma membrane. It is interesting to note that one part of the molecule, the amino acid coronamic acid (XIV), is structurally closely related to the ethylene precursor 1-aminocyclopropane-1-carboxylic acid (ACC).

3 Growth Regulating Substances of Plant Origin

3.1 General Remarks

Both extensive screening and intensive research on plant hormones and growth regulation have uncovered a high number of naturally occurring plant constituents, which affect plant growth and differentiation [cf. 3, 12, 13, 19, 29, 49]. The following report will only describe briefly some more recently published growth regulating substances. The most outstanding group is represented by the turgorines, hormone-like substances apparently regulating seismonastic, nyctinastic and photoastic movements in plants. They have been detected and fully described by Schildknecht [42, 43]. Some other examples are the terpenoid lactones [20, 22], *Eucalyptus* inhibitors [4, 5], triacontanol [10, 17, 24, 27, 37, 38], special flavanols [36] and iridoids [1, 7].

3.2 Recently Isolated Growth Regulating Substances of Restricted Distribution

A new sesquiterpenoid having plant growth inhibitory activity, (+)-vitrenal (XV), has been isolated from the liverwort *Lepidozia vitrea* [30, 31].

Studies on the endogenous inhibitors in *Avena sativa* led to the isolation of three active compounds, identified as 1,3,4-pentane-tricarboxylic acid (XVI), 3,4-dimethoxyacetophenone (XVII) and dimethoxy-butylbenzene (XVIII) [28]. In germination tests, using different plant species, the substances inhibited root growth and in part coleoptile length.

p-Hydroxybenzoic acid has been isolated from *Pinus densiflora* pollen and shown to inhibit pollen germination and pollen tube growth [25].

A new heterocyclic seed germination inhibitor was isolated from leaves of the African plant *Triumfetta rhomboida* and structurally elucidated to be 4-hydroxy-isoxazole (triumferol, XIX) [26].

S-containing heterocyclic plant growth inhibitors, like asparagusic acid and related compounds, are known to occur in *Asparagus officinalis.* New growth inhibitors of a similar type have been isolated from *Raphanus sativus* and structurally elucidated as 2-thioxothiazolidine-4-carboxylic acid (XX) [14] and raphanusanin (XXI) [16]. Raphanusanin inhibits hypocotyl growth of etiolated radish and lettuce seedlings at concentrations $> 1.5 \times 10^{-6}$ M and is thought to be involved in regulation of light-induced inhibition of hypocotyl growth. Another light-promoted growth inhibitor has been isolated from sunflower seedlings and identified as caprolactam [15]. As a potent inhibitor dihydroactinidiolide was isolated from *Eleocharis coloradensis* [48]; geniposid and genipin gentiobiosid were found to be germination and growth inhibitors in fruits of *Gardenia jasminoides* [47].

3.3 Jasmonic Acid and Related Substances

(–)-Jasmonic acid (JA) and its methyl ester (JA-Me) are known to be native plant growth regulators possessing interesting physiological properties [46]. By means of physical methods and a radioimmunoassay [23], the occurrence of JA and/or JA-Me in a variety of plants had been detected [32]. A continued screening done by radioimmunoassay led to the detection of JA in 159 further plant species belonging to 142 families: 116 Spermatophyta, 15 Pteridophyta, 6 Bryophyta, 4 Mycophyta and 1 Phycophyta (Sembdner et al. unpubl.). Thus, in all plant materials tested JA-equivalents could be detected in widely differing amounts. Therefore, it can be concluded that JA is ubiquitously distributed within the plant kingdom.

During more detailed studies on the JA formation in fruits of *Vicia faba* [33] and the fungus *Botryodiplodia theobromae* [34], also its stereoisomer (1 R, 2 S)-JA [(+)-2-iso-JA] and some related compounds have been identified (Fig. 1).

This group of substances, occurring in plants, are biogenetically related to each other. The biosynthesis of JA in plants has been shown to follow a prostaglandine-like route starting from linolenic acid [49]. The first transformation steps are catalyzed by lipoxygenase, hydroperoxide cyclase and a reductase respectively, followed by three β-oxidation steps. By use of extraction procedures, excluding or minimizing isomerization processes, it could be shown that by the fungus *Botryodi-*

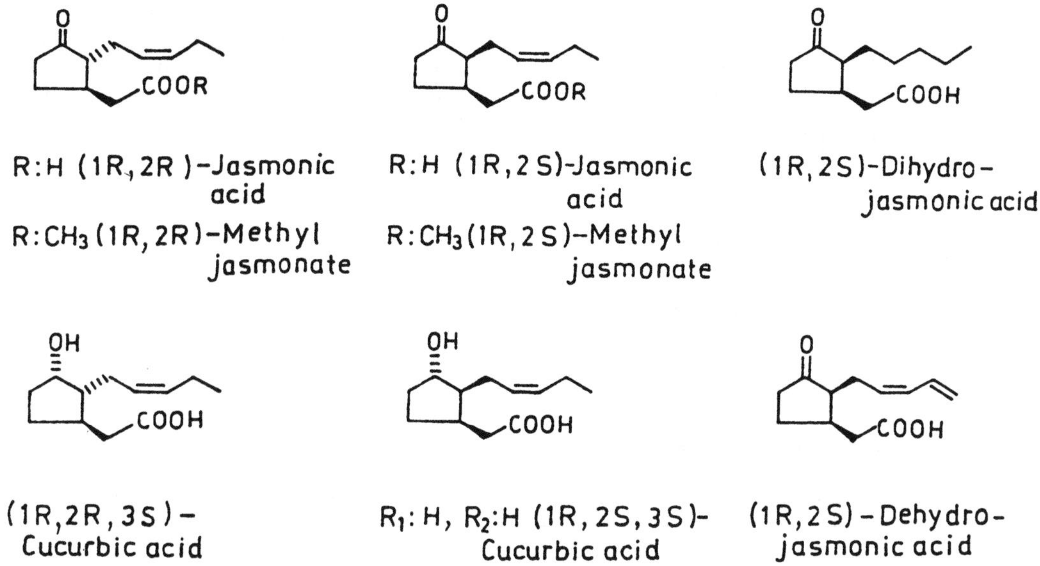

Fig. 1. Structures of naturally occurring jasmonic acid analogues

plodia exclusively (+)-2-iso-JA is biosynthesized and (–)-JA is formed by isomerization, whereas in broad bean pericarp a special ratio of (–)-JA and (+)-2-iso-JA occurred [33, 34].

The radioimmunoassay has been used to study the distribution of JA within the broad bean plant in relation to physiological development. In developing flowers the endogenous levels of JA were found to be very high, varying in a range of about 10–30 $\mu g\ g^{-1}$ fresh wt. As dilution curves of this immunoreactive material gave no parallelism to the (–)-JA standard curve in RIA, the presence of substances other than JA had to be assumed. Separation of the material by TLC prior to RIA showed the existence of a polar fraction with high immunoreactivity in addition to a small amount of JA (Fig. 2). Separation of the polar fraction by HPLC produced two immunoreactive compounds (Fig. 3) which were further investigated. Recently they have been structurally elucidated to be N [(–)-jasmonoyl]-S-tyrosine [6] and N [(–)-jasmonoyl]-S-tryptophan, respectively [Brückner et al., unpubl.].

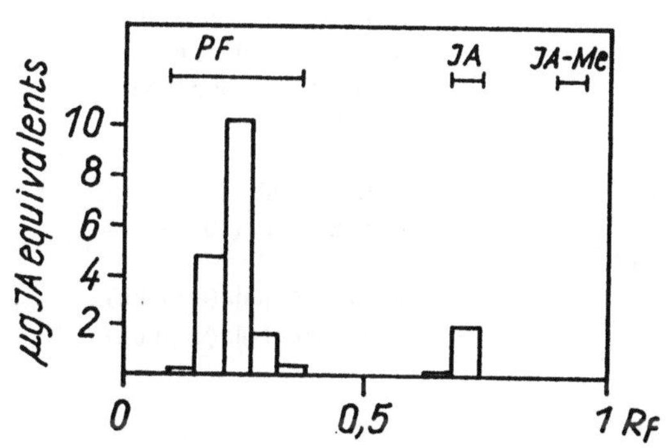

Fig. 2. TLC separation of JA-immunoreactive material of an acidic chloroform extract from broad bean flowers. Solvent systems: chloroform ethyl acetate acetic acid = 14:6:1; PF = polar fraction

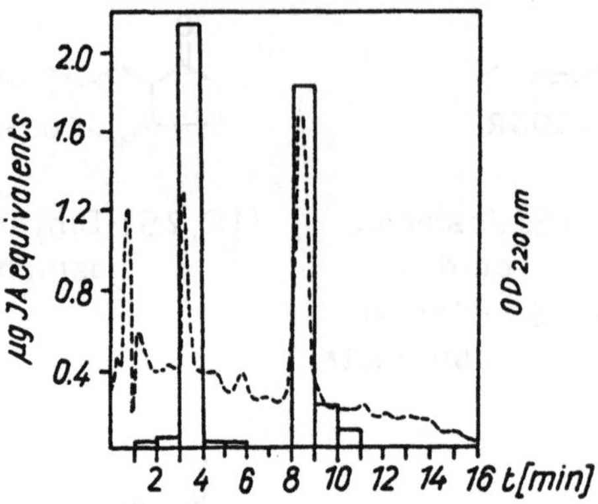

Fig. 3. HPLC separation of the immunoreactive "polar fraction" (cf. Fig. 2). Millichrom OB4: MeOH/AcOH (0,2%) = 6:4, 2x62 mm. SiRP 18 (5 μm); ----- OD 220 nm

Table 1. Physiological effects of jasmonic acid

Process influenced/Plant system	Effect	References
Length growth		
Wheat seedlings	Inhibition	Dathe et al. (1981) [a]
Dwarf rice seedlings (±GA_3)	Inhibition	Yamane et al. (1980) [a]
		Yamane et al. (1981) [a]
		Sembdner et al. (1983) [a]
Lactuca hypocotyl	Inhibition	Yamane et al. (1981) [a]
Sugarcane cuttings	Stimulation	Liebisch et al. (unpubl.)
IAA-induced curvature growth		
Avena coleoptiles	No influence	Yamane et al. (1981) [a]
Root growth		
Wheat seedlings	Inhibition	Sembdner and Klose [46]
Lactuca sativa	Inhibition	Yamane et al. (1981) [a]
Adventitious root formation		
Wheat seedlings (+auxin)	Stimulation	Klose and Sembdner (unpubl.)
Vigna radiata	Stimulation	Zimmerman and Vick [50]
Growth of axillary buds		
Linum usitatissimum	Inhibition	Sebanek and Jandakova [45]
Callus growth		
Glycine max (+cytokinin)	Inhibition	Ueda and Kato (1982) [a]
Seed germination		
Avena sativa	Inhibition	Satler and Thimann (1981) [a]
Lactuca sativa	Inhibition	Yamane et al. (1981) [a]
Amaranthus caudatus	No effect	Sembdner et al. (1983) [a]
Agrostemma githago	Inhibition	Herrmann and Sembdner (unpubl.)
Pollen germination		
Camellia sinensis	Inhibition	Yamane et al. (1981) [a]
Impatiens balsamina	Inhibition	Yamane et al. (1982) [a]
Lilium longiflorum	Stimulation/ Inhibition	Sembdner et al. (unpubl.)
Petunia hybrida	Inhibition	Sembdner et al. (unpubl.)

Table 1 (cont.)

Process influenced/Plant system	Effect	References
Leaf senescence		
Avena sativa (+cytokinin)	Promotion in light and dark	Ueda and Kato (1980) [a]
Pelargonium zonale		Satler and Thimann (1981) [a]
Hordeum vulgare	Promotion	Weidhase et al. (unpubl.)
Stomata closure		
Avena sativa	Promotion	Satler and Thimann (1981) [a]
Leaf abscission		
Vigna sp.	Promotion	Curtis [8]
Carotene pigment formation		
Fruits of *Lycopersicon esculentum*	Inhibition of lycopene accumulation. Stimulation of β-carotene accumulation	Saniewski and Czapski (1983) a
Respiration		
Avena sativa	Promotion	Satler and Thimann (1981) [a]
Content of reducing sugars and proteins		
Avena sativa	Reduction	Satler and Thimann (1981) [a]
Ethylene biosynthesis		
Vicia faba (±-auxin)	No effect	Bergner and Sembdner (unpubl.)
Fruits of *Lycopersicon esculentum*	Stimulation	Saniewski and Czapski [41]

[a] Reviewed by Sembdner and Klose [46]; bibliography of literature see therein

Exogenously applied JA and JA-Me exhibit a variety of physiological effects in different bioassays. Published results are summarized in Table 1. Further physiological studies done in our laboratory are paralleled by investigations on the endogenous JA levels with the aim to finally answer the question whether JA plays an endogenous hormonelike role in the regulation of these processes.

4 Conclusions

Screening and plant hormone research have uncovered a high number of growth regulating substances of plant and microbal origin. Most of them are secondary metabolites affecting plant growth and development after exogenous application, but do not have an endogenous regulatory role like plant hormones. Nevertheless, they can be important for agrochemical use or as models for syntheses of new plant growth substances. Some groups, like the brassinosteroids, turgorines and jasmonic acid, are of increasing interest as they possess plant hormonelike properties. Jasmonic acid and related compound are widely distributed in plants; their physiological potencies, biosynthesis and metabolism are being studied intensively. Whether jasmonic acid has an endogenous regulatory role, like a classical plant hormone, has to be confirmed by further investigations.

References

1. Adam G, Khoi NH, Bergner C, Lien NT (1979) Phytochemistry (Oxf) 18:1399
2. Barber J, Garson MJ, Staunton J (1981) J Chem Soc Perkin I 1981:2584
3. Bearder JR (1980) In: MacMillan J (ed) Hormonal regulation of development I. Encyclopedia of plant physiology, new series. Springer, Berlin Heidelberg New York, vol 9, p 9
4. Bolte ML, Bowers J, Crow WD, Paton DM, Sakurai A, Takahashi N, Ujiie M, Yoshida S (1984) Agric Biol Chem 48:373
5. Bolte ML, Crow WD, Takahashi N, Sakurai A, Ujiie M, Yoshida S (1985) Agric Biol Chem 49:761
6. Brückner C, Kramell R, Schneider G, Knöfel HD, Sembdner G, Schreiber K (1986) Phytochemistry (Oxf)
7. Cameron DW, Feutrill GI, Perlmutter P, Sasse JM (1984) Phytochemistry (Oxf) 23:533
8. Curtis RW (1984) Plant Physiol (Bethesda) Suppl 75:127
9. Cutler HG (1984) In: Ory RL, Rittig FR (eds) Bioregulators: chemistry and uses. ACS Symp Ser No 257
10. Eriksen AB, Selldin G, Skogen D, Nilsen S (1981) Planta (Berl) 152:44
11. Gnanamanickam SS, Starratt AN, Ward WB (1982) Can J Bot 60:645
12. Gross D (1975) Phytochemistry (Oxf) 14:2105
13. Gross D (1980) Z Chem 20:397
14. Hase T, Koreeda M, Hasegawa K (1983) Phytochemistry (Oxf) 22:1275
15. Hasegawa K, Knegt E, Bruinsma J (1983) Phytochemistry (Oxf) 22:2611
16. Hasegawa K, Shiikara S, Iwagawa T, Hase T (1982) Plant Physiol (Bethesda) 70:626
17. Henry EW, Gordon CJ (1980) J Exp Bot 31:1297
18. Hirota A, Sakai H, Isogai A (1985) Agric Biol Chem 49:731
19. Hoad GV (1983) In: Crozier A (ed) The biochemistry and physiology of gibberellins, vol 2. Praeger, New York, p 57
20. Hoffmann HMR, Raba J (1985) Angew Chem 97:96
21. Ishihara A, Shiraishi K, Sato H, Sakamura S, Nishiyama K, Sakai R, Furusaki A, Matsumoto T (1977) J Am Chem Soc 99:636
22. Kalsi PS, Sood VB, Masih AB, Gupta D, Talwar KK (1983) Phytochemistry (Oxf) 22:1387
23. Knöfel HD, Brückner C, Kramell R, Sembdner G, Schreiber K (1984) Biochem Physiol Pflanz 179:317
24. Knowles NR, Ries SK (1981) Plant Physiol (Bethesda) 68:1279
25. Kobayashi S, Ariga M, Ozawa T, Imagawa H (1984) Agric Biol Chem 48:389
26. Kusumi T, Chang CG, Wheeler M, Kubo I, Nakanishi K, Naoki H (1981) Tetrahedron Lett 22:3451
27. Lewak S, Skowronska E (1982) Physiol Plant 55:183
28. Lohaus E, Blas I, Schäfer W, Rüdiger W (1982) Z Naturforsch Sect C 37:802
29. Mandava NB (1979) In: Mandava NB (ed) Plant growth substances. ACS Symposium Ser 111, p 135
30. Matsuo A, Uto S, Nozaki H, Nakayama M (1980) Chem Commun 1980:1220
31. Matsuo A, Uto S, Nozaki H, Nakayama M (1984) J Chem Soc Perkin Trans I 1984:215
32. Meyer A, Miersch O, Büttner C, Dathe W, Sembdner G (1984) J Plant Growth Regul 3:1
33. Miersch O, Meyer A, Vorkefeld S, Sembdner G (1986a) J Plant Growth Regul (in press)
34. Miersch O, Preiss A, Sembdner G, Schreiber K (1986b) Phytochemistry (Oxf) (in press)
35. Murao S, Hayashi H (1983) Agric Biol Chem 47:1135

36. Ohigashi H, Minami S, Fukui H, Koshimizu K, Mitzutani F, Sugiura A, Tomana T (1982) Agric Biol Chem 46:2555
37. Ries S, Wert U (1982) J Plant Growth Regul 1:117
38. Ries S, Wert V, Biernbaum JA, Gibson T, Bradley WJ (1983) J Am Soc Hortic Sci 108:917
39. Sakai R (1980) Ann Phytopath Soc Jpn 46:499
40. Sakai R (1981) Ann Phytopath Soc Jpn 47:35
41. Saniewski M, Czapski J (1985) Experientia (Basel) 41:256
42. Schildknecht H (1983) Angew Chem 95:689
43. Schildknecht H (1984) Endeavour (Oxf) new ser 8:113
44. Schreiber K (1979) Adv Pesticide Sci 2:336
45. Sebanek J, Jandakova B (1984) Biochem Physiol Pflanz 179:351
46. Sembdner G, Klose C (1985) Biol Rundsch 23:29
47. Shimomura H, Sashida Y, Nakada H, Yamamoto A, Kawakubo Y, Kawasaki J (1983) Plant Cell Physiol 24:123
48. Stevens KL, Merill GB (1981) Experientia (Basel) 37:1133
49. Vick BA, Zimmermann DC (1984) Plant Physiol (Bethesda) 75:458
50. Zimmermann DC, Vick BA (1983) Plant Physiol (Bethesda) Suppl 72:108

III. Hormones: Mechanism

Plant Growth Substances and the Ionic Permeability of Membranes

M.F. HIPKINS and J.R. HILLMAN[1]

1 Introduction

The concept that plant growth substances might exert their primary effects by altering membrane properties was proposed by Veldstra [20]. In reviewing the effects of growth substances on ion transport Van Steveninck [19] posed a series of questions on their mode of action with respect to ion transport, including (1) whether growth substances directly affect membrane permeability or conductance through interaction with the membrane structure, and (2) whether growth substances affect metabolic processes which involve the establishment of ion gradients. A number of authors have attempted to answer these questions, in particular for ABA [21] and IAA [16, 18]. In searching for a simpler experimental system for ABA, Lea and Collins [11] abandoned biological membranes and used artificial phospholipid membranes. They found transient fluctuations in conductance which they associated with the formation of multimeric channels. Their data, therefore, suggest that ABA has an effect on membrane conductance which is independent of receptor molecules, indicating that ABA has a mode of action analogous to uncouplers (compounds which make membranes permeable to H^+) or ionophores (compounds which increase membrane permeability to certain ions or groups of ions). The evidence from biological systems on whether ABA has ionophoretic properties is conflicting. It has previously been suggested that ABA was an uncoupler [3] or might have properties like those of an uncoupler [13]. However, Boyer [1] found no evidence for uncoupling of chloroplasts by ABA. Moreover, MacRobbie [12] found ABA stimulated ^{86}Rb flux through the plasmalemma, but did not believe the membranes became "leaky" (see also Kasamo [10]). But Markhart et al. [14] suggested an interaction with the lipid phase of the membrane. More recently, ABA has been suggested to interact with phosphatidylethanolamine to render phospholipid vesicles permeable to water, urea and erythritol [17]. Interest in IAA has centered mainly on its receptors [16], but since it can cross artificial membranes readily [2] and is a weak acid, it might exhibit ionophoretic properties.

We have set out to investigate whether ABA or IAA exerts an ionophoretic effect on biological membranes. We have chosen simple photosynthetic membrane vesicles as an experimental system because electrogenic ion fluxes through such membranes may be monitored readily via electrochromic band shifts of the photosynthetic pigments [8].

1 Department of Botany, University of Glasgow, Glasgow G12 8QQ, Great Britain

Plant Growth Substances 1985
(ed. by M. Bopp)

The electrochromic band shift of photosynthetic membranes is associated with the light-driven reactions of photosynthesis and results from the specific spatial arrangement of components involved in photosynthetic electron transport within the functional membrane, in accordance with the chemiosmotic hypothesis [15]. In chloroplasts, the primary electron donors of Photosystem II (P680) and Photosystem I (P700) are located towards the inside of the thylakoid membrane, whilst the primary electron acceptors (pheophytin for Photosystem II, chlorophyll for Photosystem I) are located towards the outside. Thus, when light causes the oxidation of the donor and reduction of the acceptor, a transmembrane electric field across the thylakoid membrane is established (Fig. 1).

The existence of the electric field can be detected spectrophotometrically because the membrane is pigmented, and the absorption spectrum of the photosynthetic pigments is affected by the field, resulting in a small band shift. The time course of the absorption change typically consists of a fast absorption change (complete in a few nanoseconds) followed by a slower decay associated with ion movements across the membrane, primarily H^+ flux through the ATPase. Any treatment of the membranes, which leads to increased ionic permeability, induces the decay of the change to accelerate [9]. Analogous electrochromic band shifts occur in the chromatophores of photosynthetic bacteria. They are most easily demonstrated in chromatophores: closed vesicles formed from pinched-off invaginations of the intra-cytoplasmic membrane, and containing all the components of the bacterial cyclic photosynthetic electron transport chain [7].

Analysis of the decay of the electrochromic band shift in chloroplasts and chromatophores is a rapid and sensitive means of determining whether compounds alter the ionic permeability of the membranes.

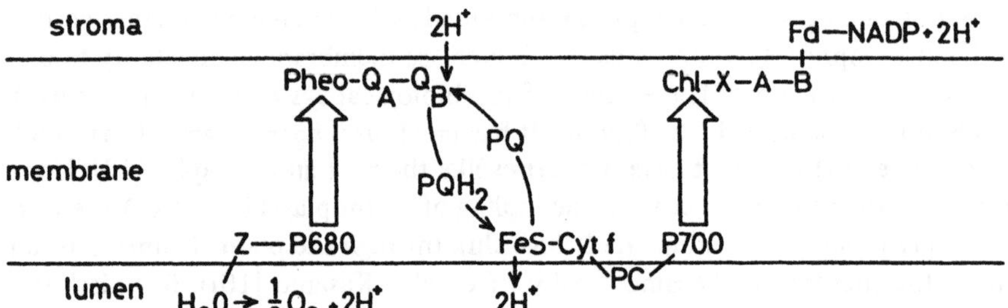

Fig. 1. Simplified schematic diagram of the spatial arrangement of the photosynthetic electron transport chain in the thylakoid membrane of chloroplasts. The diagram omits the "Q-cycle". P680, P700 = reaction centre chlorophylls of Photosystem II and Photosystem I respectively; Pheo = pheophytin; $Q_A Q_B$ = secondary electron acceptors of Photosystem II; PQ = plastoquinone; FeS = Rieske iron sulphur centre; cyt f = cytochrome *f*; PC = plastocyanin; Chl = chlorophyll; X,A,B = bound iron-sulphur centres; Fd = ferredoxin; Z = unknown intermediate(s)

2 Materials and Methods

Chloroplasts from peas [6] and spinach chloroplasts were used. An aliquot of chloroplasts was osmotically shocked in 2% (v/v) suspending medium to which 5 mM $MgCl_2$ had been added. The osmotic balance of the medium was then restored by the addition of an equal volume of double-strength suspending medium.

Chromatophores from *Rhodopseudomonas sphaeroides* strain 2.4.1 and *Rps. capsulata* strain Z1 were resuspended either in 100 mM KCl, 20 mM MES pH 6.8 (2.4.1); or in 50 mM KCl and 50 mM Tris pH 8 [6]. Both were stored in the refrigerator until used. For spectrophotometric measurements they were diluted with the same buffer and treated with 6 μM antimycin A to inhibit the slow rise of the electrochromic band shift [7] and consequently to simplify the kinetic analysis.

Flash-induced absorption changes were measured and analysed kinetically as described previously [5, 6].

3 Results

The rate of decay of the electrochromic band shift, both in bacterial chromatophores and in chloroplasts, is strongly influenced by the ionic permeability of the functional photosynthetic membrane. Adding the "classical" uncoupler, carbonyl cyanide-*m*-chlorophenyl hydrazone (CCCP), to chromatophores from *Rps. sphaeroides* strain 2.4.1 altered the electrochromic band shift measured at 523 nm (Fig. 2). The rate of decay was markedly accelerated, with increasing concentrations of CCCP. This observation was confirmed by kinetic analysis (Table 1), where the decay of the absorption change was fitted to the sum of two first-order phases. In the control, the first (faster) phase of decay comprised 20% of the total extent of the change; when CCCP was added at a concentration of 24 nM, the percentage was increased to 24%, and the proportion of the change that decays with the faster half-time increased steadily as the

Table 1. Kinetic analysis of the decay of the electrochromic band shift in *Rps. sphaeroides* chromatophores and the effect of CCCP and valinomycin on the parameters of the decay[a]

	First phase		Second phase	
	Extent	Half-time (ms)	Extent	Half-time (ms)
Control	148 ± 15	26 ± 7	594 ± 14	388 ± 19
2.4×10^{-8} M CCCP	175 ± 36	49 ± 17	565 ± 41	413 ± 51
2.4×10^{-7} M CCCP	356 ± 32	34 ± 5	410 ± 38	202 ± 19
2.4×10^{-6} M CCCP	596 ± 33	13 ± 1	75 ± 8	475 ± 164
1.7×10^{-9} M Valinomycin	173 ± 16	23 ± 5	584 ± 12	428 ± 21
1.7×10^{-8} M Valinomycin	376 ± 23	15 ± 2	315 ± 14	162 ± 9

[a] The data were fitted to the sum of two first-order decays by a least-squares procedure. The errors are those of the least-squares fit

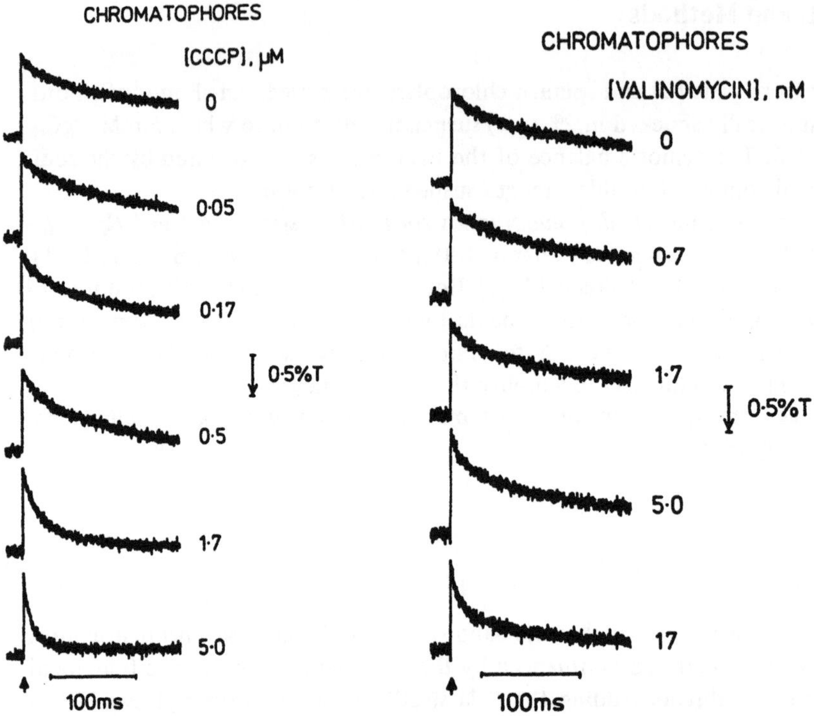

Fig. 2. The effect of CCCP and valinomycin on the decay of the electrochromic band shift in chromatophores from *Rps. sphaeroides* strain 2.4.1. Measurement wavelength 523 nm. The exciting flash was fired at the time indicated by the *upward arrow*. An upward deflection is a transmission decrease

concentration of CCCP was increased. This acceleration is expected since CCCP adds another route for the decay of the transmembrane field by making the chromatophore membrane more permeable to H^+ [4]. Similar effects are found with other "classical" uncouplers. In addition, Fig. 2 shows that valinomycin (a potassium ionophore, which makes biological membranes permeable to K^+) also accelerated the decay of the electrochromic band shift. Kinetic analysis of the decays in the presence of valinomycin shows a similar pattern to that seen with CCCP (Table 1). Valinomycin had very similar

Table 2. Kinetic analysis of the decay of the electrochromic band shift in spinach chloroplasts and the effect of valinomycin on the parameters of the decay. (See footnote to Table 1)

(Valinomycin), M	First phase		Second phase	
	Extent	Half-time (ms)	Extent	Half-time (ms)
0	787 ± 95	5.2 ± 1.4	4884 ± 49	204 ± 7
1.0×10^{-9}	992 ± 121	13.2 ± 2.0	4737 ± 131	124 ± 5
3.0×10^{-9}	1860 ± 257	7.2 ± 1.7	3478 ± 271	41 ± 3

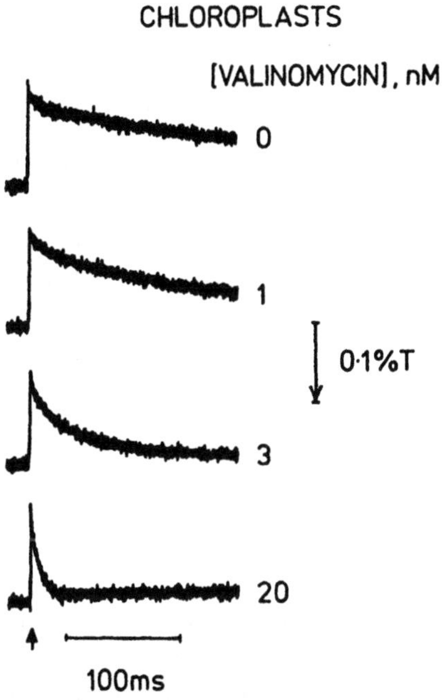

Fig. 3. The effect of valinomycin on the decay of the electrochromic band shift in isolated spinach chloroplasts. Measurement wavelength 518 nm. 10 μM methyl viologen was added as electron acceptor

two-fold effects on spinach chloroplasts (Fig. 3; Table 2): the decay was analysed as the sum of two first-order decays, and the half-times of the decays were decreased and the extent of the faster phase was increased at the expense of the slower phase.

In contrast, ABA did not accelerate the rate of decay of the electrochromic band shift in chromatophores of *Rps. capsulata* (measured at 502 nm; Fig. 4) or isolated pea chloroplasts (measured at 518 nm). The kinetic analysis is shown in Table 3. For

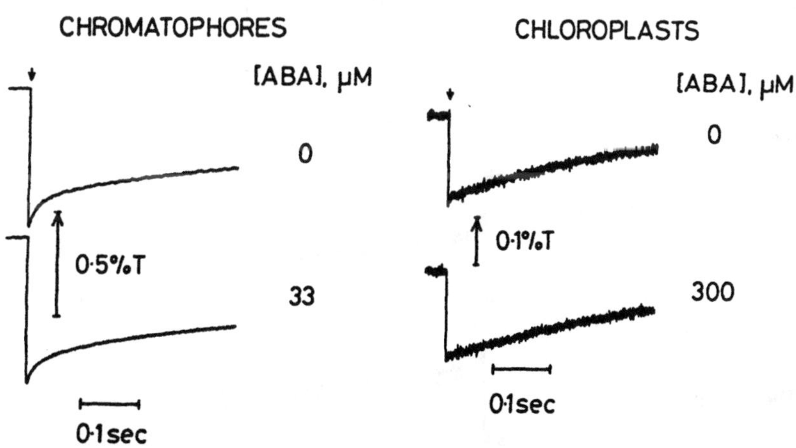

Fig. 4. The effect of ABA on the decay of the electrochromic band shift. *Left:* chromatophores from *Rps. capsulata* strain Z1, measurement wavelength 502 nm. *Right:* pea chloroplasts, measurement wavelength 518 nm. The chloroplasts were suspended in a medium containing 0.3 M sucrose, 50 mM KCl, 7.5 mM $MgCl_2$, 0.5 mM $CaSO_4$, 20 mM tricine, pH 7.6 with 10 μM methyl viologen. A downward deflection is a transmission decrease

Table 3. Kinetic analysis of the decay of the electrochromic band shift in *Rps. capsulata* chromatophores and pea chloroplasts and the effect of ABA on the parameters of the decay

Material	(ABA), M	First phase		Second phase	
		Extent	Half-time (ms)	Extent	Half-time (ms)
Chromatophores	0	381 ± 14	15 ± 2	1256 ± 8	749 ± 51
	3.3×10^{-5}	337 ± 14	17 ± 3	1361 ± 9	779 ± 58
Chloroplasts[a]	0	–	–	4386 ± 23	127 ± 2
	3.3×10^{-4}	–	–	4566 ± 23	144 ± 2

[a] In this case the decay was fitted to one first-order phase

chromatophores, the rate of decay of both phases was slightly retarded by ABA. For chloroplasts, where the data were fitted to a single phase of decay, there was again a slight slowing down induced by ABA.

The effect of IAA on the rate of decay of the electrochromic band shift was also investigated (Fig. 5). Figure 5 indicates that neither in chromatophores nor in chloroplasts did IAA appreciably alter the decay of the band shift; this is clearly shown by the kinetic analyses (Table 4). In chromatophores the decay half-times were slightly

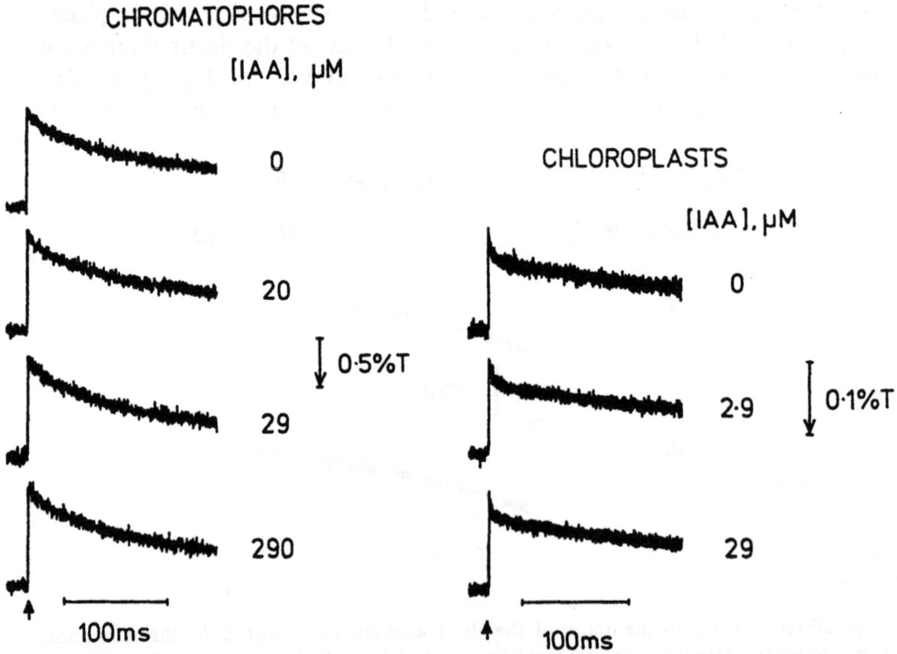

Fig. 5. The effect of IAA on the decay of the electrochromic band shift. *Left:* chromatophores from *Rps. sphaeroides* strain 2.4.1, measurement wavelength 523 nm. *Right:* spinach chloroplasts, measurement wavelength 518 nm. An upward deflection is a transmission decrease

Table 4. Kinetic analysis of the decay of the electrochromic band shift in *Rps. sphaeroides* chromatophores and spinach chloroplasts and the effect of IAA on the parameters of the decay

Material	(IAA), M	First phase		Second phase	
		Extent	Half-time (ms)	Extent	Half-time (ms)
Chromatophores	0	148 ± 15	26 ± 7	594 ± 14	388 ± 19
	2.0×10^{-5}	153 ± 26	37 ± 14	577 ± 31	420 ± 43
	2.0×10^{-4}	186 ± 60	66 ± 24	537 ± 64	416 ± 72
Chloroplasts	0	1411 ± 315	2.3 ± 1.3	5390 ± 77	225 ± 13
	2.9×10^{-6}	1655 ± 155	2.5 ± 0.5	4885 ± 41	241 ± 9
	2.9×10^{-5}	1243 ± 141	2.4 ± 0.6	5210 ± 36	237 ± 7

increased by IAA, whilst the small increase in the extent of the faster phase of decay was within the standard deviation of the kinetic analysis. For chloroplasts, the data suggested that there was no appreciable trend towards either faster decays or an increased proportion of the decay having a faster half-time.

4 Discussion

Our data suggest that in chloroplasts and chromatophores, ABA and IAA do not facilitate electrogenic ion fluxes. In our experimental system these compounds do not have properties like CCCP or valinomycin, as has been suggested for artificial lipid membranes [11]. Because of the reported specific interaction between ABA and phosphatidylethanolamine, it has been suggested that no response to ABA would be expected in chloroplasts since this lipid is absent from thylakoid membranes [17]. In contrast, chromatophore membranes (at least in *Rps. sphaeroides*) contain phosphatidylethanolamine as a major component of the lipids, and still no effect of ABA was seen (Table 3).

Our experimental system is unable to detect electroneutral exchange of, e.g., H^+ for K^+ as is seen with compounds like nigericin. However, even if nigericin-like action were to occur, the thermodynamic equivalence of proton concentration gradients and transmembrane electric fields [15] means that uncoupling would not necessarily take place.

Moreover, we cannot rule out the possibility that ABA or IAA makes membranes permeable to an ion that was not present in the media. Bearing these points in mind, we conclude that ABA and IAA do not have the ability to produce generalised effects on the ionic permeability of membranes in vivo, but that any effects are likely to be associated with a specific receptor.

5 Conclusion

We have used the light-induced electrochromic band shift in isolated, broken chloroplasts and chromatophores from photosynthetic bacteria to investigate whether ABA or IAA alters the ionic permeability of biological membranes. In this experimental system we found that neither growth substance modified electrogenic ion movements across the membranes. These data are in contrast with previous work using ABA and artificial phospholipid membranes. We conclude that ABA and IAA have no general, non-specific effect on the ionic permeability of membranes though effects might be seen in membranes possessing suitable receptors.

References

1. Boyer J (1973) Phytopathology 63:466
2. Gutnecht J, Walter A (1980) J Membr Biol 56:65
3. Hemberg T (1978) Physiol Plant 43:65
4. Heytler PG (1963) Biochemistry 2:357
5. Hipkins MF (1978) Biochim Biophys Acta 502:161
6. Hipkins MF, Hillman JR (1981) Z Pflanzenphysiol 104:217
7. Jackson JB, Dutton PL (1973) Biochim Biophys Acta 325:102
8. Junge W (1977) Annu Rev Plant Physiol 28:503
9. Junge W, Witt HT (1968) Z Naturforsch Teil B 23:244
10. Kasamo K (1981) Plant cell Physiol 22:1257
11. Lea EJA, Collins JC (1979) New Phytol 82:11
12. MacRobbie EAC (1981) J Exp Bot 32:563
13. Mansfield TA (1976) J Exp Bot 27:559
14. Markhart AH, Fiscus EL, Naylor AW, Kramer PJ (1979) Plant Physiol (Bethesda) 64:611
15. Mitchell P (1966) Biol Rev 41:445
16. Rubery PH (1981) Annu Rev Plant Physiol 32:569
17. Stillwell W, Hester P (1984) Phytochemistry (Oxf) 23:2187
18. Stoddart JL, Venis MA (1980) In: MacMillan J (ed) Encyclopaedia of plant physiology (new ser), vol 9. Springer, Berlin Heidelberg New York, p 445
19. Van Steveninck RFM (1976) In: Lüttge U, Pitman MG (eds) Encyclopaedia of plant physiology (new ser), vol 2B. Springer, Berlin Heidelberg New York, p 307
20. Veldstra H (1944) Enzymologia 7:137
21. Walton DC (1980) Annu Rev Plant Physiol 31:453

Ca^{2+}: Calmodulin Phytohormone-Linked Plant Senescence Control

Y.Y. LESHEM, M. FREUD-SILVERBERG, J. WURZBURGER, G. BAR-NES, Z. MALIK, and Y. LANGSAM [1]

1 Introduction

Further insight into the understanding of the regulatory role of Ca was the discovery some 2 decades ago of the calcium-binding protein calmodulin (CM) and the subsequent demonstration that several key enzymes, including some phospholipases [22] are dependent on Ca or a Ca:CM complex for activation. In both mammals and plants, Ca concentration on the cell's interior is several orders of magnitude greater than that on its exterior [4], while CM gradients manifest an opposite distribution – Ca being primarily in the cystosol, very sparse in the membrane and virtually absent from the cell's exterior [15]. Moreover, it is claimed [17, 18] that peroxidized polyunsaturated fatty acids (PUFA) and phosphatidic acid may serve as endogenous Ca *ionophores*, i.e. without lysing membranes they may transfer Ca from the cell's exterior to the interior, cause breakdown of the transmembrane Ca gradient and equilibrate concentration on either side.

In a recent paper [13] we presented data which we believe to be the first report that in plant systems, under certain conditions, Ca may not only fail to halt senescence, but that upon forming a part of a CM: phospholipase-A_2 (PLA_2) complex, may actually hasten the process. This has since been corroborated [9, 20] and appears to involve membrane deterioration as evidenced, inter alia by: (1) a membrane-bilayer lipid-phase transition from a juvenile liquid crystalline configuration to a senescence-associated solid gel one; (2) marked increase of ethylene, closely paralleled by ESR-detected oxy-free radical – in particular superoxide (O_2) – signals; (3) an increase of microviscosity ($\bar{\eta}$) and phase-transition temperatures; (4) an increase of free fatty acid: sterol ratios, which by some workers [2] is regarded as a major determinant of membrane fluidity; (5) a marked increase of lipoxygenase (LOX), which specifically catalyzes the oxidation of unsaturated fatty acids containing the *cis, cis*-1,4 pentadiene system, including linoleic and linoleic acids; (6) shedding in the form of "ghosts" – initially of the outer, then of the inner chloroplast membrane, leaving naked chloroplasts, which subsequently undergo total ultrastructural disruption. In this research, electron microscopy provides visual evidence of these dismantling processes.

Before presenting experimental data, both reviewing earlier work and providing new evidence, the Ca:CM:PLA_2 pathway as possibly suggested by our findings is herewith detailed.

1 Department of Life Sciences, Bar-Ilan University, Ramat-Gan 52 100, Israel

Plant Growth Substances 1985
(ed. by M. Bopp)

1.1 The Basic Hypothesis

An initial perturbation or change, the nature of which is still speculative, allows passage of external Ca across the membrane into the cytosol, or entry into the cytosol of Ca from Ca-containing organelles. This sets a "cascade effect" into motion – incoming Ca combines with cytosolic CM, activates PLA_2, which in return releases the 2-positioned PUFAs; these subsequently being catabolized by LOX to the hydroperoxidized products. The latter may now function as specific endogenous Ca ionophores, which enable more Ca to penetrate, which in turn combines with more CM, activates more PLA_2 and so on ad mortem. With increased cycling, membranes become progressively senescent: this is manifested in phase changes – wide angle X-ray diffraction patterns changing from liquid crystalline to solid gel. A further parameter contributing to membrane breakdown is phospholipid de-esterification by the Ca:CM:PLA_2 complex and the marked increase with ageing of the free fatty acid:sterol ratios. The physiological implication of the liquid → gel-phase change is that fatty acid tails in the bilayer become "frozen" and completely lose their motional freedom. As a result, the membrane becomes more rigid, loses pliability since microviscosity increases and embedded proteins are no longer able to move. In event of stress this frozen state precludes membrane adaptation – thus leaks are formed and overall function impaired. Gel-phase membranes are also more prone to phospholytic enzyme action.

Phase change is linked to ethylene production since conversion of precursor 1-aminocyclopropane-1-carboxylic acid (ACC) to ethylene is believed to be a free-radical-associated process and as mentioned above, the PLA_2-released PUFAs are lipoxygenized in a free-radical-generating process. ESR studies showed that the PLA_2: CM:CA activity is accompanied by a marked increase of an $O_2^{\overline{\cdot}}$ free-radical signal, which in parallel experiments was also reflected by increased ethylene evolution. The specificity of the $O_2^{\overline{\cdot}}$-ESR signal is further indicated by its reversal by superoxide dismutase. Support of the above hypothesis is the reversal of the CM-promoted effect by CM antagonists, such as fluphenazine hydrochloride (FP), CZ and W_7 and the simulation of senescence by application of exogenous Ca-ionophore 23187. Figure 1 schematically represents the Ca-intrusion stage.

1.2 Growth Substance-Associated Controlling Mechanisms

By interaction with anti-oxidative mechanisms, cytokinins, α-tocopherol (vitamin E), and ascorbic acid (vitamin C) may, on the one hand, check these effects at the LOX-located point of attack, either by serving as direct free-radical scavengers or by incipient prevention of free-radical formation [10, 12]. With regards to polyamines, which are known to possess anti-senescent effects, it appears that the higher polymerized species may exert a Ca-extruding effect [7]. On the other hand, enhanced ethylene evolution appears to be a clear-cut manifestation of the Ca intrusion into the cytosol [9, 13, 20], this process being CM-mediated.

In the present paper a hypothesis is presented, in which (1) at least one facet of plant senescence is mediated by phospholipid catabolism; (2) Ca intrusion into the cytosol may be the triggering mechanism, and (3) this overall ethylene-associated

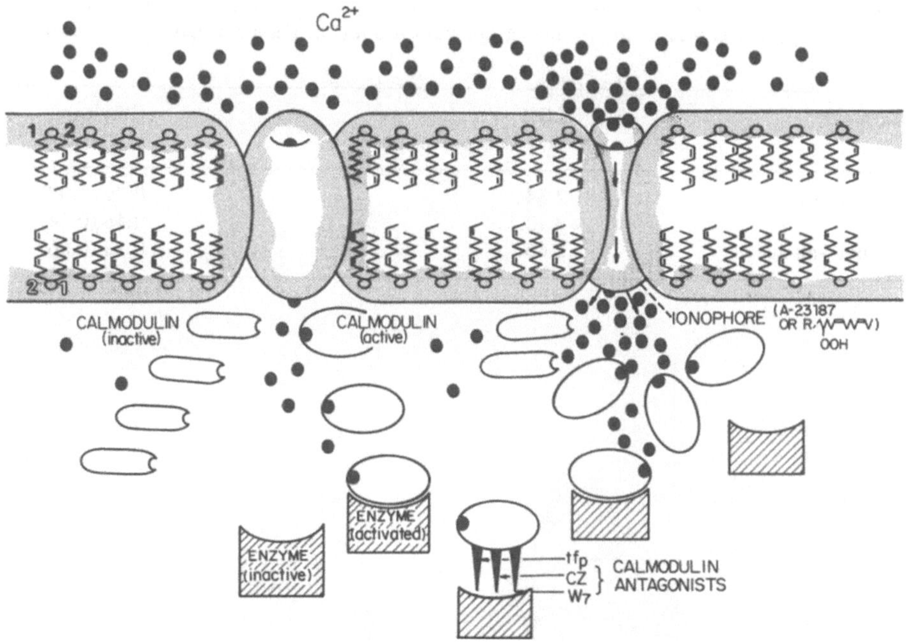

Fig. 1. Dynamics of transmembrane Ca intrusion into the cytosol. The numbers *1* and *2* in the phospholipids schematically represent *sn*-1 and *sn*-2 long-chain fatty acids respectively

process is partially controlled by, and contributes to, the biosynthesis of several plant regulators.

2 Experiments and Results

The experiments were performed with peas (*Pisum sativum* cv. Alaska or Greenfeast). Most of the methods have been described in previous papers [12–14, 16, 21]. Details are given in the legends of the Figures and Tables if necessary.

2.1 Lipoxygenase and Ethylene: Effect of Calcium and n-Propyl Gallate

Table 1 shows that both LOX and ethylene are increased by $CaCl_2$ application. This trend is significantly reversed by the free-radical scavenger *n*-propyl gallate.

2.2 X-Ray Diffraction Patterns of Microsomal Membranes Incubated with Calcium, Calmodulin, and Phospholipase A_2

Results presented in Fig. 2 show that at zero hours (A) the liquid-crystalline phase centering around a Bragg spacing of 4.6 Å is the prevalent one, while after 2 h (B),

Table 1. Effect of application of 10^{-3} M $CaCl_2$ and 10^{-3} M *n*-propyl gallate on lipoxygenase and ethylene in ageing pea foilage discs – incubation was for 8 h (I.B. Ferguson and Y.Y. Leshem, unpulished data)

Treatment	LOX $\mu l\ O_2$/mg prot./min^{-1}	Ethylene nl/8 discs/h^{-1}
H_2O-Control	2.34	0.29 (0.09)
$CaCl_2$	3.23	0.46 (0.14)
n-Propyl Gallate	1.66	0.23 (0.06)

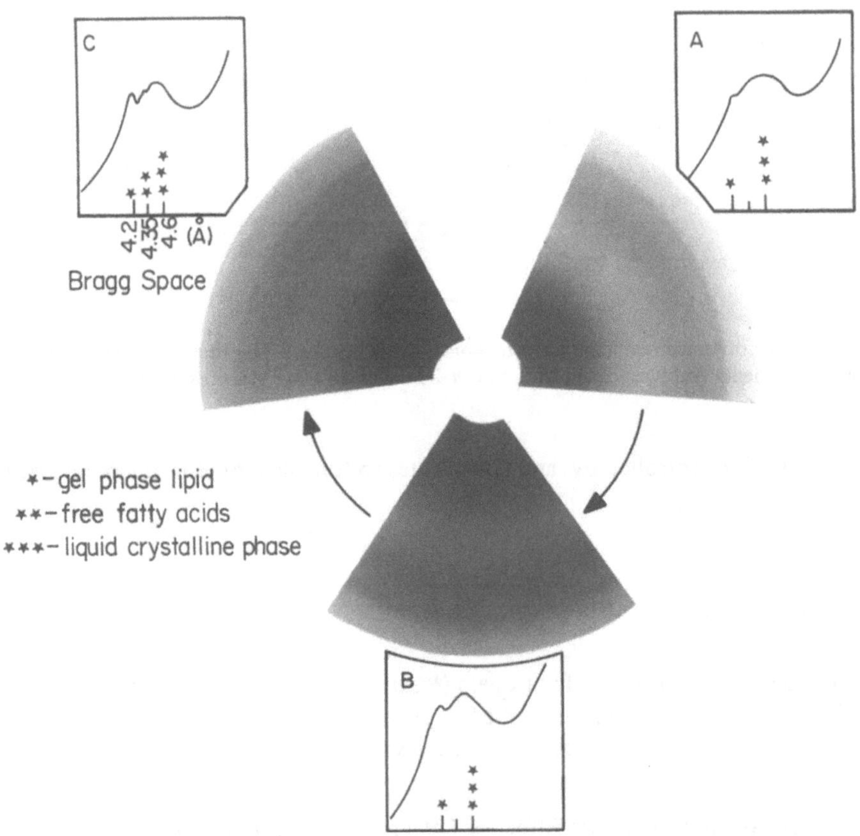

Fig. 2. Effect of Ca metabolism on wide-angle X-ray diffraction patterns and densitometer scans of microsomal membranes. Each treatment contained 10 mg microsomal protein in 2 ml Epps buffer, pH 8.5, 50 μM $CaCl_2$ and when added, 12 units PLA_2 and 23 units calmodulin. Means of four replicates. (Adapted from Sridhara and Leshem [20])

there is a significant increase of the sharply demarcated solid-gel phase band at 4.2 Å. After 20 h (C), in addition to the gel-phase band, a further one appears at 4.35 Å; this probably reflecting packing of free fatty acids and indicating advanced senescence. These free fatty acids contribute to the increase of free fatty acid:sterol ratios discussed herewith. The accompanying densitometer scans represent graphically the same data.

2.3 Effects on Free Fatty Acid: Sterol Ratios and Ethylene

Results shown in Table 2 expressed as percent relative change from those of microsomes at zero time (Serial No. 1) taken after 30 min clearly show that while PLA_2 (Serial No. 2) causes a marked increment above the zero control, highest ratios were observed by the optimal combination of Ca, CM, and PLA_2 (Serial No. 3); the effect being reversed by the CM-inhibitor FP (Serial No. 4). Detectable differences in ethylene only developed somewhat later and typical values after 20 h are also given in Table 2. These show that ethylene closely follows the general trend of free fatty acid: sterol ratios. The implications of this correlation will be dealt with in Sect. 3.

Table 2. Effect of Ca, calmodulin and phospholipase A_2 on free fatty acid:sterol ratios and ethylene evolution of pea foilage membranes

Serial No.	Treatment	Free fatty acid:sterol (Δ)[a]	Ethylene[b] nl ml^{-1}
1	Microsomes + Ca	0	1.74 (0.20)
2	1 + PLA_2	269 (20)	2.24 (0.30)
3	2 + CM	421 (21)	3.08 (0.19)
4	3 + FP	304 (43)	1.42 (0.47)

[a] Treatment procedure according to Sridhara et al. [20]
[b] Treatment procedure according to Leshem et al. [13]

2.4 Ultrastructural Changes in Chloroplast Membranes Induced by Ca, Calmodulin, and PLA_2

Figure 3 (left) indicates that upon being treated with Ca, like a moulting insect, the chloroplast sheds the outer of its two enveloping membranes in the form of a "ghost", thus leaving a still intact, but dishabilled and gaunt plastid retaining its original form. That the envelope is indeed a monolayer is shown in the transmission electron micrograph (Fig. 3, right), indicating progressive stages of the shedding process. Figure 4 indicates that senescence is markedly affected by treatment with an admixture of Ca, CM, PLA_2, and ionophore. Both outer membranes are shed, the internal ultrastructure dismantled and finally, small microsomal vesicles and membrane fragments remain.

2.5 Effects of Polyamines on Ca Extrusion from Liposomes

Figure 5 indicates that at physiological concentrations (10^{-3} M and below), the two tetraamines, spermine and to a lesser degree agmatine, induced Ca extrusion. The triamine spermidine was only active at the high non-physiological concentration of 10^{-2} M and the two diamines, putrescine and cadaverine, were devoid of activity.

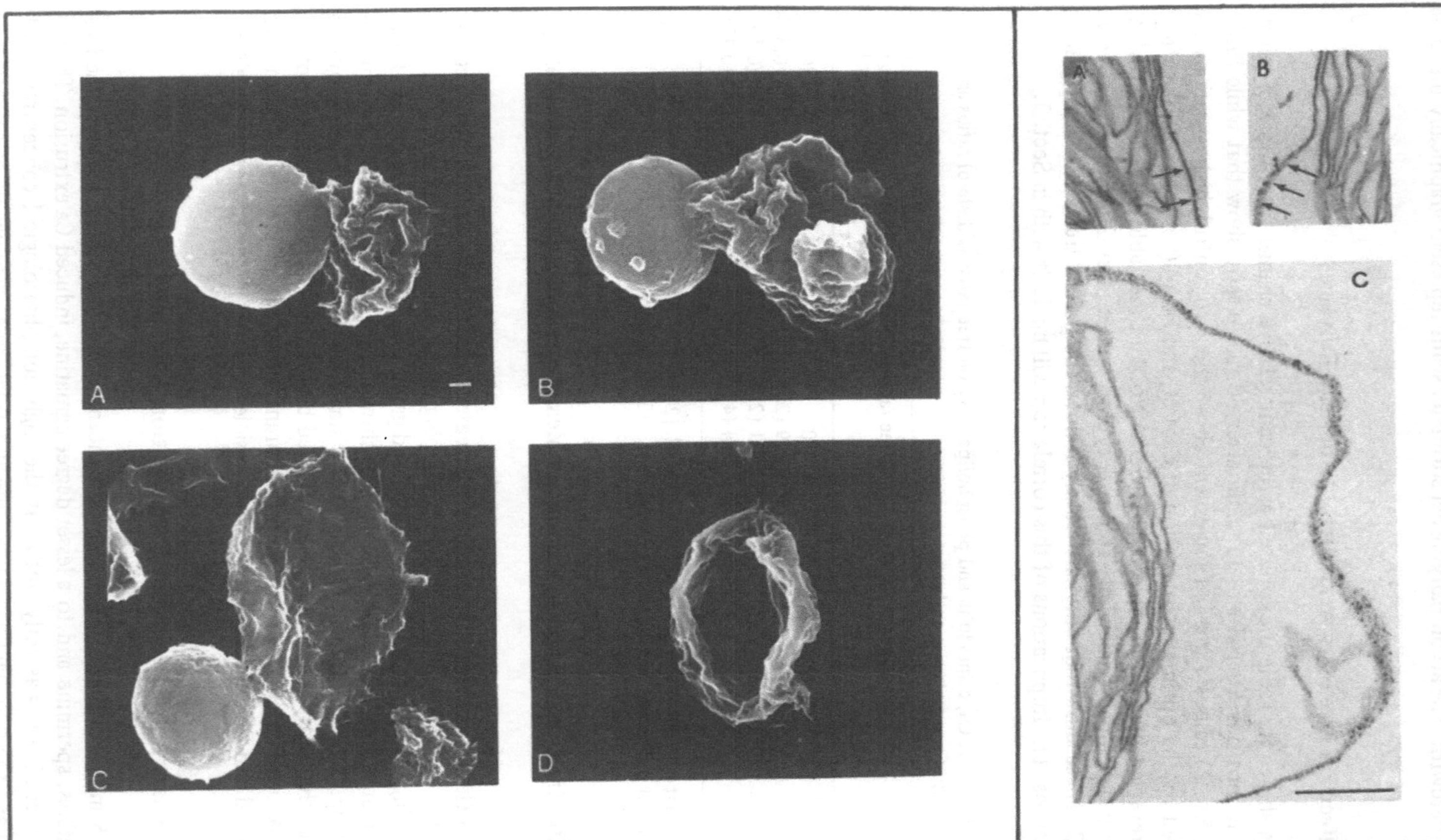

Fig. 3. *Left:* Scanning electron micrographs of pea (*Pisum sativum* cv. Sparkler) chloroplasts indicating outer membrane shedding during Ca-induced senescence. **A–C** represent progressive stages of envelope shedding. **D** shows the detached "ghost". In **B** note "blebs" on chloroplast surface. These probably indicate initial stages of shedding of the inner membrane. *Right:* Transition electron micrographs of the same chloroplasts. **A–C** represent progressive stages of shedding of the outer monolayer of the originally bilamellar structure. *Bar* indicates 1 μ

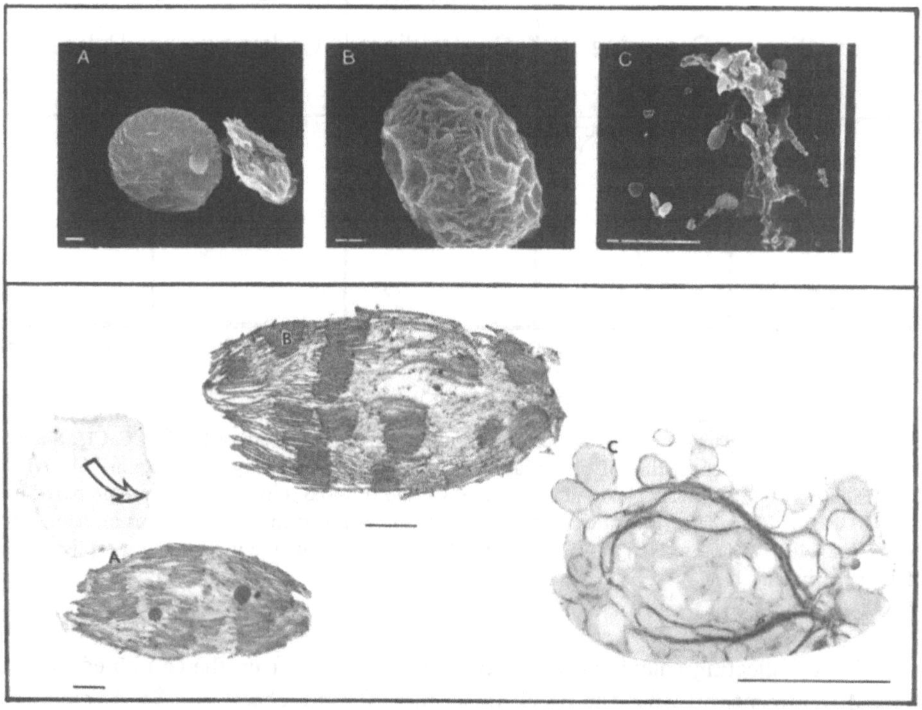

Fig. 4. *Top:* Ultrastructural changes in chloroplast membranes induced by treatment with a "senescence cocktail" containing 10^{-3} M $CaCl_2$, 100 units each of calmodulin and PLA_2 and 50 μM Ca-ionophore A23187. **A** Chloroplast denuded of both enveloping membranes. **B** Successive stage of breakdown of internal structure. **C** Terminal senescence in which ultrastructure is completely disrupted with the formation of microsomelike vesicles, probably obtained from closing of PLA_2-digested membrane fragments.
Bottom: Transmission electron micrographs of the same "senescence cocktail" treatment. **A** Membrane-denuded chloroplast accompanied by its "ghost" (*arrow*): note that the ghost is a single membrane. **B** Chloroplast lacking its enveloping membranes in the process of disruption. **C** Monolayered microsomal vesicles typical of "terminal senescence". *Bar* indicates 1 μ

3 Discussion

The data presented in Tables 1 and 2 and Fig. 2 indicate that Ca significantly enhances several parameters which are intricately associated with plant senescence – LOX, ethylene, membranal free fatty acid:sterol ratios, transition from liquid crystalline to gel phase, and that the free-radical scavenger *n*-propyl gallate reversed some of these changes (Table 1). This observation confirms the free-radical involvement reported in senescence [10] and indicates that Ca adds to the free-radical pool since scavengers and/or agents, which prevent incipient free-radical formation, such as cytokinins, α-tocopherol and the SOD inhibitor allopurinol, have a similar LOX-inhibiting effect as the *n*-propyl gallate [10, 11]. Moreover, as shown previously [9, 13], these changes are also blocked by CM inhibitors, such as FP, and the more specific calmidazolium and the naphthalene sulphonamide compound W_7, while on the other hand Ca-ionophore 23187 enhances ageing [13].

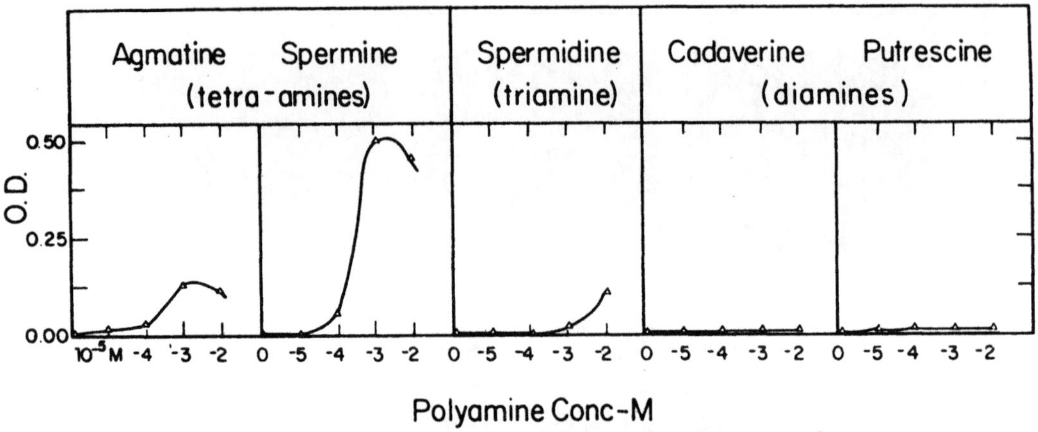

Fig. 5. Effect of polyamines on Ca extrusion from liposomes encapsulating 10^{-3} M $CaCl_2$. Single-bilayered liposomes, consisting of phosphatidylcholine and phosphatidylethanolamine (4:1 v:v), encapsulating 10^{-3} M $CaCl_2$, cleansed of external Ca and 0.2–0.3 μ in diameter were prepared by the method of Ehrenberg [3]. Ca concentration is expressed as relative OD units as measured by tetramurexide (TMX) dye [16] at 482 nm after 10 min. Each value is a mean of four replicates in three experiments

Table 2, indicating a maximal free fatty acid:sterol ratio from the combined PLA_2: Ca:CM treatment, provides the biochemical basis for the observed 4.35 Å free fatty acid band seen in Fig. 2. This implies that Ca triggered de-esterification of membrane phospholipids and that ethylene increase go hand in hand with phase transition from liquid to gel, which is clearly seen in Fig. 2.

The EM scanning and transmission micrographs (Figs. 3 and 4) of treated chloroplasts usually demonstrate the effects of intruding Ca. Figure 3 shows that the outer membrane is shed as a ghost and the chloroplast becomes somewhat bloated and that "blebs", indicating initial prying open of the inner membrane, also appear on the remaining inner membrane surface. Upon addition of PLA_2, CM, and Ca-ionophore A-23187, complete disintegration ensues (Fig. 4), the senescence sequence being the shedding of the inner membrane leaving a dismantled organelle, which then loses its ultrastructure and finally, only microsomal vesicles remain.

Figure 5 indicates that the higher tetra-amines cause Ca extrusion from liposomes – the physiological implication being that Ca extrusion could be a means of combatting senescence.

One possible explanation of the tetra-amine-inducted Ca extrusion is that in keeping with animal studies, we are dealing with a cation-exchange reaction, whereby polyamines bind to anionic sites in the plasma membrane and completely displace bound Ca [7]. This released Ca may in turn activate Ca^{2+} ATPase pumps or induce cation-mediated physical changes in other phospholipid membrane components [6] and these may stabilize membranes; stabilizing and anti-senescing effects of polyamines being well-documented [1, 6, 19].

In terms of applied horticulture, the present scheme is apparently problematic since several reports indicate that exogenous Ca application, especially to fruit, may check senescence. However, Ferguson and Watkins observed that in Ca-treated apples, the cation's buildup is primarily in the *free space* [5]. This implies that upon application to non-senescing tissue, exogenous Ca *increases* the transmembrane gradient by ion concentration buildup on the membrane exterior where, as shown by Legge et al. [8], it may induce membrane stabilization. However, upon advent of even a minute Ca leak or ionophoretic intrusion into the cytosol the "cascade effect", outlined in Sect. 1, is set into motion. Thus, a solution to the conundrum, reconciling the apparently contradictory reports, may be that the compartmentalization of Ca is crucial. When externally situated in intact cells, it maintains senescence deferral; however, upon entry into the cytosol it exerts the opposite effect.

4 Conclusions

Intrusion of Ca into the cytosol appears to be a pivotal regulatory mechanism in plant catabolism. Even a minor inequilibrization of transmembrane Ca gradients sets a "cascade" effect into motion, whereby Ca, combined with cytosolic CM, activates PLA_2. This activation promotes plant senescence in at least two ways: (1) the Ca:CM:PLA_2 complex is now able to impair integrity by attack on the *sn*-2 acyl (usually PUFA) membranal phospholipid chain – this involving increased leakiness, phase transition to a gel state and de-esterification of PUFA; (2) the released PUFAs induce LOX activity – the latter previously being unable to catabolize PUFAs in intact phospholipids – and thus produce hydroperoxides and a series of oxy-free radicals, which either directly attack any neighbouring molecules and/or enhance conversion of ACC to ethylene, the latter adding further momentum to the cascade. Cytokinins, anti-oxidants and scavengers, including vitamin E and C, serve to slow down this effect and appear to work in concert with SOD dismutase. It is suggested that further experimentation on Ca channel-blocking agents, calmodulin antagonists and Ca extruders may provide a rewarding mode of approach to senescence control.

References

1. Ben Arie R, Lurie S, Matoo AK (1982) Plant Sci Lett 24:239
2. Borochov A, Faiman-Weinberg R (1984) What's New Plant Physiol 15:1
3. Ehrenberg B (1982) FEBS Lett 140:219
4. Ferguson IB (1984) Plant Cell Environ 7:474
5. Ferguson IB, Watkins CB (1981) Aust J Plant Physiol 8:259
6. Jones GP, Paleg LG (1984) Biochemistry 23:1521
7. Koenig H. Goldstone A, Lu CY (1983) Nature 305:530
8. Legge RL, Thompson JE, Baker JE, Lieberman M (1982) Plant Cell Physiol 23:161
9. Leshem YY (1984) In: Siegenthaler PA, Eichenberger W (eds) Structure, function and metabolism of plant lipids. Elsevier, Amsterdam, p 181
10. Leshem YY (1984) Can J Bot 62:294
11. Leshem YY, Barness G (1982) In: Wintermans J, Kuiper P (eds) Biochemistry and metabolism of plant lipids. Elsevier, Amsterdam, p 275
12. Leshem YY, Grossman S, Frimer A, Ziv J (1979) In: Appelqvist LA, Liljenberg C (eds) Advances in the biochemistry and physiology of plant lipids. Elsevier, Amsterdam, p 193

13. Leshem YY, Sridhara S, Thompson JE (1984) Plant Physiol 75:329
14. McKersie BD, Thompson JE, Brandon JK (1976) Can J Bot 54:1074
15. Muto S (1982) FEBS Lett 147:161
16. Onishi ST (1978) Anal Biochem 85:165
17. Putney JW, Poggioli P, Weiss SJ (1981) Philos Trans Roy Soc Lond B Biol Sci 296:37
18. Serhan C, Anderson D, Goodman E, Dunham P, Weisman G (1981) J Biol Chem 256:273
19. Slocum RD, Kaur Sawnhey R, Galston AW (1984) Arch Biochem Biophys 235:283
20. Sridhara S, Leshem YY New Phytol (in press)
21. Thompson JE, Pauls KP, Chia LS, Sridhara S (1983) In: Thompson WW, Mudd B, Gibbs M (eds) Biosynthesis and function of plant lipids. Am Soc Plant Physiol, Rockville, p 173
22. van den Bosch H (1982) In: Hawthorne JN, Ansell GB (eds) Phospholipids. Elsevier, Amsterdam, p 313

Phytohormones and Gene Expression

B. PARTHIER [1]

1 Introduction

In 1966 Jan van Overbeck wrote: "We are beginning to learn that the site of action of hormones ... is close to the gene" [26]. He, the plant physiologist, must have been impressed by the progress in animal hormone research at that time, but also showed a feeling for research trends. At present, when indeed biologists are going to study plant genes in detail, the bewildering multiplicity of phytohormone effects contrasts with our ignorance about the molecular mechanisms involved in the various physiological responses.

Usually two sites of hormone (plant growth substance) action at the subcellular/molecular level are discussed: an early ("primary") effect at the cell membranes and a subsequent ("secondary") effect at the genetic level. The situation becomes still more complicated as the observed responses may be the result of actual concentration ratios between various compounds of synergistic and antagonistic actions. The ratios can change during the development of cells or tissues, but alteration in the "sensitivity" of the biological material towards phytohormones is another prerequisite for a specific response [25].

In this survey four selected phytohormone responses are discussed under the aspects of their action in gene expression (Table 1). Each of the examples is considered to represent a physiological model process which is regulated by the counteraction of

Table 1. Plant growth substances interacting in physiological responses and in gene expression (transcription)

Physiological process	Action in mRNA synthesis	
	Protagonistic	Antagonistic [a]
1. Amylase synthesis in aleurone cells	Gibberellins	Abscisic acid
2. Embryogenesis and seed development	Abscisic acid	Cytokinins
3. Cell enlargement	Auxins	(Cytokinins, ethylene)
4. Plastid development	Cytokinins	(Abscisic acid)

[a] In parentheses: action in transcription not confirmed

1 Institut für Biochemie der Pflanzen, Akademie der Wissenschaften der DDR, 4020 Halle (S.), Weinberg 3, GDR

Plant Growth Substances 1985
(ed. by M. Bopp)

dominating growth substances. Although there is no doubt that their fundamental action concerns nucleic acid metabolism [11], little is yet known about the molecular interaction between hormones as chemical signals and the plant cell targets involved in gene expression. Amount (biosynthesis) or function (activity) of enzymes or other polypeptides involved in gene activation, transcription, gene product modification, translation or post-translational steps are reported to change upon hormone addition. This diversity of effects easily suggests multiple sites and modes of phytohormone actions rather than a unique event that regulates plant gene expression. It may, however, reflect the limited sensitivity of available methods. This field is now rapidly improving due to the use of recombinant DNA techniques and subtle detection methods for gene products, which enable us to analyse hormone actions in gene expression with greater sensitivity and in more detail.

2 Animal Hormone Action

Rapid progress in the molecular action of animal hormones has pointed out that steroid hormones control transcription of specific genes in the nucleus after having been bound to highly specific receptor molecules. The hormone-receptor complex is then transported to the site of action in the chromatin, e.g. as described for oestradiol action in differentiating uterine tissue. The expression of the uteroglobin gene in rabbits, or ovalbumine and lysozyme genes in chicken, is subject to a tissue-specific regulation of transcription by animal steroid hormones. This hormonal control at the DNA level is explained by the occurrence of regulatory sequences, which are located variably distant to the promoters of steroid hormone-controlled genes [4]. These consensus sequences, which are recognised by hormone-binding receptor proteins, can be located even several thousands of nucleotides upstream from the respective promoter region [1]. Although a definitive proof of the functional relevance of the in vitro binding studies are still lacking, such regulatory sites can be considered as cryptic enhancers that are activated through receptor binding [1].

A first experimental approach to the relevant biological function is made with in vitro recombined genes [13].

3 Gene Expression Modulated by Phytohormones

Looking down from the summits of animal hormone action to the plains of phytohormone effects, one first tries to discover similarities. From the structural and biosynthetic point of view, gibberellins are related to animal steroid hormones and thus appear to be good candidates for similar regulatory functions.

3.1 GA-Controlled α-Amylase Synthesis

In cereal aleurone cells, external GA controls de novo synthesis and secretion of hydrolytic enzymes, prominently α-amylase isoenzymes, and thus, GA regulates the mobilisation of endosperm reserves for the growing embryo during the germination process. Already the classic studies of Varner's and Jacobsen's laboratories have shown that the GA-promoted increase in α-amylase biosynthesis is associated with corresponding changes in the levels of translatable mRNA [9, 10]. Time-course studies of the GA-induced rates of amylase biosynthesis demonstrate identical curves for the synthesis of the enzyme in vivo and in vitro. Using cloned α-amylase cDNA as a hybridisation probe for aleurone RNA, other authors reported [18] that 10^{-6} M GA in the incubation medium causes an increase of amylase mRNA as early as 1 h after GA addition. The enhancement lasts for several hours and then declines rapidly, which can be attributed to mRNA degradation. (RNase is likewise induced by GA in this system.) There is no stored α-amylase mRNA in mature aleurone cells [18].

ABA suppresses the GA-induced α-amylase synthesis at the level of mRNA concentration if assayed by in vitro translation and by RNA blots with the respective cDNA probes [12, 18]. These results may indicate regulation of amylase gene expression by the two antagonistic phytohormones which could switch the transcription process on or off. On the other hand, the actual amylase mRNA concentration could be the result of GA-induced synthesis and ABA-promoted degradation. Evidence is provided for a transcriptional control by both hormones with isolated nuclei from treated aleurone protoplasts [12].

The mechanism of GA (and ABA) action in the transcription process is unclear. Many reports on phytohormone stimulated increase in RNA polymerase activity [28] may indicate effects of secondary nature because of the time lag in response after hormone addition. In vitro addition of isolated RNA polymerases to assay mixtures are not convincing as long as reports with well-characterized RNA products are missing.

Together with calcium ions, GA seems to be involved also in the stability and secretion of α-amylase. While the influence of Ca^{2+} (10–20 mM) on the release of amylase is readily explained through an effect on secretion at the plasma membrane rather than ion exchange phenomena in the cell wall [17], the role of GA remains obscure unless the regulatory correlations between enzyme synthesis and excretion are better understood. Recently, it was shown by pulse-labelling and DNA-RNA hybridisation [6] that neither GA nor Ca^{2+} is required for synthesis, stability and excretion of α-amylase isoenzymes 1 and 2, but the transcription of mRNA for isoenzymes 3 and 4 is specifically controlled by GA. Ca^{2+} is only involved in the post-translational processing (excretion?) of these isoenzymes.

To summarise (Fig. 1), it is highly probable that GA and ABA control the synthesis of aleurone cell enzymes by enhancing mRNA transcription, whereas the steady-

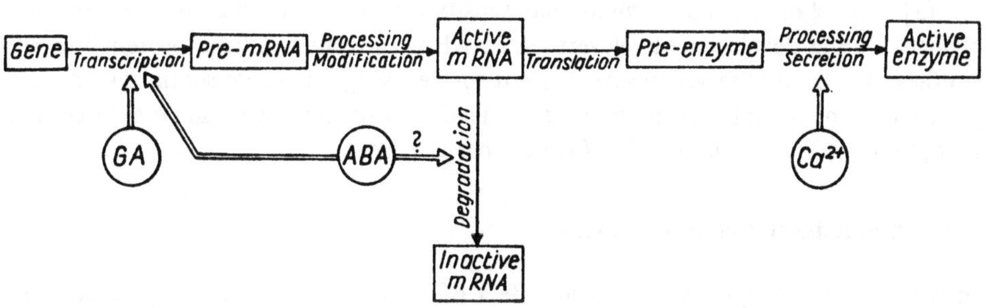

Fig. 1. Possible sites of actions of GA and ABA in α-amylase gene expression

state concentration (turnover) of the mRNAs is regulated complementary with other substances (ABA?). The sequential appearance of enzymes in the medium (or in vivo transport into the endosperm) may be mastered by changes in concentrations of localised Ca^{2+} in the aleurone cells.

3.2 ABA-Dependent Embryogenesis

Embryogeny and seed germination are normally separated by a period of developmental arrest that is controlled by phytohormones, particularly by abscisic acid. ABA levels rise sharply and then fall during seed development; the highest concentration occurs in the embryo when its dry weight increases rapidly. According to a model for cotton seeds proposed by Dure [7], ABA synthesised in the ovule is transported into the embryo, where it inhibits the translation of specific mRNA of so-called germination enzymes. Embryo-specific genes are turned off, resulting in the prevention of precocious germination, which is characterised by radicle elongation and the emergence of root hairs. More details about this process were obtained with developing seeds of *Brassica napus* by Crouch et al. [5], exploring storage protein synthesis as a biochemical marker for the study of ABA-controlled embryogenesis and germination.

If immature embryos are excised from the seeds and are cultured in vitro in a hormone-free medium, precocious germination is observed, but the accumulation of the two major storage proteins, cruciferin and napins, is prevented unless exogenous ABA (10^{-6} M) is added. Embryogenesis is now inhibited, thus indicating a concurrence between embryonic and germination-specific processes during seed development. Using storage protein cDNA as hybridisation probes, the pattern of storage protein mRNA has been studied during normal in vivo development and in cultured embryos controlled by external ABA. Only in the latter system did ABA show a rapid and specific effect on the levels of storage protein mRNA. Interestingly, the amount of napin mRNA, which appears prior to cruciferin mRNA, is found unaltered during the development of isolated cultured embryos, whereas cruciferin mRNA increases by a factor of ten. The authors suppose [5] that if the mRNA level is already high, ABA maintains the level; but the hormone enhances mRNA transcription, if the initial level is low (e.g. cruciferin mRNA). This is in accordance with reports [8] demonstrating that proteins characteristic for late embryogeny are synthesised in precociously germinating cotton embryos only in the presence of ABA.

In conclusion, ABA modulates storage protein gene expression in developing embryos, but the hormone is probably not directly involved in developmental specificity [5]. One should keep in mind that sensitivity of the embryos to the hormone changes during development. Nevertheless, ABA-controlled storage protein mRNA synthesis has been observed in several plant species, suggesting an essential role of this phytohormone in embryogenesis and seed development. Its effect can be antagonised by cytokinins, but not by GA (M. Crouch, pers. comm.).

3.3 Auxin-Induced Cell Enlargement

In the third model system the classic effect of auxins, cell enlargement, can be measured via elongation of isolated pea epicotyls or soybean hypocotyls. Of the hypotheses put forward to explain this auxin response, only that of gene activation can be taken into consideration. This idea is supported by studies, in which rapid and selective changes of poly-A^+ mRNA or of translation products have been visualised in response to auxin addition. Few polypeptide spots appear within 30 min after auxin addition ("early responsive mRNA", cf. [23, 29]). This response in gene expression is rather close to the fastest elongation response (15 to 20 min). Thus, gene expression is suggested to be directly correlated to cell enlargement, a view supported by inhibitors of gene expression. A cascade response to IAA is discussed, in which translation products of the early responsive genes might play a regulatory role for late responsive ones. The most striking feature of both early and late responses is the high selective nature of gene-product synthesis.

These data are confirmed by means of DNA clones complementary to two auxin-responsive mRNAs in the elongating region of soybean hypocotyl [27], and to three respective mRNAs in pea epicotyl [22]. RNA blot analyses reveal a rapid decrease of relevant mRNAs from excised sections incubated without auxin, but auxin addition results in the appearance of hybridisable mRNA within 20 min [22]. This expression of auxin-controlled genes is very specific and only dependent on auxins; neither CK, GA, ABA, ethylene (inhibitors of elongation) nor fusicoccin (induces H^+ secretion and promotes cell elongation) nor stress factors can trigger auxin-specific mRNA sequences [22, 27]. The possibility cannot be excluded that these growth substances act in gene expression steps subsequent to transcription.

It was also confirmed by the use of cDNA clones that auxin-dependent cytoplasmic pH change is not the driving force for mRNA induction [22]. Rapid kinetics, specificity of the auxin-induced mRNAs and the inhibition of cell enlargement by α-amanitin suggest that the auxin-modulated synthesis of certain proteins is involved in cell enlargement, provided they are characterised by a rapid turnover.

3.4 Cytokinin-Stimulated Plastid Development

In the preceding examples hormone-induced modulation is connected with nuclear gene expression. It would be interesting to know whether phytohormones also influence the expression of genes not localised in the nucleus, but in plastids. Earlier results have shown that exogenous cytokinins accelerate plastid development to etioplasts in dark-grown and to chloroplasts in light-grown plants or cotyledons [19].

The situation is complicated by the fact that the biogenesis of the organelles requires both plastid and nuclear gene expression. In order to facilitate the experimental system, RuBPCase has been used as a plastid marker enzyme. It consists of two types of subunits; the large one (LS) is encoded and synthesised within the plastid; the small one (SS) is encoded in the nucleus and synthesised in the cytosol. It is then transported into the plastids by vectorial processing.

During germination of pumpkin seeds in darkness, after a lag phase, the activity of RuBPCase in the cotyledons shows a steep increase, corresponding to the transformation of proplastids to much larger etioplasts. Detached cotyledons, which are deficient in endogenous CK and thus sensitive to exogenous CK, respond with organ enlargement and increase in the activities of various cytoplasmic and plastid enzymes, among them RuBPCase [16]. Similar increases of enzyme concentration and [^{14}C]-leucine incorporation into enzyme protein indicates de novo synthesis. It is suppressed by inhibitors of gene expression [16], but also by ABA [14]. Cell-free translation of total or poly-A$^+$ RNA from CK-treated cotyledons in a wheat germ system reveals changes in several polypeptides, among them increases of the precursors of RuBPCase SS and chloroplast light-harvesting protein, while a few non-identified polypeptides disappear ([16], and unpublished results). Only a few translatable RNA species change in abundance in response to CK action, in contrast to numerous alterations found in the population of total soluble proteins [2].

Translation in a cell-free *E. coli* system of chloroplast RNA (poly-A$^-$ RNA) from CK-treated cotyledons results in neither qualitative nor quantitative differences of the polypeptide pattern, if compared with water-treated samples. However, per cotyledon total plastid mRNA increases three-fold. This result is confirmed by quantitative dot hybridisation with cloned RuBPCase LS gene [16]. It is important to mention that illuminated cotyledons with or without CK treatment give similar results. The coaction of hormone and light in chloroplast formation [24] is additive and complementary for RuBPCase synthesis under varying light intensities [15].

We suggest therefore that cytokinins (and ABA) do not control plastogenesis in cotyledons via selective transcription of plastid genes. This may occur in CK-dependent cell cultures [3, 20]. The observed CK-dependent increase of the whole set of plastid RNA-translation products points to an involvement of plastid RNA polymerase. Like other selectively CK-stimulated plastid constituents, it consists of nuclear-encoded subunits. Thus, the hormonal control of cotyledon development is superimposed by CK-promoted plastid differentiation through activated nuclear genes that code for organelle polypeptides. While CK is the dominating regulator in etiolated tissues, in light the master control on gene expression rates is exerted by phytochrome [21]. We can only speculate whether and by which way the hormone action is connected with the signal transduction chain of the phytochrome system.

4 Conclusions

Each of the "classic" plant growth substances seems to be involved in one or more steps of gene expression. Here, I have reviewed the transcription process only, because many new data convincingly support the old idea that hormones induce RNA synthesis, as indicated by mRNA quantification with cell-free translation systems and "Northern" blot analysis of mRNA hybridised to cDNA probes. The examples cited in this review provide evidence that modulation of the transcription of specific genes belongs to a fundamental phytohormone action. Several criteria favour this view, especially in the cell enlargement response to auxins that site and mode of plant hormone action in gene expression resemble that of animal hormones: rapid response, specificity, and blockade by transcription inhibitors.

In spite of the progress made many important questions remain to be answered:

1. What is the molecular mechanism of phytohormone-triggered transcription? Amplification of the transcription rate or switching on genes by promoter selection? Activation of RNA polymerases, of DNA topoisomerases? Modification or stabilisation of mRNA (see also Fig. 2)?

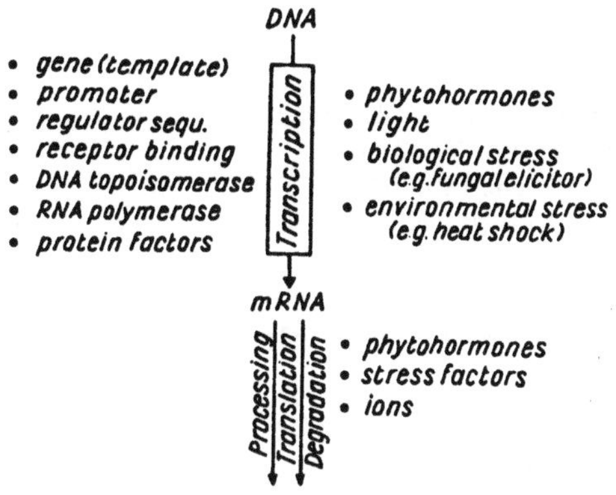

Fig. 2. Effectors eliciting an induction (stimulation) of gene expression (*right side*), and possible targets for the inductors (*left side*) in the transcription process

2. How can we discriminate, at the molecular level, between antagonistic effects of different hormones in the same cell, in seemingly the same step of gene expression?
3. Which mode of regulation secures that a single hormone in the same cell and at the same time induces certain gene products, but suppresses other ones?
4. What is the structure, the localisation, the amount, and the specificity of phytohormone receptors? Does binding of hormone-receptor complexes to regulatory DNA sequences play a pivotal role in gene expression modulation?
5. Are hormone-induced changes in gene expression causally correlated with the physiological response?

Finally, induction of gene expression is not a special activity or ability of hormones. Bioactive factors, such as light, heat shock, and other environmental and biological stressors can elicit mRNA synthesis, too (Fig. 2). With regard to this special viewpoint, the "hormone concept" has a more general meaning. In addition, plant growth substances can exert direct and specific effects beyond the level of gene expression. This point leads back once more to Van Overbeck, who asked [26]: "Why couldn't there be several sites of primary hormone action, just as there are several doors that can be opened by one key?" The future will show whether this last question can withstand for another 20 years, then in the challenging age of gene analysis.

References

1. Cato ACB, Geisse S, Wenz M, Westphal HM, Beato M (1984) EMBO J 3:2771
2. Chen CM, Leisner SM (1985) Plant Physiol (Bethesda) 77:99
3. Colijn CM, Sijmons P, Mol JNM, Kool AJ, Nijkamp HJJ (1982) Curr Genet 6:129
4. Compton JG, Schrader WT, O'Malley BW (1983) Proc Natl Acad Sci USA 80:16
5. Crouch ML, Tenbarge K, Simon A, Finkelstein R, Scofield S, Solberg L (1985) In: Van Vloten-Doting L, Groot GSP, Hall TC (ed) Molecular form and function of the plant genome. Plenum, New York, pp 555
6. Deikman J, Jones RL (1985) Plant Physiol (Bethesda) 78:192
7. Dure LS (1975) Annu Rev Plant Physiol 26:259
8. Dure LS, Galau GA (1981) Plant Physiol (Bethesda) 68:187
9. Higgins TJV, Zwar JA, Jacobsen JV (1976) Nature 260:166

10. Ho DTH, Varner JE (1974) Proc Natl Acad Sci USA 71:
11. Jacobsen JV (1977) Annu Rev Plant Physiol 28:537
12. Jacobsen JV, Beach LR (1985) Nature 316:275
13. Jost JP, Geiser M, Seldran M (1985) Proc Natl Acad Sci USA 82:988
14. Klyachko NL, Ananiev E, Kulaeva ON (1979) Physiol Veg 17:607
15. Lehmann J, Parthier B (1985) Biochem Physiol Pflanz 180:703
16. Lerbs S, Lerbs W, Klyachko NL, Romanko EG, Kulaeva ON, Wollgiehn R, Parthier B (1984) Planta (Berl) 162:289
17. Moll BA, Jones RL (1982) Plant Physiol (Bethesda) 70:1149
18. Muthukrishan S, Chandra GR, Albaugh GP (1983) Plant Mol Biol 2:249
19. Parthier B (1979) Biochem Physiol Pflanz 174:173
20. Seyer P, Lescure AM (1984) Plant Sci Lett 36:59
21. Sharma R (1985) Photochem Photobiol 41:747
22. Theologis A, Huynh TV, Davis RW (1985) J Mol Biol 183:53
23. Theologis A, Ray PM (1982) Proc Natl Acad Sci USA 79:418
24. Tong Z, Kasemir H, Mohr H (1983) Planta (Berl) 159:136
25. Trewavas AJ, Cleland RE (1983) Trends Biochem Sci 8:354
26. Van Overbeck J (1966) Science 153:721
27. Walker JC, Legocka J, Edelman L, Key JL (1985) Plant Physiol (Bethesda) 77:847
28. Wollgiehn R (1982) In: Parthier B, Boulter D (ed) Encyclopedia Plant Physiology 14B: Nucleic acids and proteins in plants II. Springer, Berlin Heidelberg New York, p 125
29. Zurfluh LL, Guilfoyle TJ (1982) Plant Physiol (Bethesda) 69:332

Enzymes of Auxin and Cytokinin Biosynthesis Encoded in Ti Plasmids*

J. SCHRÖDER, I. BUCHMANN, and G. SCHRÖDER[1]

1 Introduction

Infection of dicotyledonous plants and many gymnosperms with *Agrobacterium tumefaciens* leads to tumorous overgrowths, which are commonly called crown galls. When cells from such galls are transferred without contamination by Agrobacteria to simple media for plant tissues, they grow and divide without exogenous supply of auxins and cytokinins. This is in contrast to most normal plant tissues. Hormone-independent, unorganized proliferation in culture has long been the most characteristic property of crown gall cells.

Many experiments have shown that hormone independence is the result of gene expression from the T-DNA, that is, the part of Ti plasmids transferred from Agrobacteria to plant cells during the infection, and recent reviews provide excellent descriptions of the system [6, 19, 23]. Genetic experiments with Ti plasmids mutagenized in the T-DNA region indicate that three T-DNA genes are major determinants in hormone independence (see for example: [7, 9, 10, 13, 20]. The following is a brief summary of some important points emerging from such studies (cf. Fig. 1):

1. The three genes 1, 2, and 4, together in plant cells, induce hormone-independent, fast-growing tumors. Typically, they form unorganized tissues or teratomas in culture. No other T-DNA gene is required for hormone independence.
2. Mutations in genes 1 and/or 2 lead to slow-growing tumors, which typically sprout shoots in culture. They are cytokinin-independent, and supplementation with auxins restores the fast-growing wild-type tumor phenotype, and shoot formation is repressed.
3. Mutations in gene 4 induce slow-growing tumors. They often form roots in culture, and they are auxin-independent. Rooting is suppressed by cytokinin supplementation, and the fast-growing phenotype of wild-type tumors is restored.

It should be noted that this summary is a generalization from many results; if the data are considered in detail, it becomes obvious that the type of test tissue or test plant can affect the responses. Nevertheless, the overall picture clearly suggested that the three T-DNA genes together cause auxin- and cytokinin-independent, unorganized growth, that genes 1 and 2 together act "auxinlike", and that gene 4 acts "cytokininlike". This report summarizes recent advances in understanding the functions of these

* The following contributions (pp. 177–196) are part of the workshop *Tumor Formation*

1 Institut für Biologie II, Universität Freiburg, Schänzlestr. 1, 7800 Freiburg, FRG

Plant Growth Substances 1985
(ed. by M. Bopp)

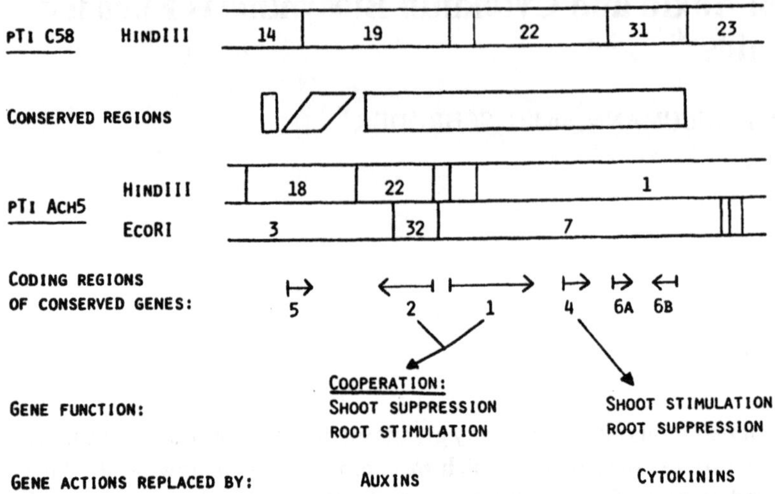

Fig. 1. "Hormone genes" in the T-regions of Ti plasmids and their functions. A part of the T-region from nopaline plasmid pTiC58 and the TL-region from octopine plasmid pTiAch5 are shown. *Lines with arrows* indicate the protein-coding regions of those genes which are conserved in both types of Ti plasmids. See text for explanation of the functions assigned to genes 1, 2, and 4

T-DNA genes, and it also discusses briefly genes for hormone biosynthesis, which are encoded in another part of the Ti plasmids.

2 Strategy for Analysis of T-DNA "Hormone Genes"

The genetic data could be explained by at least three models or their combinations:

a) T-DNA gene products activate plant genes for hormone biosynthesis.
b) T-DNA genes code for enzymes of plant growth regulator biosynthesis.
c) T-DNA gene products activate hormone responses without changing hormone concentrations.

All of these models are difficult to test with plant cells since the contribution of plant genes and proteins cannot be evaluated properly. Plant genes for hormone biosynthesis are not characterized on the DNA level; plant enzymes specifically involved in auxin and cytokinin biosynthesis are insufficiently characterized; and the molecular basis of hormone-controlled differentiation processes is poorly understood. In view of these problems, it is not surprising that most of the published reports concentrated on investigating model (b). It is also characteristic that the crucial experiments for analysis of the tumor-gene function were not performed with crown gall cells, but with the T-DNA genes expressed in *E. coli*. Basically, the strategy involved the following steps:

1. Expression of the T-DNA genes into protein in *E. coli*.
2. Tests whether extracts from such cells catalyze steps in auxin or cytokinin biosynthesis.

3. Return to crown gall cells after establishing the function of the T-DNA proteins in *E. coli*.

It should be noted that successful application of this approach requires that T-DNA genes are correctly expressed into protein in prokaryotic cells and that the proteins are functional in *E. coli*. Another, often not explicitly stated assumption in most experiments was that T-DNA-encoded proteins catalyze steps in biosynthesis of auxins and cytokinins, which are already known in plants or bacteria.

3 Auxin Genes of the T-DNA

Indole-3-acetic acid is considered to be the main auxin in plants. If T-DNA-encoded proteins are involved in its biosynthesis, it would be expected that at least one of them utilizes substrates, which are present in all plant cells, and that the enzyme either directly produces the auxin or initiates a pathway to indole-3-acetic acid which proceeds with the second T-DNA protein and/or with plant enzymes.

Figure 2 summarizes the most often discussed reactions in auxin biosynthesis (see [26]). Although all of the necessary enzyme activities have been described in extracts from plants or from bacteria, it has rarely been possible to show unambiguously that the enzyme activities measured in vitro are specifically involved in

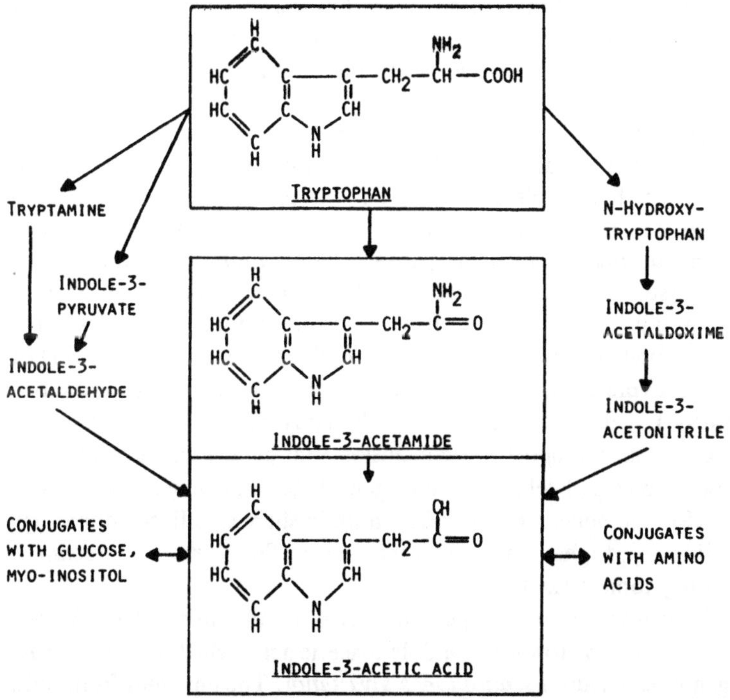

Fig. 2. Pathways to the auxin indole-3-acetic acid. The most often discussed reactions are depicted (see [26] for a detailed discussion)

indole-3-acetic acid biosynthesis in vivo. The one exception is probably auxin formation in the bacterium *Pseudomonas savastanoi*; in this case, it has been demonstrated convincingly that the auxin is synthesized from tryptophan through the intermediate indole-3-acetamide [8, 12]. The same pathway has been claimed for higher plants [22], but the conversion of tryptophan into indole-3-acetamide was later interpreted as artificial in vitro reaction of peroxidases.

T-DNA genes 1 and 2 are weakly expressed into protein in *E. coli* [24]. After recloning of the genes to higher expression of the proteins, it was investigated whether *E. coli* extracts, containing the proteins, catalyzed any of the reactions outlined in Fig. 2. Obviously, some of the substances are normal intermediates of tryptophan metabolism in *E. coli*, and especially tryptophan was quickly converted into several other components in the absence of T-DNA genes. In these cases, it was difficult to see differences between controls and extracts containing the T-DNA gene products. However, in one case the difference was striking: Extracts containing the protein encoded in gene 2 (mol. wt. 49000) very efficiently catalyzed the hydrolytic conversion of indole-3-acetamide into indole-3-acetic acid, while controls showed no such reaction. The reaction product was identified by HPLC chromatography and mass spectrometry [25], and the results were confirmed by independent experiments [27]. It was then also shown that this enzyme activity was present in crown gall cells and that it was absent in a habituated cell line, which was hormone-independent, but free of T-DNA [25]. Detailed studies on the substrate specificity of the T-DNA-encoded protein showed that indole-3-acetamide was not the only substrate; the enzyme also hydrolyzed phenylacetamide, naphthylacetamide, indole-3-acetonitrile, and esters of the auxin with glucose and myo-inositol [11]. These data suggest that the enzyme may have a general function in producing active auxins from less active precursors or storage forms.

The results indicated that one of the two T-DNA "auxin genes" codes for an enzyme of auxin biosynthesis. What then is the function of gene 1, the other auxin gene? Kinetic data suggest that indole-3-acetamide is the preferred substrate for the enzyme encoded in gene 2 in vitro [11]. The presence of this substance in normal plants is doubtful; and in analogy with the pathway in *Pseudomonas savastanoi*, it seemed reasonable to assume that gene 1 catalyzes the reaction from tryptophan to indole-3-acetamide. In this case, however, it proved difficult to demonstrate the enzyme activity after expression of the gene in *E. coli*. Despite expression cloning of the gene to high protein expression, the enzyme activity was very low (L. Thomashow, personal communication), and *E. coli* enzymes metabolizing tryptophan tend to mask the activity. Lines of evidence support the conclusion that gene 1 encodes a tryptophan monooxygenase: Crown gall cells containing gene 1, but not gene 2, accumulate indole-3-acetamide [28], and gene 1 DNA and its protein share significant homology with the corresponding tryptophan monooxygenase from *Pseudomonas savastanoi*, including a FAD-binding domain [29].

Taken together, the available results suggest that genes 1 and 2 of the T-DNA code for a two-step pathway from tryptophan to indole-3-acetic acid, which is also realized in the phytopathogenic bacterium *Pseudomonas savastanoi*. The enzymes from both sources are related, but they are not identical (T. Kosuge, personal communication), and it seems possible that they share common ancestor proteins. So far, the same pathway has not been convincingly shown in plant cells. It is an attractive, but unproven hypothesis, that *Agrobacteria* introduce via the T-DNA genes a "foreign" pathway, which is not controlled in plant cells.

4 Cytokinin Gene of the T-DNA

The genetic data predict that a single gene of the T-DNA is sufficient for cytokinin effects in crown gall cells. If the function of the protein is in hormone biosynthesis, one would expect that it utilizes substrates available in all plant cells, and that the enzyme activity either directly produces a cytokinin from inactive precursors or initiates a pathway to cytokinins, which proceeds with plant enzymes. Figure 3 summarizes in a very simplified scheme the reactions mostly discussed in plants [14].

Gene 4 is easily expressed into protein in *E. coli* [24]. Several independent investigations showed that extracts, containing the protein, synthesize N6-(Δ2-isopentenyl)-5'AMP from free 5'AMP and Δ2-isopentenylpyrophosphate, while control extracts possess negligible or no enzyme activity. This was demonstrated with the T-DNA gene from an octopine Ti plasmid [1], a nopaline Ti plasmid [3], and with the genes from both sources [4]. Most of the published experiments were performed after gene ex-

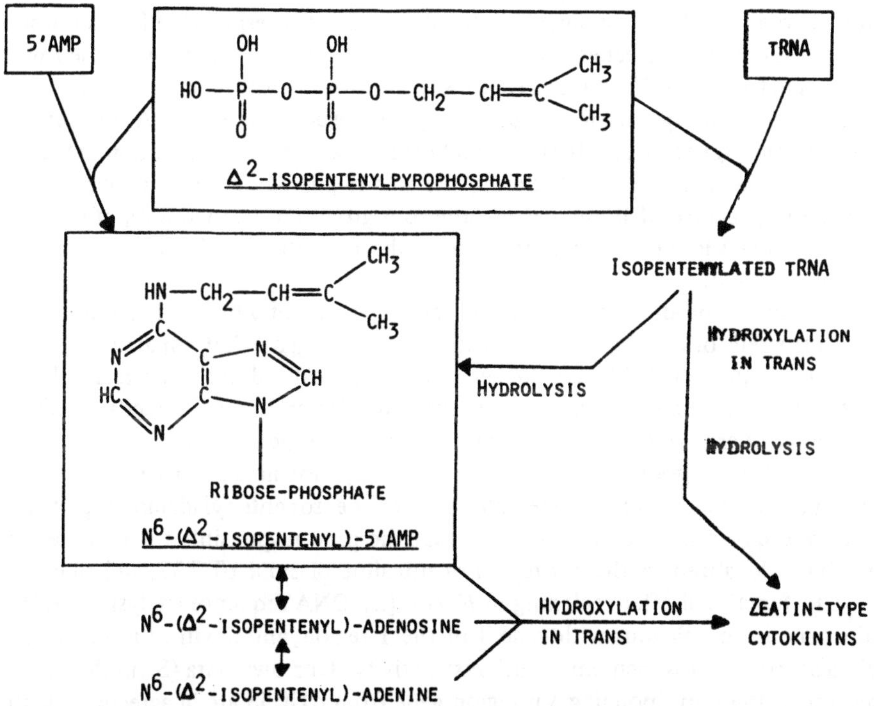

Fig. 3. Biosynthesis of cytokinins. The most often discussed pathways are depicted in a simplified scheme: Synthesis by isopentenylation of free 5'AMP *(left)*; synthesis by isopentenylation of adenine residues present in tRNAs *(right)*, followed by hydrolytic liberation of cytokinin-active molecules from the RNA (see [14] for a review)

pression in *E. coli*, and identification of the product was by HPLC [1, 3], and additionally by mass spectrometry [4]. The protein encoded in the cytokinin gene was also demonstrated in extracts from crown gall cells with the help of an antiserum prepared against a decapeptide derived from the DNA sequence of the gene. The fraction of the extracts containing the immunoreactive protein (mol. wt. 27000) also contained isopentenyltransferase activity, and it was possible to show that this activity was specifically inhibited by the antiserum [4].

These results indicate that gene 4 from the T-DNA codes for a prenyltransferase active in cytokinin biosynthesis in transformed plant cells. Such enzyme activities have been described before in crown gall cells [17], in cytokinin-independent plant cells [5], and also in the phytopathogenic bacterium *Corynebacterium fascians* [18]. Since the genes for these enzymes have not been isolated, a comparison on the DNA and the protein sequence level are not yet possible. It is likely, however, that the activity in crown gall cells represented expression of the T-DNA gene.

5 Hormone Genes in the Vir-Region of Ti Plasmids

T-region and Vir-region (vir from *vir*ulence) are physically separated parts of Ti plasmids (see for review: [6, 19, 23]), and they have different functions. The T-region is transferred to and stably maintained in the plant cells; it is responsible for tumorous growth, but it contains no genes which are necessary for its own transfer. These functions are encoded in the Vir-region. Established crown gall cell lines do not contain sequences of the Vir-region. Presumably, the genes responsible for DNA transfer are expressed in the *Agrobacteria*. It is not excluded, however, that Vir-region DNA sequences are present in plant cells during the initial phases of tumor induction. In view of the importance of hormones in tumorous growth, it is, therefore, of interest that genes of the Vir-region also play a role in the formation of the plant hormones auxin and cytokinin.

Agrobacteria do produce indole-3-acetic acid regardless of whether Ti plasmids are present in the cells or not. In at least one case the data suggest that nopaline plasmids affect auxin production [15], and the gene has been localized in the Vir-region [16]. So far there is no direct evidence that the gene codes for an enzyme directly involved in auxin biosynthesis, and this question requires further experiments.

Agrobacteria also produce cytokinins. Interestingly, nopaline strains excrete trans-zeatin, while octopine strains excrete cytokinins of the isopentenyladenine type [21]. Akiyoshi and co-workers recently showed that the gene responsible for trans-zeatin production is localized in the Vir-region of nopaline plasmid pTiT37, and that the gene is also functional after recloning in *E. coli* [2]. DNA sequence analysis revealed that the gene is homologous with gene 4 of the T-region, and in vitro the Vir-region protein also possesses isopentenyltransferase activity. Our own data (Schröder et al., in prep.) with the corresponding Vir-region gene from pTiC58 are in agreement with these results. In addition, we have made a detailed comparison of the two isopentenyltransferases, and the results suggest that the enzymes have very similar properties under most assay conditions. In view of these in vitro data, it appears somewhat puzzling that the Vir-region gene leads to trans-zeatin production in vivo, and the mechanism of the hydroxylation step remains to be clarified.

The functional importance of Vir-region-encoded hormone biosynthesis is not quite clear. The genes may be classified as "virulence genes" if increased hormone production by *Agrobacteria* during the infection participates in conditioning the plant cells for gene transfer [24].

6 Conclusion

Ti plasmids of *Agrobacterium tumefaciens* contain at least four genes coding for enzymes of hormone biosynthesis. Three of the genes are localized in the T-DNA region, which is transferred from the *Agrobacteria* into plant cells. Two genes encode enzymes for a two-step pathway from tryptophan through indole-3-acetamide to indole-3-acetic acid; the third gene encodes an isopentenyltransferase active in cytokinin biosynthesis. These three genes are responsible for hormone-independent growth of crown gall cells. The fourth Ti plasmid hormone gene is localized in the Vir-region, the part of the Ti plasmids responsible for transfer of the T-DNA into plant cells. The protein has isopentenyltransferase activity in vitro. In vivo it leads to accumulation of trans-zeatin in the medium. Vir-region-encoded hormone production by *Agrobacteria* may be a factor in the virulence of these bacteria.

Acknowledgments. Supported by Deutsche Forschungsgemeinschaft (SFB 206).

References

1. Akiyoshi DE, Klee H, Amasino RM, Nester EW, Gordon MP (1984) Proc Natl Acad Sci USA 81:5994
2. Akiyoshi DE, Regier DA, Gordon MP (1985) Nucleic Acids Res 13:2773
3. Barry GF, Rogers SG, Fraley RT, Brand L (1984) Proc Natl Acad Sci USA 81:4776
4. Buchmann I, Marner FJ, Schröder G, Waffenschmidt S, Schröder J (1985) EMBO J 4:853
5. Chen CM In: Wareing PF (ed) Plant growth substances 1982. Academic Press, New York, p 155
6. Hille J, Hoekema A, Hooykaas P, Schilperoort RA In: Verma DPS, Hohn T (eds) Plant gene research: genes involved in microbe-plant interactions. Springer, Berlin Heidelberg New York, p 287
7. Hoekema A, Pater BS, Fellinger AJ, Hooykaas P, Schilperoort RA (1984) EMBO J 3:3043
8. Hutcheson SW, Kosuge T (1985) J Biol Chem 260:6281
9. Inzé D, Follin A, Van Lijsebettens M, Simoens C, Genetello C, Van Montagu M, Schell J (1984) Mol Gen Genet 194:265
10. Joos H, Inzé D, Caplan A, Sormann M, Van Montagu M, Schell J (1983) Cell 32:1057
11. Kemper E, Waffenschmidt S, Weiler EW, Rausch T, Schröder J (1985) Planta (Berl) 163:257
12. Kosuge T, Heskett MG, Wilson EE (1966) J Biol Chem 241:3738
13. Leemans J, Deblaere R, Willmitzer L, De Greve H, Hernalsteens JP, Van Montagu M, Schell J (1982) EMBO J 1:147
14. Letham DS, Palni LMS (1983) Annu Rev Plant Physiol 34:163
15. Liu ST, Kado CI (1979) Biochem Biophys Res Commun 90:171
16. Liu ST, Perry KL, Schardl CL, Kado CI (1982) Proc Natl Acad Sci USA 79:2812

17. Morris RO, Akiyoshi DE, MacDonald EMS, Morris JW, Regier DA, Zaerr JB (1982) In: Wareing PF (ed) Plant growth substances 1982. Academic Press, New York, p 175
18. Murai N (1981) In: Guern J, Péaud-Lenoel C (eds) Metabolism and molecular activities of cytokinins. Springer, Berlin Heidelberg New York, p 17
19. Nester EW, Gordon MP, Amasino RM, Yanofsky MF (1984) Annu Rev Plant Physiol 35:387
20. Ream LW, Gordon MP, Nester EW (1983) Proc Natl Acad Sci USA 80:1660
21. Regier DA, Morris RO (1982) Biochem Biophys Res Commun 104:1560
22. Riddle VM, Mazelis M (1965) Plant Physiol (Bethesda) 40:481
23. Schell J, Van Montagu M, Holsters M, Zambryski P, Joos H, Herrera-Estrella L, Depicker A, Hernalsteens JP, De Greve H, Willmitzer L, Schröder J (1984) In: Kumar A (ed) Eucaryotic gene expression. Plenum, New York, p 141
24. Schröder G, Klipp W, Hillebrand A, Ehring R, Koncz C, Schröder J (1983) EMBO J 2:403
25. Schröder G, Waffenschmidt S, Weiler EW, Schröder J (1984) Eur J Biochem 138:387
26. Sembdner G, Gross D, Liebisch HW, Schneider G (1980) In: MacMillan J (ed) Hormonal regulation of development, vol 9. Springer, Berlin Heidelberg New York, p 281 (Encyclopedia of plant physiology)
27. Thomashow LS, Reeves S, Thomashow MF (1984) Proc Natl Acad Sci USA 81:5071
28. Van Onckelen H, Rüdelsheim P, Inzé D, Follin A, Messens E, Horemans S, Schell J (1985) FEBS Lett 181:373
29. Yamada T, Palm CJ, Kosuge T (1985) In: Szalay AA, Legocki RP (eds) Advances in molecular genetics of the bacteria-plant interaction. Cornell University Publishers, Ithaca, p 180

Cytokinin Biosynthetic Genes and Enzymes from *Agrobacterium tumefaciens* and other Plant-Associated Prokaryotes

R.O. MORRIS, G.K. POWELL, J.S. BEATY, R.C. DURLEY, N.G. HOMMES, L. LICA, and E.M.S. MACDONALD[1]

1 Introduction

Phytohormone involvement in the action of gall-forming phytopathogenic bacteria such as *Agrobacterium tumefaciens* or *Pseudomonas savastanoi* has received much attention recently. Genes, which specify production of auxins and cytokinins, are present in both organisms [23] and, in the case of *A. tumefaciens*, a set of these genes is trans-

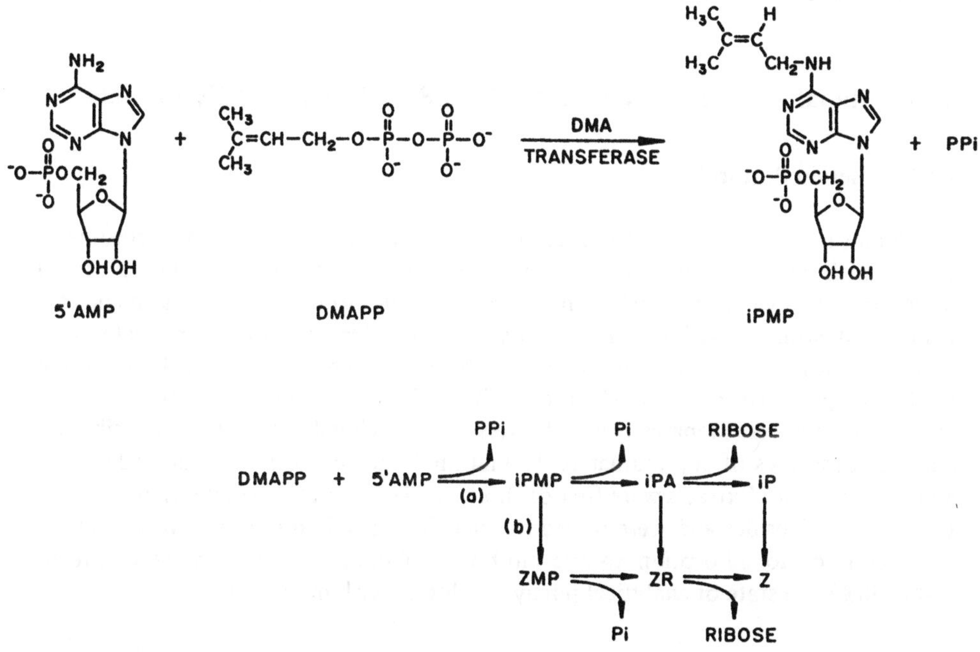

Fig. 1. Putative cytokinin biosynthetic pathway: (a) DMA transferase; (b) Oxidase. *Abbreviations:* DMAPP = dimethylallylpyrophosphate; iPMP = *iso*-pentenyladenosine-5′-phosphate; iPA = *iso*-pentenyladenosine; iP = *iso*-pentenyladenine; ZMP = ribosylzeatin-5′-phosphate; ZR = ribosylzeatin; Z = zeatin

1 Department of Agricultural Chemistry, Oregon State University, Corvallis, OR 97331, USA

Plant Growth Substances 1985
(ed. by M. Bopp)

ferred to the plant [8] where they determine tumor morphology and endogenous hormone content [1, 12]. Knots incited by *P. savastanoi* differ from galls incited by *A. tumefaciens* in that there is apparently no transfer of bacterial DNA to the host plant (Kosuge T., personal communication). Nevertheless, *P. savastanoi* secretes high levels of indole-3-acetic acid [9] and has been reported to secrete cytokinins [26].

Chen [5] and Letham et al. [17] have suggested that cytokinins are synthesized by the pathway illustrated in Fig. 1. Key enzymes in this pathway are (1) DMA transferase (Dimethylallylpyrophosphate: 5'AMP dimethylallyl transferase) [6] which attaches the dimethylallyl side chain to the exocyclic amino group of 5'AMP to give *iso*-pentenyl adenosine-5'-phosphate (iPMP); and (2) an oxidase [7] which hydroxylates the 4"-methyl group of the isoprenoid side chain to give the *trans*-zeatin series of cytokinins (zeatin, ribosylzeatin, and ribosylzeatin-5'-phosphate).

In this chapter, we describe the application of immunoaffinity chromatography, HPLC-RIA, and molecular cloning to characterize two genetic loci in *A. tumefaciens*, which control cytokinin biosynthesis. The *tmr* locus is part of the Ti plasmid T-region; it encodes a DMA transferase. It has substantial sequence homology with a second locus, *tzs*, which is present only on nopaline Ti plasmids. *Tzs* also encodes a DMA transferase and can be expressed in *E. coli* with production of zeatin. Finally, the secretion of cytokinins by *P. savastanoi* is discussed. Plasmid-borne cytokinin biosynthetic genes are present and have been cloned and expressed in *E. coli*.

2 Immunoaffinity Extraction and HPLC-RIA Analysis of Cytokinins

2.1 Extraction Protocol

Cytokinins in plant or bacterial extracts may be purified by immunoaffinity chromatography on immobilized anti-cytokinin antibodies using the rapid extraction protocol of MacDonald and Morris [19]. Monoclonal antibodies to both ribosylzeatin and *iso*-pentenyladenosine have been prepared by immunization with the appropriate cytokinins conjugated to bovine serum albumin [28]. When immobilized to cellulose, these antibodies provided a matrix which effectively absorbed cytokinins from crude extracts. Columns were prepared from mixtures of monoclonal anti-cytokinin antibodies and had capacities of approximately 100 ng ml^{-1} matrix for *iso*-pentenyladenosine and ribosylzeatin. Extracts were first dephosphorylated to convert cytokinin 5'-phosphates to nucleosides and were passed through DEAE-cellulose to remove interfering material prior to adsorption to the antibody column. Cytokinins were recovered (60%-80%) in a state of analytical purtiy by elution with methanol.

2.2 Analysis by HPLC-RIA

Analysis of immunoaffinity-purified extracts by HPLC-RIA [20] allowed identification and quantitation of most cytokinins. For example, analysis of the cytokinins in wheat endosperm gave an HPLC trace, which exhibited almost no UV-absorbing peaks other than those of the cytokinins themselves. Estimation of the amounts of zeatin and ribosylzeatin either by peak integration or by RIA gave values in good agreement, indicating that peaks were > 90% homogeneous. Cytokinin identities and levels were confirmed by mass spectrometry [10]. Most cytokinins can be assayed by this procedure, but the 4″-O-glucosides do not interact with available antibodies and were not detected. They may be measured after treatment with β-glucosidase.

3 Cytokinin Biosynthetic Genes in Agrobacterium tumefaciens

Early studies indicated that crown gall tumors contain elevated cytokinin levels [11, 21]. Phytohormone levels were found to be influenced specifically by the presence of the *tms* and *tmr* loci from *A. tumefaciens* [1]. Preliminary data [22], since confirmed by Akiyoshi et al. [2], suggested that DMA transferase activity was high in crown gall tumors, but only when a functional *tmr* locus was present. *Tmr* was, therefore, a logical candidate for the structural gene for DMA transferase. In order to determine whether this was true, a DMA transferase assay was developed, and applied to cloned *tmr* in *E. coli*.

3.1 DMA Transferase Assay

DMA transferase may be assayed rapidly and easily by immunological means [13, 22]. Dimethylallylpyrophosphate and [^{3}H]-5′-AMP were incubated with the enzyme; the product, *iso*-pentenyladenosine-5′-phosphate (iPMP), was treated with alkaline phosphatase and the resulting *iso*-pentenyladenosine was collected on small columns of immobilized anti-iPA antibody. After elution, the amount of [^{3}H]-iPA was estimated by HPLC. Figure 2 illustrates a typical assay.

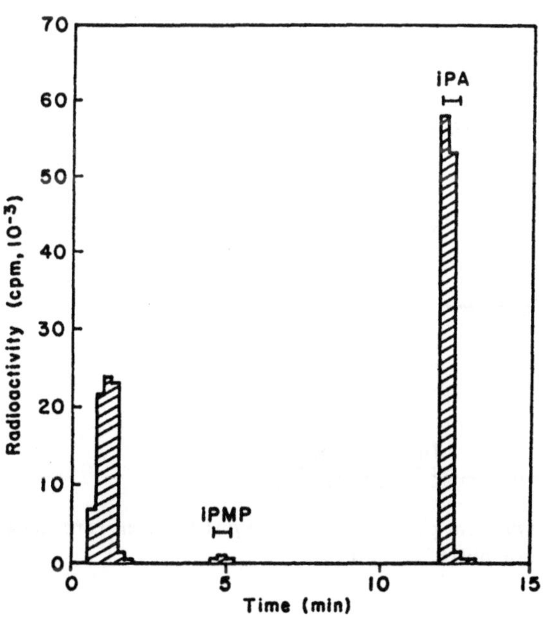

Fig. 2. DMA transferase assay. Enzyme was prepared from 2 gm tobacco crown gall tissue (line 1590): Incubation: 37 °C/2 h. Product collected on octadecyl silica, eluted, treated with alkaline phosphatase and subjected to immunoaffinity chromatography. Cytokinin standard retention times are indicated by ⊢⊣

3.2 Cloning tmr from A. tumefaciens

Sequence studies of pTiI5955 T-DNA [3] indicated the presence of a 0.7 kb open reading frame in the *tmr* locus. A restriction fragment encompassing the equivalent coding region from the closely related plasmid pTiAch5 was ligated into the *E. coli* plasmid pUC18 as indicated in Fig. 3. Transformants (pIP182 and pIP192), bearing *tmr*, expressed DMA transferase activity, whereas controls [HB101 or HB101 (pUC18)] did not (Table 1). Similar results have been obtained by Barry et al. [4]. The *tmr* locus specifies DMA transferase and it is likely that cytokinin production in crown gall tumors is a reflection of the expression of the bacterial gene.

Table 1. DMA transferase in *E. coli* bearing cloned *tmr*

Strain	DMA transferase[a]
HB101	102
HB101 (pIP182)	2954
HB101 (pIP192)	10169

[a] Activity: cpm (iPA + iP) synthesized/g cell extract/h

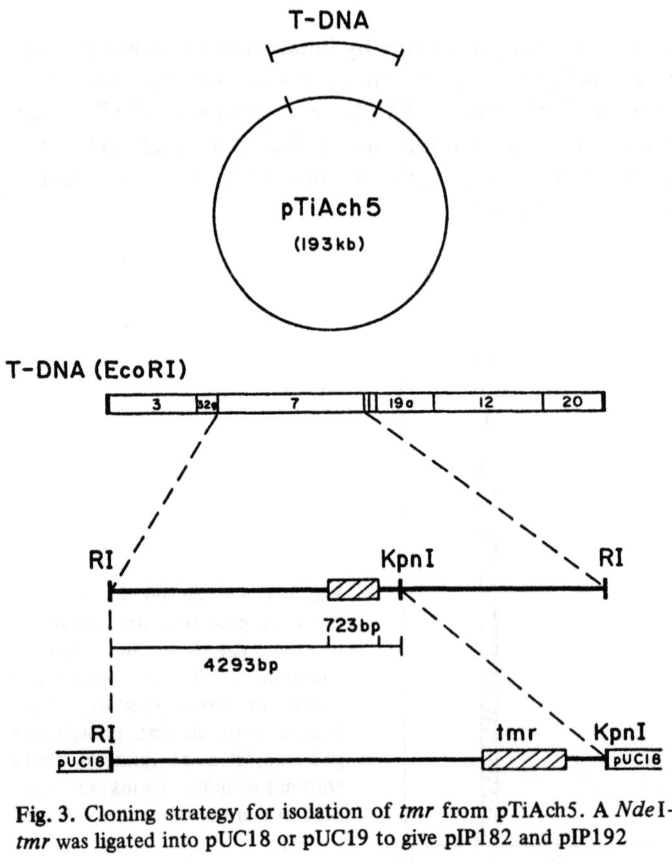

Fig. 3. Cloning strategy for isolation of *tmr* from pTiAch5. A *Nde*I–*Kpn*I fragment containing *tmr* was ligated into pUC18 or pUC19 to give pIP182 and pIP192

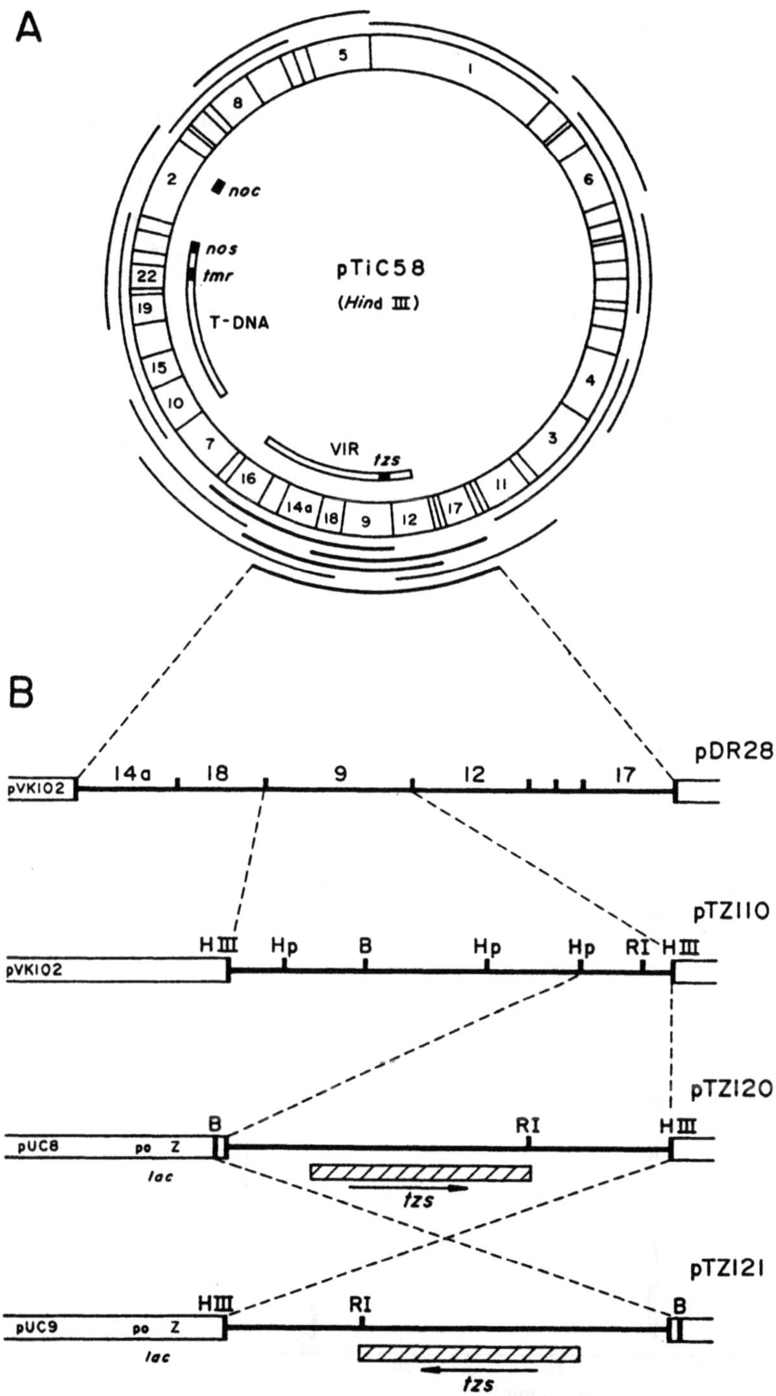

Fig. 4A,B. Cloning strategy for isolation of *tzs* from pTiC58. **A** Production of zeatin by pTiC58 library clones (——zeatin production; ——no production). Map data for the T-DNA and *vir*-regions are from Lemmers et al. [16] and Lundquist et al. [18]. **B** Construction of pTZ120 and pTZ121

3.3 Cloning tzs from A. tumefaciens

Some *A. tumefaciens* strains carry a second gene, *tzs*, which causes secretion of zeatin into the culture medium [14]. It is present only on nopaline Ti plasmids [24]. In order to map *tzs* to the Ti plasmid and to characterize it, a library of restriction fragments of the nopaline plasmid pTiC58 was prepared in the cosmid vector pVK102 [15] and transferred to cured *A. tumefaciens* strain NT1 or to *E. coli* HB101 (Fig. 4). Assay of cytokinin production by representative transformants, containing fragments spanning the pTiC58 plasmid, indicated that only those which contained *Hin*d III-9 were capable of producing cytokinins. Subcloning a 1.4 kb fragment of *Hin*d III-9 into pUC18 or pUC19 (Fig. 4b) gave two recombinants, pTZ120 and pTZ121, which both conferred upon *E. coli* the ability to secrete zeatin (Fig. 5; Table 2). Apparently, *E. coli* RNA polymerase can recognize the native *A. tumefaciens* promoter.

Tzs maps outside the T-region (Fig. 4A). Since there is no evidence to suggest that this portion of the Ti plasmid is integrated permanently into the host plant, the role

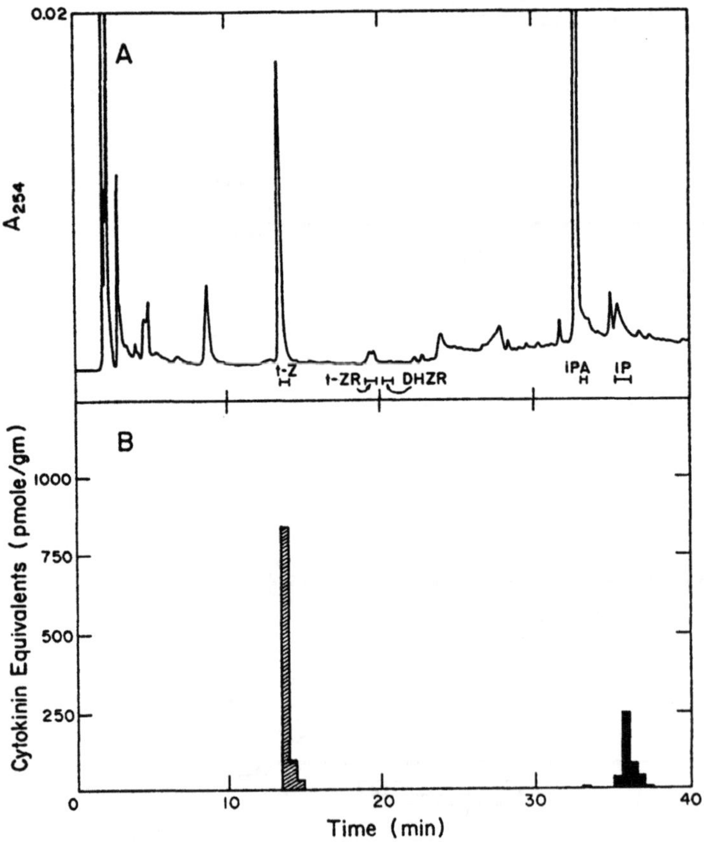

Fig. 5A,B. Secretion of zeatin by HB101 (pTZ121). Analysis of mid-log phase culture filtrate after immunoaffinity chromatography. **A** Absorbance at 254 nm. **B** RIA with anti-tZR antibody ▨ and anti-iPA antibody ■

Table 2. Cytokinin production and DMA transferase in *E. coli* bearing *tmr* or *tzs*

Strain	Insert	Zeatin[a]	iP[a]	DMA transferase[b]
HB101	None	11	120	180
HB101 (pIP192)	*tmr*	35	197	6300
HB101 (pTZ121)	*tzs*	19900	8700	13000
HB101 (pCK101)	*tmr* and *tzs*	110000	14900	3400

[a] ng l^{-1} mid-log phase
[b] Activity: cpm (iPA + iP) synthesized/g cell extract/h

of *tzs* in tumorigenesis remains to be established. It should, however, be noted that *tzs* falls very close to, or perhaps within, the virulence region as determined by Lundquist et al. [18]. *Tzs* may play a role in the initial stages of tumor formation.

3.4 Sequence Homology Between tmr and tzs

The nucleotide sequence of *tzs* was determined using the dideoxy technique [25]. The 1.4 kb fragment contained an open reading frame of 726 nucleotides, which is illustrated in Fig. 6. The peptide corresponding to this sequence should have a mol. wt. of 27598 daltons. Maxicell experiments (not shown) indicated that a peptide of 26800 daltons was produced.

The nucleotide sequences of the coding regions of *tmr* and *tzs* were very similar (Fig. 6). Homology was greater than 54% overall with regions of substantially higher homology scattered throughout. Identities were greatest for the first and second bases of each triplet and as expected, amino acid sequence homology was even greater than nucleotide sequence homology. Comparison of the amino acid sequences indicated that homology was 70% at the N-terminus and decreased towards the C-terminus. Because there was such a high degree of structural similarity, it seemed likely that there might also be functional similarity. In fact, *tzs* does encode a DMA transferase (Table 2). This result was consistent with the observation that expression of *tzs* in *E. coli* produced not only zeatin, but also *iso*-pentenyladenine (Fig. 5B). It appears, therefore, that *tzs* encodes a DMA transferase, but that its expression in *E. coli* results in the production of zeatin. The possibility was considered that it might also possess hydroxylase activity.

3.5 Expression of tmr and tzs

Because zeatin production by cloned strains containing *tzs* was low (19 ng ml^{-1}), it seemed likely that DMA transferase activity was limiting the amount of *iso*-pentenyladenosine-5′-phosphate, *iso*-pentenyladenosine, or *iso*-pentenyladenine available for hydroxylation. In order to determine whether this was the case, the *tmr* and *tzs* genes were assembled on a single replicon, pCK101, with both genes in the correct orientation for maximum transcription. Zeatin secretion by strains bearing pCK101 was substantially greater than that by strains bearing pTZ121 or pIP192 (Table 2).

```
dma
ATGGACCTGCATCTAATTTTCGGTCCAACTTGCACAGGAAAGACGACGACCGCGATAGCTCTTGCCCAGCAGACAGGGCTTCCAGTCCTTTCGCTTGATCGGGTCCAATGCTGTCCTCAACT
:::   :: ::::: :: : ::: :: :::::::  :: :: :::   :  :::::     : :: ::  : :: :::    :: ::  :: : :::::::: :: :::::::::::::: :
ATGTTACTCCATCTCATCTACGGACCGACTTGCAGCGGCAAAACGGACATGGCGATCCAAATCGCACAAGAAACCGGGTGGCCGGTGGTTGCCCTTGATCGTGTGCAATGCTGTCCTCAAAT
tzs

ATCAACCGGAAGCGGACGACCAACAGTGGAAGAACTGAAAGGAACGACGCGTCTCTACCTTGATGATCGGCCTCTGGTGGAGGGTATCATCGCAGCCAAGCAAGCTCATCATAGGCTGATCG
 : :: :: :::::: ::::     :      ::: :: ::  ::::  : :  : ::  : :::   :: :: ::    ::::: ::: : :  ::  ::   :: :::: : : :: ::
CGCGACAGGTAGCGGAAGACCTTTGGAATCGGAATTGCAATCAACGCGGAGAATATATTTGGATTCCCGCCCCCTCACCGAGGGCATCCTTGACGCTGAGAGTGCCCATCGTCGACTCATAT

AGGAGGTGTATAATCATGAGGCCAACGGCGGGCTTATTCTTGAGGGAGGATCCACCTCGTTGCTCAACTGCATGGCGCGAAACAGCTATTGGAGTGCAGATTTTCGTTGGCATATTATTCGC
 :: ::: ::   :   :: :: : :  :: :::::::: ::::: :: :: :  ::: : ::::: ::::::::   ::      : ::::::  : : :::::  :::::: : :  ::
TCGAAGTGGATTGGCGGAAGTCCGAAGAGGGTCTTATTCTCGAGGGCGGGTCGATTTCGCTTCTCAATTGCATGGCTAAAAGTCCGTTTTGGAGATCGGGTTTTCAATGGCATGTCAAGCGG

CACAAGTTACCCGACCAAGAGACCTTCATGAAAGCGGCCAAGGCCAGAGTTAAGCAGATGTTGCACCCCGCTGCAGGCCATTCTATTATTCAA---GAGTTGGTTTATCTTTGGAATGAACC
:     :    ::    :: :::: : :      ::::::     : :::  : : :::::   :   :  : :: :   :      :          :::::::   : :: :::::  : ::
CTACGTCTTGGGGATTCGGACGCCTTTCTCACCCGAGCCAAGCAACGCGTTGCGGAAATGTTTGCCATCCGGGAAGATCGCCCCTCGTTGTTGGAGGAGTTGGCGGAACTCTGGAACTACCC

TCGGCTGAGGCCCATTCTGAAAGAGATCGATGGATATCGATATGCCATGTTGTTTGCTAGCCAGAACCAGATCACGGCAGATATGCTATTGCAGCTTGACGCAAATATGGAAGGTAAGTTGA
:         : :: ::: :: :::: ::::: :::::::: :  :: ::    ::::: :: :  :: : :: :           : :     : :::: :::      : : :   :: : :
TGCCGCTCGACCGATTTTGGAAGATATCGACGGATATCGCTGCGCAATTCGTTTTGCGCGCAAACACGATCTCGCAATCAGCCAGTTGCCAAATATTGATGCAGGGCGGCACGTAGAGCTCA

TTAATGGGATCGCTCAGGAGTATTTCATCCATGCGCGCCAACAGGAACAGAAATTCCCCCAAGTTAACGCAGCCGCTTTCGACGGATTCGAAGGTCATCCGTTCGGAATGTATTAG
:  : :  :: ::: : :: ::: :     ::::::: :   ::::: :    : :: :: ::        : ::  :     :          :   : :             : :
TAGAGGCCATAGCTAATGAATATCTTGAACATGCGCTCTCGCAGGAGCGCGATTTTCCTCAGTGGCCAGAAGATGGCGCAGGACAGCCTGTTTGCCCGGTCACGCTGACGCGAATTCGG
```

Fig. 6. Homology between *tmr* and *tzs*. Homology was determined using the NUCALN program of Wilbur and Lipmann [29]. Sequence of *tmr* (= dma) coding region from Barker et al. [3]

4 Cytokinin Biosynthetic Genes in Pseudomonas savastanoi

4.1 Cytokinin Production by P. savastanoi and Deletion Mutants

Preliminary RIA of unfractionated culture filtrates showed that *P. savastanoi* produced substantially more zeatinlike cytokinins than did *A. tumefaciens* C58. The olive-specific strain EW1006 produced extremely high levels. In order to identify the

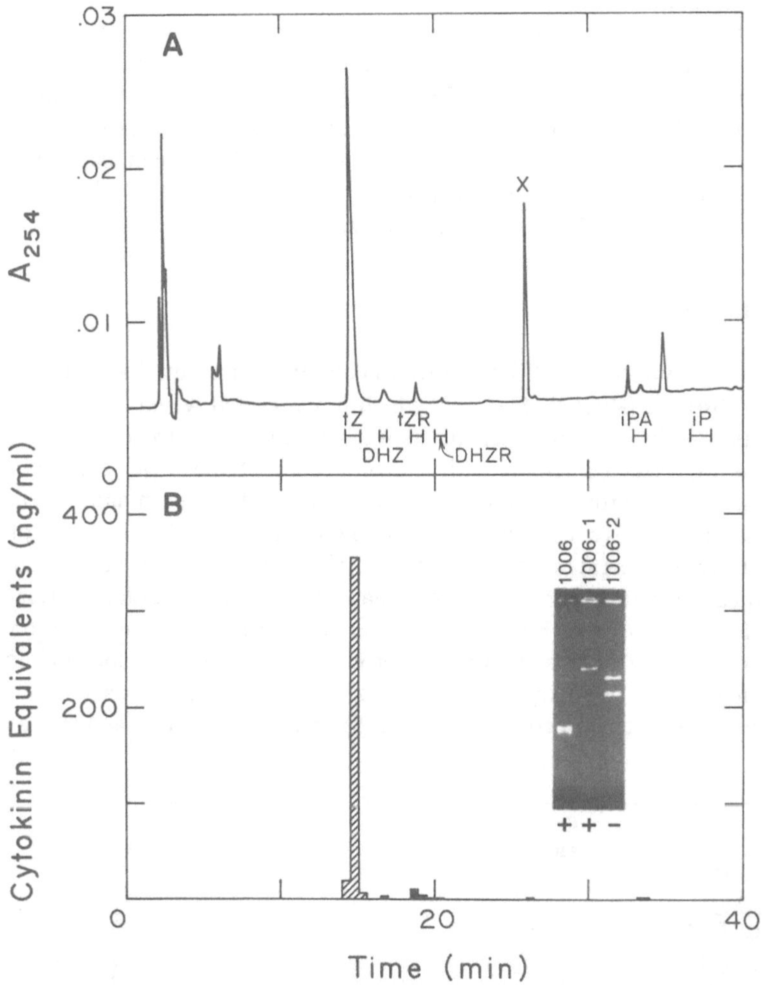

Fig. 7A,B. Cytokinins secreted by *P. savastanoi* strain EW1006. One ml of supernatant from a mid-log phase culture of EW1006 was purified by immunoaffinity chromatography and fractionated by HPLC-RIA. **A** Absorbance at 254 nm; peak X: ribosyl-1″-methylzeatin. **B** RIA with anti-tZR antibody ▨ and anti-iPA antibody ■. *Inset*: Plasmids present in strain EW 1006 and deletion mutants.

1006 Plasmids: 105 kb and 81 kb

1006-1 Plasmids: 105 kb

1006-2 Plasmids: 81 kb and 65 kb (deletion of 105 kb)

(+) Produces zeatin; (–) no detectable zeatin produced

cytokinins, a culture filtrate from EW1006 was analyzed by immunoaffinity chromatography, HPLC, and RIA. The results are shown in Fig. 7.

Four UV-absorbing peaks cross-reacted with anti-ZR antibody and one with anti-iPA antibody. Those cross-reacting with anti-tZR antibody had retention times corresponding to zeatin (14.5 min), dihydrozeatin (16.5 min), ribosylzeatin (18.8 min), and ribosyldihydrozeatin (20.6 min). Small, but significant, amounts of *iso*-pentenyladenosine were detected at 33.3 min. Mass spectrometry of the permethylated derivatives confirmed the identities in all cases.

In order to determine whether cytokinin biosynthesis was plasmid-encoded, spontaneous deletion mutants of EW1006 were isolated and assayed for their ability to produce cytokinins. As illustrated in Fig. 7B (inset), zeatin secretion was plasmid-encoded. Wild-type EW1006 secreted high levels of zeatins, as did a deletion mutant (EW1006-1) from which the smaller of the two plasmids (81 kb) had been lost. However, EW1006-2, which retained the 81 kb plasmid, but from which 40 kb of the 105 kb plasmid had been deleted, no longer secreted zeatin.

4.2 Ribosyl-1″-Methylzeatin

In addition to known cytokinins, EW1006 filtrates contained substantial amounts of a UV-absorbing substance (Peak X, Fig. 7A), which cross-reacted weakly with anti-tZR antibody. Its polarity was intermediate between that of ribosylzeatin and *iso*-pentenyladenosine and was consistent with the properties of the novel cytokinin ribosyl-1″-methylzeatin recently identified by Surico et al. [27]. Permethylation gave a derivative whose spectrum is shown in Fig. 8. An obvious molecular ion was present at m/z=435 and fragment ions at m/z=404, 390, 362, 286, 260, 246, 230, and 188. It was almost identical to that of permethylribosylzeatin except that homologous fragment ions were increased in mass by 14 amu indicative of the presence of an extra methyl group. The position of this extra methyl group was assigned by inspection of the fragment ion pattern. The tricyclic ion at m/z=174 resulting from side-chain cleavage (with loss of side chain carbons 3, 4, and 5) and loss of ribose; and ions at

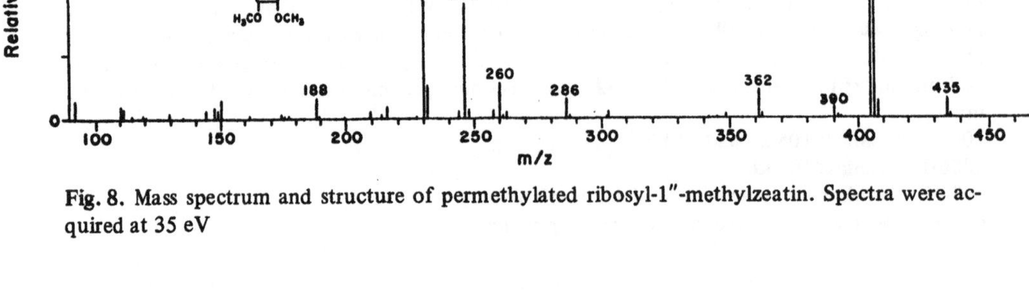

Fig. 8. Mass spectrum and structure of permethylated ribosyl-1″-methylzeatin. Spectra were acquired at 35 eV

m/z=148 and 119 in the mass spectrum of underivatized peak X, were consistent only with the methyl group being on the 1″-position of the side chain. Peak X is, therefore, ribosyl-1″-methylzeatin as proposed by Surico et al. [27].

4.3 Cloning and Expression of Cytokinin Biosynthetic Genes from P. savastanoi

Plasmid DNA from EW1006 was cloned into pRK290 and a functional recombinant, pPS001, containing a 13.5 kb fragment from the 105 kb plasmid, was isolated. The size of the insert was reduced to 4.5 kb by further restriction digestion to give pPS003, which was assayed for cytokinin production. Analysis by HPLC gave peaks at 15.1 min and 20.8 min, which corresponded in retention times to zeatin (16 ng ml^{-1}) and ribosylzeatin (5 ng ml^{-1}), respectively. Permethylation of these fractions gave derivatives whose mass spectra were identical to those of authentic permethylzeatin and permethylribosylzeatin. No evidence was found to suggest that ribosyl-1″-methylzeatin was produced.

5 Conclusion

Agrobacterium tumefaciens contains two plasmid-encoded genes which specify cytokinin biosynthesis. The T-DNA gene, *tmr*, is expressed only when transferred to the host plant and is responsible for elevated tumor cytokinin content. The other gene, *tzs*, is expressed primarily in the bacterium. The two genes exhibit a remarkable degree of sequence homology. When present in *E. coli*, both express DMA transferase and both cause secretion of zeatin. In the case of *tmr*, zeatin secretion is very low, although significantly above background. Zeatin production in *E. coli*, containing either *tmr* or *tzs*, may be a consequence of the presence of oxidases which do not originate from *A. tumefaciens*.

Acknowledgments. This work was supported by grants PCM 83-03371 from the National Science Foundation and 83-CRCR-1-1249 from the Competitive Research Grants Office, Science and Education Administration, U.S. Department of Agriculture.

References

1. Akiyoshi DE, Morris RO, Hinz R, Mischke BS, Kosuge T, Garfinkel DJ, Gordon MP, Nester EW (1983) Proc Natl Acad Sci USA 80:407
2. Akiyoshi DE, Klee H, Amasino RM, Nester EW, Gordon MP (1984) Proc Natl Acad Sci USA 81:5994
3. Barker RF, Idler KB, Thompson DB, Kemp JD (1983) Plant Mol Biol 2:335
4. Barry GF, Rogers DG, Fraley RT, Brand L (1984) Proc Natl Acad Sci USA 81:4776
5. Chen CM (1982) In: Wareing PF (ed) Plant growth substances 1982. Academic Press, London, p 155
6. Chen CM, Melitz DK (1979) FEBS Lett 107:15
7. Chen C, Leisner SM (1984) Plant Physiol (Bethesda) 75:442

8. Chilton MD, Drummond MH, Merlo DJ, Sciaky D, Montoya AL, Gordon MP, Nester EW (1977) Cell 11:263
9. Comai L, Kosuge T (1980) J Bacteriol 143:950
10. Durley RC, MacDonald EMS, Sayavedra-Soto LA, Morris JW, Morris RO (1985) Plant Physiol (submitted)
11. Einset JW (1980) Biochem Biophys Res Commun 93:510
12. Garfinkel D, Simpson R, Ream L, White R, Gordon M, Nester EW (1981) Cell 27:143
13. Hommes NG, Akiyoshi DE, Morris RO (1985) In: Law J, Rilling H (eds) Methods Enzymol 110:340
14. Kaiss-Chapman RW, Morris RO (1977) Biochem Biophys Res Commun 76:453
15. Knauf VC, Nester EW (1982) Plasmid 8:45
16. Lemmers M, De Beuckeleer M, Holsters M, Zambryski P, Depicker A, Hernalsteens JP, Van Montagu M, Schell J (1980) J Mol Biol 144:353
17. Letham DS, Tao GQ, Parker CW (1982) In: Wareing PF (ed) Plant growth substances 1982. Academic Press, London, p 143
18. Lundquist RC, Close TJ, Kado CI (1984) Mol Gen Genet 193:1
19. MacDonald EMS, Morris RO (1985) In: Law J, Rilling H (eds) Methods Enzymol 110:347
20. MacDonald EMS, Akiyoshi DE, Morris RO (1981) J Chromatogr 214:101
21. Miller CO (1974) Proc Natl Acad Sci USA 71:334
22. Morris RO, Akiyoshi DE, MacDonald EMS, Morris JW, Regier DA, Zaerr JB (1982) In: Wareing PF (ed) Plant growth substances 1982. Academic Press, London, p 175
23. Nester EW, Kosuge T (1981) Annu Rev Plant Physiol 35:531
24. Regier DA, Morris RO (1982) Biochem Biophys Res Commun 104:1560
25. Sanger F, Nicklen S, Coulsen AR (1977) Proc Natl Acad Sci USA 74:5463
26. Surico G, Sparapano L, Lerario P, Durbin DR, Iacobellis NS (1975) Experientia (Basel) 31:929
27. Surico G, Evidente A, Iacobellis NS, Randazzo G (1985) Phytochemistry (Oxf) 24:1499
28. Trione EJ, Krygier BB, Banowetz GM, Kathrein JM (1985) J Plant Growth Regul 4:101
29. Wilbur WJ, Lipman DJ (1983) Proc Natl Acad Sci USA 80:726

The Evolution of Polar Transport Models, and some Possibilities for the Regulation of Auxin Carriers*

P.H. RUBERY[1]

1 Historical Sketch of the Evolution of Polar Auxin Transport Models

In 1928, Went used the *Avena* curvature bioassay to measure transport of the newly discovered "Wuchsstoff" (later identified with IAA), from agar donor to receiver blocks, through short (2–4.2 mm) sections of oat coleoptile segments [48]. With these methods, the transport appeared exclusively basipetal and fulfilled Paál's 1919 prediction [33] of an internally-secreted, basally-moving, correlation carrier in phototropically-stimulated coleoptiles. Although quantitative velocity measurements (10 to 15 mm h^{-1}, see [14, 25] for discussion of "velocity") had to await Van der Weij's work [47], Went considered that simple diffusion could not explain the speed of movement and that protoplasmic streaming was responsible. Although Van der Weij had argued in 1932 [47] that the temperature insensitivity of transport velocity (see also [21]) excluded such a role, it was over 40 years before experiment [6] and theory [29] showed that streaming made only a small contribution to the velocity of polar auxin transport. Indeed, Van der Weij anticipated other contemporary ideas [47]: he coined the term "polar diffusion" (cf. [13]) and suggested (in translation) that "pure diffusion occurred within each cell, with polarity being conferred by a polar permeability [4] of the boundary between two cells so that auxin can only pass in a basal direction". A layer of unstirred cytoplasm adjacent to the cell wall was suggested as the intracellular transport route. Seeking to minimise transport barriers, Went had [48] suggested that Wuchsstoff did not penetrate the vacuole during transport – a concept whose consequences have recently been considered [30]. It is unclear what was envisaged for the hypothesised [48] diffusive step through the intervening cell walls, whose permeability properties were not then recognised as greatly different from plasma membranes. Went [48] indeed suggested that there was no protoplasmic discontinuity at the junction of young adjacent cells, and his opinion that permeability changes would not significantly influence transport velocity is consistent with a supposed plasmodesmatal type of route. His conclusion [48] that transport velocity is proportional to cell length has been reversed by recent analyses [15, 29] of the chemiosmotic theory of polar transport, to be discussed below.

In common with Went [49], Van der Weij favoured transport polarity being fundamentally due to an overall electrical polarity of the coleoptile (apex negative), perhaps resulting in "cataphoresis" of weakly acidic Wuchsstoff molecules [47, 49]. One

* The following contributions (pp. 197–246) are part of the workshop *Auxin Transport*

1 Biochemistry Department, University of Cambridge, Cambridge, Great Britain

Plant Growth Substances 1985
(ed. by M. Bopp)

attractive feature of this electrical theory of polarity was that it could explain "uphill" polar transport, although it could not account per se for transport specificity. However, a few years later Clark [7] showed that electrical and transport polarity were unconnected since the electrical polarity could be inverted by environmental manipulation (light, gravity, applied voltage), whereas IAA-transport polarity could not. Thus, Clark concludes [8] that "a possible asymmetry in membrane permeability theoretically linked with polar transport (accumulation, secretion?)" is not related to inherent potential differences.

Together with the availability of radioactive IAA, the discovery that TIBA [32] and NPA [31] were inhibitors of polar auxin transport allowed further progress towards the understanding of its mechanism. The role of NPA was inferred from its effects on coleoptile growth in interaction with IAA. A model of auxin transport was proposed [31] involving surface diffusion (to explain the velocity; see also [8]), with NPA binding strongly to proteinaceous adsorption sites on the plasma membrane and so blocking sorption-desorption migration of IAA from one site to another.

These new tools enabled Hertel and Leopold [21] convincingly to suggest that a specific and locally distributed auxin secretion across the plasma membrane was responsible for the polarity of transport. The transport velocity was still considered dependent on protoplasmic streaming. The key observation was that TIBA stimulated the net uptake of [^{14}C]IAA by maize coleoptile cylinders [20, 21] - indicating that TIBA was inhibiting an efflux rather than an influx component of the overall uptake. Because of the "uphill" characteristics of polar auxin transport and its sensitivity to disruption of energy metabolism [13, 24], the proposed secretion and, less critically, the uptake of auxin were seen as "active transport" processes requiring energisation of a carrier. This "polar secretion hypothesis" [21] thus retained Went's [48] concept that both intracellular movement (accelerated by protoplasmic streaming) and diffusion from cell to cell across the wall were intrinsic parts of the complete polar transport system, which persists when plasmodesmatal connections are broken [44]. The main conceptual advance was to see polarity in terms of asymmetric transport-catalyst distribution rather than as a consequence of a driving-force gradient along the tissue. However, Raven [36, 37] later suggested that electrophoresis could generate polar membrane protein distribution (cf. [16]).

Because of difficulties in distinguishing unidirectional influx and efflux in multicellular systems, Hertel and Leopold [21] suggested that single cells might be a fruitful system. Indeed, these figured strongly in subsequent developments, which lead to the realisation that primary active transport was unnecessary for auxin movement across cell membranes [35, 42, 43]. Early work had shown that auxin could easily diffuse across membranes in its neutral undissociated form (IAAH; pK = 4.7) by virtue of its hydrophobic character and consequent lipid solubility [1, 46]. As for any weak acid, the IAAH concentration increases as the pH is lowered (50% dissociation at the pK) and auxin uptake is thus strongly pH-dependent, reflecting the "titration" of the less permeant IAA anion. However, the significance of these simple facts for polar auxin transport was slow to be recognised. Rubery and Sheldrake [42, 43] and Raven [35] re-emphasised that the relative anion impermeability means that plant cells, whose cytoplasm is exposed to a more acidic external environment (cell wall in contact with any incubation medium), will accumulate auxin. This is because IAAH concentrations tend to equalise across the membrane, but the ratio of IAA^- to IAAH levels would be higher at the higher pH. The anions are effectively trapped in the more basic compartment. Thus, an IAA anion carrier in the plasma membrane would tend to lessen this accumulation by catalysing passive efflux of IAA^- ions down their concentration gradient, reinforced by the usual negative electrical membrane potential. Stimulation of [^{14}C]IAA uptake into cultured cells by TIBA, which could be blocked by excess nonradioactive IAA, was interpreted as inhibition of such a saturable carrier [43], whose preferential distribution at the basal ends of cells in tissues was suggested to account for transport polarity [43]. Such polar permeability would allow a cell taking up auxin from a source to bring its cytoplasmic auxin concentration to effective equilibrium with a higher cell wall auxin concentration at the base than at the apex of the cell. This is due to the auxin accumulation ratio being lower at the basal end, with its higher density of auxin efflux sites. The concentration difference becomes amplified as auxin moves along a file of individually polar cells, to produce an overall uphill auxin gradient whose dependence on metabolic energy reflects the ATP-dependent maintenance of the pH and electrical gradients necessary as driving forces for transmembrane auxin transport (see Fig. 2 in [40] and [13, 24]. This secondary energisation of polar transport contrasts with the more direct application of energy envisaged by the polar secretion hypothesis [21] and is reflected in the designation "chemiosmotic polar diffusion hypothesis" adopted by Goldsmith when discussing this work in her 1977 review [13].

Both the auxin efflux carrier and also an uptake carrier [42, 43], which may be an IAA^-/H^+ symport [38], were subsequently detected in segments of a variety of organs capable of polar transport [2,9,12,39,45]. They have additionally been demonstrated in microsomal membrane preparations presumed to include a population of outside-out plasma membrane vesicles [22]. Recently, some direct immunocytochemical evidence for a basally-located auxin efflux system has been obtained [23] in pea stem cells using a fluorescent monoclonal antibody to an NPA-binding protein assumed to be closely associated with the carrier. A role for *saturable* carriers is shown by a saturability of overall polar transport [14, 21], which loses polarity if overloaded [14], and is underpinned by mathematical analysis [29]. This gives quantitative support to the chemiosmotic hypothesis, including a need for a saturable component to achieve the characteristic auxin distribution profiles found along the transporting pathway. It is now accepted that diffusion, given a biologically specified polarity with appropriate permeability coefficients for IAAH and IAA^-, can explain observed polar transport velocities [6, 15, 29]. The symport carrier may augment IAAH diffusion [40].

2 Some Thoughts on Regulation of Auxin Transport

TIBA is itself polarly transported and displays reciprocal transport interactions with IAA in hypocotyl sections [12]. A two-site model has been proposed to resolve the apparent paradox of TIBA being a *noncompetitive* inhibitor and yet a substrate of the polar transport system [12].

In contrast, NPA is not subject to polar transport, and there is considerable in vitro evidence for a specific NPA-binding site [41], which excludes physiological concentrations of IAA (see [40]). This could represent a regulatory subunit of the IAA^- efflux carrier. Recent work in my laboratory (D.A. Morris and P.H. Rubery 1985, unpubl.) with protein synthesis inhibitors suggests that in zucchini hypocotyl segments the NPA target turns over (half-life ca. 1.5–2 h) more rapidly than the auxin efflux catalyst. Possibly the latter may have a basal activity and be uniformly distributed, but is capable of augmented activity if associated with a regulatory subunit whose localised insertion or removal at the plasma membrane could generate the asymmetry critical for polar transport. Control could be accomplished in the longer-term by turnover and in the short-term by a natural occurring NPA analogue. The endorphin precedent springs to mind [18, 25], with phenolic compounds [26] and "Kartoffelfaktor" [18] as possible candidates.

Milborrow and Rubery [27] recently pointed out that in tissues with a *gradient* of carriers, the asymmetry at cellular junctions needed for polar transport would arise if in each cell the carrier density were uniform, but decreased in a regular way down a file of cells. In such a model, transport polarity could be controlled developmentally, by carrier synthesis and turnover, rather than by vectorial processes in individual cells.

The growing appreciation that Ca^{2+} is an important regulator in plants [17, 37] is reflected in several models linking Ca^{2+} and auxin. Hertel, pursuing his long-expressed idea [21] that auxin transport and action are intimately related, suggested [19] that auxin carriers, on the tonoplast as well as the plasma membrane, could catalyse counter-transport of IAA^- and Ca^{2+} down their respective electrochemical gradients. This would increase cytoplasmic calcium concentration, which seems a priori more suitable for signal transduction than an auxin-related calcium *decrease* suggested by some indirect evidence [17]. Dela Fuente [10, 11] draws heavily on the role of Ca^{2+} in animal stimulus-secretion coupling to propose that sufficiently high intracellular auxin levels trigger membrane depolarisation to allow gated calcium influx, which then activate the IAA^- efflux carrier via calmodulin [10]. The resulting jump in efflux is proposed in some unspecified way to accelerate Ca^{2+} entry, which is eventually attenuated by autoactivation of calcium efflux pumps [11]. This proposed sequence lacks detailed experimental support and there are some difficulties related to the nonspecific effects of permeant weak acids on membrane potentials [3]. In the context of shoot geotropism, Pickard [34] proposes that cytoplasmic Ca^{2+} increases when gravity-sensitive calcium channels open and activates IAA efflux by calcium-stimulated protein phosphorylation, increased IAA-sensitive proton pumping then ultimately leading to lateral separation of Ca^{2+} and H^+ across the shoot.

Clearly, all of these ideas must be tested against hard data. Functional reconstitution in sealed membranes of purified auxin transport proteins and related control elements must be an important goal. To show what *can* be achieved, I conclude with some elegant work on the acetylcholine receptor (AchR), which is a ligand-gated ion channel. It has four types of subunits, of known sequence and conformation in the membrane [5, 28]. Site-directed mutagenesis has enabled specific amino acid substitutions and sequence deletions in the α-subunit to be constructed. The effects of these deliberately created changes on Ach binding, gating, and channel specificity have been assessed by injecting mutant α-subunit mRNAs (copied from cDNAs) into frog oocytes together with normal mRNAs for the other subunits. The cells not only translate the mRNAs (the products can be immunoassayed), but insert them into their plasma membranes so conferring Ach sensitivity on the oocyte, which can be measured by pharmacological and electrophysiological methods. A map was constructed to show the structure-function correlations deduced for the membrane-spanning receptor. The fact that [28] has 14 authors shows the scale of team-work required.

References

1. Albaum HG, Kaiser S, Nestler HA (1937) Am J Bot 24:513
2. Astle MC, Rubery PH (1983) Planta (Berl) 157:53
3. Bates GW, Goldsmith MHM (1983) Planta (Berl) 159:231
4. Brauner L (1930) Ber Dtsch Bot Ges 48:109
5. Brisson A, Unwin PNT (1985) Nature 315:474
6. Candé WZ, Goldsmith MHM, Ray PM (1973) Planta (Berl) 111:279
7. Clark WG (1937) Plant Physiol (Bethesda) 12:737
8. Clark WG (1938) Plant Physiol (Bethesda) 13:529
9. Davies PJ, Rubery PH (1978) Planta (Berl) 142:211
10. Dela Fuente RK (1984) Plant Physiol (Bethesda) 76:342
11. De Guzman CC, Dela Fuente RK (1984) Plant Physiol (Bethesda) 76:347
12. Depta H, Rubery PH (1984) J Plant Physiol 115:371
13. Goldsmith MHM (1977) Annu Rev Plant Physiol 28:439
14. Goldsmith MHM (1982) Planta (Berl) 155:68
15. Goldsmith MHM, Goldsmith TH, Martin MH (1981) Proc Natl Acad Sci USA 78:976
16. Goldsworthy A, Rathore KS (1985) J Exp Bot 36:1134
17. Hepler PK, Wayne RO (1985) Annu Rev Plant Physiol 36:397
18. Hertel R (1981) Biochem Physiol Pflanz 176:495
19. Hertel R (1983) Z Pflanzenphysiol 112:53
20. Hertel R, Leopold AC (1962) Naturwissenschaften 16:377
21. Hertel R, Leopold AC (1963) Planta (Berl) 59:535
22. Hertel R, Lomax TL, Briggs WR (1983) Planta (Berl) 157:193
23. Jacobs M, Gilbert SF (1983) Science 220:1297
24. Kaldewey H (1984) In: Scott TK (ed) Hormonal regulation of development II. The functions of hormones from the level of the cell to the whole plant. Encyclopedia Plant Physiol (New Ser), vol 10, p 80
25. Katekar GF, Navé J-F, Geissler AE (1981) Plant Physiol (Bethesda) 68:1460
26. Marigo G, Boudet AM (1977) Physiol Plant 41:197
27. Milborrow BV, Rubery PH (1985) J Exp Bot 36:807
28. Mishina M, Tobimatsu T, Imoto K, Tanaka K-i, Fujita Y, Fukuda K, Kurasaki M, Takahashi H, Morimoto Y, Hirose T, Inayama S, Takahashi T, Kuno M, Numa S (1985) Nature 313:364
29. Mitchison GJ (1980) Proc R Soc Lond B Biol Sci 209:489
30. Mitchison GJ (1981) Proc R Soc Lond B Biol Sci 214:69
31. Morgan DG, Söding H (1958) Planta (Berl) 52:235
32. Niedergang-Kamien E, Skoog F (1956) Physiol Plant 9:60
33. Paál Á (1919) Jahr Wiss Bot 58:406
34. Pickard BG (1985) Annu Rev Plant Physiol 36:55
35. Raven JA (1975) New Phytol 74:163
36. Raven JA (1979) New Phytol 82:285
37. Raven JA, Rubery PH (1982) In: Smith H, Grierson D (eds) Molecular biology of plant development. Blackwell, Oxford, p 28
38. Rubery PH (1978) Planta (Berl) 142:203
39. Rubery PH (1979) Plant Sci Lett 14:365
40. Rubery PH (1980) In: Skoog F (ed) Plant growth substances 1979. Springer, Berlin Heidelberg New York, p 50
41. Rubery PH (1981) Annu Rev Plant Physiol 32:569

42. Rubery PH, Sheldrake AR (1973) Nature New Biol 244:285
43. Rubery PH, Sheldrake AR (1974) Planta (Berl) 118:101
44. Sheldrake AR (1979) Planta (Berl) 145:113
45. Sussman MR, Goldsmith MHM (1981) Planta (Berl) 150:15
46. Sutter E (1944) Ber Schweiz Bot Ges 54:197
47. Van der Weij HG (1932) Recl Trav Bot Neerl 29:381
48. Went FW (1928) Recl Trav Bot Neerl 25:1
49. Went FW (1932) Jahr Wiss Bot 76:528

Roles of Transport and Binding in the Specific ΔpH-Dependent Accumulation of Auxin by Zucchini Membrane Vesicles

K.A. CLARK and M.H.M. GOLDSMITH[1]

1 Introduction

The growth of plants is coordinated in part by the polar transport of the endogenous auxin indoleacetic acid (IAA). Studies on the pH-dependent accumulation of auxin by cells and tissues suspended in buffered auxin solutions [4, 13, 15–17, 19, 21, 22] have led to the realization that this process involves some of the same elements as polar transport in isolated tissues and intact plants [2, 5, 6]. The available evidence suggests a model for the ΔpH-dependent uptake and transport of IAA (Fig. 1) that includes: (1) permeation of the neutral lipophilic form of the weak acid (IAAH) through the lipid bilayer; (2) a saturable symport of the auxin anion (IAA^-) with protons; and (3) a passive efflux of IAA^- via a specific saturable anion channel (carrier) sensitive to napthylphthalamic (NPA) [20] and 2,3,5-triiodobenzoic acids (TIBA), herbicides known to stimulate cellular auxin accumulation at the expense of polar auxin transport [3, 5, 19, 21, 22]. Localization of the IAA^- efflux sites at the base of cells is believed to be the physical basis for polar transport [8].

Plant cells and tissues are multicompartmented systems whose complexity makes it difficult to measure either the transmembrane auxin fluxes or the magnitude of the pH and voltage gradients postulated to be responsible for polar transport in vivo. Furthermore, measurements are usually limited to initial rates of uptake by cells or tissues. Because of the intervention of auxin metabolism, steady-state accumulations are usually not obtainable. Confronted with these difficulties, membrane vesicles

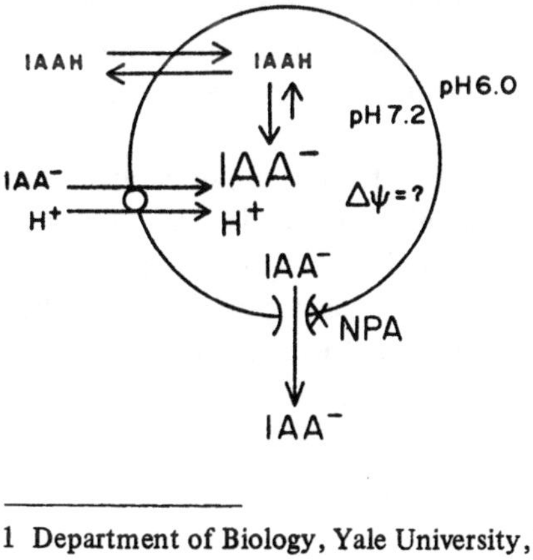

Fig. 1. Processes postulated to account for the ΔpH-dependent accumulation of IAA by cells or membrane vesicles as described in the text. Vesicles suspended in pH 6 buffer have a ΔpH of 1.2; the voltage gradient across the vesicle membrane, Δψ, has not been measured

1 Department of Biology, Yale University, New Haven, CT 06511, USA

Plant Growth Substances 1985
(ed. by M. Bopp)

isolated from plant tissues offer a single compartment that is relatively free of metabolic reactions and can be used to test the hypotheses that (1) as expected for permeation of a lipophilic weak acid, the amount of auxin accumulated depends on the magnitude of the pH gradient, and that (2) NPA stimulates net uptake by blocking a specific efflux of auxin anions via a saturable site.

Hertel et al. [7] observed that microsomal vesicles isolated from zucchini hypocotyls accumulate IAA in a ΔpH-dependent fashion. Microsomal vesicles display some of the same properties as auxin transport in vivo, including saturation by high auxin concentrations and stimulation of uptake by TIBA and NPA [7, 9]. Lomax [10], however, reported the surprising result that the observed auxin accumulation exceeds that predicted by the pH gradient. She suggested that a two-proton symport of IAA, driven by a negative membrane potential, could provide an additional driving force for auxin accumulation. We also find that the auxin accumulation by zucchini microsomal vesicles is greater than predicted from the ΔpH, but we have not been able to detect directly any membrane voltage, and now report a different explanation for this result.

2 Materials and Methods

We have modified the protocol of Hertel et al. [7] (cf. [1]). Hypocotyls (subapical 2 cm) from 6-day-old, dark-grown zucchini (*Cucurbita pepo* L. cv. Black Beauty) seedlings were homogenized in isotonic medium buffered at pH 8 (250 mM sorbitol, 50 mM HEPES/BTP, 3 mM EGTA, 1 mM PMSF, 1 mM DTT, 1 mM KCl, and 0.1% BSA). The temperature was 0°–4 °C for this and subsequent procedures unless stated otherwise. Microsomal membrane vesicles were isolated by differential centrifugation (8000 × g supernatant centrifuged 40 min at 50000 × g). Vesicles were washed and repelleted once, and then a transmembrane pH gradient was established by resuspending the vesicles in osmoticum buffered at pH 6 (330 mM sorbitol, 10 mM MES/BTP, and 1 mM KCl). Osmolarity was adjusted to other desired values by adding concentrated sorbitol. We used [^{14}C]-butyric acid as a pH-probe and measured its uptake simultaneously with that of [^{3}H]-IAA. [^{3}H]-IAA and [^{14}C]-butyric acid (BA) were added 6 min after adjusting the osmolarity. Uptake was monitored by sampling aliquots (0.1 ml) in triplicate at appropriate intervals. Vesicles were separated from the radioactive incubation medium by filtration (Millipore HAWP 0.45 μM). Radioactivity was determined by scintillation counting, and tritium counts were corrected for spillover of [^{14}C].

The ΔpH-dependent uptake was calculated from the difference in radioactivity before and after the pH gradient was collapsed by incubating vesicles for 30 min at 12 °C in presence of carbonyl cyanide m-chlorophenyl hydrazone (CCCP) (5 μM) plus valinomycin (5 μM). About one-third of the total [^{3}H]-IAA and a somewhat smaller fraction of the [^{14}C]-BA dpm remaining on the filter after treatment of the vesicles with CCCP were independent of the pH gradient.

3 Results

Based on the accumulation of [^{14}C]-BA, zucchini vesicles suspended in pH 6 buffer have a transmembrane pH gradient of about 1.2 units. The vesicles appear to be sealed because their pH gradient persists for more than 1 h (Fig. 2) and varies with pH_0 in a predictable fashion [1]. Also, as we shall see below, the vesicles are osmotically active. The magnitude of the pH gradient measured with [^{14}C]-BA agrees well with measurements made under similar conditions by electron spin resonance [11, 12].

In the absence of any anion permeability, IAA and BA should be accumulated to the same extent as both of these weak acids have almost the same pK. After 30 min, accumulation of IAA exceeds that of BA by two- to three-fold (Fig. 2). Furthermore, although NPA does not affect the accumulation of BA (and thus ΔpH), it increases the net accumulation of IAA still further. BA accumulation peaks within 1 min followed by a gradual decline of about 37% over the following 90 min, reflecting a slow decay of the pH gradient. The initial rates of IAA and BA accumulation are similar (Fig. 2, inset), but it takes 20 to 30 min longer for IAA to reach its higher steady-state level of accumulation.

Concentrations of weak acids high enough to overcome the buffering capacity of the solution within the vesicles should dissipate ΔpH and decrease the uptake of both IAA and butyric acid nonspecifically [14]. In our experiments, this happens when the external concentrations of either the IAA or butyric acid are above 10 μM. Figure 3 shows that increasing the concentration of IAA in the range 0.1 to 10 μM does not affect ΔpH, as butyric acid uptake remains constant, but dramatically lowers the uptake of [^{3}H]-IAA both in the presence and absence of NPA. The competition between [^{3}H]-IAA and micromolar concentrations of cold IAA is specific, as it occurs with IAA and other auxin analogs, but not with benzoic or butyric acids [1].

Our data shows that the accumulation of [^{3}H]-IAA in either the presence or absence of NPA exceeds that expected from ΔpH. Because the accumulation of IAA in the

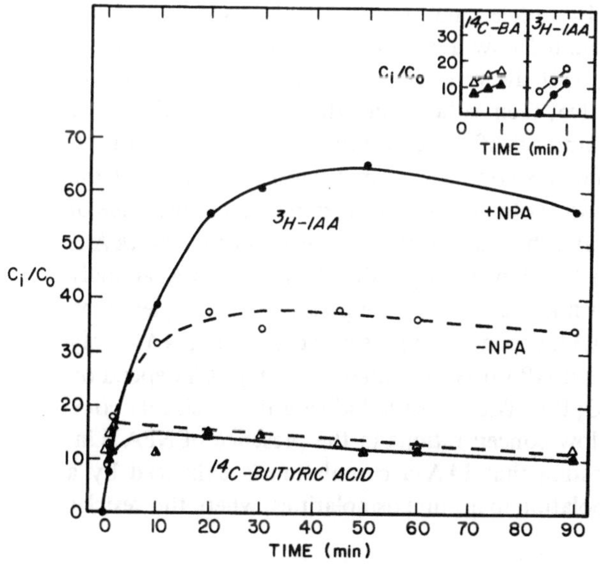

Fig. 2. The ΔpH-dependent accumulation of [^{3}H]-IAA exceeds that of the pH probe, [^{14}C]-butyric acid. NPA, when present, is 5 μM. The accumulation of each acid is expressed as the ratio of concentration of radioactivity inside the vesicles (C_i) to that in the external medium (C_o) [1]

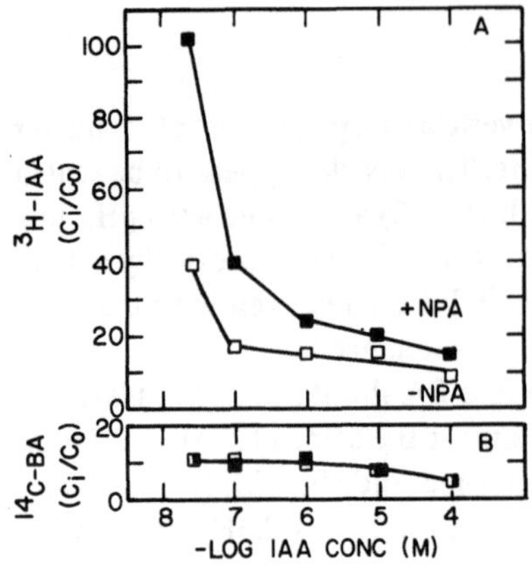

Fig. 3

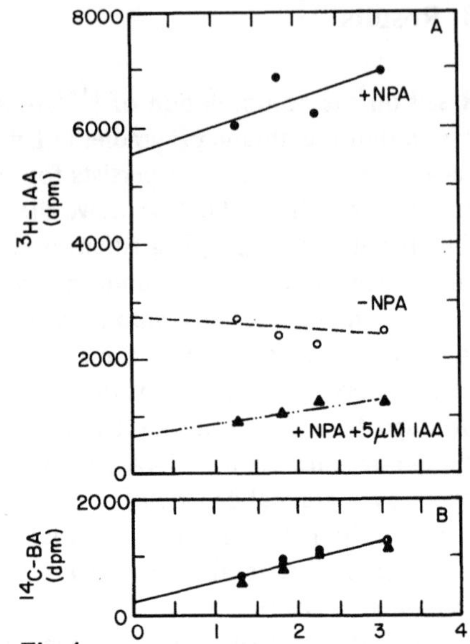

Fig. 4

Fig. 3. The ΔpH-dependent accumulation of [^{3}H]-IAA and [^{14}C]-BA. Increasing concentrations of cold IAA up to 1 μM affect only the accumulation of [^{3}H]-IAA and not that of [^{14}C]-BA, whereas still higher concentrations affect uptake of both IAA and BA nonspecifically. When present, NPA is 5 μM; 30 min uptakes

Fig. 4. Variation of ΔpH-dependent uptake of [^{3}H]-IAA and [^{14}C]-butyric acid with the reciprocal of the external osmolarity. As described in the text, the results with [^{14}C]-BA indicate that the vesicles are osmotically active; 30 min uptakes. When present, NPA is 5 μM

absence of NPA already exceeds that predicted for a weak acid with an impermeable anion, the stimulatory action of NPA cannot be attributed solely to decreasing the efflux of IAA^-. An alternative hypothesis is that a specific saturable binding contributes to the accumulation of IAA by vesicles and presumably also by cells.

To test this hypothesis, we determined how auxin accumulation varies as a function of the external osmotic strength. If the vesicles contain a fixed amount of osmotic solute and are osmotically active, their volume and thus the accumulation of BA should vary linearly with the reciprocal of the concentration of external osmoticum (sorbitol). Figure 4B shows that this condition is met, with about 80% of the uptake of BA dependent on the osmotic volume and uninfluenced by the presence of either NPA or 5 μM IAA. By contrast, the uptake of low concentrations of IAA is much greater than expected at all vesicle volumes and, in the absence of NPA, actually increases somewhat as the vesicles shrink (Fig. 4A). In the presence of NPA, less than 20% of the total accumulation of IAA is dependent on vesicle volume (Fig. 4A). Only in the presence of 5 μM cold IAA and NPA does the uptake of [^{3}H]-IAA depend on the vesicle volume similarly to [^{14}C]-BA. These results indicate that a sizable component of accumulation of IAA at low concentrations in the presence of NPA is independent of vesicle volume, suggesting that IAA accumulation is enhanced by a saturable binding site. Even on extrapolation to infinite osmolarities, where the vesicles would have virtually no internal volume, NPA still stimulates IAA uptake (Fig. 4A). At least in this situation, it seems unlikely that NPA could be affecting auxin fluxes across the vesicle membranes.

Returning to the model in Fig. 1, we find no direct evidence for the postulated saturable H^+/IAA^- symport in vesicles. Instead, the evidence suggests participation of a saturable ΔpH-dependent binding of IAA. Preliminary calculations indicate that in the presence of NPA about 200 molecules of IAA per μm^2 of internal vesicle surface area would have to be bound to account for these results. This is not an unreasonable number of receptor sites per μm^2, although it is about 80-fold higher than the amount of auxin-binding protein detected in plasma membrane fractions [9, 18]. The affinity of the saturable site for IAA appears not to be altered by the presence of NPA (Fig. 3A). This observation indicates that the same site is responsible for the saturable accumulation in both the presence and absence of NPA.

4 Conclusions

1. The saturable, NPA-stimulated accumulation of [^{3}H]-IAA by zucchini microsomal membrane vesicles is severalfold greater than expected for the ΔpH-dependent uptake of a weak acid.
2. As indicated by the accumulation of butyric acid, NPA does not affect the pH gradient.
3. This accumulation of IAA depends on the presence of a pH gradient and is reversible on collapse of the ΔpH. It, therefore, most likely occurs within the vesicles.
4. A sizable fraction (70–80%) of the IAA accumulated inside the vesicles is not dissolved in the intravesicular solution, but rather interacts with a specific, saturable, NPA-stimulated binding site.
5. When the binding site is saturated by cold auxin or other auxin analogs, the accumulation of [^{3}H]-IAA is in accord with ΔpH, as determined by [^{14}C]-BA accumulation. Thus, the ΔpH-dependent saturable binding of IAA inside the vesicles is sufficient to account for accumulation of IAA in excess of that predicted by ΔpH.

Acknowledgment. This research is supported in part by a grant from the United States Department of Energy 85ER13337.

References

1. Clark KA, Goldsmith MHM (1985) In preparation
2. Davies PJ, Rubery PH (1978) Planta (Berl) 142:211
3. Depta H, Rubery PH (1984) J Plant Physiol 115:371
4. Edwards KL, Goldsmith MHM (1980) Planta (Berl) 147:457
5. Goldsmith MHM (1982) Planta (Berl) 155:68
6. Goldsmith MHM, Goldsmith TH, Martin MH (1981) Proc Natl Acad Sci USA 78:976
7. Hertel R, Lomax TL, Briggs WR (1983) Planta (Berl) 157:193
8. Jacobs M, Gilbert SF (1983) Science 220:1297
9. Jacobs M, Hertel R (1978) Planta (Berl) 142:1

10. Lomax TL (1983) Ph.D. Thesis, Stanford University
11. Lomax TL, Mehlhorn RJ (1985) Biochim Biophys Acta 821:106
12. Lomax TL, Mehlhorn RJ, Briggs WR (1985) Proc Natl Acad Sci USA 82:6541
13. Raven J (1975) New Phytol 74:163
14. Rottenberg H (1979) Methods Enzymol 55F:547
15. Rubery PH (1977) Planta (Berl) 135:275
16. Rubery PH (1978) Planta (Berl) 142:203
17. Rubery PH (1979) Planta (Berl) 144:173
18. Rubery PH (1981) Annu Rev Plant Physiol 32:569
19. Rubery PH, Sheldrake AR (1974) Planta (Berl) 118:101
20. Sussman MR, Goldsmith MHM (1981a) Planta (Berl) 151:15
21. Sussman MR, Goldsmith MHM (1981b) Planta (Berl) 152:13
22. Thomson KSt, Hertel R, Muller S, Tavares JE (1973) Planta (Berl) 109:337

Active Auxin Uptake by Specific Plasma Membrane Carriers

T.L. LOMAX [1]

1 Introduction

The directed movement of auxin plays a major regulatory role in the development of plants. The chemiosmotic hypothesis proposes that both cellular accumulation and polar transport of IAA (indole-3-acetic acid) are driven by transmembrane pH and electrical gradients [2, 13]. It has been proposed that a carrier(s) may be necessary to sustain reasonable rates of cell-to-cell transport and to provide the saturability, specificity, and polarity that are features of in vivo auxin transport through a variety of tissues [13, 14]. These sites have been proposed to include both a saturable, specific IAA/proton uptake symport and a specific, perhaps electrogenic, anion efflux carrier, which can be blocked by specific, noncompetitive transport inhibitors, such as TIBA (2,3,5-triiodobenzoic acid). Hertel [4] has recently suggested that the efflux carrier is instead a symport with auxin leaving the cell as $IAAH/H^+$.

Until recently, no in vitro system has existed with which to test models of auxin transport. Now, evidence has been presented for such a system: tightly-sealed zucchini membrane vesicles, which have a pH gradient of the same direction as found in intact plants [5, 8] and which provide an excellent system for studying the various transport components of the plasma membrane without the complications of other cellular components [3, 12]. The vesicles are well-suited for the study of in vitro IAA transport, defined here as $[^{14}C]$-IAA accumulation that is sensitive to the electrogenic H^+-carrier FCCP (carbonyl-cyanide p-trifluoromethoxyphenylhydrazone) and that can be stimulated by auxin transport inhibitors such as TIBA [5, 9].

To understand the dynamics of auxin transport and to establish whether the transport is an active or simply passive accumulation down a pH gradient, direct measurements both of vesicle volume and of the size of the pH gradient were needed to compare with the measurements of auxin uptake. Recently, we have demonstrated that the nitroxide spin-probe ESR technique can be used to obtain such measurements from zucchini membrane vesicle preparations identical to those used for auxin uptake studies [8]. Using the nitroxide spin-probe technique to provide direct measurements of the magnitude and persistence of pH gradients across the vesicles obtained with ESR together with $[^{14}C]$-IAA uptake measurements, results were obtained, which are consistent with active, electrogenic transport of IAA into sealed vesicle space via

1 Carnegie Institution of Washington, Department of Plant Biology, 290 Panama Street, Stanford, CA 94305, USA. Current address: Department of Botany, University of Washington, Seattle, WA 98195, USA

Plant Growth Substances 1985
(ed. by M. Bopp)

IAAH/H^+ cotransport [9]. Here, evidence is presented that this accumulation of [^{14}C]-IAA is facilitated auxin influx via saturable, specific, plasma membrane carriers, as predicted by Rubery and Sheldrake [13], Sussman and Goldsmith [14], and Goldsmith [2] rather than the simple, nonmediated accumulation of a weak acid across a pH gradient.

2 Results

Membrane particles from etiolated zucchini (*Cucurbita pepo* cv. Dark Green) were prepared and uptake of [2-^{14}C]-IAA was measured as previously described [5, 9]. Vesicle volume and pH gradient determinations were made by electron spin resonance and ΔpH was calculated as described by Lomax and Mehlhorn [8].

Benzoic acid (BA), like IAA, is a lipophilic weak acid, which has a pH and structure similar to that of IAA, but which is inactive as an auxin. In membrane vesicles from zucchini hypocotyls, the uptake of [^{14}C]-IAA is saturable by low concentrations of unlabeled IAA, but is not inhibited by excess benzoic acid until quite high concentrations are reached (see Table 1 and [7]). The half-maximal level of IAA saturation of these vesicles appears to be approximately $2-4 \times 10^{-7}$ M. In comparison, the uptake of [^{14}C]-BA is inhibited by neither IAA nor BA in the submillimolar range (see Fig. 12 in ref. [7]). This result provides evidence that while accumulation by diffusion is certainly a component, IAA accumulation in these vesicles is via a specific, saturable mechanism.

If the IAA transport site(s) described above represents the physiological transport site, the auxin specificity of in vitro uptake should display a pattern similar to that measured for in vivo polar auxin transport. When the relative effectiveness of various auxin analogs and antagonists in inhibiting both in vitro and in vivo [^{14}C]-IAA transport are compared, the order of effectiveness of these auxin analogs in inhibiting polar auxin transport closely parallels their inhibition of the association of [^{14}C]-IAA to membrane vesicles in zucchini (Table 1). The single exception is 1-NAA (a very effective inhibitor of auxin transport in vivo, but only a weak competitor for IAA association in vitro); this same discrepancy has also been observed with suspension-cultured crown gall cells [13]. Although caution should be taken in the interpretation of absolute competition values, the similarity in relative effectiveness of the analogs in inhibiting both in vitro and in vivo auxin transport substantiates the relevance of

Table 1. Activity of auxin analogs as competitors for in vitro vs in vivo IAA transport

	2-NAA	IAA	2,4-D	1-NAA	BA	PAA
IAA uptake inhibition, pK_D [a]	7.1	6.4	5.4	4.7	4.5	4.4
Polar transport inhibition, pC_{50} [b]	4.5	4.0	3.5	5.6	3.0	3.0

[a] pK_D = −log concentration necessary for a 50% reduction of [^{14}C]-IAA association by membrane vesicles

[b] pC_{50} = −log conc. for 50% inhibition of polar auxin transport by hypocotyl segments (from [6])

the membrane vesicle system to polar transport by intact cells, and provides further proof as to the existence of specific and saturable uptake carriers.

In relation to studies of cellular auxin transport [2, 3, 13, 14], the in vitro auxin accumulation by membrane fractions from zucchini hypocotyls can only be understood if it occurs in outside-out plasma membrane vesicles. Jacobs and Hertel [6] had tentatively localized their "site III" auxin binding at the plasma membrane. To confirm that this was the only membrane fraction capable of pH-dependent [^{14}C]-IAA uptake, the membranes were separated on 1%–12% Dextran gradients (Fig. 1), which does not disturb the sealed integrity of the vesicles by subjecting them to osmotic shock.

FCCP-sensitive IAA accumulation is found in a single peak, which coincides with NPA binding considered to be a marker for the plasma membrane, and is clearly separated from the cytochrome c oxidase (mitochondria) and cytochrome c reductase (endoplasmic reticulum) activity as well as from the major peak of Cl^--stimulated

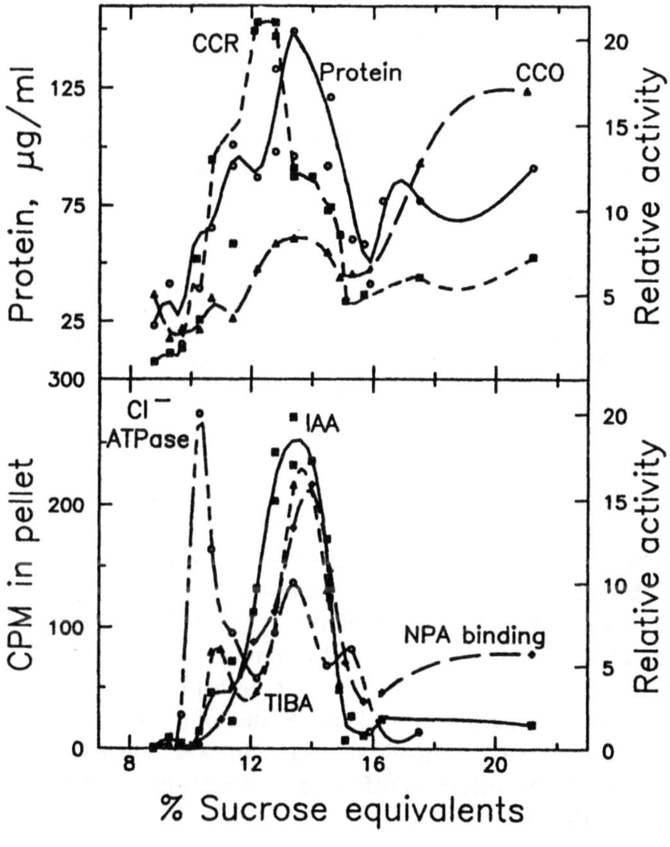

Fig. 1. Localization of the auxin transport sites. A vesicle resuspension was prepared as for the in vitro uptake assay, layered onto a linear gradient of 1%–12% Dextran T-70 in 250 mM sucrose, 50 mM Tris base, pH 7.6 with HCl, and spun for 2 h at 75000 × g. Fractions were then repelleted to remove the Dextran. Assays for marker enzymes are described in Mandala et al. [10]. [^{14}C]-TIBA accumulation and [^{3}H]-NPA binding were measured as previously described in [6]. *CCR* cytochrome c reductase; *CCO* cytochrome c oxidase

ATPase activity, which is considered to be associated with the tonoplast [10]. In other experiments (data not shown), the [^{14}C]-IAA uptake co-sedimented with binding of the fungal toxin fusicoccin (FC), a plasma membrane marker, but was separated from latent IDPase and glucan synthetase I, both markers for the Golgi. pH-sensitive TIBA accumulation both by the tonoplast and at the region of maximum IAA transport indicates that vesicles of both tonoplast and plasma membrane origin are sealed in these *Cucurbita* membrane preparations (Fig. 1). The fact that IAA is selectively accumulated by plasma membrane vesicles provides further evidence for the specific nature of the in vitro auxin transport system.

To quantitate the sealed vesicle volume and the magnitude and duration of the pH gradient across the vesicle membranes, and to analyze further kinetic data and concentration ratios for the accumulation of auxin, ESR spectroscopy of nitroxide spin probes provided a convenient, accurate, and sensitive method. The volume of sealed vesicles within a solution can be determined by quantitating the ESR signal of a membrane-permeable yet water-soluble nitroxide probe in the presence and absence of an impermeable paramagnetic quenching agent. The ratio of quenched to unquenched signal is a linear function of the total intravesicular volume, which is sealed and inaccessible to the quencher. A weak-acid nitroxide probe partitions preferentially into basic environments. This difference can be used to calculate the ΔpH [8, 11].

These ESR techniques were used to determine both the vesicle volume and the size of the pH gradient in zucchini vesicles made with either 50 mM or 200 mM Tris buffer at pH 7.9, and resuspended in pH 5.5 buffer. In both cases the pH gradient decays slowly and linearly and, therefore, represents a negative exponential decline in H^+ gradient (Fig. 2 in ref. [9]). The size of the pH gradients which we measured agrees well with measurements which have since been made by Clark and Goldsmith [1] using [^{14}C]-butyric acid. The duration of auxin accumulation in vesicles loaded with 50 mM buffer (approx. 4 h) is in good accord with the time that would be necessary for complete decay of the pH gradient at the rate estimated here, 0.003 pH units min^{-1}. With the greater buffer strength (200 mM), more protons are available to maintain the pH gradient, prolonging the IAA accumulation up to 6 h.

However, [^{14}C]-IAA uptake by vesicles prepared with 200 mM buffer was half again as large than for those prepared with 50 mM buffer [9]. This is the predictable result of the greater volume of vesicles with a higher internal osmolarity. The sealed vesicles made with 50 mM buffer were found by ESR to constitute about 0.6% of the total volume of the membrane resuspension or 7.5 $\mu l\ mg^{-1}$ protein. Those prepared with 200 mM were half again as large, 0.9% of the total volume or 11.3 $\mu l\ mg^{-1}$ protein. The 50 mM buffer was in osmotic balance with the external buffer while the osmotic concentration of the 200 mM buffer was higher than that of the external buffer (600 m Osmol vs 350 m Osmol), which caused the vesicles to swell. Thus, the difference in vesicle size when loaded with different buffer strengths explains the observed differences in total IAA accumulation.

3 Discussion

Our experiment argues against the suggestion by Clark and Goldsmith [1] that a considerable portion of the pH-dependent [^{14}C]-IAA association is binding rather than accumulation within the vesicle due to electrogenic uptake. If the association was binding, one would not expect it to increase exactly as the internal volume of the vesicle increases. The magnitude of the pH gradient is the same in both cases [9] so the total IAA association is not merely pH-dependent; it would be necessary to invoke a binding dependent upon buffer strength, which appears unlikely.

Using the vesicle volume data from the ESR determinations, the internal IAA concentration within the sealed vesicles was calculated to be 4–5 μM with internal buffer concentrations of either 50 or 200 mM Tris [9]. This represents at least a 25-fold accumulation of auxin over the outside concentration. With the measured pH gradient, the maximal accumulation of IAA by passive diffusion should be only five times the external concentration of IAA. Thus, internal IAA concentrations were enhanced at least five-fold over the equilibrium distribution. This difference brought us to suggest that active transport of IAA by these zucchini membrane vesicles is possible and that the active uptake is via cotransport of IAA^- with two protons [9]. An attractive analogous system is that provided by amino acid or sugar cotransport with two protons. This suggestion is substantiated by the demonstration that [^{14}C]-IAA uptake can be increased by conditions (valinomycin plus K^+ internally) which should produce an outside-positive membrane potential [4, 5, 9]. When the electrogenic protonophore FCCP is added to these vesicles, it should reduce the pH gradient only to the point where an equal, but opposite, membrane potential is reached, totally dissipating the proton motive force, but leaving a considerable ΔpH ESR measurements have shown that this is indeed the case [9]. However, although a pH gradient exists, the ionophore-sensitive IAA accumulation is reduced to zero [9], indicating that the uptake is dependent upon the proton motive force rather than simply a pH or binding phenomenon.

4 Conclusion

Our experimental results have demonstrated that auxin uptake can occur via a specific, electrogenic $IAAH/H^+$ symport. This result does not exclude other kinds of IAA accumulation in excess (cf. [1]).

References

1. Clark KA, Goldsmith MHM (1986) This volume
2. Goldsmith MHM (1982) Planta (Berl) 155:68
3. Goldsmith MHM, Goldsmith TH, Martin MH (1981) Proc Natl Acad Sci USA 78:976
4. Hertel R (1986) This volume
5. Hertel R, Lomax TL, Briggs WR (1983) Planta (Berl) 157:193
6. Jacobs M, Hertel R (1978) Planta (Berl) 142:1
7. Lomax TL, Briggs WR (1983) Carnegie Inst Wash Yearbook 82:30
8. Lomax TL, Mehlhorn RJ (1985) Biochim Biophys Acta 821:106
9. Lomax TL, Mehlhorn RJ, Briggs WR (1985) Proc Natl Acad Sci USA 82:6541
10. Mandala S, Mettler IJ, Taiz L (1982) Plant Physiol (Bethesda) 70:1743
11. Mehlhorn RJ, Candau P, Packer L (1982) Methods Enzymol 88:751
12. Reinhold L, Kaplan A (1984) Annu Rev Plant Physiol 35:45
13. Rubery PH, Sheldrake AR (1974) Planta (Berl) 118:101
14. Sussman MR, Goldsmith MHM (1981) Planta (Berl) 151:15

Two Comments on Auxin Transport: The Uptake/Efflux-Mechanism and the Problem of Adaptation

R. HERTEL[1]

1 Uptake/Efflux: Role of Protons, Transmembrane Potential, Symports, and a Possible Binding

A "simple" transport scheme of auxin accumulation in the cytoplasm and auxin "secretion" from each cell consists of two specific molecular transmembrane processes at the plasmalemma: (1) an accumulation of IAA into the cytoplasm driven by a pH gradient and by the electrical membrane potential, and (2) an efflux of IAA through polarly distributed exit carriers. Figure 1 illustrates the mechanism at a cell or at an outside-out plasma membrane vesicle. The exit carriers are inhibited non-competitively by phytotropins and TIBA [3].

The chemiosmotic mechanism for auxin transport was originally proposed by Rubery and Sheldrake [14] as shown on the left in Fig. 1.

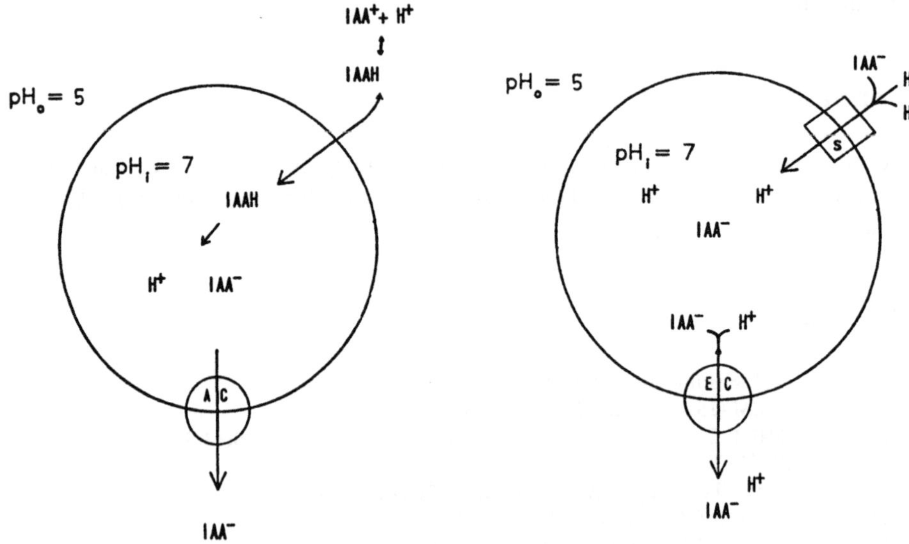

Fig. 1. Two chemiosmotic models. *S* electrogenic symport co-translocating 2 H^+ with 1 IAA^-; *AC* anion carrier; *EC* efflux carrier for IAA^- and H^+

1 Institut Biologie III der Universität Freiburg, Schänzlestr. 1, 7800 Freiburg i.Br., FRG

Plant Growth Substances 1985
(ed. by M. Bopp)

As a first approximation, this scheme is well supported by experiments with tissue segments [15] and with isolated plasma membrane vesicles [7].

However, IAA uptake is not a simple ion trap mechanism, but is specific and can be saturated by IAA [10]. Also, estimates of IAA uptake into the vesicles yielded accumulation ratios higher than expected from the H^+ gradients [8, 11], and furthermore, uptake is increased by an – outside positive – membrane potential [7, 8, 11].

Also, with transport tests in tissue segments it was shown [5, 6] that auxin accumulation was electrogenic, but surprisingly polar transport – with efflux limiting – was *not* electrogenic. This latter observation suggests that the notion of an anion (IAA^-) carrier should be dropped: auxin may leave the cell or the vesicle non-electrogenically, not as an anion, but as IAA^- plus one H^+. This efflux carrier would thus be a symport; in Fig. 1, right side, the original chemiosmotic scheme is modified and "lifted by one order of protons".

The important question to be settled experimentally is whether the rate of "normal" IAAH diffusion is slow enough not to interfere with the proposed saturable, specific accumulation/efflux mechanisms.

An alternative to accumulation of auxin into the "free" vesicle volume is proposed by Clark and Goldsmith [2]: a NPA-stimulated binding is held responsible for much of the observed specific IAA association. This hypothesis, however, appears unlikely for the following reasons: (1) this binding would have to be voltage-dependent [8, 11], and (2) the deviation observed with IAA accumulation from the expected value when the vesicles were shrunk [2], is also seen with the potential probe TPP^+ (Benning and Hertel, unpubl.); this suggests that the transmembrane potential increases upon shrinking (Donnan potential?), which could lead to the observed excess IAA accumulation at high osmotic strength without the necessity to invoke a binding.

2 More Complex Kinetics and the Problem of Adaptation. Effects of 3,4,5-TIBA (Triiodobenzoic Acid). NPA-Binding Kinetics

Auxin stimulates its own transport [4,9]. This polar transport, controlled at the efflux site, does not display a simple linear or saturation function, but a bell-shaped optimum curve [8] – as do many biological processes. Furthermore, auxin transport seems to adapt in response to stimulation [8].

In order to approach an understanding of these striking kinetic features, the plausible properties of the efflux carrier were computer modelled [8]. The function of the exit carrier is proposed to be a gating cycle of an integral membrane protein with two different conformations.

Whenever the relevant output is the number of cycles or transitions, rather than the state, then certain cyclic reaction schemes (transport mechanisms), by thermodynamic necessity, produce optimum functions [8].

Sensory adaptation (desensitation) is the decrease in sensitivity towards a stimulus following the exposure to this stimulus. De-adaptation or re-sensitation is the return to full sensitivity upon lowering the stimulation.

On a molecular level, adaptation could mean (see model in [8]) that the protein is accessible to a (covalent) modification – methylation or phosphorylation – only in one of the two conformations. The second conformation should be demethylated or dephosphorylated. The (covalent) modifying reactions, i.e. adaptation and de-adaptation, have to be slow (15–20 min) compared with the gating cycle (seconds or less).

The inhibition of transport (auxin efflux) by NPA, 2,3,5-TIBA and 3,4,5-TIBA appeared to be independent of the IAA donor concentration, indicating non-competitive, indirect interactions with the carriers [3].

It was also noted [3] that transport inhibition by high 3,4,5-TIBA concentrations was only partial, whereas with NPA and 2,3,5-TIBA it was almost total. If these three compounds act via a "common site in vivo", then 3,4,5-TIBA might displace the other inhibitors. Total inhibition by NPA or 2,3,5-TIBA was partially alleviated by increasing concentrations of the "soft" inhibitor 3,4,5-TIBA [3].

Figure 2 shows a time-dependent change in sensitivity (= adaptation) of IAA transport towards the inhibitor 3,4,5-TIBA. (Similar experiments had first been performed by W. Petersen and B.G. Kang in our lab; unpubl.) IAA export from maize coleoptile segments was followed over 3 h after a 1 h pretreatment with labelled auxin.

The slow recovery of the inhibited system (Fig. 2, +T) is not due to a pile-up of auxin, since pretreatment without IAA leads to the same recovery (data not shown). Newly applied TIBA acts strongly (+T at 120 min in Fig. 2), and increasing TIBA does not increase the inhibition (+4 × T).

Studies [3] on binding kinetics of phytotropins and cross-competition with [^{3}H]-NPA and [^{14}C]-2,3,5-TIBA have suggested that the efflux-carrier complex bears these different, but interacting binding areas: one site for auxin to be translocated and two regulatory sites with affinity for the inhibitors. Of the latter, one has a high affinity

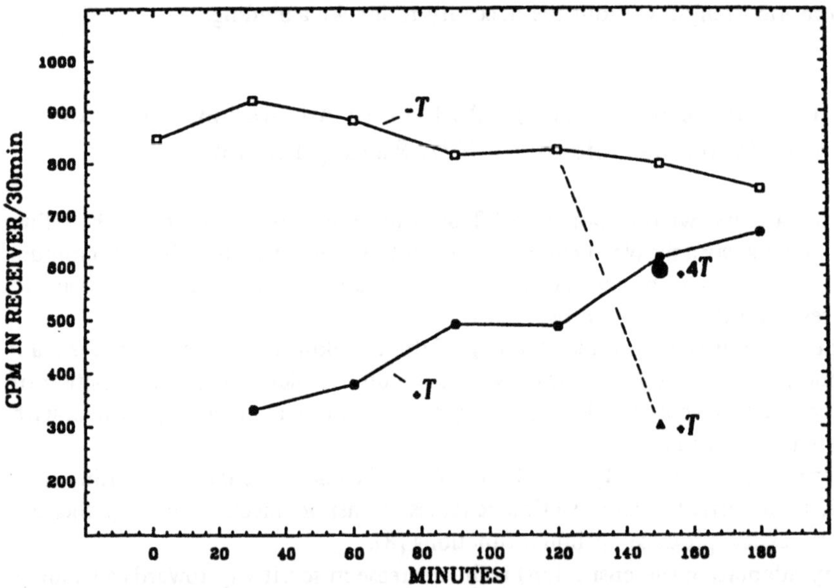

Fig. 2. Transport of [^{3}H]-IAA and inhibition by 3,4,5-TIBA in corn coleoptile segments (conditions see [9]; two segments per blocklet, 3.5 mm long; 5 × 10^{-7} M labelled IAA in donors). The segments were transfered to fresh receivers every 30 min. One set of receivers contained 2.5 × 10^{-5} M 3,4,5-TIBA (+ T)

for phytotropins, such as NPA (K_d = 3 × 10^{-8} M) and a low affinity of 2,3,5-TIBA ($K_d \geq 10^{-5}$ M), while the other one has a different affinity pattern (K_d = 3–5 × 10^{-7} M for both NPA and TIBA).

Some kinetic evidence supports this suggestion for two different interacting NPA-binding sites. A complicated time course is seen when studying on- and off-kinetics of labelled NPA at its specific and saturable receptor. Very slow phases of association, "trapping" of labelled NPA at the binding complex by shifting the pH from 5 to 8 [12] and a stimulation of the dissociation rate by adding unlabelled NPA (Michalke, unpubl.) have been found. With solubilized binding sites (0.3% CHAPS or 1.5% octyl-glucoside, incubated in binding buffer, pH 5, with tritiated NPA; quickly diluted into pH 7.8 buffer and filtered through Whatman glass filters and GF-B coated with poly-ethylenimine [1]) very similar – surprising – kinetic features were observed.

3 Conclusion

Are the complex, auxin transport kinetics due to many interacting elements, such as proteins and carriers and calcium etc., or else, are the biochemical elements already sophisticated and complex systems? The latter might be true – see the complex kinetics of the acetylcholin receptor [13]. A detailed analysis, e.g. of the efflux carrier will be relevant to reconstruct "physiological" kinetics.

Complicated and teleonomic, meaningful biological functions – and auxin transport certainly is such a function – might be realized in multiple biochemical processes on one single molecule, such as a protein (carrier) complex.

References

1. Bruns RF, Lawson-Wendling K, Pugsley TA (1983) Anal Biochem 132:74
2. Clark KA, Goldsmith MHM (1986) This volume
3. Depta H, Eisele K-H, Hertel R (1983) Plant Sci Lett 31:181
4. Goldsmith MHM (1982) Planta (Berl) 155:68
5. Goldsmith MHM, Rose R (1983) Plant Physiol (Bethesda) 72:Suppl 145
6. Hasenstein K-H, Rayle D (1984) Plant Physiol (Bethesda) 76:65
7. Hertel R, Lomax TL, Briggs WR (1983) Planta (Berl) 157:193
8. Hertel R (1983) Z Pflanzenphysiol 112:53
9. Hertel R, Flory R (1983) Planta (Berl) 82:123
10. Jacobs M, Hertel R (1978) Planta (Berl) 141:1
11. Lomax TL, Mehlhorn RJ, Briggs WR (1985) Proc Natl Acad Sci USA 82:6541
12. Michalke W (1982) In: Marme' D et al. (eds) Plasmalemma and the tonoplast. Elsevier, Amsterdam, p 129
13. Prinz H, Maelicke A (1983) J Biol Chem 258:10273
14. Rubery PH, Sheldrake AR (1974) Planta (Berl) 118:101
15. Sussman MR, Goldsmith MHM (1981) Planta (Berl) 151:15

Further Characterization of the Presumptive Auxin Transport Carrier Using Monoclonal Antibodies

M. JACOBS and T.W. SHORT [1]

1 Introduction

A key prediction of the chemiosmotic hypothesis of polar auxin movement [3, 9, 10] is that the carrier by which auxin anion molecules leave a transporting cell is located in differentially greater numbers at the cell's basal end. In order to visualize the auxin anion carriers (AAC), we have made monoclonal antibodies (Abs) against the naphthylphthalamic acid (NPA) receptor in etiolated pea stem tissue and used those antibodies in indirect immunofluorescence experiments [4]. Our presumption was that the receptor for NPA, a specific auxin transport inhibitor believed to act at the basal efflux site [1, 11], would be the same protein as, or closely associated with, the AAC. Our results have led us to the general conclusion that within the resolving power of our technique, the preponderance of sites in etiolated pea third internode tissue is at the basal ends of cells sheathing the vascular bundles (Fig. 1).

Here, we report further work we have done utilizing the anti-NPA receptor monoclonal Abs: more specific localization of the fluorescent labeling within the cells sheathing the vascular bundles in pea; examination of the effects of the anti-pea Abs on labeling, auxin transport, and NPA binding in other species; and preliminary characterization of the size of the protein recognized by our Abs in pea.

2 Materials and Methods

2.1 Plant Material

Pea third internodes (*Pisum sativum*, cv. Alaska), maize coleoptiles (*Zea mays*, WF9 X Bear 38) and zucchini hypocotyls (*Cucurbita pepo*, cv. Gray) were sown and grown as described in previous publications [5, 6].

2.2 Chemicals, Radiochemicals, and Antibodies

Chemicals were purchased from Sigma Chemical Co (St. Louis, MO, USA), [^{3}H]-NPA was purchased from Research Products International Corp. (Elk Grove, IL, USA),

1 Department of Biology, Swarthmore College, Swarthmore, PA 19081, USA

Plant Growth Substances 1985
(ed. by M. Bopp)

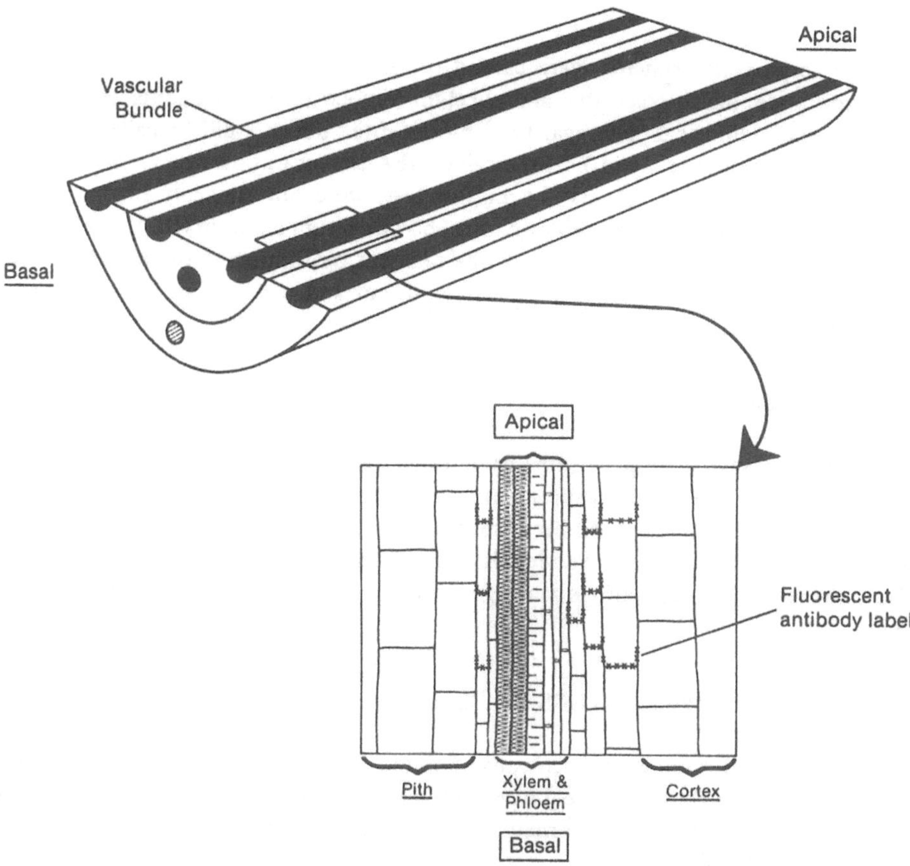

Fig. 1. Median longitudinal section of an etiolated pea third internode segment. Enlargement indicates fluorescence labeling pattern obtained using Abs active in inhibiting NPA binding in the screening assay

and [^{14}C]-IAA was purchased from Amersham Corp. (Arlington Heights, IL, USA). Secondary antibodies used in immunofluorescence and immunoprecipitation experiments were purchased from DAKO/Accurate Chemicals (Westbury, NY, USA) and Sigma. Monoclonal antibodies were produced as described previously [4]. Unlabeled NPA was a gift from R. Hertel.

2.3 Indirect Immunofluorescence

Experiments were performed as described previously [4] on median longitudinal sections of pea and zucchini. Maize coleoptiles were sectioned longitudinally through the region of the vascular bundle on either side of the coleoptile. Amyloplasts were visualized by immersing sections for 20 min in I_2KI stain after normal treatment in primary and secondary Abs.

2.4 Auxin Transport

Auxin transport experiments were carried out on 5 mm subapical segments of pea third internode, maize coleoptile and zucchini hypocotyl. The segments were placed between two cylindrical plain agar (2%) blocks, the "donor" and "receiver". Donor IAA concentration was 7×10^{-6} M, receiver NPA concentration when it was present was 10^{-8} M or 5×10^{-7} M. Basal transport tests were run vertically for maize and zucchini, horizontally for pea. Ab or control bovine serum albumin (BSA, 0.2%) solutions were applied to the basal end of the segment by dipping it twice into the appropriate solution before mounting between donor and receiver agar blocks. Transport time was 3 h.

2.5 NPA Binding

NPA-binding assays were performed as described previously [4] except that the final concentration of the plant tissue membrane in the assay tube was halved. General membrane preparations from different species were prepared in exactly the same way. Ab or control treatments were added directly to the assay tubes before spinning. Aliquots of membrane preparations were assayed for protein content according to Lowry [8].

2.6 Immunoprecipitation and SDS Gel Electrophoresis

In immunoprecipitation experiments, an etiolated pea third internode membrane preparation (5000–80000 × g pellet) was solubilized with 0.3% Triton X-100 and mixed with the appropriate clonal supernatant and Tris-buffered saline "2X" (20 mM Tris, pH 7.5, 0.3 M NaCl). The clonal supernatant was purified prior to use on a column of CM-Affigel Blue (BioRad Laboratories). The mixture was incubated on a shaker at 4 °C for 4 h, rabbit-anti-mouse IgG was added, then that mixture incubated for an additional 2 h. Protein A - agarose beads (Sigma), were then added and the incubation was allowed to proceed overnight. The protein A bead-Ab complexes were then pelleted, the supernatants discarded, and the pellets washed twice with Tris-buffered saline "1X" (10 mM Tris, pH 7.5, 0.15 M NaCl). The immunoprecipitates were boiled in SDS application buffer (0.125 M Tris-HCl, pH 6.8, 4% SDS, 20% glycerol, 10% B-mercaptoethanol) to separate the antigen-Ab complexes, the agarose beads were re-pelleted, and the supernatants applied to a 12% discontinuous polyacrylamide gel. Gels were run for 1.5 h at 20 mA, then an additional 3.5 h at 30 mA and stained with Coomassie Brilliant Blue.

3 Results and Discussion

3.1 Immunofluorescence and Amyloplast-Containing Cells

Further observation of longitudinal sections of dark-grown pea third internode tissue using indirect immunofluorescence has established a positive correlation between cells specifically labeled at their basal ends using positive monoclonal Abs and amyloplast-containing cells, which in pea form a "starch sheath" around the vascular bundles. In sections treated only with Abs, amyloplasts and the cells that contain them are nonetheless easily visible under phase-contrast illumination. In about 80%–90% of the cases of basal end fluorescent labeling observed, a check of the same section under phase-contrast indicated that the labeled cells were either starch sheath cells themselves or immediately adjacent to them. Figure 2 shows the results of an experiment in which median longitudinal sections of pea third internodes were treated first with the primary and secondary Abs, then with an I_2KI stain for starch that leaves amyloplasts dark blue or black. The dark amyloplasts of the starch sheath show up even in the fluorescence image of the section, and a strong correlation between the bright, specific labeling and amyloplast-containing cells is evident.

This correlation raises an interesting possibility if the fluorescently-labeled cells indicate the preferred auxin transport pathway in pea stems. That is, amyloplasts, displaced in response to shifts in the stem orientation within a gravitational field, could interact directly with the auxin transport carriers of the main transport stream *in the same cells* and have a proportionately greater and more rapid effect on patterns of auxin transport in the stem as it adjusts to a new orientation.

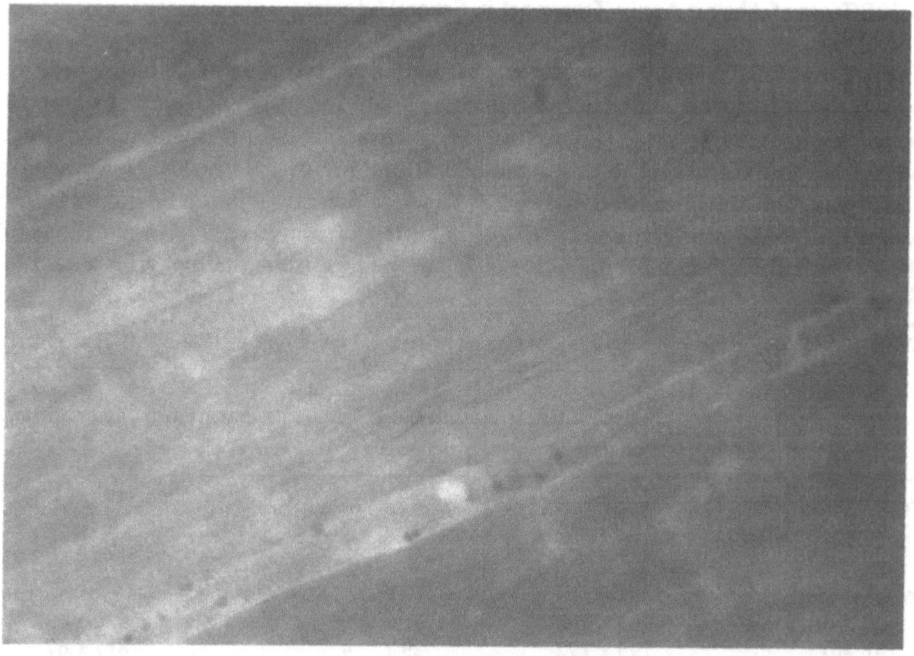

Fig. 2. Fluorescent antibody and I_2KI labeling of tissue associated with a pea stem vascular bundle. Longitudinal section prepared as described in text and illuminated here with UV light to activate fluorescein fluorescence (× 190). Under bright field, mature xylem and phloem tissue would be visible in the *upper left* corner with a single layer of starch sheath cells stretching across the *lower right*. Basal end of section is at *lower left*. In color, the bright amyloplast-containing cells are yellow (the fluorescein dye color), while the *upper left* portion of the image is gray to blue-gray

We have tried rhodamine-conjugated secondary Abs in indirect immunofluorescence experiments in an attempt to cut down the background fluorescence that exists at fluorescein wavelengths and help to more closely examine the relation between amyloplast-containing and fluorescently-labeled cells. However, we find the background fluorescence employing rhodamine just as high, if not higher, than with fluorescein in etiolated pea tissue. We have also made protoplasts from pea stem tissue: healthy protoplasts can be generated, some containing amyloplasts and with only a portion of the cells labeled with the fluorescein-conjugated secondary Ab, as one might expect from observations of tissue sections. However, the labeling we have seen so far is not specific to our positive as compared to our control Abs. It may be that with the cell surface in a protoplast suspension so much more easily accessible to large molecules in the medium, a potential low-affinity recognition of NPA sites by the fluorescein moiety of the secondary Ab complex [7, 11] is contributing to a high background in both control and experimental treatments.

We have checked for fluorescent labeling several times in zucchini hypocotyl and maize coleoptile tissue using both individual monoclonal Abs and oligoclonal mixtures of positive monoclonals as the primary Ab treatment. No specific labeling, of amyloplast-containing cells or others, is seen.

3.2 Effects of Abs on Auxin Transport in Several Species

Table 1 shows the results of experiments carried out to test the effects of the positive Abs on both basipetal auxin transport *and* the inhibition of basipetal auxin transport by NPA in several species. The mixture of positive Abs did not significantly affect basal transport when applied as a basal end dip in any of the species. However, for each tissue, once a concentration of NPA that only partially inhibited basal transport was used, the Ab mix did reduce NPA's inhibition of transport. The implications of these results are rather interesting. Not only can the positive Abs affect NPA action in

Table 1. Antibody effects on basal auxin transport and on its inhibition by NPA in different species

	Receivers		
	Plain agar	5×10^{-7} M NPA	10^{-8} M NPA
Pea			
Control[a]	132 ± 29[b]	7 ± 2	70 ± 9
Ab Mix[c]	161 ± 31	8 ± 2	190 ± 50
Zucchini			
Control	903 ± 155	35 ± 3	600 ± 12
Ab Mix	795 ± 192	29 ± 4	817 ± 67
Maize			
Control	3670 ± 192	30 ± 2	–
Ab Mix	3568 ± 167	651 ± 281	–

[a] 0.2% BSA: concentration chosen to match protein concentration in clonal supernatants
[b] Data shown are cpm in receiver ± standard error of the mean (SEM)
[c] Ab mix included clone 8, 8 ascites, and either 24 or A8-8 (all positive clones)

different unrelated species, but also in each case they affect NPA action without affecting basal auxin transport. This agrees quite nicely with the suggestions that have been made recently [2, 11] that the NPA site and the auxin transport site may be two separate portions of the AAC complex.

3.3 Effects of Abs on NPA Binding in Several Species

Table 2 shows the results of an experiment testing the effects of two of the positive Abs on NPA binding to membrane preparations from several different species. It is, of course, this binding assay in pea that acts as the screening assay for our selection of monoclonals to the NPA receptor. Not surprisingly, in pea, both positives, A8-8 and 8, inhibit NPA binding, with A8-8 more efficient. In zucchini and maize, both are equally inhibitory, with more than 50% inhibition. Thus, the clones that are positives in pea seem to be able to recognize the NPA receptors of other species by this measure as well as in the transport tests.

Table 2. Antibody inhibition of NPA binding in different species

	A [^{3}H]-NPA	B [^{3}H]-NPA + 10^{-5} M Unlabeled NPA	C NPA specific Binding (A–B)	D NPA specific Binding (%Contr.)
Pea				
Control[a]	415 ± 15[b]	279 ± 4	136	100
Clone 8	312 ± 8	221 ± 5	91	67
Clone A8-8	273 ± 21	227 ± 2	46	34
Zucchini				
Control	548 ± 14	268 ± 3	280	100
Clone 8	363 ± 11	237 ± 16	126	45
Clone A8-8	352 ± 5	221 ± 5	131	47
Maize				
Control	667 ± 14	360 ± 8	304	100
Clone 8	445 ± 10	302 ± 3	143	47
Clone A8-8	460 ± 11	323 ± 7	137	45

[a] 0.2% BSA: concentration chosen to match protein concentration in clonal supernatants
[b] Data shown are cpm in assay pellet ± SEM; μg protein per assay approximately the same in pea, zucchini, and maize

3.4 Immunoprecipitation Experiments

In preliminary attempts to characterize the protein recognized in pea by our positive monoclonal Abs, we combined proteins solubilized from a general membrane preparation with a mixture of several of our positive Abs (A8-8, 8, and 8 ascites). Employing a rabbit-anti-mouse IgG and protein-A-conjugated agarose beads as described in Sect. 2, we precipitated the positive Ab-pea protein complexes, released them from the pellet and ran them on an SDS gel. The results of two such experiments are shown in Figs. 3 and 4. In Fig. 3, a band at 77 kD appeared more and more prominently as the concentration of positive Ab was increased in the precipitations shown in lanes 3, 4, and 5. Since the precipitation run with control Ab in lane 6 showed no such band, we focused on the 77 kD protein in further experiments as potentially interesting. In Fig. 4, a later immunoprecipitation result with finer resolution, we see the 77 kD band again: here, the positive lane (No. 2) was again an oligoclonal mixture of positive Abs, but the control lane (No. 3) shows what precipitated when we used BSA at a protein concentration equivalent to that of the clonal supernatants. One can discern the BSA band at 66 kD in lane 3; the double band near it in the positive lane results from the monoclonal Abs used in the immunoprecipitation themselves. That is also the case for the band visible in lane 2, but not in lane 3 at ca. 200 kD. However, the lower molecular weight bands at the bottom of the gel in lane 2 that are at least considerably darker than the same area in lane 3 may be, as the 77 kD band appears to be, specific to our positive Abs. We are currently performing immunoprecipitation experiments with protease inhibitors present to see if the lower molecular weight bands are affected (e.g., disappear). Hertel's group, using one of our positive Abs (clone 8), has probed western blots of general solubilized pea and zucchini membrane protein preparations run on SDS gels and reports the development of a band at 77 kD on each

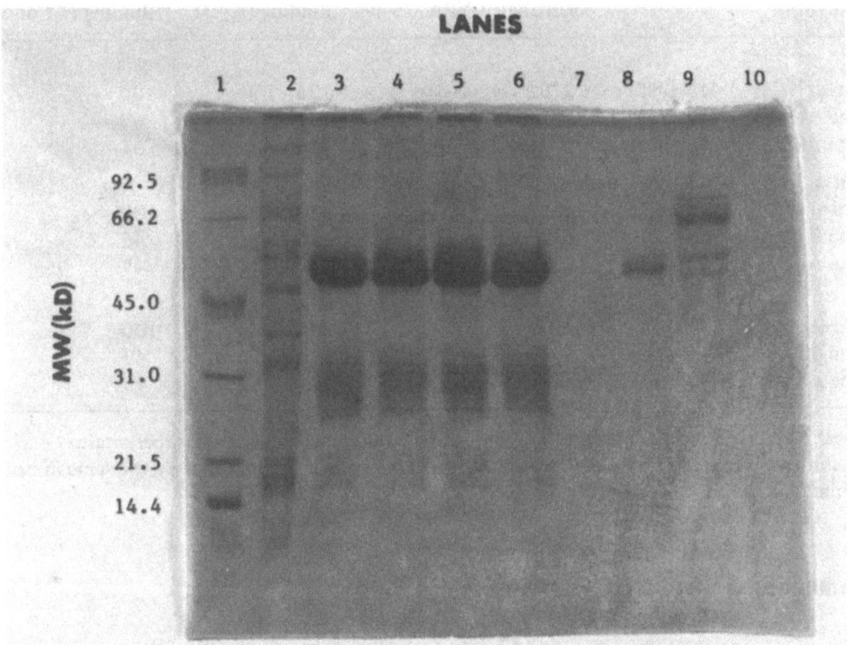

Fig. 3. SDS polyacrylamide gel electrophoresis (PAGE) of immunoprecipitated solubilized pea membrane proteins. Lane 1: low MW standards. Lane 2: total solubilized pea membrane proteins. Lanes 3–5: solubilized proteins precipitated by positive monoclonal mixture, quantity of antibodies increasing from lane 3 (90 μg) to lane 4 (900 μg) to lane 5 (2700 μg). Lane 6: solubilized proteins precipitated by control antibodies. Lane 7: protein A beads alone, boiled in application buffer. Lane 8: rabbit-anti-mouse antibody alone. Lane 9: oligoclonal mix alone

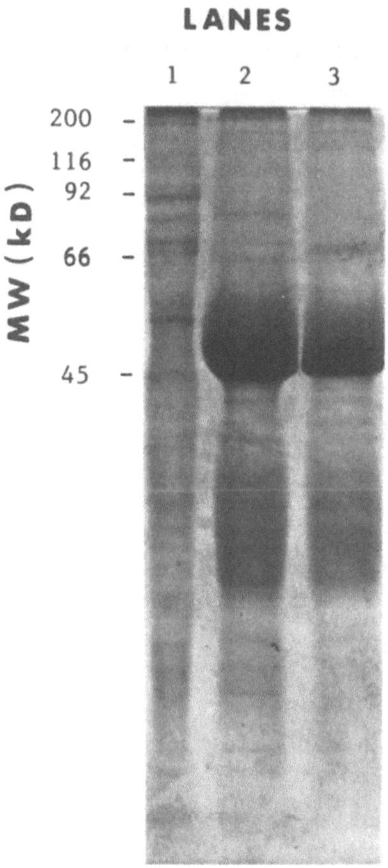

Fig. 4. SDS PAGE as in Fig. 3. Lane 1: total solubilized pea membrane proteins. Lane 2: proteins precipitated by oligoclonal mix of positive Abs. Lane 3: proteins precipitated when 0.2% BSA replaces the oligoclonal mix

blot (Hertel, pers. comm.). These immunoblot experiments, if they are corroborated, will strengthen the case for the 77 kD protein as the major NPA-binding moiety in peas and other species.

4 Conclusions

A strong correlation between cells labeled with monoclonal antibodies to the presumptive auxin transport carrier and amyloplast-containing cells has been observed in pea stem tissue. In an examination of the effectiveness of the anti-pea Abs in different species, we have evidence that they are: (1) capable of recognizing the NPA receptor in general membrane preparations from those species (Table 2); (2) presumably capable of affecting the NPA receptor "in vivo" (in the cells of a transporting segment from those species) (Table 1); yet (3) apparently only capable of specifically labeling cells in longitudinal sections from pea, not maize or zucchini. The protein recognized by the Abs in pea appears to be of molecular weight 77000; other smaller proteins that could represent subunits or merely protease breakdown products may be recognized by the positive Abs as well. We are looking forward to using the knowledge we now have of the NPA receptor/AAC to purify it and study its behaviour in artificial membranes.

Acknowledgments. We are grateful to the National Science Foundation for a grant (PCM 8314844) that supported this research and to Joanna Miller and Phillip Kloeckner for assistance with some of the experiments. We are also grateful to Scott F. Gilbert and David Kravitz for their help in monoclonal antibody production.

References

1. Cande WZ, Ray PM (1976) Planta (Berl) 129:43
2. Depta H, Rubery PH (1984) J Plant Physiol 115:371
3. Goldsmith MHM (1977) Annu Rev Plant Physiol 28:439
4. Jacobs M, Gilbert SF (1983) Science 220:1297
5. Jacobs M, Hertel R (1978) Planta (Berl) 142:1
6. Jacobs M, Taiz L (1980) Proc Natl Acad Sci USA 77:7242
7. Katekar GF, Geissler AE (1975) Plant Physiol (Bethesda) 56:645
8. Lowry OH, Rosebrough NJ, Farr AL, Randall RJ (1951) J Biol Chem 193:265
9. Raven JA (1975) New Phytol 74:163
10. Rubery PH, Sheldrake AR (1974) Planta (Berl) 118:101
11. Sussman MR, Goldsmith MHM (1982) Planta (Berl) 152:13

The Requirement for Calcium and Boron in Auxin Transport

R.K. DELA FUENTE, P.M. TANG, and C.C. DE GUZMAN[1]

1 Introduction: The Need for Calcium in Auxin Transport

The requirement for Ca in auxin transport was first observed in sunflower hypocotyl tissue with Ca deficiency induced by chelation of tissue Ca. The longer the hypocotyl segments were incubated in EDTA solution, the more Ca was lost to the medium and the greater was the inhibition of auxin transport [4]. These results were confirmed with sunflower seedlings grown in Ca-deficient Hoagland solution [3]. Auxin transport was found to be linear with the log concentration of Ca in the root medium from 10^{-6} to 10^{-3} M. The most significant finding in these experiments, regardless of the method used to induce Ca deficiency, was that the inhibition of auxin transport can be reversed to the control, nondeficient level by the simple incubation of the hypocotyl segments in Ca solution, prior to the measurement of auxin transport [3, 4].

The requirement for Ca cannot be replaced by the more common divalent cations, such as, Sr, Mg, Mn, Fe, Co, or Ni. The trivalent cation La at 1 mM or less, seemed to be able to replace Ca [3, 4], but more recent data (Table 1) show that this effect of

Table 1. Effect of some cations in auxin transport in calcium-deficient sunflower hypocotyl segments[a]

Seedling growing medium		Segment incubation medium (2 h)		
0–3 day (mM)	3–4 day	Pipes only	1 mM of same ion	1 mM Ca
		Auxin transport (cpm 4 h^{-1})		
0.1 Ca	0.1 Ca	1743 ± 50		2575 ± 16
0.1 Ca	1.0 Ca	2660 ± 110	3179 ± 91	3088 ± 49
0.1 Ca	1.0 La	956 ± 42	1094 ± 58	1978 ± 57
0.1 Ca	1.0 Mg	1082 ± 38	1902 ± 86	3013 ± 53
0.1 Ca	1.0 Sr	1725 ± 27	1868 ± 32	2741 ± 44

[a] The seedlings were in Hoagland solution with 0.1 mM Ca for the first 3 days. To reduce displaceable Ca in some treatments, the plants were transferred for 24 h to solutions without Ca, but in the presence of the element being tested to substitute for Ca. Thereafter, auxin transport was determined by the classical apical agar donor containing 1-^{14}C(IAA) and a blank agar receiver at the basal end of 20, 5 mm hypocotyl segments, replicated three times

1 Department of Biological Sciences, Kent State University, Kent, OH 44242, USA

Plant Growth Substances 1985
(ed. by M. Bopp)

La may be due to its more efficient means of displacing Ca from adsorption sites, relative to other ions.

2 The Possible Function of Calcium in Auxin Transport

One of the more widely known functions of Ca in living cells is the maintenance of membrane integrity [5, 6]. In general, when the Ca present in the environment of plant and animal cells is low, the rate of leakage of endogenous substances is high. The reverse is true in the case of IAA, the lesser the Ca content of the tissue, the lower the rate of IAA efflux [3, 4]. This indicates, a priori, the presence in the plasmalemma of specific, IAA-carrier molecules, the operation of which requires the presence of Ca. It is very likely that this Ca-IAA secretion process has a common ancestry with that of the Ca-requiring secretion processes found in animal cells, known as "stimulus-secretion coupling" [10], or even to the more general concept of "second messengers" [7]. An external signal, chemical or physical, modifies membrane permeability causing the downhill flow of Ca into the cytoplasm. The rise in Ca and the concomitant steps that follow, initiates the secretion of the substance in question. Both IAA-transport flux and velocity were found to be less in low Ca hypocotyl, which is consistent with the view that the secretion process is an all-or-nothing response at a certain level of Ca in the environment of the cells [3].

3 Consequences of IAA Transport on Calcium Movement

If our hypothesis is correct that influx of Ca triggers auxin secretion, and since IAA is secreted at the basal end of the cell, then it is likely that Ca enters the cell through the basal plasmalemma. Furthermore, since the cell maintains a low concentration of Ca in the cytoplasm, then exit or efflux of Ca would likely occur in higher proportion on all sides relative to the basal end. Using 20 mm hypocotyl segments, we found a higher efflux of Ca at the apical end, relative to the basal end [1]. It was also found that the acropetal movement of Ca was promoted by auxins IAA and α-NAA, but not by the nonauxin β-NAA. In addition, the auxin-enhanced Ca efflux was found to be inhibited by the auxin-transport inhibitor, TIBA, suggesting that auxin has to be transported in order for acropetal Ca movement to occur [1]. Like auxin transport, acropetal Ca movement was also inhibited by cyanide and low temperatures, and was not affected by reversing the morphological orientation of the segment relative to gravity [2].

4 The Need for Boron in Auxin Transport

It is possible that the inhibition of auxin transport may not be the direct result of a lack of Ca, but rather of its effect on the tissues. This problem was pursued with the aid of another essential element, boron, whose deficiency symptoms in plants resemble that of Ca. Both Ca and B are relatively immobile ions in plant systems [8, 9]. Withdrawal of B or Ca from the root medium leads to necrosis of root and shoot meristems. It was deduced that if the inhibition of auxin transport is a direct result of the lack of Ca, then B-deficient hypocotyls should not exhibit an inhibition of auxin transport, unless B itself is also involved.

Experiments with B-deficient seedlings showed auxin transport was just as much inhibited as in the Ca-deficient hypocotyl [11–13]. This suggested (1) that auxin-transport inhibition may be due to the effect of the deficiency on the tissue, and only indirectly to the lack of the essential element itself, or (2) that B itself may also possess site(s) involved in auxin transport. The first possibility is supported by the finding that both Ca- or B-deficient hypocotyls also exhibited a higher rate of K^+ efflux, as well as a higher rate of respiration, relative to nondeficient hypocotyl segments [11, 12]. Further experiments showed that incubation of B- or Ca-deficient hypocotyl in B or Ca solution, respectively, reversed membrane permeability to the level of the nondeficient hypocotyl [11–13]. This suggested that both elements possess sites in the plasmalemma that control its permeability. The reversal of permeability with short incubation also suggests that both Ca and B may be component parts of their respective sites.

A second site for each element was also found. The second site for B was indicated by the finding that auxin transport was not reversed in B-deficient hypocotyl incubated in B solution, even though membrane permeability was reversed to the control level [11, 13]. However, the transfer of B-deficient seedlings to B solutions prevented further decline in auxin transport [12], suggesting that the formation of the functional second site for B probably requires an intact plant, or a much longer period of incubation of the hypocotyl segment in B solution. The lack of reversal of auxin transport to the control level, in B-deficient hypocotyl incubated in B solution, gives some hint that B itself may not be an essential component of its second site.

On the other hand, the incubation of Ca-deficient hypocotyl in Ca solution completely reversed both the high rate of solute leakage as well as the inhibition of auxin transport [11–13]. We believe the second site for Ca is involved in the operation of the IAA-carrier site. The two Ca sites can be distinguished from each other by the property of the first site to remain functional even in the absence of Ca, to a certain extent, when other polyvalent cations are present [11]. As seen earlier, the second site is highly specific for Ca.

5 Conclusion

The sites for Ca were identified in plant materials with normal B nutrition. Likewise, the B sites were found in plants with normal Ca content and deficient only in B. Thus, the auxin-transport process requires the presence of the two immobile essential elements, Ca and B, each of which could control auxin transport in at least two sites.

Acknowledgments. Supported by National Science Foundation Grants PCM 78-04920 and DMB 84-16619.

References

1. De Guzman CC, Dela Fuente RK (1984) Plant Physiol (Bethesda) 76:348
2. De Guzman CC, Dela Fuente RK (1986) Plant Physiol (Bethesda) (in press)
3. Dela Fuente RK (1984) Plant Physiol (Bethesda) 76:342
4. Dela Fuente RK, Leopold AC (1973) Plant Physiol (Bethesda) 51:845
5. Leopold AC, Poovaiah BW, Dela Fuente RK, Williams RJ (1973) In: Plant growth substances 1973. Hirokawa, Tokyo, p 780
6. Manery JF (1966) Fed Proc 25:180
7. Rasmussen H (1981) Calcium and cAMP as Synarchic Messengers. Wiley, New York
8. Raven JA (1979) New Phytol 79:465
9. Raven JA (1980) New Phytol 84:231
10. Rubin RP (1982) Plenum, New York
11. Tang PM (1983) Thesis, Kent State University, Kent, Ohio, USA
12. Tang PM, Dela Fuente RK (a) (1986) Plant Physiol (Bethesda) (in press)
13. Tang PM, Dela Fuente RK (b) (1986) Plant Physiol (Bethesda) (in press)

Cellular Patterns Determined by Polar Transport

T. SACHS [1]

1 Introduction

The topic of this chapter is the control of the patterned placement, or the relative positioning, of various cell types. Polar transport means that the sides of cells are unequal and that the orientation of the cells within a tissue is coordinated. This transport is, therefore, an expression of pattern at a cellular level. Polarity at a cellular level, furthermore, can also be seen microscopically in the arrangement of specialized cells in vascular strands that connect the shoots with the roots. It is asked below whether the transport and the structural expressions of polarity are related and could be used to study one another. This leads to an hypothesis concerning the control of differentiation of cells in longitudinal rows.

2 Polarity Expressed by Differentiation

Vöchting's [10] original evidence that the plant axis is polar was based on the localization of new organs on cuttings. In many, though not all, cases this localization is independent of environmental cues. New organ localization can now be accounted for, at least partially, by the differences between the parts of the cutting that arise from continued polar transport [9]. It is often forgotten, however, that Vöchting's [11] evidence for tissue polarity went further: he showed that plants that had part of the shoot-root orientation of their axis inverted, by grafts or an appropriate suppression of new organs, survive, but are unable to continue their development. Transport through mature xylem and phloem is not polar, but Vöchting found that the development of the vascular system is abnormal in the inverted plant: it does not form new connections between the shoot and the root, but is rather concentrated in the form of tumors, at the base of the shoots. This effect of polarity raises the question whether vascular differentiation is dependent on auxin and its polar transport. Auxin induces the differentiation of all types of vascular tissues in wounded plants and tissue culture (Fig. 1A; reviewed by Sachs [7]). This effect of auxin is unique in its ability to *orient* the differentiation of the new tissues when it is applied locally and in its ability to replace the effect of known organs, especially expanding leaves (Fig. 1B). Since it is

1 Department of Botany, The Hebrew University of Jerușalem, 91904 Jerusalem, Israel

Plant Growth Substances 1985
(ed. by M. Bopp)

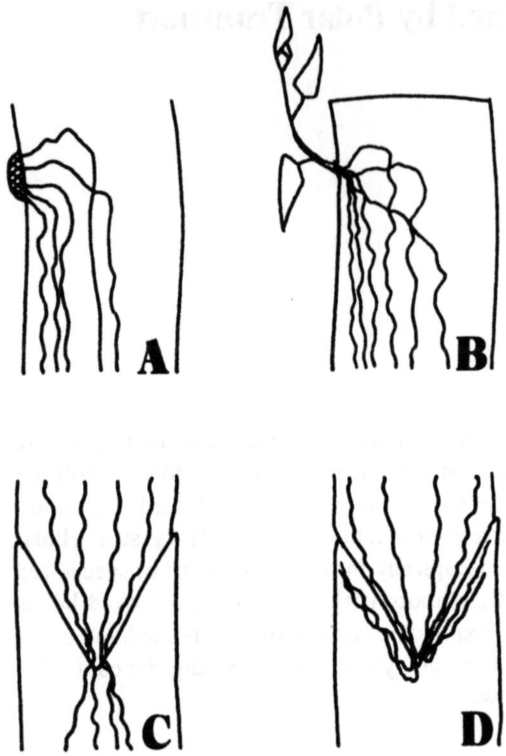

Fig. 1A–D. Induction of vascular differentiation and its relation to polarity. Original vascular tissues not marked; *wavy lines* show the course of new vessels. **A** Cut stem to which a source of auxin was applied laterally. **B** Growth of a lateral bud after shoot decapitation. **C, D** Grafts; only in **D** was the lower member inverted before being grafted. The differentiation in **D** does not connect the shoot with the root

known that auxin is produced by such leaves, it must be part of the signal by which they control the formation of their vascular contacts with the plant. In other words, auxin acts as an natural inducer of differentiation.

3 Stability and Lability of Polarity

This raises the question whether vascular differentiation, which can be observed with a microscope, follows tissue polarity as expressed by auxin transport. A strict effect of polarity, even at the expense of the functional relations between the shoot and the root, is easy to see in grafts (Fig. 1C,D) and in wounded plants (Fig. 2A,B). However, it is also easy to find vascular differentiation at all possible angles to the original axis of the organ: near buds growing after they are released from apical dominance (Fig. 1B), in plants wounded in various ways (Fig. 2C), near a source of auxin (Fig. 1A), and even in double grafts where a part of the axis is inverted, provided this inverted region is short [5]. This raises an apparent contradiction: polarity is both determined and labile [8]. A survey of all the various experiments suggests that differentiation along a new axis occurs only when polar transport is blocked in some way, usually by cuts. New differentiation then occurs along the expected path of the diffusion of auxin from where it accumulates, as a result of polar transport, to where it can be removed in the direction of the root.

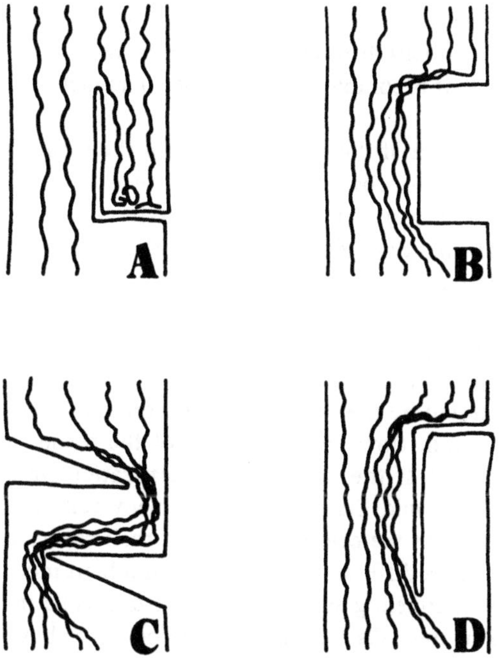

Fig. 2A–D. Regeneration of vascular tissues in wounded plants. Only new vessels are shown by the *wavy lines*. **A** Differentiation continues into a flap of tissue pointing downwards. This differentiation serves no functional purpose. **B** If the flap of tissue in **A** is removed, differentiation around the wound occurs readily. This differentiation connects the shoots and the roots. **C** New regenerative differentiation occurs readily in transverse tissue bridges, when they are the only shoot-root connection. **D** Little or no new differentiation occurs in tissue flaps that have direct connections to the roots, but not the shoots

This suggests that auxin movement and tissue polarity are linked by a positive feedback relation: the movement of auxin polarizes the tissue and this polarization orients and enhances auxin movement [7, 8]. Such a positive feedback would account very well for the stability of polarity as long as polar transport can continue and for the lability of this very same polarity when it can not. The critical assumption of the hypothesis is that the movement of auxin can induce a new tissue polarity. For polarity expressed by vascular differentiation, the evidence for this is in direct microscopic observation. There is now evidence that a new polarity of auxin transport is also induced where it is expected on the basis of vascular differentiation [2]. Horizontal tissue bridges of bean plants, as shown in Fig. 2C, were allowed to differentiate new vascular tissues and were then cut and tested for auxin transport. Movement of radioactivity from [^{3}H] indoleacetic acid in the new shoot-to-root direction was twice that from the root to the shoot. The capacity to transport auxin, furthermore, appears in such tissue bridges long before the transport of other substances, which is as expected if auxin transport is an early stage in the differentiation of the tissue bridges along their new axis [1].

4 Control of Pattern by Canalized Flow

The hypothesis that vascular differentiation is an expression of polarity and depends on the transport of auxin makes it possible to consider how the vascular patterns are determined. The overall form of the vascular system suggests that differentiation occurs along the flow of auxin, from the young tissues of the shoot to the root tips. This flow pattern is seen most clearly in regenerative differentiation around wounds (Fig. 2B). Root tips, furthermore, appear to orient differentiation towards themselves by acting as sinks for auxin flow in a way that is complementary rather than identical with that of the shoot tips (Fig. 2D, [5]).

The outstanding cellular aspect of the pattern of the vascular system is the arrangement of the specialized cells in discrete strands or rows. It should then be asked why, if differentiation occurs along the flow of auxin, it is limited to certain cells. This limitation could not depend on a predetermination of the cells by early events, since wounds and auxin application induce differentiation of any of the cells of a competent tissue. The precise course of new xylem vessels is, furthermore, quite variable, even in intact plants. It is the inductive flow itself, therefore, that must become canalized to the cells that later differentiate into mature vascular channels. An hypothesis as to how this canalization could occur is suggested by the previous conclusion, that there is a positive feedback relation between auxin flow and the ability of the tissue to transport this flow in a polar fashion [6, 7]. If the induction of polarization were gradual, then the inductive signals (auxin) would first pass through many cells and would become confined to discrete channels as the cells specialize and their capacity for polar transport increases. Each cell along a differentiating file would inhibit most of its neighbors from differentiating by channeling signals through itself, while inducing further flow and differentiation in the cells that are morphologically above and below.

This hypothesis, first suggested some time ago [6] has withstood the test of mathematical modelling [4] and is supported by indirect evidence of various types. The evidence has been summarized in some detail elsewhere [7] and for reasons of space its major lines will only be mentioned here. (1) The induction of differentiation is, as required by the hypothesis, gradual and requires a flow of auxin during a period that is long relative to that of the entire process of differentiation, about 2 days. (2) Vascular strands act as a sink for auxin flow and, unless they are well supplied with auxin, attract new strands towards themselves. (3) The transport of auxin has been shown to be most rapid in the vascular system. In maturing stems it has even been found to be canalized. It occurs first in many cells and later becomes confined to the vascular region [3]. (4) The cells neighboring vascular strands often have forms that indicate "partial induction" – as though they transported the critical signals, and became elongated, but were then prevented from further differentiation by the specialization of the neighboring vascular system.

5 Conclusions

It was shown above that polar transport has major consequences for cell differentiation; these can be readily seen microscopically. On the basis of this differentiation, it can be suggested that tissue polarity not only controls auxin flow, but is also induced by a new flow due to diffusion. The resulting gradual feedback between polarity and flow would account not only for stable and labile expressions of tissue polarity, but also for the canalization of both flow and differentiation to discrete rows of cells.

This hypothesis, perhaps the most detailed suggestion concerning a major cellular pattern, differs from other proposed controls in stressing the directional properties of both induction and differentiation. It shows that the course of signals that act across the entire plant could be locally modified to patterns at a cellular level. Finally, differentiation does not appear to be necessarily separate from the specification of the special location of cells: on the contrary, it is the gradual feedback between differentiation and the distribution of differentiation-inducing signals, rather than a global developmental plan, that results in the dependable though variable patterns found in the biological world.

References

1. Gersani M (1986) J Exp Bot 36:1809
2. Gersani M, Sachs T (1984) Differentiation 25:205
3. Lachaud S, Bonnemain JL (1984) Planta (Berl) 181:207
4. Mitchison GJ (1980) Proc R Soc Lond B Biol Sci 207:79
5. Sachs T (1968) Ann Bot (Lond) 32:391
6. Sachs T (1969) Ann Bot (Lond) 33:263
7. Sachs T (1981a) Adv Bot Res 9:151
8. Sachs T (1981b) In: Schweiger HG (ed) International cell biology 1980–1981. Springer, Berlin Heidelberg New York, p 489
9. Sachs T (1984) In: Barlow PW, Carr DJ (eds) Positional controls in plant development. Cambridge University Press, Cambridge, p 193
10. Vöchting H (1878) Über Organbildung im Pflanzenreich. I. Cohen, Bonn
11. Vöchting H (1892) Über Transplantation am Pflanzenkörper. Laup'schen Buchhandlung, Tübingen

Auxin Transport and Shoot Tropisms: The Need for Precise Models

R.D. FIRN[1] and S. TAMIMI[1,2]

1 Introduction

Over half a century has passed since the Cholodny-Went model of plant tropisms was outlined [10] and since that time the model has had both its forceful advocates and its critics. The purpose of the present chapter is not to go over old arguments because such arguments have obviously failed to resolve the dispute. Instead, the present chapter will attempt to lead the arguments into new areas in the hope that greater progress will result. We shall concentrate on one aspect of the C-W model, the involvement of auxin movement in the induction of differential growth causing shoot gravitropism, and we will try to show that existing knowledge about shoot tropisms demands that new ideas and approaches be brought to this area of research. It will be argued that the proponents of the C-W model have failed to develop the model with the result that it is now too vague and it no longer provides the necessary theoretical framework on which to base future research. Taking into account some known constraints, a revised model will be outlined in order to illustrate the very significant changes which would be needed to the original Cholodny-Went model, if it were to attempt to account for some known facts. This revised model is presented in the hope that it will provoke others to refute it and replace it with a more precise, satisfactory one.

2 Some Constraints on Models of Shoot Tropisms Involving Auxin Movement

2.1 Auxin in the Basipetal Transport Stream Seems Not to Control Elongation

"The involvement of transportable auxin with a number of phenomena, particularly tropisms, is so universally acknowledged that to raise the question at all may appear unrewarding" [7].

The original formulation of the C-W theory demanded that auxin in the basipetal transport system (BATS) controlled elongation, indeed, it was a central tenet of the

1 Department of Biology, University of York, York, Y01 5DD, Great Britain
2 Present address: Department of Biology, University of Jordan, Amman, Jordan

Plant Growth Substances 1985
(ed. by M. Bopp)

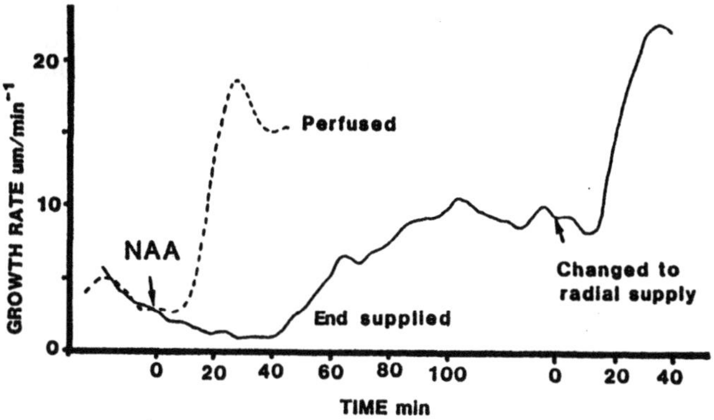

Fig. 1. The inability of the BATS to supply a saturating dose of NAA to the auxin responsive cells in sunflower hypocotyl sections. The length of sections were monitored continuously before and after addition of NAA to the apical cut surface or to the bathing medium. (Redrawn from Firn [4])

model. However, the evidence that auxin in the BATS controls elongation depends greatly on old correlative studies on shoot tropisms – a classic circular argument. Some recent tests we have conducted on the role of the BATS seem to suggest that Mer's scepticism was not unreasonable. First, it is evident that the BATS is not very efficient at providing auxin to the elongation limiting cells (ELC) in hypocotyls. If NAA is supplied to the apical cut surface of sunflower hypocotyl sections, high (unphysiological?) concentrations need to be used to produce a significant response [4]. Furthermore, the response is slower and less reproducible than if the NAA is supplied radially via the cuticle and the BATS, even when supplied with very high concentrations of NAA, cannot provide a saturating dose to the ELC (Fig. 1). Such results made it difficult to understand how small changes in the amount of auxin in the basipetal transport flow could bring about dramatic changes in the rate of hypocotyl elongation. Consequently, further studies were undertaken to assess the role of the BATS in intact seedlings.

To extend our knowledge of the role of the BATS in supplying auxin to the cells in the elongation zone, studies were made of the effect of locally inhibiting the BATS in order to reduce the flow of auxin into the elongation zone via the BATS. A small ring of lanolin containing a BATS inhibitor (the morphactin CFM) was applied just below the apical hook of etiolated sunflower seedlings in which hypocotyl elongation was being monitored. Despite a > 90% inhibition of the BATS, the elongation rate of the hypocotyl was not reduced during the next few hours (Fig. 2) [9]. Such results would seem to suggest that auxin flowing into the hypocotyl via the BATS is not rate limiting to cell elongation in the short-term. Recently, equivalent studies have been conducted on maize coleoptiles with very similar results (A. Parsons, unpubl.) Obviously, more studies of this type are needed and they should include an analysis of the endogenous IAA following the relevant treatments, but it is already apparent that the role of the BATS in controlling short-term growth is questionable.

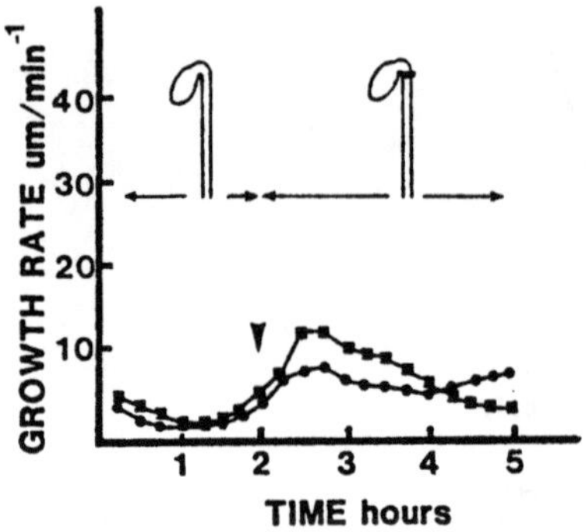

Fig. 2. The effect of blocking the BATS just beneath the apical hook on the elongation of sunflower hypocotyls (averaged data from three plants). (Redrawn from Tamimi and Firn [9])

2.2 Lateral Trans-Organ Movement of Auxin is Not the Cause of Differential Growth

It may seem a provocative thing to say, but nearly all studies of gravitopically-induced lateral auxin movement have been studies of a process which is demonstrably unnecessary to the initial induction of differential growth. It is now quite clear from studies on semicylinders that the initial gravitropic response of coleoptiles or hypocotyls does not depend on IAA movement over a large distance, either longitudinally or laterally. An "upper" or a "lower" semicylinder in isolation both show initial gravitropic responses, which are similar to those found in the equivalent parts of intact organs [3, 5]. Consequently, measurements of IAA movement, whether endogenous or exogenous, between the upper and lower halves of such gravistimulated organs are measurements of a process which is not essential to the initial growth rate changes causing gravitropism. While it has been shown repeatedly over the last 5 decades that very small amounts of auxin can be redistributed across gravistimulated organs (usually a few percent) it must now surely be recognised that any important movement (if indeed there is any important IAA movement) must be occurring locally. The extent to which studies of lateral trans-organ movements of IAA have dominated research on plant tropisms is undoubtedly due to the fact that workers were being guided by a vague, inaccurate model. Had the model been modified to recognise that lateral trans-organ IAA movement is not essential, it is likely that a completely different experimental approach would have been adopted.

2.3 Morphactins Inhibit Some Tropisms in Which Auxin Movement is Unlikely to be Involved

The fact that compounds, which inhibit auxin transport, such as the morphactins, abolish tropistic responses has been used as evidence that auxin transport must be involved in tropistic responses. However, it is becoming apparent that these inhibitors abolish tropistic responses even in organs in which it is not usually believed that auxin transport is involved. The best example of such an inconsistency is grass node gravitropism, which is sensitive to these inhibitors yet is generally accepted not to involve lateral IAA movement [1]. More controversial are hypocotyl and coleoptile gravitropism, hypocotyl and coleoptile phototropism towards continuous unilateral light and root gravitropism. In all these cases there are growing doubts that the differential growth in such organs are solely accounted for by lateral IAA movement [6, 11]. Thus, the ability of a compound to inhibit tropistic responses and to inhibit the BATS does not prove that the BATS is directly linked with tropistic responses. It is possible that tropistic responses and the BATS share a common reaction or process susceptible to morphactin inhibition. It could even be argued that morphactins disturb the tropistic responses of organs and as a consequence the BATS is inhibited. No matter how these compounds act, they offer interesting opportunities for probing the BATS and tropistic response mechanisms if used imaginatively.

3 Revising Models of Shoot Gravitropism in the Light of the Above Constraints

What kind of models could one propose in the light of the factors discussed above, if one was determined to retain a role for auxin movement in tropistic responses? In order to focus our discussion of this point, we shall concentrate on shoot gravitropism although similar precise models are also needed for phototropism.

3.1 A Possible Model

The model is based on the idea that the cells involved in lateral IAA movement are distinct from the cells involved in the BATS. This is a significant departure from traditional thinking, but it has a number of attractive features. In particular, it would explain why so many careful investigations have shown that gravi- or photostimulation has such a small effect on the BATS. A close examination of any study of longitudial and lateral IAA movements in stimulated organs always shows that an IAA molecule in such organs has a much greater chance of moving longitudinally than laterally [2, 8]. If the BATS were redirected by gravi- or photostimulation, one would surely predict a greater disturbance of longitudinal IAA movement. It is proposed in this new model that the cells involved in lateral IAA movement (the lateral transport cells) are situated between the cells involved in the BATS and the ELC. This separation of the ELC and the BATS being necessary because of the factors discussed in Sect. 2.1, and one could imagine that the ELC were the peripheral cell layers [5] and the cells involved in the BATS were near the vascular tissues. In the model, tropistic stimulation of the lateral transport cells would activate the lateral transport mechanism, hence control the flow of IAA from the BATS to the ELC; lateral trans organ IAA movement not be required (see Sect. 2.2). It would be necessary to postulate that the half-life for IAA in the ELC was very short in order to explain the dramatic reduction in elongation at the upper side of a gravistimulated organ. In this model, the morphactins would act by interfering with lateral IAA movement in the lateral transport cells and with the BATS in the cells in which that process occurs. However, the model has some drawbacks. It would be difficult to explain the finding that a gravitropic response can be measured in organs in which the elongation has been stimulated by radial IAA application [5], as in this case, the applied IAA would presumably swamp any gradients of endogenous IAA.

The model could be modified. It could be suggested that the lateral transport cells are always actively moving IAA leaking from the BATS back into the BATS, hence stopping any leak of IAA into the ELC. Tropistic stimulation would simply increase or decrease lateral IAA movement to generate the necessary IAA gradients. This modified model is handicapped by the fact that morphactins do not mimic IAA when applied to hypocotyl sections (which is what would be predicted if they simply blocked the IAA carriers and allowed IAA in the BATS to flow out to the ELC).

While this simple model has obvious deficiencies, in our opinion it serves as an example of the type of detailed model which those studying the involvement of auxin transport in tropisms should be attempting to produce. A model, which is not precise, becomes so flexible that it offers very little real guidance in designing experiments or evaluating results. Many, who subscribe to the Cholodny-Went model of plant tropisms, no doubt have in mind a clearer view of the model than is usually presented in textbooks - that auxin movement is involved in tropisms in some unspecified way. However, it seems remarkable that 50 years of research has resulted in a less defined model rather than a more detailed one.

4 Conclusions

Scientific models are only useful if they result in progress - stimulating thought and providing a fruitful framework for experimentation. A good model not only survives attempts at disproof, it leads to new questions which, when answered, increase our knowledge. It is our contention that the Cholodny-Went model has either been served poorly by its advocates or it is a model which no longer can be considered good. It has not led us to an understanding of which cells preceive light or gravity. It has not led us to any workable model as to how light or gravity could influence the BATS at a cellular or molecular level. Indeed, it is hard to think of any question, which has arisen from studies guided by the original Cholodny-Went model, which has resulted in significant new understanding. The time has come to make it a predictive model by giving it a new precision or to seek alternative explanations of plant tropisms. In the present chapter, we have tried to do the former, but we are conscious of the possibility that the latter might be more profitable.

References

1. Bridges IG, Wilkins MBU (1973) Planta (Berl) 112:191
2. Cane AR, Wilkins MBU (1969) Plant Physiol (Bethesda) 44:1481
3. Carrington CMS, Firn RD (1983) J Exp Bot 34:283
4. Firn RD (1982) Plant Physiol (Bethesda) 70:1568
5. Firn RD, Digby J (1977) Aust J Plant Physiol 4:337
6. Firn RD, Digby J (1980) Annu Rev Plant Physiol 31:131
7. Mer C (1969) New Phytol 68:275
8. Shaw S, Gardner G, Wilkins MBU (1973) Planta (Berl) 109:11
9. Tamimi S, Firn RD (1985) J Exp Bot 36:955
10. Went FC, Thimann KV (1937) Phytohormones. MacMillan, New York
11. Wilkins MBU (1984) In: Wilkins MBU (ed) Advanced Plant Physiology. Pitman, London, p 163

Auxin and Root Gravitropism: The State of Our Knowledge

T.J. MULKEY and M.A. VAUGHAN [1]

1 Introduction

During the past 10 years, increasing emphasis has been placed on the study of the involvement of hormones and calcium in growth and gravitropic response of roots. The establishment of gradients of calcium ions across the tip of roots appears to be essential for the initiation of asymmetric growth associated with the gravitropic response [7, 8, 10] and may be the signal transducer which links the root cap-associated sensor of gravitational orientation [9] and the hormonal control of asymmetric growth, which occurs during gravicurvature (GC).

2 Hormonal Involvement in Root Gravitropism

Indole-3-acetic acid (IAA) and abscisic acid (AbA) are central to the controversy concerning hormonal control of root graviresponse. The Cholodny-Went hypothesis of gravitropism [1, 19] proposes that GC of roots results from gravity-induced lateral transport of auxin on the lower side of horizontal roots. The most widely accepted and studied alternate hypothesis of GC is the "root cap inhibitor" model, which substitutes AbA for IAA as the inhibitory hormone-affecting growth on the lower surface of horizontal roots [20]. The role of AbA as the root cap inhibitor has been discounted by the work of numerous workers [4, 12, 17, 18]. The role of IAA as the hormone involved in GC is supported by the work of several workers [2, 3, 11].

3 Interactions of Auxin and Ethylene

Two patterns of gravitropic response have been noted for roots of *Zea mays*. The type I gravitropic response, which occurs in about 30% of the roots examined, is characterized by an initial upward curvature followed by downward bending of the roots (Fig. 1). In these roots, complete (90°) curvature does not occur. Type II gravitropic response, which occurs in about 70% of the roots examined, is characterized by downward bending without an initial upward curvature (Fig. 2). These roots over-

1 Department of Life Sciences, Indiana State University, Terre Haute, IN 47809, USA

Plant Growth Substances 1985
(ed. by M. Bopp)

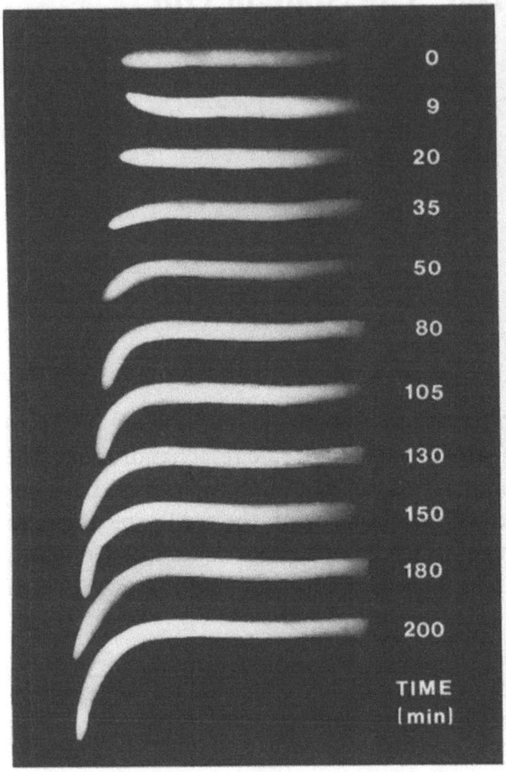

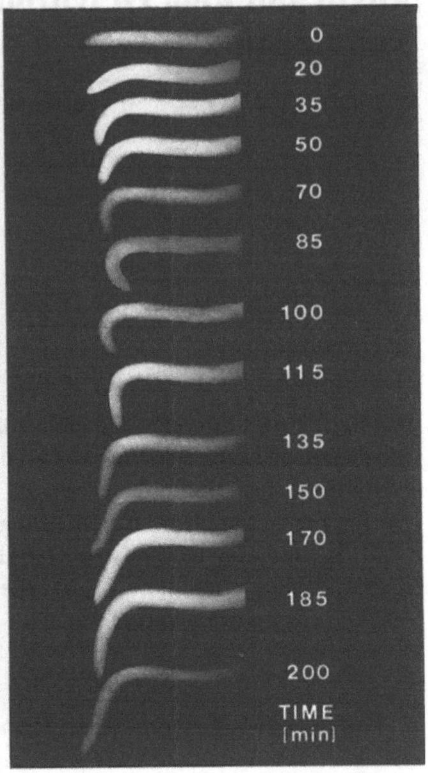

Fig. 1 **Fig. 2**

Fig. 1. Kinetics of type I curvature of intact primary roots of maize. The type I gravitropic response is characterized by an initial upward curvature evident 9 min after the root is placed horizontally. Significant downward curvature is apparent after 35 min (T.J. Mulkey and M.L. Evans)

Fig. 2. Kinetics of type II curvature of intact primary roots of maize. The type II gravitropic response is characterized by significant downward curvature within the first 20 min and an overshoot of 90° (85 min). This is followed by an oscillatory readjustment of the angle of curvature (T.J. Mulkey and M.L. Evans)

shoot 90° curvature and undergo an oscillatory search for vertical orientation. These two types of response have been reported by Hild and Hertel [6] for maize coleoptiles. They suggest that the differences in patterns of GC results from varying sensitivity of the tissue to auxin. Thus, type I roots would be less sensitive to auxin and the initial downward flux of auxin promotes growth in the lower hemisphere of the root and initiate upward curvature. Subsequent accumulation of auxin inhibits growth of the lower hemisphere of the root and induces downward curvature. Conversely, type II roots are more sensitive to auxin and the initial movement of auxin to the lower portion of the root immediately inhibits growth in the lower hemisphere of the root.

It has been shown that the inhibitory effects of auxin on root growth are correlated with auxin-induced ethylene formation [14]. We have found that application of inhibitors of ethylene (1-canaline, aminoethoxyvinylglycine, silver thiosulphate, aminoethoxyacetic acid) block GC and slight upward curvature of roots occurs. This suggests accumulation of auxin triggers ethylene formation within the lower hemisphere of the root and ethylene inhibits cellular elongation and induces GC. The upward curvature of roots treated with ethylene inhibitors results from auxin-induced promotion of growth in the lower half of the root in the absence of inhibition of growth by ethylene.

4 Interaction of Calcium and Auxin

Calcium movement across tips of roots has been shown to be essential for the induction of GC [7, 8, 10]. Lateral transport of indoleacetic acid is promoted by the presence of a calcium gradient [7]. Inhibitors of IAA transport/activity (triiodobenzoic acid, naphthalpthalmic acid, and morphactins) inhibit GC [11] and inhibit gravity-induced calcium gradient formation across the tip/cap of maize roots [9].

Using a computer-based video image analysis system, we have examined the kinetics of gravicurvature induced by exogenously applied calcium. An examination of growth rates along the upper (Fig. 3) and lower (Fig. 4) surfaces of roots indicates that the rate of elongation and region of elongation are similar between control roots and decapped roots which have 1.5 mm^3 agar blocks containing 1 mM calcium chloride placed asymmetrically at the root tip. High concentrations of calcium (10 mM) significantly altered the elongation rate and position of maximal elongation. As would be expected, the rate of curvature (Fig. 5) is similar between control and decapped, 1 mM $CaCl_2$-treated roots. Since exogenously applied calcium establishes a trans-root

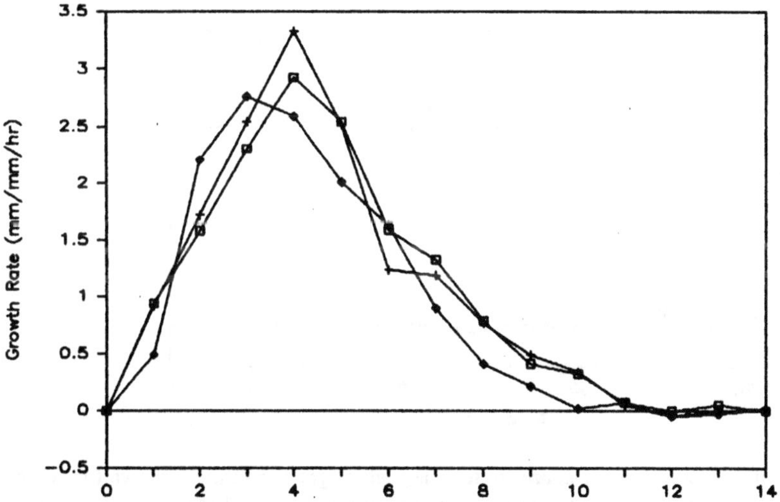

Fig. 3. Elongation rates of regions along the upper surface of maize roots responding to gravity. Comparison of elongation rates of control roots (□) and of decapped roots treated with 1 mM $CaCl_2$ (+) or with 10 mM $CaCl_2$ (◇) asymmetrically applied to the tip in 1.5 mm^3 agar blocks. Growth rates shown are average rates 50 min after exposure to the gravitational/calcium stimulus

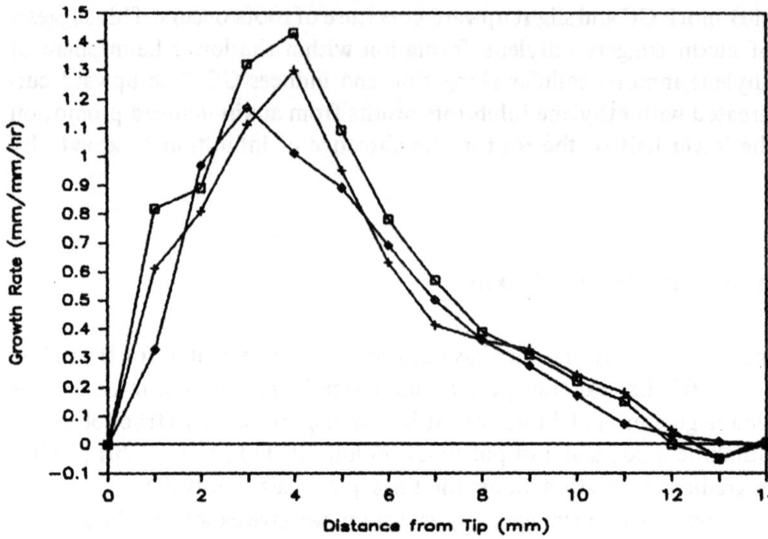

Fig. 4. Elongation rates of regions along the lower surface of maize roots responding to gravity. Comparison of elongation rates of control roots (□) and of decapped roots treated with 1 mM $CaCl_2$ (+) or with 10 mM $CaCl_2$ (◇) asymmetrically applied to the tip in 1.5 mm^3 agar blocks. Growth rates shown are average rates 50 min after exposure to the gravitational/calcium stimulus

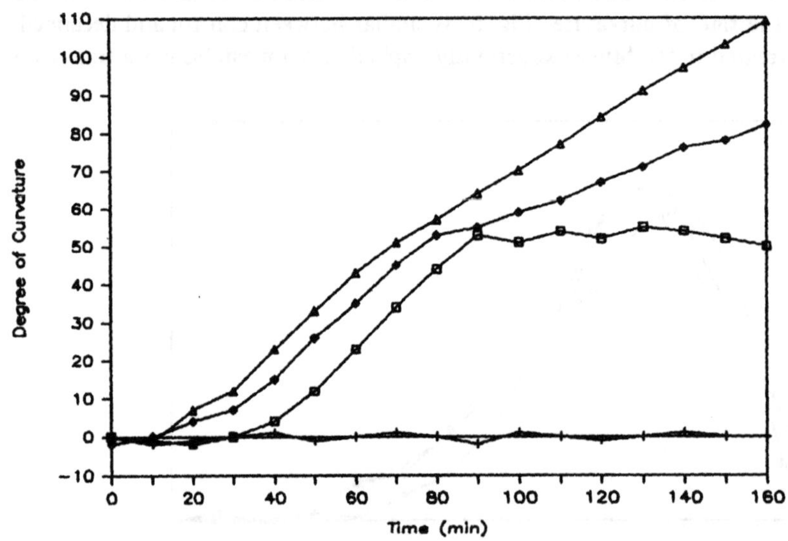

Fig. 5. Kinetics of gravicurvature of maize roots. The angle of gravicurvature is shown for decapped roots (+), for control roots (□), and for decapped roots treated with 1 mM $CaCl_2$ (◇) or with 10 mM $CaCl_2$ (△) asymmetrically applied to the tip in 1.5 mm^3 agar blocks

calcium gradient more quickly than the normal gradient develops, curvature is initiated earlier in the calcium-treated, decapped roots. Another notable difference is that exogenous calcium maintains the trans-root calcium gradient and prolongs curvature of the roots past vertical orientation. A definite and unexplained decrease in the rate of curvature occurs as the root tip passes through vertical orientation with respect to the gravitational stimulus.

Another calcium/auxin interaction has been suggested by Raghothama et al. [16]. These workers found that inhibitors of calmodulin activity (phenothiazine and napthalene sulfonamide drugs) accentuate auxin activity in *Avena* coleoptiles. In roots of *Zea mays*, we have found no promotion of growth by auxin (10^{-10} M to 10^{-6} M) in roots pretreated with phenothiazine drugs (Stelazine and chlorpromazine) or napthalene sulfonamide drugs (W5, W7, W13). This difference in effect of calmodulin inhibitors on auxin-induced growth may be another difference in the response of roots to auxin [14] compared to the response of stems to auxin.

Other calcium antagonists do effect root growth and GC. Hasenstein and Evans [5] have reported promotion of root elongation by auxin in maize roots treated with calcium chelators. Lee et al. [8] have shown that calcium chelators inhibit GC. We have shown that pharmacologic calcium entry blockers (verapamil, diltiazem, nifedipine, cinnarizine) promote growth of maize roots [15], but inhibit GC [13]. These data suggest an interaction of calcium with auxin in roots which is not modulated by calmodulin or dependent on inhibition of root elongation by calcium antagonists.

5 Conclusions

The role of plant hormones in relation to root elongation and more specifically, asymmetric elongation associated with gravicurvature, has been discussed. The evidence presented supports a role for auxin in modulating asymmetric growth, which results in gravicurvature. The inhibition of cellular elongation along the lower hemisphere of a horizontally-oriented root appears to be due to auxin-induced ethylene. Inhibitors of auxin action and of ethylene synthesis/action block gravicurvature. Calcium ions are involved with the transduction of the signal between the sensor at the root cap and the elongation zone. Within the elongation zone, calcium affects both the lateral transport and the growth-promoting activity of auxin.

Acknowledgments. This work was funded in part by an Indiana State University Grant-in-Aid of Research to T.J.M., a Sigma Xi Grant-in-Aid of Research to M.A.V., and NSF/ASPP Travel Grants to T.J.M. and to M.A.V.

References

1. Cholodny N (1926) Jahrb Wiss Bot 65:447
2. Davies PJ, Doro JA, Tarbox AW (1976) Physiol Plant 36:333
3. Gaither DH, Ableles FB (1975) Plant Physiol (Bethesda) 56:404
4. Hartung W (1976) Planta (Berl) 128:59
5. Hasenstein KH, Evans ML (1985) Plant Physiol (Bethesda) 77:80

6. Hild V, Hertel R (1972) Planta (Berl) 108:245
7. Lee JS, Mulkey TJ, Evans ML (1983a) Plant Physiol (Bethesda) 73:874
8. Lee JS, Mulkey TJ, Evans ML (1983b) Science 220:1375
9. Lee JS, Mulkey TJ, Evans ML (1984) Planta (Berl) 160:536
10. Moore R (1985) Am J Bot 72:785
11. Mulkey TJ, Evans ML (1982) J Plant Growth Regul 1:259
12. Mulkey TJ, Evans ML, Kuzmanoff KM (1983) Planta (Berl) 157:150
13. Mulkey TJ, Evans ML, Lee JS, Kuzmanoff KM (1983) Plant Physiol (Bethesda) 72:63
14. Mulkey TJ, Kuzmanoff KM, Evans ML (1982) Plant Physiol (Bethesda) 70:186
15. Mulkey TJ, Vaughan MA (1985) Plant Physiol (Bethesda) 77:105
16. Raghothama KG, Mizrahi Y, Poovaiah BW (1985) Plant Physiol (Bethesda) 79:28
17. Smith J, Ho THD (1982) Plant Physiol (Bethesda) 69:339
18. Suzuki T, Kondo N, Fujii T (1979) Planta (Berl) 145:323
19. Went FW (1926) Pro Kon Akad Wetensch Amsterdam 30:10
20. Wilkins MB, Gibbons GSB, Shaw S (1970) In: Carr DJ (ed) Plant growth substances 1970. Springer, Berlin Heidelberg New York, p 717

IV. Hormones: Effects

Are Angiosperm Hormones Present in, and Used as Hormones by, Algae?*

W.P. JACOBS[1]

1 Introduction

Most studies of hormones in plants have focused on angiosperms. The farther down the presumed evolutionary tree one looks, the fewer actual chemical identifications of putative hormones one sees. Despite the number of algal extracts showing auxin-like or gibberellinlike activities when tested in angiosperm bioassays, remarkably few of the chemicals responsible have been identified. Even fewer reports demonstrate that chemicals that act as hormones in angiosperms also actually function as endogenous *hormones* in algae.

The intriguing question for those of us interested in biochemical evolution is: Did the angiosperm hormones only appear as chemicals after the vascular plants evolved on land, or did the vascular plants evolve hormonal uses for chemicals that were already present, but not acting as hormones in lower forms? It is that question which we shall be exploring today, particularly as it relates to auxins and gibberellins. (Such questions have been raised previously about the evolution of the cytokinins zeatin and iso-pentenyladenine [19], about abscisic acid and lunularic acid [25], and in a more general form by Leopold and Noodén [21], p. 16).

The evolutionary tree[2] of Fig. 1 lists various lower plant groups in which chemicals known as angiosperm hormones have been identified. The method of choice was gas chromatography-mass spectrometry (GC-MS). No gibberellin has been chemically identified from any alga apparently, despite the scattered reports of gibberellinlike activity in bioassays (summary [4]). When the auxinlike activity reported in various algal extracts was calculated as though it were due to indole-3-acetic acid (IAA), the levels were typically far below those reported in angiosperms. A search for IAA in the green alga *Caulerpa* resulted in no detectable IAA [8]. Among the various algal groups, only the brown algae and the green algae contain species from which auxins have been identified by MS. In the brown algae, Abe et al. [1] found IAA by MS in extracts of *Undaria*, with a yield much less than that of angiosperms. By other criteria,

* The following contributions (pp. 249–274) are part of the workshop *Evolutionary Aspects*

1 Biology Department, Princeton University, Princeton, N.J., USA

2 This evolutionary tree is merely one of many that have been proposed. [The names of the various plant groups (e.g., Psilopsida) are similarly subject to argument.] Since we know how variously interpreted can be the results of our "controlled" laboratory experiments, it is scarcely surprising that there are so many views about the presumed evolutionary happenings of millions of years ago. For recent examples, see Philos Trans Roy Soc Lond (B) 309:1–342 (1985)

Plant Growth Substances 1985
(ed. by M. Bopp)

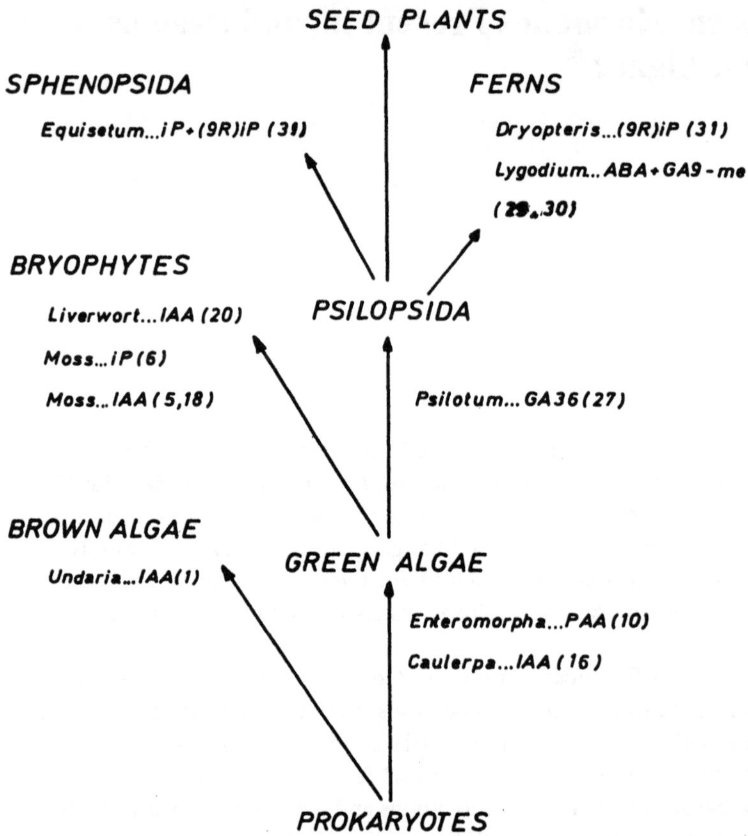

Fig. 1. A possible evolutionary tree, showing the major lower plant groups in which seed-plant hormones have been identified by mass spectrometry or other sophisticated methods. *Abbreviations:* IAA = indole-3-acetic acid; PAA = phenylacetic acid; iP = iso-pentenyladenine; (9 R) iP = 9-ribose-iso-pentenyladenine; ABA = abscisic acid; GA9-me = methylated gibberellin-9; GA36= gibberellin-36

they concluded that phenylacetic acid (PAA) and its p-hydroxy derivative were also present [2]. In green algae, MS was used to identify PAA in extracts of *Enteromorpha* [10].

2 Hormones in Caulerpa

From the evolutionary point of view, particular interest centers on the green algae, because they are usually considered to be the most likely ancestors of the Bryophytes and vascular plants. If one were to select a green alga to investigate as a possible source of angiosperm hormone, it would be sensible to pick an alga large enough and differentiated enough to be likely to benefit from the coordinating effects of hormonal action. It would be even better if the alga were already known to show correlative effects in its development. *Caulerpa* is such a green alga. *C. prolifera*, the most often investigated species, can be as long as several meters. It develops three different organ-like structures (leaflike blades, a rhizomelike structure, and rhizoids), each with a different orientation to the environment. The coordinated development of the intact coenocytic organism, as well as the striking polarity of regeneration of excised portions of the "leaves" and "rhizomes" (e.g. [9, 12, 17]), suggests hormonal control.

2.1 Indole-3-Acetic Acid

IAA added to *Caulerpa* species stimulated "leaf" development [8, 24], had no effect on rhizoid or rhizome regeneration [12], and decreased rhizome elongation [24].

Encouraged by these results with added IAA, we have recently reinvestigated whether IAA is present in *Caulerpa* as an *endogenous* auxin [16]. Each of the three coauthors used a different extraction, purification, and detection procedure on samples of *C. paspaloides* (Fig. 2). Falkenstein used the *Avena* section bioassay of activity at the Rf of IAA, combined with an estimate of 32% recovery from her internal standard, to calculate that her original 3.3 kg sample contained 3.1 mg IAA/kg fresh weight. Hamilton used high performance liquid chromatography and an electrochemical detector to measure IAA. Correction for his 41% recovery revealed an average of 0.7 mg IAA/kg in his original samples. (A single determination on *C. prolifera* showed 0.09 mg IAA/kg.) Finally, acid-base partitioning and high performance liquid chromatography followed by capillary GC-MS of the methylated and silylated fractions confirmed that IAA was present in the *C. paspaloides* extract (Fig. 3). Their conclusion was that IAA is present in *C. paspaloides* at approximately 1 mg kg^{-1} fresh weight, with an error of ± one order of magnitude. Contrary to the earlier reports on algae, the low estimates

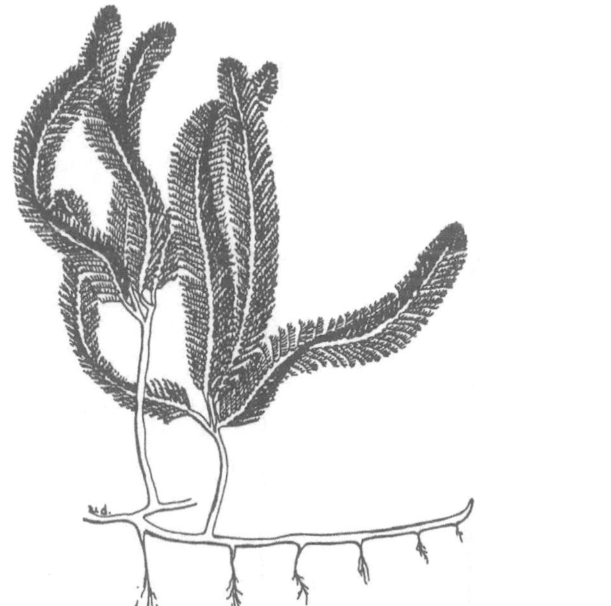

Fig. 2. Habit drawing of *Caulerpa paspaloides*

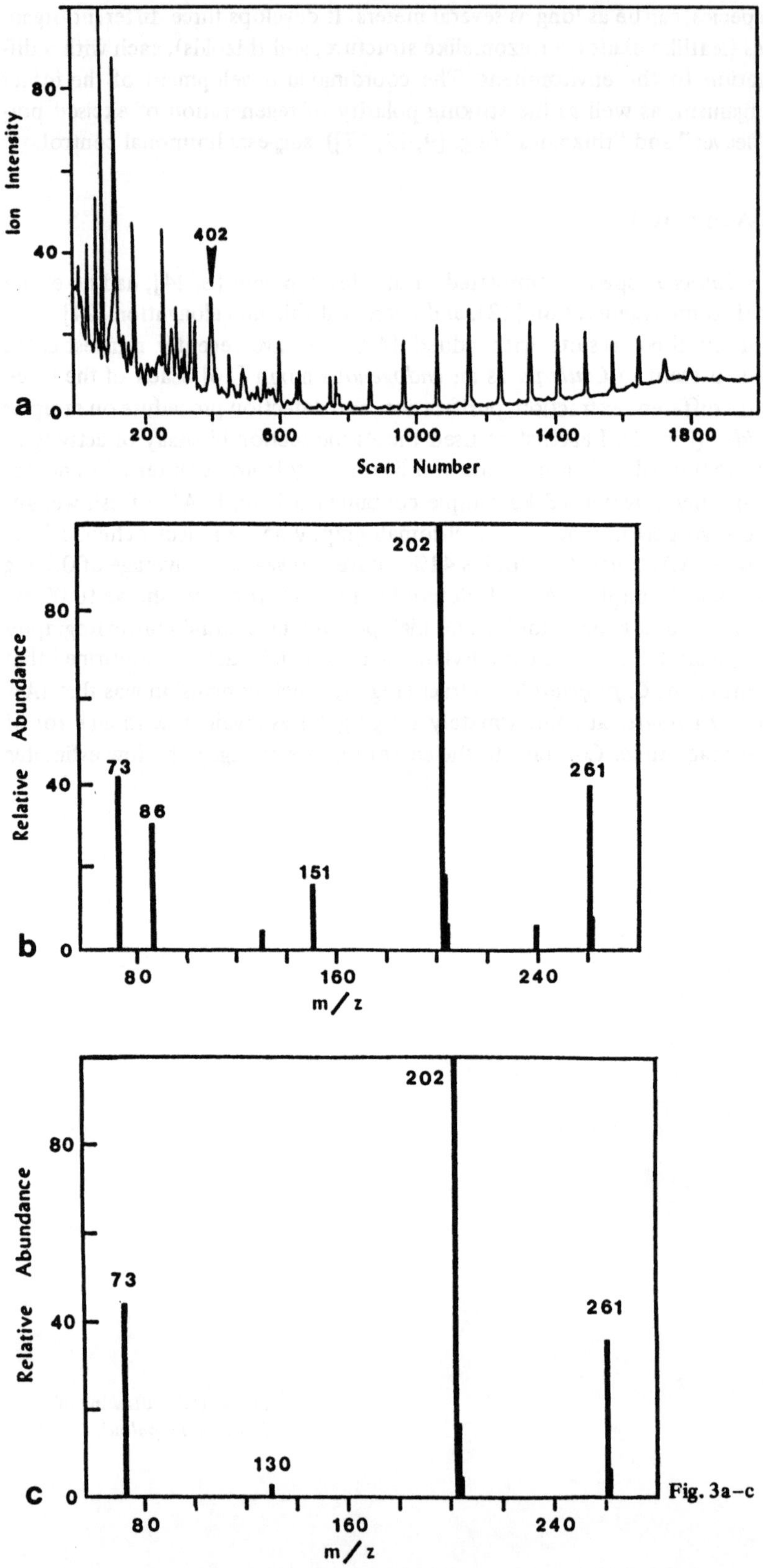
Ion Intensity
80
40
0
402
a
200
600
1000
1400
1800
Scan Number
Relative Abundance
80
40
0
202
73
86
151
261
b
80
160
240
m/z
Relative Abundance
80
40
0
202
73
130
261
c
80
160
240
m/z

Fig. 3a–c

of which were mostly based on bioassays of extracts that had *not* been corrected for losses in processing, this concentration is in the range that others have recently reported in angiosperms.

2.2 Gibberellins

Augier [3] in his survey of various algal groups, found gibberellinlike activity from several zones of his chromatograms when extracts of *Caulerpa prolifera* were tested in the lettuce-hypocotyl and *Avena*-leaf bioassays. In collaboration with Prof. J. MacMillan, I have begun an attempt to identify the chemicals responsible for this activity. Using the Tan-ginbozu dwarf-rice assay (which responds to more known gibberellins than any other bioassay), we found two zones of gibberellinlike activity in the acidic-ethyl acetate fraction of *C. paspaloides* after separation by high performance liquid chromatography (Fig. 4). Preliminary investigation of scans from a smaller aliquot that had been methylated and silylated, then subjected to GC-MS, has revealed no known gibberellin. We do not know if this is due to inadequate sample size or to previously undiscovered sources of gibberellinlike activity in the alga. (These investigations are, of course, continuing.)

C. prolifera reacts to gibberellic acid (added in hormonal concentrations to the culture medium) in a way that contrasts intriguingly with the alga's response to IAA. Gibberellic acid consistently stimulated rhizome elongation and the rate of rhizoid initiation, while having no such effects on leaf development [15]. This effect of added gibberellic acid on rhizoid initiation suggests that endogenous gibberellinlike substance(s) may be involved in controlling rhizoid development and in mediating the

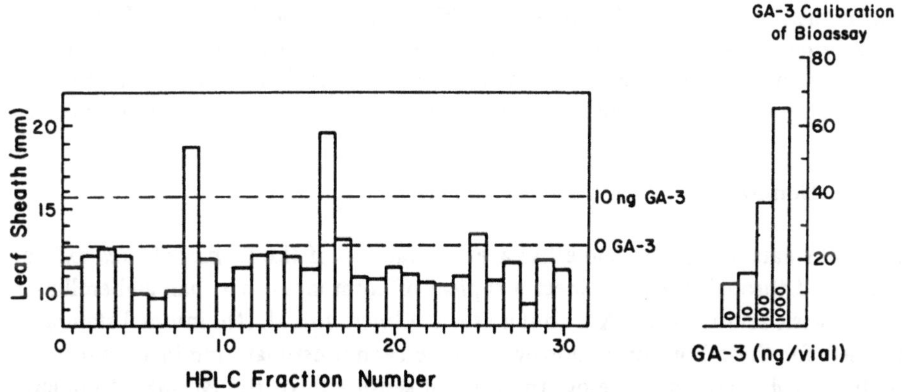

Fig. 4. Gibberellinlike activity in a *C. paspaloides* extract after high performance liquid chromatography, as estimated by the Tan-ginbozu dwarf-rice bioassay

Fig. 3a–c. Evidence from capillary GC-MS for the occurrence of IAA in *C. paspaloides* [16]. **a** Total ion current trace of a sample from the acid-ethyl acetate fraction, with *n*-alkanes added as retention standards. Scan 402 (*arrow*) had m/z 202, characteristic ion from indoles, as a major component. **b** The mass spectrum of putative IAA-MeTMS (scan 402) after subtraction of scan 398 as background. **c** The mass spectrum of authentic IAA-MeTMS after GC-MS as for the sample a and b fast morphogenetic effect of gravity on rhizoid initiation that was recently described by Jacobs and Olson [14]. When *C. prolifera* was turned upside down by rotating it 180° around the rhizome axis, the next rhizoid cluster to form near the rhizome tip was formed in the normal time interval, but on the new underside (on the opposite side of the rhizome axis from the rhizoid clusters that had developed before inversion).

It is not necessary to hypothesize that the endogenous gibberellin like substances in *Caulerpa* become concentrated on the lower side of the inverted rhizome, because amyloplasts have been shown to collect on the lower side before the next rhizoid clusters forms and at the site where it will form [22]. Reaction of a generally distributed gibberellin with the asymmetrically distributed amyloplasts could presumably lead to the localized formation of rhizoids. If gibberellins and amyloplasts are connected in such a reaction system, the parallel with gibberellins and the starch granules of barley seeds is obvious.

2.3 Distribution of Hormones

But *are* the hormones of *Caulerpa* generally distributed? The question is of particular interest because the large and complex plant body of *C. prolifera* is a single giant, multinucleate cell – a feature which was the salient attraction for most developmental biologists who have investigated the alga (cf. [13]). Lacking the microscopic compartments provided by the myriad cells of more conventional organisms, *Caulerpa* might either produce "organ"-forming substances locally for immediate local use, or might have the substances generally distributed with the local response depending on other local conditions. The present evidence, although scanty, favors the latter hypothesis. IAA is apparently present throughout the plant judging by the quick and completely nonpolar distribution of [^{14}C] added as [^{14}C] IAA to *C. prolifera*'s "leaf" tips [7]. Also favoring the second hypothesis is the fact that the local effects of added hormones (IAA on "leaves", gibberellic acid on rhizoids and rhizomes) resulted even when the hormones were added in the culture medium surrounding the entire plant.

2.4 Other Hormones

The GC-MS data from the acid-ethyl acetate fraction of the *C. paspaloides* extract was searched through for mass spectra typical of hormones other than gibberellins – such as abscisic acid and 4-Cl-IAA – but no such MS were found. (No internal standards were used for these hormones, however, so we cannot estimate minimal concentrations that might have been below the level of sensitivity of the GC-MS.) Lunularic acid was isolated from a liverwort [28] and shown to be involved in photoperiodic control of *Lunularia*'s dormancy. On the basis of GC retention times and a flame ionization detector, Pryce [25] reported that lunularic acid was also present in all nine algal genera that he examined and suggested that the role of lunularic acid in the lower plants was taken over by abscisic acid in more highly evolved plants. However, the validity of lunularic acid's presence in algae has been questioned by Gorham [11], who used different methods and found that his GC peaks of authentic lunularic acid and putative lunularic acid from algal extracts, although coincident under some conditions, could be separated by using different columns. The level in the evolutionary tree at which abscisic acid evolved obviously needs further study.

3 Conclusions

The current status of our information is that the angiosperm hormone IAA is present in (at least one genus of) the green algae and has hormonallike action there. Whether phenylacetic acid is also present in *Caulerpa* (or whether IAA is also present in *Enteromorpha* - see Fig. 1), we do not yet know; by analogy with angiosperms we might expect they would both be present. This occurrence of IAA as a hormone in the green algae on the presumed "main line" of plant evolution, suggests that the answer to our original question is that the Bryophytes and vascular plants did not have to evolve hormonal use for IAA (much less its synthesis, which had already been accomplished by bacteria), but had merely to adapt the hormone to their own uses.

Still unresolved is the question of when the polar movement of IAA, so famous in angiosperms, first evolved. In mosses and liverworts polar transport of IAA was demonstrated [23, 26]. Current hypotheses about polar movement of auxins through angiosperm tissues assume movement across cell membranes as a prerequisite. From such hypotheses it would be expected that a single giant cell, like *Caulerpa prolifera*, would not show polar movement of IAA - an expectation that fits Brennan and Jacob's data from [^{14}C] IAA [7]. However, more detailed evidence is needed not only from coenocytic *Caulerpa*, but also from *multicellular* green algae to determine if polar movement of hormonal IAA has already evolved in these algae.

From the point of view of comparative biochemical evolution, it is intriguing that two of the major chemicals involved in polar transmission of signals in living organisms should be the indole-based indole-3-acetic acid in higher plants and 5-hydroxy tryptamine (serotonin) in higher animals - the latter serving to transmit nerve impulses in polar fashion across synapses.

For gibberellins, the evidence is less compelling. Although *Caulerpa* reacts to added gibberellic acid as though it were a hormone, the gibberellinlike activity found in *Caulerpa* extracts has not yet been chemically identified. Also, unlike IAA, gibberellins have not been identified from bacteria. Hence, the question still remains open whether the gibberellinlike activity found in *Caulerpa* and other algae is due to chemicals with the same basic chemical structure as the gibberellins of the angiosperms. I would guess that it is.

(Note added: Tietz and Kasprik presented a poster at the meeting reporting the presence of ABA in the green alga *Stigeoclonium*, as determined by GC-MS.)

Acknowledgments. My heartfelt thanks to Prof. J. MacMillan and his co-workers at the University of Bristol for their hospitality and help with GC-MS and with the Tan-ginbozu gibberellin bioassay. Thanks also to Wendy Davis for preparing Fig. 2.

References

1. Abe H, Uchiyama M, Sato R (1972) Agric Biol Chem 36:2259
2. Abe H, Uchiyama M, Sato R (1974) Agric Biol Chem 38:897
3. Augier H (1974) Ann Sci Nat Bot Biol Veg 15:1
4. Augier H (1976) Bot Mar 19:245
5. Ashton NW, Schulze A, Hall P, Bandurski RS (1985) Planta (Berl) 164:142
6. Beutelmann P, Bauer L (1977) Planta (Berl) 133:215
7. Brennan T, Jacobs WP (1980) Ann Bot 46:129
8. Dawes CJ (1971) Phycologia 10:375
9. Dostal R (1945) Bull Int Acad Tchéque Sci 46:133
10. Fries L, Åberg S (1978) Z Pflanzenphysiol 88:383
11. Gorham J (1977) Phytochemistry (Oxf) 16:249
12. Jacobs WP (1964) Staz Zool Napoli 34:185
13. Jacobs WP (1970) Ann NY Acad Sci 175:732
14. Jacobs WP, Olson J (1980) Am J Bot 67:141
15. Jacobs WP, Davis W (1983) Ann Bot 52:39
16. Jacobs WP, Falkenstein K, Hamilton RH (1985) Plant Physiol (Bethesda) 78:844
17. Janse JM (1906) Jahrb Wiss Bot 42:394
18. Jayaswal RK, Johri MM (1983) Phytochemistry (Oxf) 24:1211
19. Kaminek M (1974) J Theor Biol 48:489
20. Law DM, Basile DV, Basile MR (1985) Plant Physiol (Bethesda) 77:926
21. Leopold AC, Noodén LD (1984) Encyclopedia Plant Physiology (new ser), vol 10, p 4
22. Matilsky MB, Jacobs WP (1983) Planta (Berl) 159:189
23. Maravolo NC (1976) Am J Bot 63:526
24. Mishra AK, Kefford NP (1969) J Phycol 5:103
25. Pryce R (1972) Phytochemistry (Oxf) 11:1759
26. Rose S, Bopp M (1983) Physiol Plant 58:57
27. Takahashi M, Yamane H, Satoh Y, Takahashi N, Iwatsuki K (1984) Phytochemistry (Oxf) 23:681
28. Valio IFM, Schwabe WW (1970) J Exp Bot 21:138
29. Yamane H, Takahashi N, Takeno K, Furuya M (1979) Planta 147:251
30. Yamane H, Sato Y, Takahashi N, Takeno K, Furuya M (1980) Agr Biol Chem 44:1697
31. Yamane H, Watanabe M, Satoh Y, Takahashi N, Iwatsuki K (1980) Plant & Cell Physiol 24: 1027

Hormonal Coordination of the Processes Leading to Cell Fusion in Algae: A Glycoprotein Hormone from Red Algae

S.D. WAALAND[1]

1 Introduction

In a number of algae and fungi, sexual reproduction, including the induction of sexuality, the meeting of gametes, and gamete fusion, have been shown to be coordinated by specialized morphogenetic substances. These substances differ widely in their chemical composition. Among the fungi, substances which induces sexual morphogenesis include oligopeptides in some Ascomycete and Basidiomycete yeasts [1, 9, 12], steroids in the watermold *Achlya* [6], and carotenoids in Mucoralean fungi [2]. In the water mold *Allomyces*, a cyclic sesquiterpenediol has been identified as a sexual attractant [5]. Among the algae, inducers of sexual morphogenesis have only been identified in the colonial green alga *Volvox*. Within this genus, a glycoprotein male inducer has been characterized from *V. carteri* [3, 10], while in *V. capensis* a single amino acid, L-glutamic acid, has been shown to trigger sexual development [11]. Among the brown algae a number of simple volatile hydrocarbons have been shown to act as species-specific sexual attractants; in some cases these may also serve as inducers of sperm release [4, 7, 8].

Specific compounds, which induce sexual morphogenesis, have not been identified in red algae. However, a somatic cell repair process has been described in a number of filamentous red algae. This process shares many characteristics with sexual reproduction in other algae, including its coordination by a specific hormone. When an intercalary cell in a filament is killed, it may be replaced by the process of cell repair. During this process, the cells on either side of a dead cell produce new, specialized cells: a repair rhizoid cell from above and a repair shoot cell from below. These two new cells grow towards each other through the lumen of the dead cell; they are attracted to each other. When these cells meet, they fuse to form a single cell, which replaces the killed cell; in this way, the integrity of the filament is restored [15]. Thus, cell repair, like sexual reproduction, involves the production of specialized cells, attraction between these cells, and their eventual fusion. In this paper, evidence will be presented which shows that cell repair is coordinated by one or more specific *morphogenetic substances*. The isolation and characterization of one of these substances from one species will be described. The biological action of this morphogenetic substance will be discussed.

1 Department of Biology, University of Puget Sound, Tacoma, WA 98416, USA

Plant Growth Substances 1985
(ed. by M. Bopp)

2 Evidence for Control of Somatic Cell Repair by a Specific Morphogenetic Substance

First, filamentous, marine red algae in the genus *Griffithsia* have two types of vegetative cells: rhizoid and shoot cells. These two cell types differ in morphology, physiology, and biochemistry (Table 1). During cell repair, a new type of cell is produced, the repair shoot cell. This cell resembles a rhizoid in morphology and pattern of growth, but is a physiologically distinct cell type (Table 1). Second, the production of a repair shoot cell by the subjacent cell during cell repair occurs more rapidly than the production of a new shoot apical cell by a decapitated apex during normal regeneration (Table 2). Both the precocious division of the subjacent cell and the continued growth of the repair shoot cell depend on the presence of a repair rhizoid from the superjacent cell [15]. These observations suggested that the production and morphogenesis of repair shoot cells might be controlled by a substance produced by the repair rhizoid. In addition, the repair shoot cell shows a strong positive chemotropic response to the repair rhizoid.

Using an artificial cell repair system, we have been able to demonstrate that any rhizoid, and only rhizoid cells produce a substance which induces decapitated filaments to produce repair shoot cells [13, 14, 17]. By testing rhizoids from one species

Table 1. Comparison of characteristics of different cell types in *Griffithsia*

	Rhizoid apical cell	Repair shoot cell	Shoot apical cell
Morphology	Elongate, pale, refractile tip	Elongate, pale, refractile tip	Dome-shaped, deep red, uniform tip
Growth pattern	Rapid tip growth	Rapid tip growth	Slow tip growth + band growth
Phototropism	Negative	Neutral	Positive
Rhodomorphin requirement	No	Yes	No
Attraction to rhizoids	No	Yes	No
Attraction to repair shoot cells	Yes	No	No

Table 2. Kinetics of apical regeneration during cell repair and normal regeneration

Species	Time to cytokinesis (h)	
	Subjacent cell (cell repair)[a]	Decapitated apex (normal regeneration)[b]
Griffithsia pacifica	5–6	16–19
Anotrichium tenue	4.5	7– 8
Antithamnion kylinii	6–9	No regeneration

[a] Production of a repair shoot cell
[b] Production of a normal shoot apical cell

against decapitated filaments of the same or other species, we were able to show that cell repair, including cell fusion, can be obtained between isolates of the same species of *Griffithsia* from widely separated localities. However, the induction of repair shoot formation and chemotropic attraction will not occur between different species. Thus, the cell repair hormone appears to be species-specific; we have called these substances "rhodomorphin" [13].

3 Isolation and Characterization of Rhodomorphin from Griffithsia pacifica

We have isolated and characterized rhodomorphin from one species, *Griffithsia pacifica* Kylin, a filamentous marine red alga in the order Ceramiales. Details of this procedure have been published [17]. Cultures of *G. pacifica* are chopped into short filaments, washed, and placed into a medium containing antibiotics and bovine serum albumin. During the next 24 h, each filament regenerates a rhizoid and each rhizoid produces rhodomorphin. The culture medium is filtered and assayed for biological activity. To assay for the presence of rhodomorphin during isolation, we have used two bioassays: (1) the induction of the formation of repair shoots by decapitated filaments and (2) the maintenance of elongation in preinduced repair shoot cells. Both assays employ a dilution series to determine units of activity. We have defined activity units in the following way: a putative hormone solution, which induces 50% of the decapitated assay filaments to form repair shoot cells (assay 1) or maintains cell elongation in 50% of the repair-shoot assay cells (assay 2) is defined to have one unit of rhodomorphin activity per milliliter [17, 19]. Assay 2 is about five times more sensitive than assay 1.

3.1 Chemical Characterization

The chemical properties of rhodomorphin from *G. pacifica* have been studied. Rhodomorphin is bound specifically by the plant lectin Concanavalin A (Con A); Con A binds molecules containing α-glucosyl or α-mannosyl linkages. Rhodomorphin activity is not bound by lectins with affinities for 1-0-methyl-B-D-galactoside, L-fucose, or N-acetyl-D-glucosamine [18]. The nature of the binding of rhodomorphin by Con A has been further clarified by the use of highly specific glycosidases. We have found that α-mannosidase, but not α-glucosidase, reduces the binding of hormone activity by Con A; neither enzyme reduces hormone activity [19]. These results suggest that the rhodomorphin molecule has α-mannosyl linkages, which are accessable to binding by Con A and to enzymatic removal. Removal of these groups does not reduce biological activity.

That rhodomorphin from *G. pacifica* contains protein is suggested from the following evidence. Rhodomorphin activity is destroyed by heating to 50 °C for 8 min. Activity is also sensitive to digestion by trypsin and proteinase K [18]. Further, we have found that exposure of rhodomorphin to dithiothreitol, a sulfhydryl reducing agent, reduces or eliminates its biological activity. This treatment is most effective in the presence of the chaotrope 6 M guanidine-HCl, an agent which disrupts noncovalent bonds and unfolds protein molecules [19]. Thus, the rhodomorphin molecule contains protein as well as carbohydrate; the protein moiety appears to have essential disulfide bonds, which are required for the correct folding of the molecule.

The molecular weight of rhodomorphin has been determined by gel filtration and by SDS-PAGE. Rhodomorphin activity elutes from gel filtration chromatography on Bio-Gel A 1.5 M in the presence of 6 M guanidine-HCl activity in a sharp peak corresponding to M_r of 14,000 [18]. We have also determined the molecular weight of rhodomorphin by SDS-PAGE in the absence of reducing agents. To localize rhodomorphin on gels, proteins were extracted from sectioned gels and the extracts were assayed for biological activity. Critical to recovery of rhodomorphin activity from SDS-gels was the treatment of solutions with 6 M guanidine-HCl; rhodomorphin renatures to complete activity after the removal of guandine-HCl [19]. Apparently, a complete unfolding of the molecule is important to its renaturation in an active form. Rhodomorphin activity was found to be run on SDS-PAGE as a diffuse, 6–10 mm band between M_r 15000–17500. Thus, active rhodomorphin solutions appear to be microheterogeneous.

Thus, we have shown that rhodomorphin from *G. pacifica* is a glycoprotein of M_r 14000–17500 as determined by gel filtration and SDS-PAGE. It has α-mannosyl groups, which are accessible to removal by mannosidase and to binding by Con A. The protein portion of the molecule is accessible to digestion by proteases and contains disulfide bonds, which are essential for correct folding of the molecule. The microheterogeneous nature of rhodomorphin as seen by SDS-PAGE may be due to small differences in glycosylation or to proteolytic degradation during isolation.

3.2 Purification

Rhodomorphin from *G. pacifica* can be purified by a series of steps. These include the following: Con A affinity chromatography, hydrophobic interaction chromatography on octyl agarose, and gel filtration chromatography in the presence of 6 M guanidine-HCl [18]. We have assessed the efficacy of our purification scheme by SDS-PAGE of activity-containing fractions [18, 19]. Using a highly sensitive silver-staining technique, about six proteins can be detected on gels of highly purified rhodomorphin solutions. However, when a solution containing 10000 units of highly purified rhodomorphin was analyzed by SDS-PAGE, no staining band could be detected in the region of the gel corresponding to M_r 15000–17500. Either rhodomorphin is not stained by the silver stain or the amount of protein associated with 10000 units activity is below the level of detection of the stain. The silver stain can detect 1 ng of protein in a tight 1 mm band. Since rhodomorphin runs as a diffuse, 10 mm band, we would expect to be able to detect 10 ng of rhodomorphin. Thus, if 10000 units of rhodomorphin are below the level of detection with silver stain, they must contain 10 ng or less proteins. Using an average M_r for rhodomorphin of 15000 for rhodomorphin, we can calculate that the maximum molarity of a solution containing 1 unit ml^{-1} activity is 10^{-13} to 10^{-14}. While this figure appears to be quite low, Starr and Jaenicke [10] have estimated that the glycoprotein sexual inducer from *Volvox carteri* is active at concentrations of 3×10^{-16} M.

4 Mechanism of Action of Rhodomorphin

Rhodomorphin controls elongation and regulates differentiation in repair shoot cells. One action of rhodomorphin appears to be to promote the localization of cell expansion at the tips of repair shoot cells. Repair shoot cells resemble rhizoids in their morphology and in their pattern of cell elongation by tip growth. Maintenance of this morphology and growth pattern requires the continuous presence of rhodomorphin. If repair shoot cells are removed from rhodomorphin, their elongation continues for 1-2 h and then ceases. The cells then widen laterally and eventually produce normal shoot apical cells at their apices [15, 16]. If hormone is readded within 8-12 h after its removal, repair shoot cell elongation can be restored. Thus, cell elongation and morphogenesis can be manipulated by removal and readdition of rhodomorphin.

In vivo, during normal cell repair, repair shoot cells elongate only until they touch and fuse with repair rhizoids, a period of 4-8 h. However, in solutions of partially purified rhodomorphin, repair shoot cells continue to elongate for more than 5 days. For the first 2-3 days in hormone, repair shoot cells maintain their hormone dependency and will cease elongation when removed from hormone. However, after 3-4 days of continuous hormone treatment up to 90%-95% of repair shoot cells become hormone-independent; they continue to elongate even after hormone is removed. In fact, the repair shoot cells, which have grown for extended periods in hormone solution, appear to have become rhizoids. Thus, rhodomorphin not only promotes the aquisition of a rhizoidlike morphology in modified shoot cells, eventually it may induce a permanent commitment to a rhizoidal pattern of cell differentiation.

In addition to its induction of precocious cell division by decapitated filaments, rhodomorphin also may induce the formation of small cells along the sides of shoot cells. During prolonged hormone treatment (3-5 days), numerous outgrowths are produced from the sides of shoot cells. While these outgrowths resemble rhizoids, they cease elongating if removed from hormone and, thus, are certainly repair shoot cells. Under natural conditions, such outgrowths are only seen when a rhizoid cell approaches the side of a shoot cell; in these cases, fusion may occur between the rhizoid and the side growth. Thus, rhodomorphin may be involved in the formation of secondary connections between cells of adjacent filaments. These connections are quite common in red algae with complex thalli.

5 Evidence for Additional Cell Repair Hormones

One final aspect of hormonal interaction during cell repair should be mentioned. This is the chemotropic attraction seen between repair shoot cells and rhizoids [13, 15]. Rhizoids are not attracted to other rhizoids and repair shoot cells are not attracted to other repair shoot cells. However, repair shoot cells are positively attracted to rhizoid cells; the source of a chemotropic agent appears to be localized close to the tip of a growing rhizoid. It is possible that this attractant is rhodomorphin. In addition, there appears to be a distinct positive chemotropism of rhizoids to repair shoot cells. This suggests that a second chemoattractant, produced by repair shoot cells, is also involved in ensuring cell fusion. To date, the chemical nature of these two putative chemoattractants has not been investigated.

6 Conclusions

Algae and fungi employ a variety of substances to ensure the success of sexual reproduction. In certain filamentous red algae the process of somatic cell repair by cell fusion is coordinated by species-specific morphogenetic substances, rhodomorphins. These hormones induce cell division, regulate cell elongation, and control cell differentiation. Rhodomorphin from *G. pacifica* has been isolated and partially purified and characterized. It is a glycoprotein, M_r 14000–17000, which has accessible α-mannosyl groups and internal sulfhydryl bonds. It appears to be active at very low concentrations in the range of 10^{-13} to 10^{-14} M. Preliminary evidence suggests that additional chemotropic agents may also be involved in ensuring somatic cell fusion.

Acknowledgment. This research was supported by National Science Foundation Grant PCM 8309407.

References

1. Betz R, Duntze W (1979) Eur J Biochem 95:469
2. Bu'Lock JD, Jones BE, Winskill N (1976) Pure Appl Chem 47:191
3. Kochert G, Yates I (1977) Proc Natl Acad Sci USA 71:1211
4. Lüning K, Müller DG (1978) J Phycol 89:333
5. Machlis L, Nutting WH, Rapport H (1968) J Am Chem Soc 90:1674
6. McMorris TC (1968) Philos Trans R Soc Lond B Biol Sci 284:459
7. Müller DG (1981) In: Lobban CS, Wynne MJ (eds) The biology of seaweeds. Blackwell, Oxford, p 661
8. Müller DG, Gassmann G, Marner F-G, Boland W, Jaenicke L (1971) Science 171:815
9. Sakagami Y, Yoshida M, Isogai A, Suzuki A (1981) Science 212:1525
10. Starr RC, Jaenicke L (1974) Proc Natl Acad Sci USA 71:1050
11. Starr RC, O'Neil RM, Miller CE III (1980) Proc Natl Acad Sci USA 77:1025
12. Stöltzler D, Kiltz H-H, Duntze W (1976) Eur J Biochem 69:397
13. Waaland SD (1975) Protoplasma 86:253
14. Waaland SD (1978) Planta (Berl) 138:65
15. Waaland SD, Cleland RE (1974) Protoplasma 79:185
16. Waaland SD, Lucas WJ (1984) Protoplasma 123:184
17. Waaland SD, Watson BA (1980) Planta (Berl) 149:493
18. Watson BA, Waaland SD (1983) Plant Physiol (Bethesda) 71:327
19. Watson BA, Waaland SD (1985) Plant Cell Physiol (in press)

On the Hormonal System of Mosses

M. BOPP, D. GERHÄUSER, and U. KESSLER[1]

1 Introduction

Bryophytes – mosses and liverworts – are on the evolutionary line of "Chlorophyta", which means that in comparison to the preceding report, it is expected that substances operating as hormones are similar to the hormones in higher plants. However, mosses are certainly not ancestors of the higher plants, therefore, we have to determine the kind of hormonal system mosses have and to compare it with angiosperms. A few aspects of the internal regulation of moss development should give a preliminary answer to this question, thus indicating the origin of hormonal regulation during evolution.

2 Response to Exogenous Substances

Specific effects of treatments with exogenous auxins or cytokinins are well-documented and a characteristic response to ethylene is very probable. Gibberellins and abscisic acid seem to be inactive, as far as specific hormonal reactions are concerned.

2.1 Auxin

Auxin has several effects on mosses. In high concentrations, it inhibits growth and bud formation of the protonema. Furthermore, it induces cell elongation in young shoots and the formation of caulonema on the protonema is auxin-dependent. If, under stress conditions, protonema remain in the chloronema stage, caulonema formation can be induced by exogenous auxin [4, 10, 11].

2.2 Cytokinin

Cytokinin accelerates the formation and enhances the number of buds. Both responses are concentration-dependent (cf. Table 1). A different cytokinin effect was found very recently. Under standard conditions, caulonema filaments are characterized by

1 Botanisches Institut der Universität, Im Neuenheimer Feld 360, 6900 Heidelberg, FRG

Plant Growth Substances 1985
(ed. by M. Bopp)

Table 1. Bud formation 24 h after kinetin treatment

Concentration mol/l	Number of buds on cell number					Total number of bud/filament
	3 4	4 6	5	6	7	
10^{-7}	0	0	0	0.05	0.1	0.15
2×10^{-7}	0	0	0.25	0.6	0.2	1.05 [a]
5×10^{-7}	0	0.1	0.7	0.8	0.2	1.8
10^{-6}	0	0.1	0.75	0.8	0.9	2.6
10^{-5}	0	0.1	1.0	1.0	1.0	3.0

[a] This corresponds to the spontaneously formed number of buds/filament. However, spontaneous buds appear much later

Table 2

Concentration	Number of branches	
	Kinetin	6BAP
10^{-14} M	23	25
10^{-12} M	52	57
10^{-10} M	60	61
10^{-8} M	65	74
Control	9	6

Number of cells with a side branch out of 50 filaments. Only the number in the 4th and 5th cell is counted (maximum of possible branches = 100) 24 h after start of the treatment

regulatory, spontaneous branching; under unfavourable conditions, however, no side branches appear. However, they can be induced with a cytokinin concentration as low as 10^{-12} mol (Table 2). This effect differs clearly from bud induction, which will not occur at concentrations below 10^{-7} mol l^{-1}.

In a normally developed protonema, the side branches of caulonema are the target cells for cytokinin [5]. Thus, auxin and cytokinin interact sequentially and the effect of auxin is the prerequisite of a cytokinin response, although even under very deviated conditions, bud formation has been found without caulonema differentiation.

2.3 Ethylene

Ethylene – and more clearly ACC, the precursor of ethylene – induces a process, which can be regarded as accelerated aging. This is expressed as the formation of cells, which divide the filaments of the protonema in short "propagation units" [11].

3 Endogenous Hormones

After demonstration of exogenous effects, the question remains, whether the three phytohormones are "endogenous regulators" of mosses. For this purpose, one has to postulate:

1. The hormones must be present in the mosses.
2. A correlation between response and the hormone content should be found. For this purpose, mutants [6, 8] or studies on hormone metabolism can be used.

3.1 Endogenous Auxins

In the protonema of Funaria and Physcomitrella, endogenous auxins were demonstrated very recently with different methods [2, 5, 9]. The total amount found in the protonema can vary depending on the determination of the extracted material (dry or fresh weight), the extraction methods and the method of determination. Therefore, quantitatively, only values found with the same method are comparable. Nevertheless, the comparison of such results [6] was in agreement with the assumption that mosses, grown under low light intensity, a condition under which they grow without caulonema, do not have enough endogenous auxin for this developmental step.

3.2 Cytokinins

Cytokinin was determined repeatedly in the mosses, at first in a callus of a moss hybrid [3]. But in recent years, in normal-growing, wild-type protonema of Funaria and Physcomitrella, isopentenyladenine (2iP) and zeatin could be demonstrated with RIA (D. Gerhäuser and E. Weiler, unpubl. data), either in the protonema itself or in the culture medium [15]. The most remarkable feature of this determination was, that in both species the content of 2iP was around ten times higher than that of zeatin, which is in contrast to higher plants, but not differing from algae. Probably, the enzyme, which hydroxylates 2iP to zeatin is less active in mosses than in the tissue of higher plants [12], so that the 2iP can be accumulated in the cells and delivered into the substrate [14].

3.3 Ethylene

Ethylene was found in the protonema of Funaria [13]. The ethylene production is stimulated by application of ACC and IAA. Therefore, the pathway of ethylene synthesis and its regulation seem to be similar to higher plants. The endogenous ethylene production increases simultaneously with the age of the protonema to the "aging process" of the protonema cells.

4 Quantitative Aspects

According to the concentration dependency of the effects, one may assume that the hormone responses depend more on the amount of the hormones than on the overall "sensitivity" of the plants, even if a clear change of sensitivity for each caulonema cell could be shown [3]. Therefore, a correlation between the internal hormone content and morphogenesis is to be expected.

Such correlations can be demonstrated by morphogenetic mutants of Physcomitrella and Funaria.

4.1 Auxin

Several mutants, without spontaneous caulonema formation, were found in both mosses. One was tested in more detail: NAR 2. Under standard conditions, this mutant never produces caulonema. But when treated with IAA, synthetic auxins or auxin precursors, the mutant looks exactly like the wild type. The internal IAA content of this mutant is reduced, which cannot be due to a lack of auxin-forming enzymes, because with all precursors, the same response of caulonema formation has been obtained. However, an enhanced IAA oxidase activity in these cells has been found [4]. If this IAA oxidase is active in the protonema, one may expect a steady state concentration of auxin too low to induce caulonema formation.

4.2 Cytokinin

A correlation between cytokinin content and the appearance of the protonema was found in several mutants of Physcomitrella by Coves and co-workers [1]. The most pronounced type was protonemata with an enormous overproduction of buds, which is how they appear in the wild type after high doses of cytokinin. In the culture substrate of such overproducers (OVE-mutants), about 100 times more 2iP was found as in the substrate of the wild type. Regardless of whether these mutants are produced by a reduced degradation or an enhanced synthesis, a clear parallelism between endogenous cytokinin production and the morphogenetic effect exists.

Therefore, for both hormones, the endogenous content is correlated to morphogenesis, thus both hormones can be the endogenous regulators of morphogenesis.

5 Cytokinin Metabolism

Hormones in mosses do not trigger the developmental step, but are prerequisites during the whole differentiation process. Therefore, changes of the internal content by metabolism are of great importance. So far, only data for the metabolism of cytokinins are available. Kinetin is very instable in mosses [7]. After a short incubation time (about 1 h), products of side chain cleavage appear. In pulse-chase experiments with kinetin, the response disappears after removing the kinetin, whereas with 2iP or

BAP, responses remain unchanged even if the cytokinin is no longer available. We have tested the cleavage process in vivo and in vitro [7]. We have extracted "cytokinin-oxidase" after the method of Kaminek et al. (pers. comm.) and studied the metabolism of [^{14}C]-kinetin and [^{3}H]-2iP in vitro. The results were then compared with the in vivo distribution of labelled substances after feeding the mosses for 24 h.

In vivo, about 35%–40% of the initial activity of kinetin could be detected in adenine and about 5%–10% in hypoxanthine. Furthermore, besides the cytokinin oxidase, a deaminase must be in the extract. In vivo-labelled products could be identified as shown in Fig. 1.

If [^{3}H]-isopentenyladenosine is used for labelling in vitro, much less adenine was found, but about 30% isopentenyladenine, 15% inosine, and 10% hypoxanthine (Fig. 2). In vivo a constant, but low level of 2iP always remains in the cells, which is in agreement with the physiological experiments with externally applied IPA.

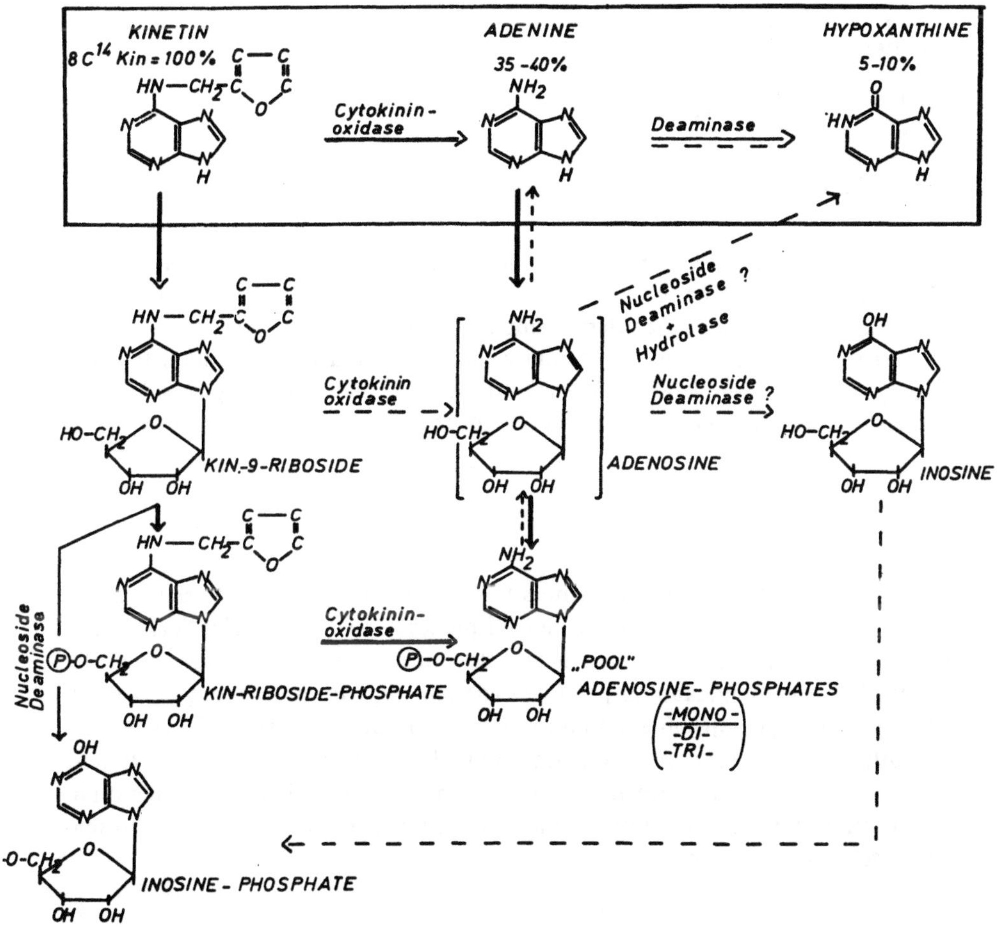

Fig. 1. The metabolism of kinetin in vitro (*box*) and in vivo. [8-^{14}C] kinetin (N^{6}-furfurylamino [8-^{14}C]-purine, 925 MBq mmol^{-1}) was added either to an enzyme preparation or fed to intact moss protonema for 24 h. Substances indicated in the scheme were labelled after 24 h

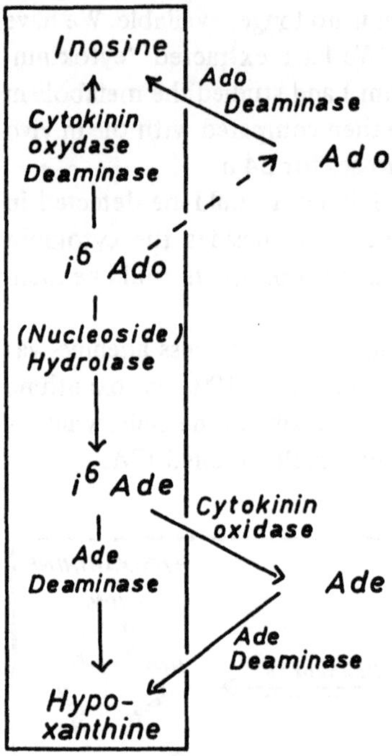

Fig. 2. The metabolism of isopentenyl adenosine (i^6 Ado) in vitro (*box*). N^6-(Δ^2-isopentenyl)adenosine (^{3}H-dialcohol) was added to an enzyme preparation up to 24 h. The substances *in the box* were found in the reaction mixture after 16–24 h. The substances *outside the box* can be found after 12–24 h in vivo treatment of intact moss protonema. Enzymes were not determined

Thus, these series of experiments also demonstrate a strong correlation between cytokinin metabolism and the effects on morphogenesis. The natural bud formation must be the consequence of the internal amount of 2iP.

With regards to the enzyme cytokinin oxidase, one must bear in mind that the moss enzyme is not completely identical with the enzyme of higher plants, because for the latter, 2iP is the better substrate [11]; for the moss enzyme, however, kinetin.

6 Conclusions

In comparing parallelisms and differences between the hormonal systems of higher plants and mosses, we found that the general processes seem to be identical in both plant groups, but in more detailed studies, clear differences can be shown, e.g. the pathway of synthesis, metabolism, and degradation, type of endogenous hormones. On the other hand, comparable morphogenetic events are regulated by the same hormone, although these processes are neither homologous nor similar in the different steps. Such events are: root formation vs caulonema differentiation, bud induction in tissue culture vs bud formation in caulonemata and leaf senescence vs tmema formation (as a form of senescence).

References

1. Ashton NW, Cove DJ, Featherstone DR (1979) Planta (Berl) 144:437
2. Ashton NW, Schulze A, Hall P, Bandurski RS (1985) Planta (Berl) 164:142
3. Beutelmann P, Bauer L (1977) Planta (Berl) 133:215
4. Bhatla SC, Bopp M (1985) J Plant Physiol 120:233
5. Bopp M (1984) Z Pflanzenphysiol 113:435
6. Bopp M, Bhatla SC (1985) In: Purohit SS (ed) Hormonal regulation of plant growth and development, vol 2. Agro Botanical Publishers, Bikaner (India), p 65
7. Bopp M, Gerhäuser D (1985) Biol Plant (Prague) 27:265
8. Cove DJ, Ashton NW (1984) In: Dyer AF, Duckett JG (eds) Experimental biology of bryophytes. Academic Press, London, p 177
9. Jayaswel RK, Johri MM (1985) Phytochemistry (Oxf) 24:1211
10. Johri MM, Desai S (1973) Nature 245:223
11. Lehnert B, Bopp M (1983) Z Pflanzenphysiol 110:379
12. McGaw BA, Horgan R (1985) Biol Plant (Prague) 27:180
13. Rohwer F, Bopp M (1985) J Plant Physiol 117:331
14. Wang TL, Cove DJ, Beutelmann P, Hartmann E (1980) Phytochemistry (Oxf) 10:1103
15. Wang TL, Horgan R, Cove DJ (1981) Plant Physiol (Bethesda) 68:735

Phytohormones and Filicinae: Chemical Signals Triggering Morphogenesis in Schizaeaceae

H. SCHRAUDOLF[1]

1 Introduction

Gametophytes of many fern species secrete specific chemicals into their substrate. These signals, named "antheridiogens" (Fig. 1), substitute for the red-light irradiation dose causing spore germination, and induce this process under dark conditions. Simultaneously, they induce premature differentiation of male sexual organs in the prothallia. The biological meaning of these combined effects is the reduction of the rate of intragametic self-fertilization, connected with the promotion of intersporophytic crossing in a bisexual gametophyte. Many biological properties of the gametophyte as well as the pheromone produced by it, become understandable only under this assumption.

SOURCES AND STRUCTURES OF ANTHERIDIOGENS

	SOURCE	STRUCTURE	ACTIVITY SPECTRUM
A_{AN}	ANEMIA SPEC.	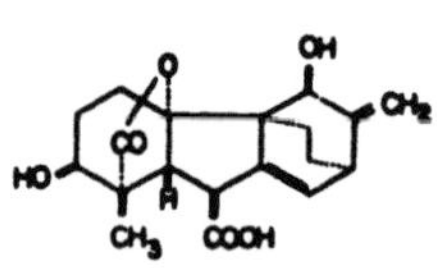	ANEMIA LYGODIUM MOHRIA
A_{LY}	LYGODIUM JAPONICUM (6)	GIBBERELLIN A_9-METHYL-ESTER	LYGODIUM
A_{PT}	PTERIDIUM AQUILINUM (1) DRYOPTERIS FILIX MAS (1) BLECHNUM GIBBUM (2) THELYPTERIS HEXAGENOPTERA (2) ONOCLEA SENSIBILIS PTERIS VITTATA	UNKNOWN	MANY POLYPODIACEAE SENSU LATO
A_{CE}	CERATOPTERIS THALYCTROIDES (3)	UNKNOWN	CERATOPTERIS SPEC.

Fig. 1. Sexual pheromones isolated from the culture media of fern gametophytes

1 Abteilung Allgemeine Botanik, Universität Ulm, 7900 Ulm, FRG

Plant Growth Substances 1985
(ed. by M. Bopp)

2 Spore Germination

One example is the extreme viability of the spores of *Anemia phyllitidis*, which show a 100% germination rate even after 20 years of dry storage. This viability persists under conditions of repeated imbibition and drying, a sequence which occurs in the spores' natural substrate. In this species, the viability of the spore is intercorrelated with an obligatory light requirement for the induction of germination. This in turn divides the natural spore population into two groups, one of which, lying on the surface of the substrate, germinates immediately after receiving optimal humidity conditions and gives rise to the typical two-dimensional growing gametophyte, which after several weeks differentiates archegonia. The second population, covered by debris, remains ungerminated, but viable by virtue of its positive photoplasticity. Due to the longevity of the spores and the thus long-lasting period of accumulation, this debris-covered population may derive from different sporophytes and, therefore, shows genetic heterogenity. This population is the target of the antheridiogen signal. At least in *A. phyllitidis*, the segregation of the pheromone starts shortly after the development of the lateral meristem, i.e. a short time before the differentiation of female sexual organs. It is still not definitely known if the time of signal segregation is coincident with the period of its biosynthesis. In any case, antheridiogen-segregating prothallia contain a compound active in bioassay, which in its physico-chemical properties is not identical with the molecule segregated.

3 Antheridiogen Effects

At the beginning of its archegonial phase, each single gametophyte is surrounded by a halo of antheridiogen (Fig. 3). In the culturing of single, female prothallia on soft agar, this halo has a radius of more than 30 cm; on natural substrate, we measured a hemisphere of high antheridiogen activity with a 10 cm radius. This spread into the substrate is favoured by the extreme biological and chemical stability of the signal molecule. It is this pheromone hemisphere, surrounding a gametophyte ready for fertilization, in which antheridiogens manifest their specific effects on debris-covered spores, that is, the combined induction of dark germination and antheridium differentiation.

Indeed, the spore densities in and on the substrate of a fern population are in the natural biotope extremely high [10]. Resulting from the spores' longevity and caused by the double action of the pheromone, the sender prothallium, ready for fertilization, after several days of pheromone segregation, is surrounded by a high number of genetically different male prothallia (Fig. 2). They contribute by their high spermatozoid production to a high probability of both fertilization and inter-sporophytic crossing.

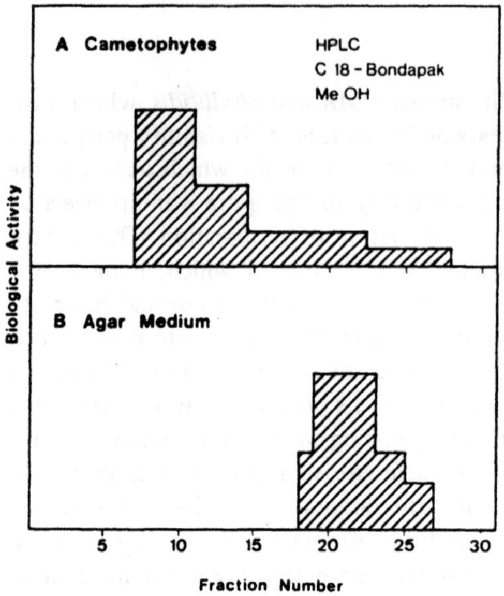

Fig. 2A,B. HPLC separation of A_{AN} isolated from female gametophytes (A), respectively agar medium of gametophyte cultures of *Anemia phyllitidis* (flow: 2 ml min^{-1}; time: 25 min, gradient: linear)

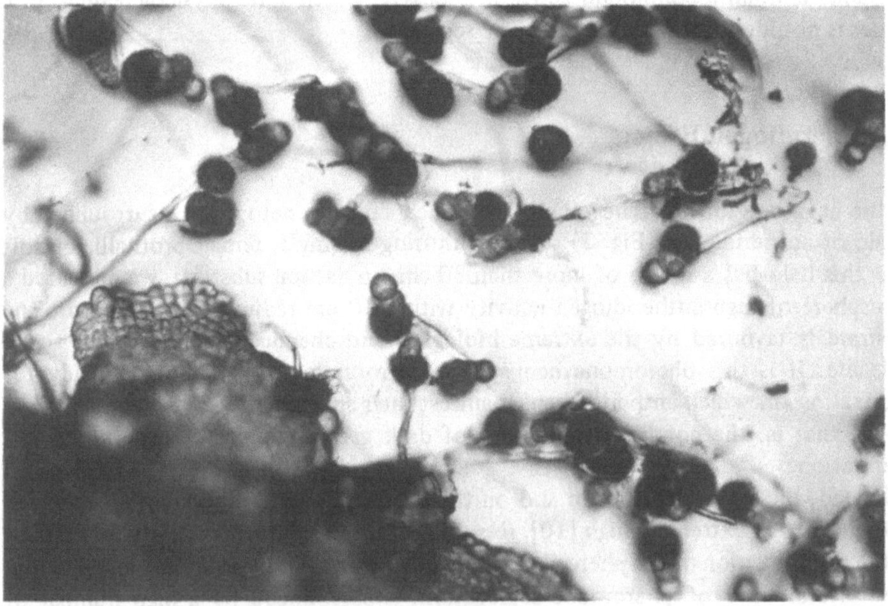

Fig. 3. Halo of antheridiogen diffusion into the agar medium of an isolated female prothallium of *A. phyllitidis*. Spores sown on the agar produce premature antheridia

4 Biological Meaning

Only an information system in which a chemical signal like the Anemia antheridiogen, produced by optimal biological timing and enforcing simultaneously dark-germination and antheridium induction, has biological meaning and a positive selection value in evolution. No homology exists between the induction of germination of a fern spore and the corresponding process in a seed or achene. The same is valid for the induction of male sexual organs in a fern gametophyte, respectively an inflorescence. But if we accept that the receptor/signal system of Schizaeaceous ferns was finally transferred during evolution to spermatophytes, whereby the pheromone changed to a hormone, then some gibberellin reactions of higher plants, which seemed so long without biological meaning and, therefore, without positive selection value, acquire then secondarily a phylogenetic significance.

This is particularly true for the phenomenon of dark induction of seed germination in higher plants by gibberellins, a reaction, which in spite of its usefulness as an experimental model, will never be realized in nature, this in default of a sufficient hormone concentration in the seeds' substrate. Perhaps the increased determination of male sexuality, caused by this hormone in unrelated taxonomical groups of spermatophytes may also be just spasmodically occurring "reminiscence" of the ancient pheromone action. The step from pheromone to hormone, accompanied by structure/activity correlations as well as the predominance of growth effects, is lastly realized in spermatophytes.

A definitive proof of an identity of the pheromone reaction in Schizaeaceae with the gibberellin system of higher plants can only be adduced if a direct comparison of the mutual binding proteins becomes possible. Nevertheless, the comparison of structure/activity relations of both systems may give some indirect knowledge about the binding molecules.

The known correlations between gibberellin structures and biological activities in the Anemia bioassay,

$$(A_7 > A_4 > A_9 \geqslant A_3 \geqslant A_1 > A_5 > A_8), \quad [11],$$

are congruent with some reactions caused by this hormone in higher plants. Especially the exceptionally high activity of Gibb A_7 finds its equivalent in the gibberellin-induced stimulation of staminate flower formation [3, 9], but also in regard to the dark stimulation of seed germination. Conversely, the Anemia-antheridiogen has biological activity in several bioassays of higher plants, such as growth stimulation of d_5-dwarf corn or dwarf peas as well as the amylase induction in barley aleuron cells [2, 14].

At least some correspondence is detectable even when nongibberellin signals are compared. Allo-gibbereic acid, but also the sesquiterpen helminthosporol induce dark germination and antheridium differentiation in Anemia gametophytes [11]. The activity of the latter compound corresponds with the gibberellin system in higher plants. Also, there are still reasonable discussions whether helminthosporol really interacts with the gibberellin receptor molecule itself [5].

Beale and MacMillan [1] reported on the activities of several derivatives of Gibb A_4 and Gibb A_9 among which especially the 2,2-dimethyl derivative of Gibb A_4 showed increased activity in bioassays as compared to the unsubstituted hormone. A corresponding increase of biological activity could be observed using the Anemia test system [12]. Even more striking is the observation that a substituted phtalimide, which proved to be active in several gibberellin assays [4, 7, 8], induced spore germination as well as antheridium differentiation in gametophytes of *A. phyllitidis* [13].

5 Conclusions

These far-reaching correspondences may additionally support the hypothesis of a close phylogenetic relation between the pheromone system of Schizaeaceae and the gibberellin reaction system of spermatophytes, a relation which certainly does not represent just a simple coincidence or analogy. Indeed, the idea that "pheromones of unicellular eucarionts could be the ancestors of hormones for all multicellular eucarionts" [6] is widely accepted among zoologists. From biological reasons, as well as from results of an indirect analysis of the mutual receptor properties, there is some probability that the primitive family of Leptosporangiatae constitutes an important step to the phylogenesis of information molecules, namely the transitional period of gibberellinlike terpenoid signals in the status of a pheromone, a status from which, in a second step, they obtained the character of a hormonal signal during evolution of conifers.

References

1. Beale MH, MacMillan J (1981) Phytochemistry (Oxf) 20:693
2. Bürcky K (1977) Z Naturforsch Sect C 32:652
3. Clark RK, Kennedy DS (1969) J Am Hortic Sci 94:131
4. Davlin RM (1981) Proc Plant Growth Regul Working Group 8:197
5. Kato J, Katsumi M, Tamura S, Sakurai A (1968) Biochemistry and physiology of plant growth substances, p 347
6. Kochert G (1978) Annu Rev Plant Physiol 29:461
7. Los M, Kust CA, Lamp G, Diehl RE (1980) J Am Hortic Sci 15:22
8. Metzger ID, Suttle IC (1982) Plant Physiol (Bethesda) Lancaster 69:24
9. Pike LM, Peterson CE·(1969) Euphytica 18:106
10. Schneller IJ (1974) Ber Schweiz Bot Ges 84:195
11. Schraudolf H (1966) Plant Cell Physiol 7:277
12. Schraudolf H (1982) Naturwissenschaften 69:286
13. Schraudolf H (1983) Z Pflanzenphysiol 109:469
14. Sharp PB, Keitt GW, Clum HH, Näf U (1975) Physiol Plant 34:101

The Control of Plant Cell Elongation by Auxin and Gibberellin

R.L. JONES [1]

1 Introduction

The purpose of this article is to provide an overview of progress in research on the effects of auxin and gibberellins (GAs) on plant cell elongation. The response of angiosperm shoots to auxins and GAs is well established, and many reviews deal with various aspects of cell elongation [2,22,34,41]. Research during the past decade has focused on the biophysical and biochemical mechanisms that underlie these responses. It is becoming increasingly clear that the biophysical basis of the growth response to both classes of hormones is similar, if not identical, and the biochemical processes that lead to changes in the properties of the cell wall may also be similar.

Our understanding of the changes that occur at the molecular level in the symplast and apoplast following hormone treatment progresses slowly. For example, although changes in the levels of mRNAs following auxin treatment have been well documented [43, 44], we do not know whether these messengers code for proteins that are directly involved in the growth response. For the elongation response to GAs, on the other hand, there has been no demonstration that specific mRNAs accumulate in tissue prior to or acompanying the growth response.

In this brief review attention will be focused on biophysical aspects of growth and on the events that occur in the apoplast of tissues that respond to auxin and GA. Biophysical measurements identify the cell wall as the site where plant growth regulators exert their growth-stimulating effects.

2 Biophysical Aspects

Plant cell expansion can be described in biophysical terms by Eq. (1) formulated by Lockhart [32] and modified by Cosgrove [8]

$$v_s = \frac{L\phi}{L + \phi} (\sigma\Delta_\pi - Y) \qquad (1)$$

where the steady-state rate of cell expansion (v_s) is a function of the hydraulic conductance of the cell or tissue (L), the extensibility of the cell wall (ϕ), the solute reflection coefficient of cell membranes (σ), the difference in the osmotic pressure

1 Department of Botany, University of California, Berkeley, CA 94720, USA

Plant Growth Substances 1985
(ed. by M. Bopp)

across the plasma membrane (Δ_π), and the yield threshold of the cell wall (Y). The parameters in Eq. (1) can be separated into those that affect the uptake of H_2O into cells, namely, L, Δ_π, and σ, and those that govern the extensibility of the cell wall, ϕ and Y. One of the goals of biophysicists has been to define the role that each of these parameters plays in growth and in particular to determine those that are influenced by treatment with plant growth substances.

Recent technical developments have led to methods that can provide accurate measurements of the water potential of cells and tissues [2, 3, 11, 12]. Cosgrove [9] has exploited the pressure probe to measure the osmotic properties of excised internode sections of pea and has concluded that the effect of auxin on cell wall extensibility alone is enough to account for hormone-stimulated elongation (Table 1). Cosgrove [9] also measured the hydraulic conductance of pea internode tissues and showed that it was an order to magnitude greater than cell-wall extensibility. He therefore argued that water supply does not normally limit elongation; rather, the absence of a significant resistance to water movement indicates that a large difference in water potential (Δ_ψ) is not required to sustain growth.

Boyer [2] and Boyer et al. [3] take issue with Cosgrove's conclusions about the role of hydraulic conductance in the control of cell enlargement. Using a new guillotine thermocouple psychrometer to determine various parameters of tissue water potential, Boyer and his colleagues argue that hydraulic conductance plays a significant role in limiting the rate at which cells can elongate. To support their contention of an effect of hydraulic conductance on enlargement, these workers point to the significant difference in the water potential between xylem (water source) and the growing tissues of soybean hypocotyls. Their measurements show a radial gradient in water potential in the stem of as much as 1 bar. Also, in contrast to the data obtained by Cosgrove [9, 12] using the pressure probe, the psychrometric measurements of exten-

Table 1. Turgor pressure, growth rate, and $T_{1/2}$ of stress relaxation in pea stems. (From Cosgrove [9])

Treatment	n	Growth rate	P_0	Y	$T_{1/2}$	ϕ_{dyn}	ϕ_{ss} [a]
		% h^{-1}	MPa		min	$MPa^{-1}h^{-1}$	
Sections preincubated in H_2O	10	2.49 (0.22)	0.61 (0.01)	0.29 (0.02)	52.2 (3.7)	0.084 (0.005)	0.074 (0.007)
Sections preincubated in IAA	8	6.70 (0.35)	0.60 (0.01)	0.29 (0.01)	18.4 (1.2)	0.24 (0.022)	0.22 (0.014)
Sections not preincubated	10	7.51 (0.43)	0.62 (0.01)	0.30 (0.01)	34.8 (3.1)	0.13 (0.016)	0.25 (0.017)

[a] In the first two treatments, stem segments were excised from the plant and pretreated with water or with 10 μM IAA before undergoing stress relaxation. In the third treatment, segments were excised directly from the plants and allowed to undergo relaxation without pretreatment. In five of the segments in this third treatment, the cut surfaces were washed and blotted dry to remove solutes released by the cut cells. P_0, cell turgor at start of stress relaxation; Y, yield threshold; $T_{1/2}$, half-time of stress relaxation; ϕ_{dyn}, apparent extensibility calculated from $\ln(2)/\epsilon \cdot T_{1/2}$; ϕ_{ss}, steady state extensibility calculated from growth rate/(P_0 – Y). The values represent means of n samples; SE are given in parentheses

sibility and hydraulic conductance made by Boyer et al. [9] show that these two parameters are approximately equal. The significance of this observation is that both water demand and water supply can modify growth.

While it is generally agreed that auxin and GAs can affect water demand by altering cell-wall extensibility, there is circumstantial evidence that auxin may also influence hydraulic conductance. Lee et al. [31] and Loros and Taiz [33] showed that the rates of plasmolysis and deplasmolysis of epidermal cells were significantly influenced by auxin. It is possible, therefore, that plant growth regulators could influence both aspects of the water relations of cells by altering either the demand for H_2O by changing wall extensibility or the supply of H_2O by modifying the hydraulic conductance of cells or tissues.

The way in which the low water potential of growing plant cells is interpreted has also led to the controversy concerning the magnitude of hydraulic conductance. Cosgrove and Cleland [9-11] argue that the low water potential of elongating tissues is a function of the presence of solutes in the apoplast and not due to a resistance to water flow from xylem to growing cells. They have found the concentration of solutes in stem segments to be about 70 mOsmol kg^{-1}, giving an estimated osmotic pressure of 2.9 bars [10]. That solutes are present in the apoplastic space of plant tissues is not disputed, but Boyer [2] questions whether these solutes are present in the apoplast at sufficiently high concentrations to influence water potential.

3 Biochemical Aspects

Although the role that hydraulic conductance may play in regulating plant cell extension is controversial, there is almost unanimous agreement that changes in cell-wall extensibility accompany elongation induced by auxins [41] and GAs [22] (but see [23] for a different view). Investigations of changes in wall extensibility have focused on two principal aspects, the chemical component(s) of cell walls that change during elongation growth and the agent(s) responsible for these changes.

3.1 Changes Accompanying Altered Wall Extensibility

Until recently it was presumed that growth-regulator treatments caused changes in the cell wall that resulted in an increase in wall extensibility. Recent experiments [6, 7, 15, 16, 28] have challenged this notion. These experiments point to the formation of covalent cross-linkages in the wall that may cause a decrease in extensibility (rigidification), and there is evidence that this rigidification may be controlled by plant growth substances. Other experiments, particularly those involving the Ca^{2+} ion, suggest that noncovalent interactions in the cell may also be important in reducing extensibility.

3.2 Covalent Cross-Linking

In a series of publications, Fry [15-18] has drawn attention to two components of the primary wall of higher plants that might be involved in the formation of inter-a)

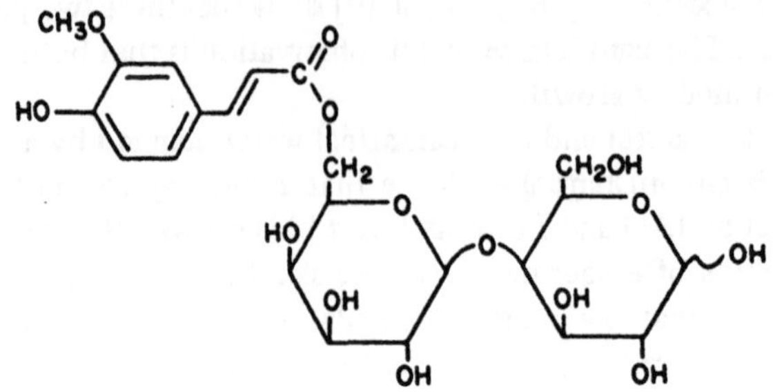

b)

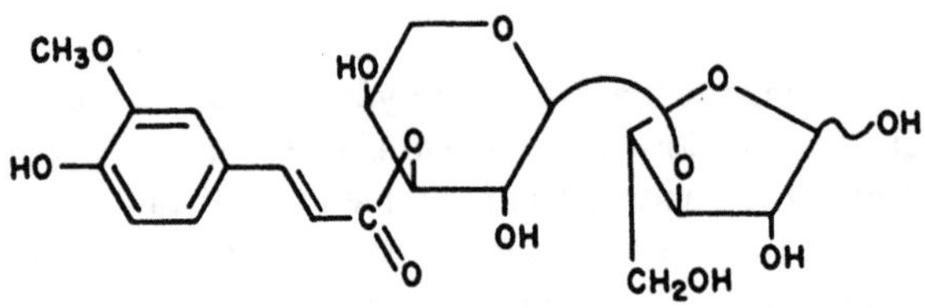

Fig. 1a,b. Ferulic acid esterified with galactopyranose (a) and arabinopyranose (b). (From McNeil et al. [34])

molecular cross-linkages. He has shown that cultured spinach cells contain ferulic acid that is esterified with arabinopyranose and galactopyranose residues of pectic polysaccharides in the wall (Fig. 1). Approximately 0.4% of the glycosyl residues of the spinach wall are esterified with ferulate, accounting for as much as 60%–70% of total cell-wall ferulate.

In a labeling study with [^{14}C]cinnamate, Fry [18] showed that radioactivity was incorporated into at least two pools of phenolic compounds in the wall. Saponification of labeled cell walls yielded free labeled *p*-coumarate and ferulate as well as ethyl acetate-soluble radioactivity. The ethyl acetate-soluble material was associated with the Driselase-resistant fraction of the cell wall. He argued that this fraction became Driselase-resistant because of the presence of cross-links between the phenolic residues.

In addition to the coupling of polysaccharides by ferulate cross-bridges, Fry [16] described a new amino acid, isodityrosine, which may also cross-link cell-wall proteins. Fry [16] has proposed that the insolubility of the principal structural protein extensin derives from the presence of phenolic cross-linkages, specifically the formation of an intermolecular diphenyl ether bridge between adjacent tyrosine molecules (Fig. 2). The protein moiety of extensin contains about 8 mol% tyrosine [34], and Fry has calculated that there are nine half-isodityrosyl residues per extensin molecule. Although some of the isodityrosyl linkages are intramolecular [13], there is indirect evidence that some of them are intermolecular, that is, diphenyl ether bridges between the tyrosine residues of adjacent extensin chains. This evidence includes the observation that extensin, which can be readily extracted from cell walls, lacks isodityrosine, whereas extensin insolubilization is paralleled by a conversion of tyrosine to isodityrosine [5, 16]. Furthermore, oxidation of phenolic cross-linkages in the cell wall by sodium chlorite is correlated with solubilization of extensin [16].

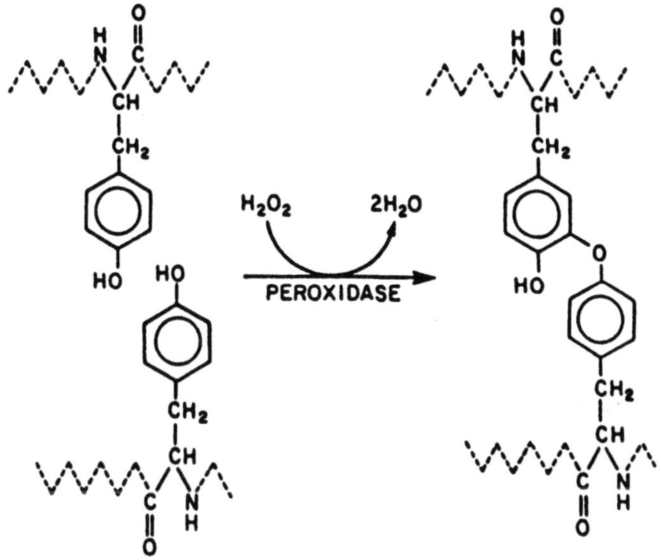

Fig. 2. A postulated mechanism for the formation of intermolecular isodityrosine residues thought to cross-link extension molecules within the cell wall. (From McNeil [34])

Because both polysaccharide and protein components of the cell wall can be shown to be insolubilized by formation of dimers between ferulate and tyrosine, respectively, attention has focused on mechanisms that may bring about this oxidative coupling. Cooper and Varner [6], Fry [17], and Lamport and Epstein [28] have proposed that peroxidases located in the cell wall catalyze the cross-linking of both ferulate and tyrosine. Furthermore, Fry has suggested that this process is hormonally regulated, since in cultured spinach cells the secretion of peroxidase from the protoplast into the cell wall is regulated by GA.

Peroxidases have long been correlated with growth in plants. In dwarf varieties of many species, including corn, pea, sorghum, tomato, and wheat, there is a strong correlation between elevated peroxidase levels and reduced shoot height [39]. Plant growth regulators have also been shown to play a role in regulating peroxidase levels in many species. In peas, for example, there are reports that auxins [36], ethylene [37], and GAs [1] all influence peroxidase activity and growth. Whereas GA and auxin promote elongation and suppress peroxidase activity in pea shoots, ethylene suppresses elongation and promotes peroxidase activity.

Using a centrifugation technique, Terry [42] and Morrow and Jones [35a] have shown a strong correlation between peroxidase activity that can be removed from the cell wall and the action of auxin and GA on elongation. In sugar cane, Terry has shown that GA treatment promotes growth by 300% over controls and inhibits cell-wall peroxidase by 53% (Table 2). Similarly, in peas, auxin-promoted elongation is paralleled by the suppression of peroxidase activity that can be removed by centrifugation (Fig. 3). Both growth promotion and inhibition of peroxidase activity can be detected by at least 30 min of auxin treatment of excised pea internode segments (Fig. 3).

Table 2. Effect of GA on elongation of sugar cane spindles and on the activities of POD, MDH, and GOT centrifuged from spindles[a]. (From Terry [42])

Treatment	Increase in length (mm)	Enzyme activity[b] ($A\ ml^{-1}\ min^{-1}\ g^{-1}$)		
		POD	MDH	GOT
GA, 6 h	2.1	27.6	6.2	1.61
H_2O, 6 h	1.1	35.3	7.8	1.94
GA/H_2O	1.9	0.78	0.79	0.82
GA, 12 h	14.2	17.7	6.2	0.96
H_2O, 12 h	4.4	39.7	9.0	1.84
GA/H_2O	3.2	0.45	0.69	0.52

[a] Results are the average of three experiments using approximately 60–70 spindles per treatment
[b] The activities of peroxidase (POD), malate dehydrogenase (MDH), and glutamate oxaloacetate transaminase (GOT) were assayed spectrophotometrically

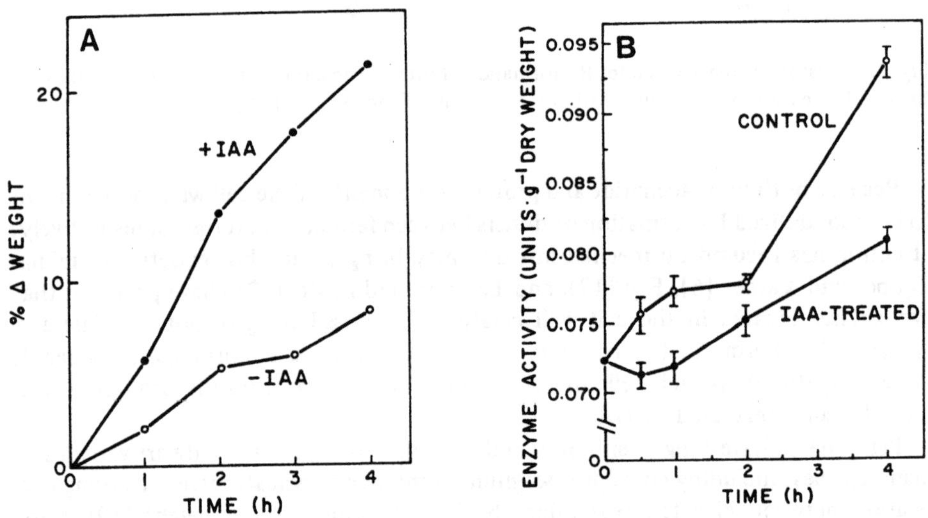

Fig. 3A,B. Growth of excised pea internode sections (A) and peroxidase activity centrifuged from sections (B) treated with (•) and without (○) 17 μM IAA. (Morrow and Jones [35a])

If peroxidases regulate cell-wall extensibility by catalyzing the oxidative cross-linking of phenols, there must also be a mechanism to ensure the production of electron donors for the formation of H_2O_2. Gross et al. [19] have proposed that malate dehydrogenase (MDH) could provide the necessary electron donor by generating $NADH_2$ for the production of H_2O_2 as shown in Eqs. (2) and (3):

$$\text{Malate} + \text{NAD} \xrightarrow{\text{MDH}} \text{oxaloacetate} + \text{NADH}_2\ ; \tag{2}$$

$$\text{NADH}_2 + \text{O}_2 \xrightarrow{\text{peroxidase}} \text{NAD} + \text{H}_2\text{O}_2\ . \tag{3}$$

Malate dehydrogenase is present in plant cell walls, and its activity can be removed by centrifugation (Table 2). Interestingly, the activity of MDH is also suppressed by GA treatment of sugar cane tissue. The regulation of peroxidase activity by auxins and GAs raises the possibility that these growth regulators themselves alter wall extensibility via their effects on peroxidase. Even if auxins and GAs affect cell-wall extensibility by reducing the extent to which oxidative cross-bridges are formed [15], this cannot be the primary mechanism by which these growth regulators act. In excised tissue segments of both monocot and dicot tissue, the effect of GA or auxin is to cause an increase in the extensibility of tissue [41]. By merely suppressing the activity of peroxidase in the cell wall, growth regulators would bring about a reduction in the rate of cell-wall rigidification and not an increase in the extensibility of the wall. If auxin or GA should play a role in suppressing the cross-linking of the cell wall in growing tissues, this role must be in addition to the one that results in an enhancement of wall extensibility. One of the roles of auxins and GAs may, therefore, be to delay the hardening or aging of the cell wall [38] and in this way prolong the duration of elongation. The actual process of elongation, however, would depend only on increased wall extensibility or enhanced hydraulic conductivity of the tissue.

3.3 Noncovalent Cross-Linking

Cations, in particular Ca^{2+}, have long been implicated in the process of cell elongation (see reviews by Hepler and Wayne [20], Taiz [41]). Interest in the role of Ca^{2+} has been rekindled in part because of the demonstration of its involvement in gravitropism [29, 30, 40] and the possibility that Ca^{2+} may play an important role in stimulus-response coupling [20]. In the coleoptile of oat, for example, Slocum and Roux [40] have shown that gravistimulation results in a redistribution of Ca^{2+}, and these workers proposed that Ca^{2+} influences the gravitropic response by affecting the extensibility of the cell wall.

The involvement of Ca^{2+} in auxin- and GA-induced elongation remains to be firmly established. Although there are many reports that show that the growth of auxin-sensitive [14] and GA-sensitive [22] tissue is affected by external Ca^{2+}, there has as yet been no satisfactory demonstration that the effects of either growth regulator on elongation are mediated directly or indirectly by Ca^{2+} [20]. Circumstantial evidence from several laboratories has implicated both auxin [5, 24] and GA [35] in the transport of Ca^{2+}.

The isolated reports of possible effects of auxin and GA on Ca^{2+} transport in elongating tissues call for a more detailed examination of this phenomenon. As Hepler and Wayne [20] point out in their recent review, the stage has been reached where the demonstration of changes in cytoplasmic and apoplastic Ca^{2+} is essential if Ca^{2+} is to be implicated in auxin- or GA-stimulated growth.

3.4 Agents Causing Altered Wall Extensibility

It is now almost unanimously accepted that increased wall extensibility brought about by growth substance treatment is a result of the action of enzymes [4, 41]. A more vexing issue concerns the role that protons play in the process of wall loosening. While most published evidence continues to favor a role for H^+, possibly in the modulation of the activity of the putative enzyme(s) that loosen the wall, recent data has again questioned the validity of the acid-growth hypothesis of auxin action [25]. Kutschera and Schopfer [25, 26] have been unable to demonstrate proton release from auxin-treated coleoptiles of *Zea mays* despite the fact that coleoptiles incubated in fusicoccin released measurable levels of protons. It is clear that the conflicting data on the role of H^+ in elongation must be resolved. Many of the controversies concern aspects of method and technique. An experimental approach utilizing protoplasts might help resolve this issue.

The existence of enzymes, whose activities are thought to be altered by pH, is also controversial. There are numerous reports of polysaccharide-degrading enzymes associated with plant cell walls [27, 41]. However, there has been no convincing demonstration linking a specific enzyme activity to auxin- or GA-induced wall loosening. An approach to this question that may hold promise is one described by Huber and Nevins [21], in which antibodies to proteins extracted from *Z. mays* cell walls were shown to inhibit auxin-induced growth of *Z. mays* coleoptiles.

References

1. Birecka JA, Galston AW (1970) J Exp Bot 21:735
2. Boyer JS (1985) Annu Rev Plant Physiol 36:473
3. Boyer JS, Cavalieri AJ, Schulze ED (1985) Planta (Berl) 163:527
4. Cleland RE (1981) In: Tanner W, Loewus FA (eds) Plant carbohydrates II. Extracellular carbohydrates. Encyclopedia of plant physiology, vol 13B. Berlin Heidelberg New York, p 256
5. Cohen JD, Lilly N (1984) Plant Physiol (Bethesda) 75:5
6. Cooper JB, Varner JE (1983a) Biochem Biophys Res Commun 112:161
7. Cooper JB, Varner JE (1983b) Plant Physiol (Bethesda) 73:324
8. Cosgrove DJ (1981) Plant Physiol (Bethesda) 68:1439
9. Cosgrove DJ (1985) Plant Physiol (Bethesda) 78:347
10. Cosgrove DJ, Cleland RE (1983a) Plant Physiol (Bethesda) 72:326
11. Cosgrove DJ, Cleland RE (1983b) Plant Physiol (Bethesda) 72:332
12. Cosgrove DJ, Van Volkenburgh E, Cleland RE (1984) Planta (Berl) 162:46
13. Epstein L, Lamport DTA (1984) Phytochemistry (Oxf) 23:1241
14. Evans ML (1984) CRC Crit Rev Plant Sci 2:317
15. Fry SC (1982a) Biochem J 203:493
16. Fry SC (1982b) Biochem J 204:449
17. Fry SC (1983) Planta (Berl) 157:111
18. Fry SC (1984) Phytochemistry (Oxf) 23:59
19. Gross GG, Janse C, Elstner EF (1977) Planta (Berl) 136:271
20. Hepler PK, Wayne RO (1985) Annu Rev Plant Physiol 36:397
21. Huber DJ, Nevins DJ (1981) Physiol Plant 53:535
22. Jones RL (1983) CRC Crit Rev Plant Sci 1:23
23. Katsumi M, Kazama H (1978) Bot Mag Tokyo Spec Issue 1:141
24. Kubowicz BD, Vanderhoef LN, Hanson JB (1982) Plant Physiol (Bethesda) 69:187
25. Kutschera U, Schopfer P (1985a) Planta (Berl) 163:483
26. Kutschera U, Schopfer P (1985b) Planta (Berl) 163:494
27. Labavitch JM (1981) Annu Rev Plant Physiol 32:385
28. Lamport DTA, Epstein L (1983) In: Proc Ann Plant Biochem Physiol Symp, 2nd edn, Columbia. University of Missouri Press, Columbia, Missouri, p 73

29. Lee JS, Mulkey TJ, Evans ML (1983) Plant Physiol (Bethesda) 73:874
30. Lee JS, Mulkey TJ, Evans MC (1984) Planta (Berl) 160:536
31. Lee YS, Stadelman EJ, Kwon YM (1980) Plant Physiol (Bethesda) 65:5
32. Lockhart JA (1965) Theor Biol 8:264
33. Loros J, Taiz L (1982) Plant Sci Lett 26:93
34. McNeil M, Darvill AG, Fry SC, Albersheim P (1984) Annu Rev Biochem 53:625
35. Moll C, Jones RL (1981) Planta (Berl) 152:450
35a. Morrow DL, Jones RL (1986) Physiol Plant (in press)
36. Ockerse R, Mumford LM (1973) Can J Bot 51:2237
37. Ridge I, Osborne DJ (1970) J Exp Bot 21:843
38. Silk WK, Jones RL (1975) Plant Physiol (Bethesda) 56:267
39. Schertz KF, Sumpter NA, Sarkisian IV, Hart GE (1971) J Hered 62:235
40. Slocum RD, Roux SJ (1983) Planta (Berl) 157:481
41. Taiz L (1984) Annu Rev Plant Physiol 35:585
42. Terry ME (1986) J Am Soc Sug Cane Technol (in press)
43. Theologis A, Ray PM (1982) Proc Natl Acad Sci USA 79:418
44. Zurfluh LL, Guilfoyle TJ (1982) Plant Physiol (Bethesda) 69:332

Mechanism of Growth Induction by Ester Compounds. Studies with Segments and Protoplasts of Oat Coleoptiles

A. HAGER, R. HAMPP, and W. MEHRLE [1]

1 Introduction

Lipophilic esters of acetic and propionic acids induce elongation growth in oat coleoptiles [7]; for naphthyl acetate see [20]. The initial increase in growth was comparable to that induced by auxins, such as IAA or 2,4 D; however, it could not be inhibited by cycloheximide as was the case with auxins. The mechanism by which esters stimulate growth was hypothesized thus: The ester permeates the plasma membrane and is hydrolyzed by endogenous esterases. The resulting acid lowers the cytosolic pH and thereby activates proton pumps that are located on the plasma membrane. Enhanced proton extrusion associated with higher pump activity causes elongation.

This mechanism is similar to that proposed for growth stimulated by permeable weak acids (e.g., propionic acid or dimethadione; [7]). Brummer et al. [1], who found the membrane potential to be increased after the addition of weak acids, also suggested an ATP-dependent H^+-pumping activity to be enhanced under these conditions. This observation is in agreement with the report of Bates and Goldsmith, who observed a rapid hyperpolarization of oat coleoptile cells after applying higher concentrations (10 mM) of undissociated benzoic and butyric acids. Additional support for this interpretation comes from experiments in which (1) maize root segments were shown to respond to the application of butyrate, trimethyl acetate, or dimethadione with a significant acidification of the cell sap, together with a rapid hyperpolarization [12], and (2) cytoplasmic pH in *Neurospora crassa* was lowered and proton extrusion was increased by the addition of butyric acid [14].

Another indication that it is the cytosolic pH which regulates proton extrusion is the "ANA" effect [5]: accumulation of lactic acid during a short period of *ana*erobiosis caused a decrease in the cellular pH of coleoptile segments. Upon transfer of the segments to normal air (restoration of respiratory ATP supply), a burst in the growth rate occurred. The period of enhanced growth declined after about 20 min, which coincided with the disappearance of lactic acid. Similarly, incubation of segments with CN^- (in the presence of oxygen) resulted in an accumulation of lactic acid and a concomitant fall in cellular pH. Removal of CN^- led to enhanced growth rates [7]. Thus, the lowering of cytosolic pH by endogenously synthesized lactic acid activated a proton extruding ATPase.

1 Universität Tübingen, Biologie I, Auf der Morgenstelle 1, 7400 Tübingen, FRG

Plant Growth Substances 1985
(ed. by M. Bopp)

In the present chapter, we report corroborative evidence for this mechanism, which was by a different experimental approach. With fluorescein as an indicator, changes of cytosolic pH upon addition of carboxyester compounds (including the diacetic acid ester, fusicoccin) were recorded on a single cell level. Furthermore, we show that incubation of coleoptile protoplasts with permeable esters results in an acidification of the incubation medium, which is due to processes inside intact protoplasts and is dependent on respiratory ATP supply.

2 Material and Methods

2.1 Isolation of Coleoptile Protoplasts

Coleoptile protoplasts were isolated from 5-day-old, dark-grown *Avena sativa* (cv. Arnold) seedlings in a medium containing 0.8 M mannitol, 1 mM $CaCl_2$, 0.5% (w/v) BSA, 10 mM ascorbic acid, 3% (w/v) Cellulase TC (Serva; Heidelberg, FRG), 0.1% (w/v) Pectolyase Y23 (Paesel, Frankfurt), adjusted to pH 5.6 with HCl. Principal steps were as in [9]. Integrity as determined by fluorescein diacetate (FDA) fluorescence was usually between 80% and 90%.

2.2 Immobilization of Protoplasts

Immobilization in an alginate matrix. A protoplast aliquot (20 μl; stained with 0.005% (w/v) FDA for 5 min and subsequently washed free from excess FDA as described above) was gently stirred into the alginate layer and sprayed with a solution of 50 mM $CaCl_2$ in 0.8 M sorbitol. After forming a solid matrix (about 5 min), the gel was rinsed twice with 0.8 M sorbitol and used for fluorimetry [15].

2.3 Determination of Fluorescence of Fluorescein in Single Protoplasts

Changes in pH-dependent fluorescence of cytosolic fluorescein were monitored with an inverted microscope [13]. Readings were taken from single, alginate-embedded cells without further treatment, or after application of different effectors (see Sect. 3).

Changes in extraprotoplast pH were monitored at 20 °C with a Knick microprocessor pH meter with the electrode mounted into a Hansatech oxygen electrode vessel. The same set up was used in experiments in which adenylates were assayed [8, 10].

2.4 Assays with Coleoptile Segments

Growth of coleoptile segments was determined according to Hager and Moser [7], and the preparation of microsomal membranes was carried out essentially as described by Hager and Biber [6]. The assay of esterase activity with membrane and protein fractions was performed with FDA (50 μM), using a Farrand fluorimeter (Mark I; excitation and emission wavelength: 490 and 520 nm, respectively).

3 Results and Discussion

3.1 Induction by Ester Compounds of Elongation Growth of Oat Coleoptile Segments

Structures of some of the compounds used are shown in Fig. 1. Application of these effectors induced growth of coleoptile segments comparable to that induced by IAA (Fig. 2). At higher concentrations of 2-naphthylacetate (0.5 mM), growth was inhibited (not shown). Possibly, this inhibition is due to the release of naphthol inside the cytosol, which interacts with respiratory phosphorylation as the cellular ATP level is decreased, too (cf. Fig. 6).

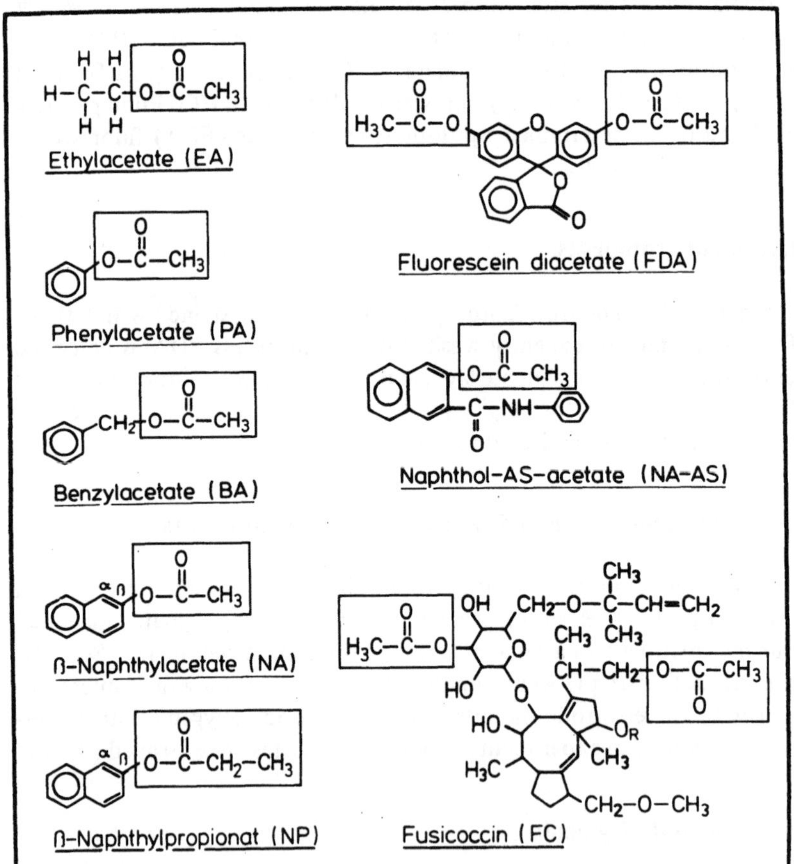

Fig. 1. Structure of esters used for growth induction or for the assay of changes in intracellular pH (FDA). Acidic moieties set free by esterases are indicated

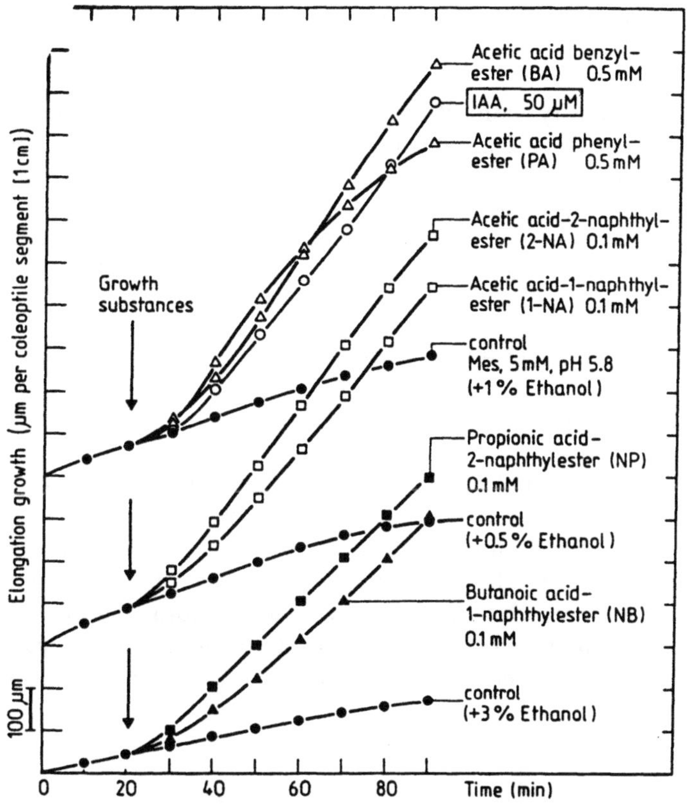

Fig. 2. Time course of the induction of elongation growth of oat coleoptile segments by lipophilic carboxyesters and IAA

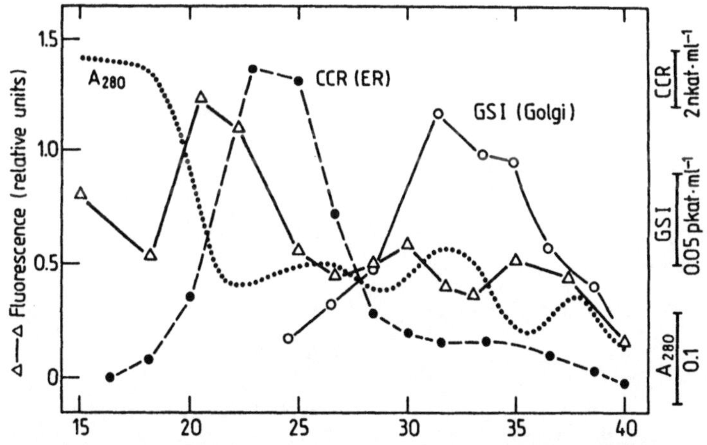

Fig. 3. Profile of esterase activity (release of fluorescein from fluorescein diacetate) and membrane markers of a microsomal membrane fraction of maize coleoptiles after centrifugation on a linear, isopycnic sucrose gradient. CCR = cytochrome c reductase (antimycin insensitive); GS I = glucan synthetase I

3.2 Intracellular Localization of Enzymes Responsible for the Cleavage of Added Esters

The association of esterase activity with different fractions obtained after density gradient centrifugation of microsomal membranes of maize coleoptiles is shown in Fig. 3. Esterase activity peaked at a low gradient density in the region of tonoplast membranes. In addition, there was some activity that coincided with markers for Golgi and plasma membranes. Thus, enzymatic activity responsible for hydrolysis of growth promoting esters, or FDA, appears to be mainly membrane-associated.

3.3 Experiments with Isolated Coleoptile Protoplasts

3.3.1 Manipulation of Cytosolic pH

Figure 4a is a phase-contrast photomicrograph of intact coleoptile protoplasts. A feature unique to this type of protoplast is the large number of vacuoles (Fig. 4b). When these protoplasts are incubated with 0.005% (w/v) FDA, illumination with blue light (Leitz filter combination I2) results in an intense fluorescence with an emission around 520 nm. This fluorescence, which is observed with intact protoplasts only, is restricted to the cytosolic compartment (Fig. 4c). It is due to the formation by cytosolic esterases of plasmalemma-impermeable fluorescein. The intensity of fluorescein

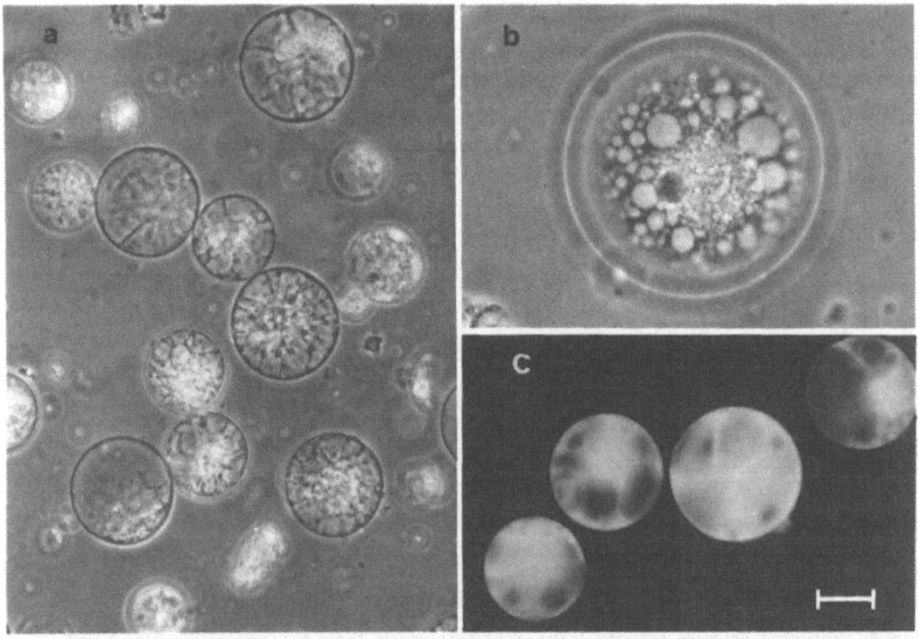

Fig. 4a–c. Photomicrographs of oat coleoptile protoplasts. A population of isolated protoplasts (**a**) and a single protoplast under phase contrast (**b**; note the large number of vacuoles); fluorescein diacetate-stained protoplasts (**c**). *Bar* represents 20 μm

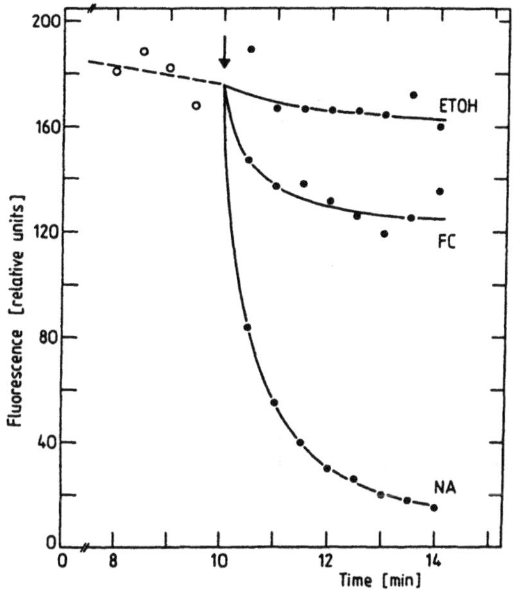

Fig. 5. Microfluorometric recordings of cytosolic fluorescein fluorescence in single oat coleoptile protoplasts, immobilized in an alginate matrix. *Open circles:* no addition. *Closed circles:* change in fluorescence upon addition (*arrow*) of (final concentrations) ethanol (ETOH; 0.1%), fusicoccin (FC; 5 μM, in ethanol), 2-nephthylacetate (NA; 50 μM, in ethanol). The decrease in fluorescence is due to cytosolic acidification. Typical examples out of a series of several independent experiments

fluorescence has been shown to be pH-dependent [7, 17, 18] and, therefore, can be used as an indicator for changes in cytosolic pH. In Fig. 5, microphotometric determinations of changes in fluorescence of single coleoptile protoplasts, immobilized in Ca-alginate, are given. Readings (about 0.8 s) were taken in 30-s-intervals. Controls showed a slight decrease in fluorescence during the incubation period, probably due to photobleaching of fluorescein or unspecific leakage. This background was slightly increased by the addition of ethanol (final concentration: 0.1%). In contrast, 2-naphthylacetate (50 μM) caused an immediate decrease in fluorescence, indicative for an acidification of the cytosol. Interestingly, addition of fusicoccin (5 μM) also resulted in decreased fluorescence, which in all trials was distinctly different from background reactions. This latter finding is noteworthy with respect to the mode of action of fusicoccin. External acidification in the presence of fusicoccin was proposed to take place via internal alkalization (e.g., [11]). However, according to our results and those of Bertl and Felle [3], fusicoccin could act in a rather different way: The diacetic acid ester, as suggested for the other esters used in this communication, could be broken down by cytosolic esterases. This would cause an acidification of the cytosol which, in turn, would activate proton extrusion by plasma membrane-located ATPases as discussed above.

3.3.2 Proton Extrusion and Intracellular ATP Level

Addition of 50 μM NA to a suspension of oat coleoptile protoplasts led to a decrease in medium pH. This acidification was considerably less obvious in the presence of antimycin A (Fig. 6), an inhibitor of the mitochondrial electron transport chain at the level of complex III. The ability of antimycin to permeate the plasma membrane of intact protoplasts is indicated by the fact that it causes an immediate decrease in

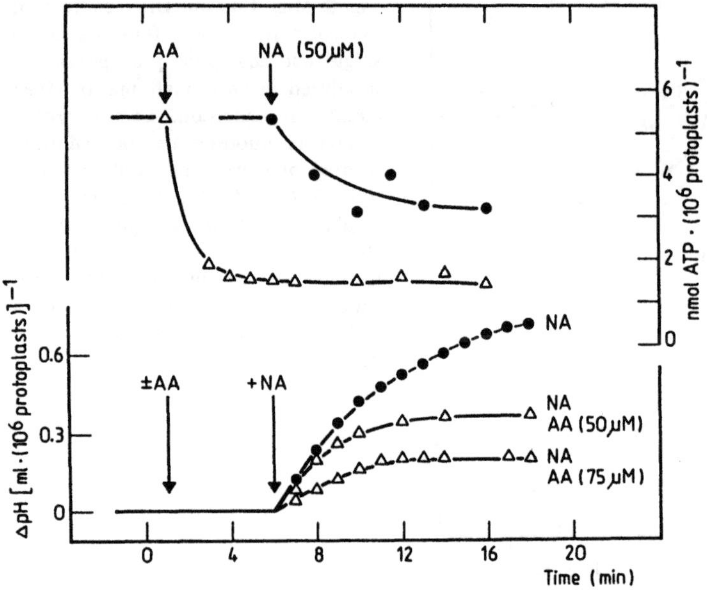

Fig. 6. Effects of 2-naphthylacetate (NA) and antimycin A (AA), singly or in combination, on acidification of the incubation medium and intracellular ATP levels of oat coleoptile protoplasts

cellular ATP (Fig. 6; see also [10]). Thus, the lowered rate of acidification of the incubation medium in the presence of antimycin is explained by assuming the involvement of a proton translocation ATPase. However, there is also a decrease in cellular ATP caused by NA itself (Fig. 6). This effect may result either from an enhanced rate of ATP consumption due to acid-stimulated proton translocation, or simply from poisoning of mitochondrial phosphorylation by naphthol.

Experiments with *Brassica* leaf protoplasts [16] indicated a release of protons upon the addition of 10 μM IAA. In contrast, fusicoccin (10 μM) caused no significant effect. In these experiments a considerable acidification of the external medium only was detectable from 10 h of incubation onwards. This event was discussed as a lag phase in proton pumping activity as a consequence of irritations during the process of protoplast isolation. More likely, however, this observation could have resulted from a pronounced loss in protoplast stability. Protoplast rupture would also lead to an acidification of the incubation medium due to the release of vacuolar constituents. As the authors did not supply any data on protoplast integrity, their results are not comparable.

4 Conclusion

As summarized in Fig. 7, lipophilic esters are taken up by both oat coleoptile segments and protoplasts isolated therefrom. Once inside the cell, these esters are hydrolyzed by relatively nonspecific esterases, leading to the formation of acids. Due to

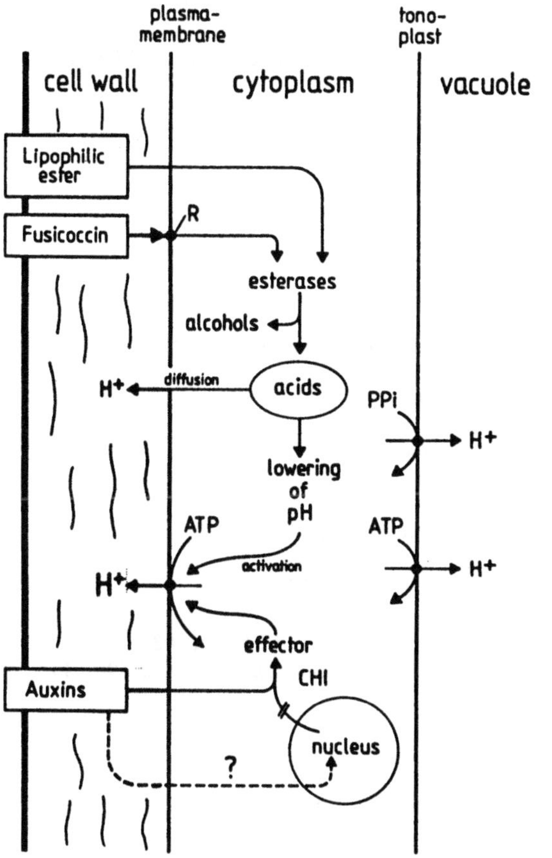

Fig. 7. Possible mechanisms of action of carboxyester compounds, including fusicoccin and auxins. CHI = cycloheximide; PP_i = pyrophosphate; R = receptor

the lowered cytosolic pH (Fig. 5), reactions are initiated that lead to proton efflux and subsequent acidification of the extracellular space (Fig. 6). The results indicate that fusicoccin could act in a similar manner.

Decreasing the cellular-ATP content through the addition of plasma membrane-permeable inhibitors of mitochondrial phosphorylation reduces external acidification. These observations can be taken as evidence that a proton-pumping ATPase is involved, although part of the acids set free from the esters may diffuse through the plasma membrane: the acidification of the external medium could not completely be abandoned by respiratory blockers.

As fusicoccin induces H^+ excretion and concomitant growth at considerably lower concentrations, compared to other carboxyesters [7], high affinity receptors [2, 4, 19] are suggested to be responsible for a concentrating mechanism.

Because of the inhibitory effect of CHI on the elongation growth of coleoptile segments, induced by native auxins, it is suggested that in contrast to the growth-inducing esters, the activation of the H^+-excreting mechanism by IAA depends on the formation of an effector molecule.

Acknowledgments. The authors are indebted for help in preparing functional coleoptile protoplasts to S. Hildenbrand, for careful technical assistance to P. Becher and Ch. Piesch, as well as for critical reading of the manuscript and help in establishing a system of computerized data aquisition (microphotometry) to Prof. W.H. Outlaw, Jr., Florida State University.

References

1. Brummer B, Felle H, Parish RW (1984) FEBS Lett 174:223
2. Beffagna N, Pesci P, Tognoli L, Marré E (1979) Plant Sci Lett 15:323
3. Bertl A, Felle H (1985) J Exp Bot 36:1142
4. Dohrmann U, Hertel R, Pesci P, Cocucci S, Marré E (1979) Plant Sci Lett 9:291
5. Hager A (1980) Z Naturforsch Sect C 35:794
6. Hager A, Biber W (1984) Z Naturforsch Sect C 39:927
7. Hager A, Moser I (1985) Planta (Berl) 163:391
8. Hampp R (1985) In: Bergmeyer J, Graßl M (eds) Methods in enzymatic analysis, vol. 7. Chemie, Weinheim, p 370
9. Hampp R, Ziegler H (1980) Planta (Berl) 147:485
10. Hampp R, Goller M, Ziegler H (1982) Plant Physiol (Bethesda) 69:448
11. Marré E (1979) Annu Rev Plant Physiol 30:273
12. Marré MT, Romani G, Marré E (1983) Plant Cell Environ 6:617
13. Mehrle W, Zimmermann U, Hampp R (1985) FEBS Lett 185:89
14. Sanders D, Hansen U-P, Slayman CL (1981) Proc Natl Acad Sci USA 78:5903
15. Scheurich P, Schnabl H, Zimmermann U, Klein J (1980) Biochim Biophys Acta 598:645
16. Schubert S, Matzke H (1985) Physiol Plant 64:285
17. Slawik J (1982) FEBS Lett 140:22
18. Slawik J (1983) FEBS Lett 156:227
19. Stout RG, Cleland RE (1980) Plant Physiol 66:353
20. Vesper MJ, Evans ML (1979) Proc Natl Acad Sci USA 76:6366

The Control of Flower Initiation and Development*

G. BERNIER and J.M. KINET [1]

1 Introduction

Flowering is generally divided into two major steps: (1) the initiation of flower primordia, and (2) the development of these primordia into mature flowers until anthesis. Despite the fact that the whole process is characterized by continuity and unity, it is necessary to make a clear distinction between the various stages, since they are not alike morphologically and do not react similarly to environmental and internal variables [5, 6, 16].

In this report we shall address the following questions: (1) Are plant growth substances (PGS) involved in the control of flower initiation and development? (2) Are PGS the primary controlling factors of these processes or are they only modulating agents acting secondarily in regard to other, more essential, but still unknown, factors? (3) Which PGS are controlling what?

A survey of recent textbooks in physiology reveals that if most physiologists recognize that PGS are of primary importance in the control of most growth and developmental processes in higher plants, they are generally reluctant to do so for flowering, at least for the step of flower initiation. One reason for that is "because answers to fundamental questions [in the field of flowering] are hard to come by and because the experiments are often long term and require rather specialized growing facilities" [32].

One may be tempted to infer then that there was almost no progress during recent years. This is not what we feel and our aim here will be to pinpoint some of the most important observations that have put a new light on this field.

2 Relationships Between Flowering and Other Developmental Processes

For analytical purposes, flowering has to be considered separately from other aspects of the plant life cycle, and because of this, the close connections that may exist between flower initiation and/or development and other developmental processes are not always fully recognized.

---

* Abbreviations see p. 301

1 Laboratoire de Physiologie végétale et Centre de Physiologie végétale appliquée (I.R.S.I.A.), Département de Botanique, Université de Liège, Sart Tilman, 4000 Liège, Belgium

Plant Growth Substances 1985
(ed. by M. Bopp)

A finding of utmost importance is that genes, which control either the transition from the vegetative to the reproductive condition or the structure and function of the flowers in various species, e.g. pea and tomato, are also involved in the control of many other processes, such as leaf morphogenesis, internode growth, branching, fruit development, senescence, etc. [22, 26].

Physiological evidence is in line with these genetic data. The leaves of the South American potato *Solanum andigenum* are known to produce a tuber-promoting material when exposed to an appropriate photoperiodic regime. Chailakhyan [8], by grafting florally-induced and non-induced scions of either the SDP Maryland Mammoth tobacco or the LDP *Nicotiana sylvestris* onto stocks of this potato, found that the induced scions, but *not* the non-induced ones, can cause tuber formation in the stocks. This suggests that at least part of the transmissible floral promoters of these two photoperiodic tobaccos and the tuber promoters of potato are identical and thus unspecific.

Flower formation is caused in the obligate SDP *Sicyos angulatus*, kept in non-inductive conditions, by grafting with day-neutral cucumber seedlings [29]. However, in these conditions the *Sicyos* receptors produce normal staminate flowers only, the pistillate flowers remaining undeveloped. Application of a cytokinin, in addition to grafting, is required to obtain the complete development of pistillate flowers. Similarly, femaleness in cucumber receptors grafted on *Sicyos* donors is markedly increased when the *Sicyos* are exposed to inductive SD rather than kept in LD. The mobile flower promoters of cucurbits are thus also involved in the control of sex expression in these plants.

In *Hordeum bulbosum*, bulb differentiation (associated with onset of dormancy) and onset of flower initiation respond in such a similar way to vernalization and/or photoperiodism that they seem to be closely linked [23].

Monocarpic senescence is usually associated with the late stages of reproductive development, i.e. the development of flowers and fruits. In the SDP *Xanthium strumarium*, however, totally disbudded plants (plants unable to produce flowers and fruits) senesce at the same rate when exposed to SD as intact fruiting plants [17]. Senescence, in this case, is apparently directly related to photoinduction of flowering.

3 The Basic Role of Correlative Influences in the Flowering Process

Examples showing that all plant parts may participate in the control of flower initiation and/or development follow.

In several biennial and perennial species, like beet, *Chrysanthemum*, orchids, etc., the dominant apical bud inhibits axillary flower initiation [6], whereas in the annual *Scrofularia arguta* it promotes axillary flowering [19]. Similar observations have been made concerning flower development [16]. Thus, whereas decapitation promotes development of lateral reproductive structures in *Impatiens*, tomato, grapevine, etc., removal of axillary shoots stimulates terminal inflorescence development under unfavorable environmental conditions in *Chrysanthemum* and *Sinapis*. Involvement of young expanding leaves is also clearly demonstrated since their removal hastens initiation in the SDP soybean, the LDP *Sinapis*, the day-neutral tomato, etc., and prevents inflorescence abortion in grapevine, *Bougainvillea* and tomato grown under conditions leading to failure of reproductive development. In general, these effects have been interpreted on the basis of participation of the apical bud and young leaves in the competition for available assimilates or by production of floral promoters or inhibitors.

Because some plants may flower well in the total absence of roots, the role of the root system is generally disregarded. This is unfortunate since evidence is accumulating showing that the roots play a critical role in many plants. Derooting is known to promote flower initiation in several photoperiodic species, such as the LDP *Scrofularia* and *Silene armeria* [6, 19]. In *Silene* the requirement for LD can be bypassed by an exposure of the root system to a high temperature [34]. In chicory, a cold-requiring plant, only the root needs apparently to be exposed to low temperature since an unvernalized apical bud grafted on a vernalized root will flower, whereas a vernalized bud grafted on an unvernalized root stays vegetative [14]. All these observations suggest that roots may produce an inhibitory factor, which is inactivated by heat in *Silene* or chilling in chicory.

In other plants, roots may, on the contrary, promote flower initiation. The rosette LDP, *Rudbeckia bicolor*, for example, is unable to flower when deprived of its root system even when its leaves are fully photoinduced [9]. Absence of roots was proposed to prevent essentially bolting and only secondarily flower initiation, but since flowering can be obtained in this plant in the total absence of stem elongation [6], it seems likely that the presence of roots is also absolutely required for flower initiation itself.

In the orchid *Aranda*, there is no absolute requirement for aerial roots for inflorescence initiation, but the number of inflorescences produced is positively correlated to the number of these roots (Goh CJ, unpublished). In other words, inflorescences are more likely to be produced in the vicinity of a greater number of aerial roots. Presence of these roots was found to enhance the flow of assimilates from leaves to shoot apices, and this might explain their promotive effect.

In the cold-requiring chinese cabbage, the effect of low temperatures can be counteracted by a high-temperature treatment (devernalization). Since this detrimental heat treatment is perceived by the root system [25], this is suggestive of production of a heat-sensitive floral promoter in the roots.

The root system is also involved in the control of flower development, at least in some species. In tomato, the flower number of inflorescences increases when root temperature is reduced [24]. This temperature effect is graft-transmissible. In cuttings of grapevine, the usual failure of inflorescences to develop to anthesis seems related to the absence of roots at planting time since, in pre-rooted cuttings, inflorescences are retained and develop normally [21]. On the other hand, root removal in the dioecious species, hemp and spinach, leads to a significant increase in the ratio of male to female plants [8].

Root effects are presumably due mainly to cytokinin production since exogenous cytokinins substitute for roots in many of the above cases, e.g. *Scrofularia, Aranda*, grapevine, spinach, hemp.

When plants are in the reproductive stage, other interorgan correlations develop. In the woody SDP, *Chamelaucium uncinatum*, for example, flower production occurs only temporarily under inductive SD and the plants finally return to vegetative growth due to the production of transmissible floral inhibitors by developing flowers [28]. In barley, cessation of spikelet initiation occurs concurrently with stamen initiation in the most advanced spikelet, a situation explained by Cottrell et al. [10] by the production of an inhibitor (GA?) in stamens.

4 Plant Growth Substances and Flowering

The above evidence strongly suggests that PGS are intimately involved in the control of flowering. Recent analytical work on this question will be reviewed now but, because of space limitation, we shall cover only a few selected cases.

4.1 Sinapis alba (LDP)

A great variety of changes have been observed in the apex of *Sinapis* plants induced to flower by a single LD [6]. Exogenous cytokinin applied once to the apex of plants kept in SD produces part of the apical changes, including a wave of cell divisions, that are normally observed after one LD, but does not cause flower formation. On this basis it was proposed that cytokinin is one of the controlling factors of flower initiation. Various lines of evidence suggested that the endogenous changes in cytokinins in induced plants are critical 16 h after start of the LD [6]. Sequential removal of induced leaves shows that the stimulus for the mitotic wave moves from these leaves to the apex at 16 h [4]. Moreover, the cytokinin levels in the leaves are higher at 16 h in induced plants than in vegetative controls [6].

Results of recent experiments confirm this observation and show further that the cytokinin fluxes are also profoundly altered by the LD. Xylem and phloem saps were collected from induced and vegetative plants during a 16-h period starting 16 h after beginning of the LD or SD. Cytokinin content is higher in root exudate of induced plants than in root exudate of plants kept in SD (Fig. 1). This increased flux of cytokinins results from a rise in cytokinin concentration, since exudate volumes collected from both induced and vegetative plants are identical. The increase is detected in the butanolic fraction only, essentially in the fractions corresponding to zeatin and zeatin riboside. Cytokinin activity in phloem sap is also markedly increased after induction. The rise is found in the butanolic fraction only, particularly in the fraction corresponding to zeatin. Although these results with phloem sap are from one experiment only, they are of special interest since they fit very well with all the other data.

These results strongly support the idea that cytokinins are one of the controlling agents of the floral transition in *Sinapis* and implicate the root system in the control of this developmental step.

4.2 Xanthium strumarium (SDP)

In *Xanthium* plants, induced to flower by a single long night, there is a dramatic decline of cytokinin activity in leaf laminae, apical bud and root exudate [33]. This situation is just opposite to what is observed in *Sinapis*, and one might be tempted then to conclude that the changes in cytokinin activity are merely dependent on day length and unrelated to flowering. Interestingly, flower initiation and decrease in

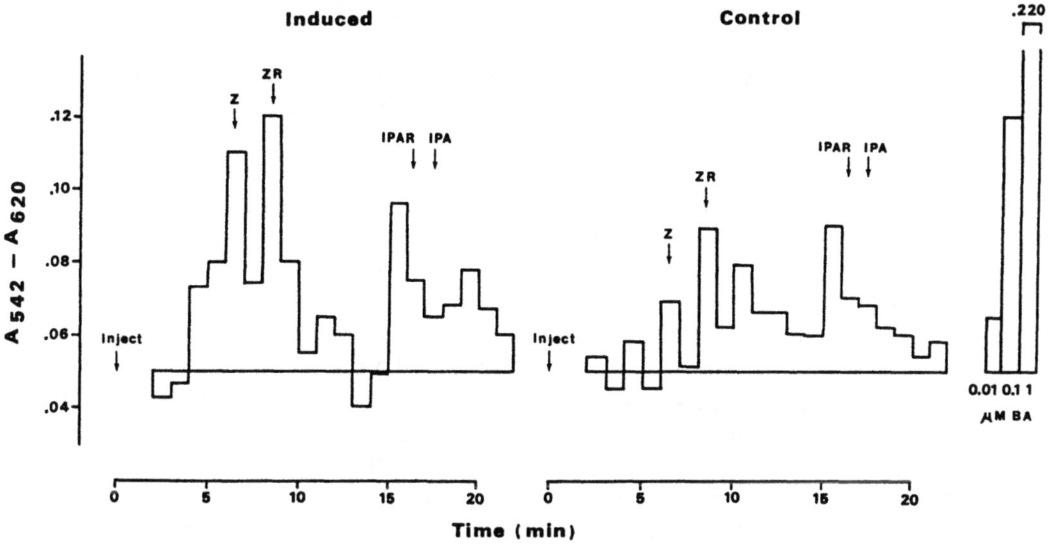

Fig. 1. *Amaranthus* bioassay of butanolic fraction of root exudates of *Sinapis alba* after analytical reversed-phase HPLC on a Nucleosil ODS (5 μm) column (100 × 4 mm), eluted with a constant or increasing concentration of acetonitrile in water (adjusted to pH 7 with triethylammonium acetate) at a flow rate of 1 ml min^{-1}. For the initial 10 min, 10% acetonitrile; for the subsequent 3 min, 10%–25% acetonitrile, linear gradient; from the 13th min, 25% acetonitrile. The root exudates were collected during a 16-h period from the 16th h after the start of a LD (induced) or a SD (control). *Arrows* indicate retention times of cytokinin standards: Z, zeatin, ZR, zeatin riboside; IPA, N^6-isopentenyladenine; IPAR, N^6-isopentenyladenosine

cytokinin levels in *Xanthium* are both nullified by interrupting the long night by a light break, suggesting that they are closely related. Thus, despite the fact that the trend of changes in cytokinin levels is opposite in *Xanthium* and *Sinapis*, these changes seem essential in the two cases.

In both species, at least a part of the stimuli generated in the photoinduced leaves is apparently rapidly transmitted to the roots where it alters the course of cytokinin production and/or release [33].

4.3 Rosette Plants

Flower initiation in rosette plants is generally accompanied by stem elongation (bolting), a process which is under the control of GA metabolism [35]. Application of GA causes stem growth in most of these plants, but flower formation in only some of them. This raises the question of the relationships between endogenous GAs and flowering.

In the LDP *Samolus parviflorus*, a growth retardant application suppresses both bolting and flowering in LD and GA overcomes these effects [2]. These observations are substantial in their implication of GAs as the sole controlling factor of flower initiation in this species, but unfortunately there is no analysis of endogenous GA levels to support this idea.

A similar situation apparently exists in other rosette LDP, such as *Arabidopsis thaliana* and *Rudbeckia bicolor* [35]. In this last species, the promotions of bolting and flower initiation by GA in SD are associated with the simultaneous stimulations of cell proliferation in the subapical tissues and the central zone of the apical meristem, respectively [13].

In other rosette LDP, such as spinach, *Silene armeria* and *Agrostemma githago*, the story is, however, entirely different [35]. Applications of growth retardants in LD reduce considerably the levels of endogenous GAs, suppress bolting, but leave flowering unaffected. Flower formation in these species is thus unrelated to GA levels. The possibility that a minor GA escaping detection would be responsible for the flowering control has sometimes been raised on the basis that different exogenous GAs may have different effects in some plants [6]. Although not entirely excluded, this possibility is not supported by recent studies, indicating that the various GAs present in a particular species form a biosynthetic pathway with only one physiologically-active GA and the others being precursors or catabolites [35].

Interestingly, these studies have also revealed that not only the production, but also the degradation of GAs is increased in LD compared to SD. Also, the sensitivity of these plants to exogenous GA is much higher under LD than under SD.

Finally, there is evidence that in some caulescent plants GAs are a part of the endogenous promoters of flowering, e.g. in *Bryophyllum* and *Pharbitis*, or are inhibitory to flower initiation, e.g. in *Fuchsia*, tomato and many woody perennials [35].

4.4 Guzmania lingulata and Other Bromeliads

Mature plants of ornamental bromeliads flower occasionally in greenhouses. Applications of AVG totally suppress this spontaneous flowering [18]. Synchronized flowering can be obtained following fumigation with ethylene or application of CEPA. In *Guzmania*, flower formation is also caused by an application of ACC or shaking the plants for 15 s (De Proft M, De Greef JA, unpublished). These two treatments result in a rapid ethylene release and, as expected, AVG prevents this release and subsequent flowering in the case of shaked plants, but not in ACC-treated plants. These data are consistent with the view that in mature bromeliads, ethylene is the only controlling factor of flowering. Since exogenous ethylene is inhibitory to flower initiation in most other plants, this type of control is unlikely of widespread occurrence.

4.5 Lycopersicon esculentum (Fig. 2)

Flowering in tomato is strongly dependent on high light intensity. There is a critical period when meiotic processes in the anthers of the first flower of the inflorescence are in progress, at the tetrad and free-microspore stages. Low irradiance during this period induces the late and complete failure of numerous inflorescences.

A constant low light regime delays flower initiation and macroscopic appearance of the inflorescence; it also causes the early abortion of the flowers before the start of meiosis.

GAs have a dual effect. When applied at the time of inflorescence initiation in plants grown under high irradiance, they induce the failure of numerous flowers [16].

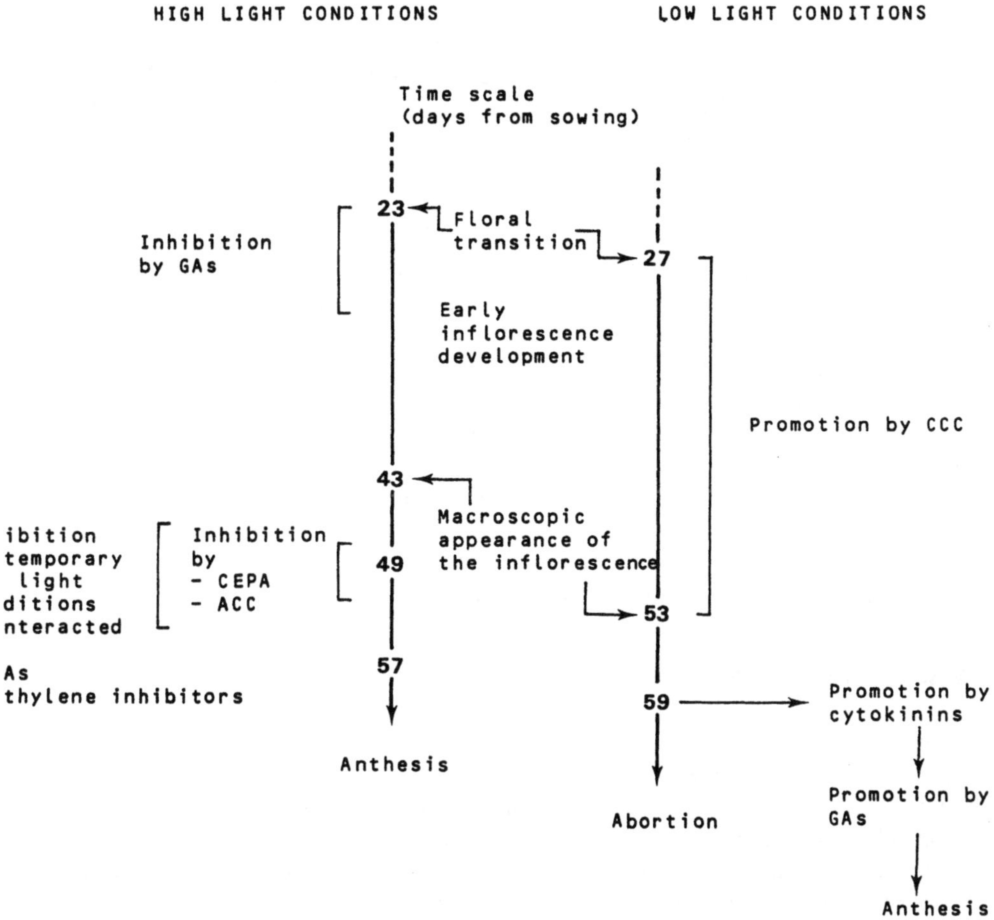

Fig. 2. Diagram summarizing the temporal interactions between light conditions and PGS for the control of inflorescence initiation and development in tomato (Computed from Abdul [1] and Kinet [16])

In contrast, application of CCC from flower initiation through macroscopic appearance of the flowers reduces the incidence of abortion in plants grown at a high temperature with little light [1]. Determinations of endogenous GA levels support the idea that normal inflorescence development is associated, during early stages, with a reduced GA concentration. Thus, early abortion occurring in plants grown under a constant low irradiance is correlated with a higher GA content than is found in inflorescences developing in favorable light conditions [16]. Moreover, the yield of diffusible GAs from the shoot tip of plants treated with CCC, which stimulates development, is greatly reduced [1].

During the late stages of inflorescence development, GAs are promotive; they prevent abortion induced by a transfer to low irradiance during meiosis. In plants grown under a constant low irradiance, they stimulate development to anthesis, but only subsequently to a cytokinin treatment, i.e. they are promotive after the inflorescence has reached a particular stage [16].

Cytokinins strongly stimulate development of inflorescences, which are aborting under a constant low light regime; these inflorescences are severely deficient in this type of regulator [16]. Cytokinins are, however, unable to induce complete development under these light conditions, just as they are ineffective in reverting the inhibitory effect of a low irradiance given during late stages of development, when endogenous cytokinin-like activity reaches a minimum. Hence, it seems that normal development is associated with high cytokinin levels, at least during an early critical stage.

Ethylene is also involved since, under a high light regime, CEPA applications induce abortion, particularly when made 8–9 days before anthesis, i.e. when low irradiance is most inhibitory [15]. At the same stage, ACC is also detrimental, whereas AVG counteracts the inhibition caused by low irradiance. This late stage of development is thus apparently controlled by a balance between the promotory GAs and the inhibitory ethylene.

These studies establish that during flowering in tomato, there is a sequence of physiological steps, each having particular requirements in PGS.

The behavior of tomato is not unique. GAs have also a dual effect in *Bougainvillea* [16]. In tulip, cytokinins and GAs promote flower development, whereas ethylene is inhibitory [11]. In a Peruvian cv. of bean, flower development is dependent on a balance between cytokinin promotion and ABA inhibition [20].

4.6 Sex Expression in Monoecious Species

The sex of cucumber flowers is dramatically influenced by PGS. The results of both exogenous treatments and analyses of endogenous levels are generally consistent with the view that ethylene and GAs are the primary feminizing and masculinizing agents, respectively, and that sex is controlled by the ratio of these two PGS [12, 16]. However, some observations are clearly at variance with this picture and indicate that the control of sex expression may differ markedly in different cucumber lines [30].

In corn, contrary to the situation in cucumber, GAs enhance femaleness. GA-like activity in the male inflorescence is lower than in the female one, and increases when reversal to female expression is induced by low irradiance [27].

5 Conclusions

The evidence reviewed above is fragmentary, but sufficient to conclude that PGS are among the primary controlling agents of flowering. Except for the cases of some rosette plants and bromeliads, in which there is apparently only one limiting factor for the whole flowering process, several regulators, either promoters or inhibitors, are usually implicated, each controlling particular events of flower initiation and development.

PGS are not the sole controlling factors of flowering. Assimilates are also involved, at least in some plants [7, 16], and possibly also oligosaccharides [31], phenolics, etc. [6]. The possible interplays between these compounds and PGS should be explored in further work.

So far, the changes in endogenous PGS at the real sites of flower initiation and development, i.e. the shoot meristems and flower primordia, remain essentially unknown. Similarly, the possible changes in meristem sensitivity to PGS in function of the developmental stage have received little attention in the past [3].

The mechanisms of control appear basically different in different species. Conceivably, these mechanisms have evolved quite independently in a wide range of plant groups. It is puzzling, however, that sometimes similar mechanisms are found in taxonomically unrelated species, like tomato and tulip. Is it due to convergent evolution of different groups?

Acknowledgments. We are grateful to Dr. J.A. De Greef and M. De Proft (Antwerpen, Belgium) and Dr. C.J. Goh (Singapore) for supplying unpublished results. Our research work was supported by a grant from the Belgian government through the programme "Action de Recherche Concertée".

Abbreviations: ABA = (±)-Abscisic acid; ACC = 1-Amino-cyclopropane-1-carboxylic acid; AVG = Aminoethoxyvinylglycine; CCC = 2-Chloroethyltrimethylammonium chloride; CEPA = 2-Chloroethylphosphonic acid; cv = cultivar; GA(s) = Gibberellin(s); LD(P) = Long day (plant); PGS = Plant growth substance(s); SD(P) = Short day (plant); STS = Silverthiosulfate.

References

1. Abdul KS, Canham AE, Harris GP (1978) Ann Bot (Lond) 42:617
2. Baldev B, Lang A (1965) Am J Bot 52:408
3. Bernier G (1984) In: Vince-Prue D, Thomas B, Cockshull KE (ed) Light and the flowering process. Academic Press, London, p 277
4. Bernier G, Bodson M, Kinet JM, Jacqmard A, Havelange A (1974) In: Sumiki Y (ed) Plant growth substances 1973. Hirokawa, Tokyo, p 980
5. Bernier G, Kinet JM, Sachs RM (1981a) The physiology of flowering, vol 1. CRC, Boca Raton, Florida
6. Bernier G, Kinet JM, Sachs RM (1981b) The physiology of flowering, vol 2. CRC, Boca Raton, Florida
7. Bodson M (1984) In: Vince-Prue D, Thomas B, Cockshull KE (ed) Light and the flowering process. Academic Press, London, p 157
8. Chailakhyan MKh (1980) Fiziol Rast 27:917
9. Chailakhyan MKh, Khlopenkova LP (1969) Dokl Akad Nauk SSSR 189:1400
10. Cottrell J, Dale JE, Jeffcoat B (1981) Plant Sci Lett 22:161
11. De Munk WJ, Gijzenberg J (1977) Sci Hortic (Amst) 7:255
12. Frankel R, Galun E (1977) In: Pilet PE (ed) Plant growth regulation. Springer, Berlin Heidelberg New York
13. Jacqmard A (1970) In: Bernier G (ed) Cellular and molecular aspects of floral induction. Longman, London, p 316
14. Joseph C, Billot J, Côme D (1982) CR Acad Sci Paris 295:417
15. Kinet JM, El Alaoui H (1984) Arch Int Physiol Biochim 92:PF16
16. Kinet JM, Sachs RM, Bernier G (1985) The physiology of flowering, vol 3. CRC, Boca Raton, Florida
17. Krizek DT, McIlrath WJ, Vergara BS (1966) Science 151:95
18. Mekers O, De Proft M, Jacobs L (1983) Acta Hortic (The Hague) 137:217
19. Miginiac E (1978) Bot Mag (Tokyo) Spec Issue I:159
20. Morgan DG, Morgan CB (1984) In: Vince-Prue D, Thomas B, Cockshull KE (ed) Light and the flowering process. Academic Press, London, p 227
21. Mullins MG (1967) J Exp Bot 18:206

22. Murfet IC (1985) In: Halevy AH (ed) Handbook of flowering, vol 4. CRC, Boca Raton, Florida, p 97
23. Ofir M (1981) Isr J Bot 30:173
24. Phatak SC, Wittwer SH, Teubner FG (1966) Proc Am Soc Hort Sci 88:527
25. Pressman E, Negbi M (1981) J Exp Bot 32:821
26. Rick CM, Robinson J (1951) Am J Bot 38:639
27. Rood SB, Pharis RP, Major DJ (1980) Plant Physiol (Bethesda) 66:793
28. Shillo R, Weiner A, Halevy AH (1984) Planta (Berl) 160:508
29. Takahashi H, Saito T, Suge H (1982) Plant Cell Physiol 23:1
30. Takahashi H, Saito T, Suge H (1983) Plant Cell Physiol 24:147
31. Tran Thanh Van K, Toubart P, Cousson A, Darvill AG, Gollin DJ, Chelf P, Albersheim P (1985) Nature 314:615
32. Vince-Prue D, Cockshull K, Thomas B (1984) In: Vince-Prue D, Thomas B, Cockshull KE (ed) Light and the flowering process. Academic Press, London, p xvii
33. Wareing PF, Horgan R, Henson IE, Davis W (1977) In: Pilet PE (ed) Plant growth regulation. Springer, Berlin Heidelberg New York, p 147
34. Wellensiek SJ (1985) In: Halevy AH (ed) Handbook of flowering, vol 4. CRC, Boca Raton, Florida, p 320
35. Zeevaart JAD (1983) In: Crozier A (ed) The biochemistry and physiology of gibberellins, vol 2. Praeger, New York, p 333

Hormonal Regulation of Plant Flowering

M. Kh. CHAILAKHYAN[1]

1 Introduction: The Florigen Hypothesis

The florigen hypothesis, underlying the hormonal plant concept, has largely evolved during the past years.

The three basic stages of this hypothesis evolution can be briefly characterized as follows. The first stage (1937–1957) was marked by the hypothesis of florigen as a system of flowering hormones of unknown nature. The second stage (1958–1974) was related to the hypothesis of florigen as a bicomponental system of flowering, consisting of two groups: gibberellins, causing formation and growth of flower stems; and hormone substances, called anthesins, causing flower formation. The third stage (1975–1984) is associated with the hypothesis of florigen as a bicomponental complementary system of flowering hormones, in which the hormones are formed as a result of both autonomous and induced regulation mechanisms [3].

Hence, it follows that the hormonal concept of plant flowering has in none of its evolutional stages been based on the concept of only one flowering hormone. The florigen hypothesis is based on numerous experiments on intact plants and differentiated models. However, the methods of plant transplantation and extraction of flowering substances directly from plants and tests of their activity on specific test objects have been most essential.

2 Plant Transplantation and Initiation of Flowering

It has been shown in many experiments that the flowering of long-day plant species under short-day conditions is induced by their grafting on the flowering plants of short-day species. On the other hand, the flowering of short-day species under long-day conditions is induced by their grafting on the flowering plants of long-day species. The flowering of receptor plants belonging to species similar to the donor ones, as well as to other species, is achieved in many cases [11].

In later experiments, the flowering occurred also at intergeneric graftings and at the graftings of long- and short-day plants on photoperiodic neutral ones [7, 15, 16]. Thus, the common nature of the hormonal systems of the photoperiodically neutral species and the short- and long-day flowering species was demonstrated.

1 Timiryazev Institute of Plant Physiology, Academy of Science of the USSR, Moscow, USSR

Plant Growth Substances 1985
(ed. by M. Bopp)

The dominating part of the graftings for flowering initiation was performed with induced, i.e. flowering donors. However, even in cases when the donors had not been induced and were in the vegetating state, the receptors started flowering. At first, we [2] performed graftings of 2-year *Hyoscyamus niger* rosette plants on stalky plants of Mammoth tobacco under long-day conditions, which resulted in stem formation in scions and flowering of the latter under the stock effect.

Other graftings involved a combination of two vegetating components of the long-short-day species, *Bryophyllum daigremontianum*, whose flowering is induced only when the long-day conditions are replaced by short-day ones. In these graftings, the stock and leaves were under long-day, and the scion with leaves was under short-day conditions, the intermediate shoots between them flowered due to gibberellins, delivered from stock leaves, and anthesins, delivered from scion leaves [5].

The results of graftings, in which the plant flowering occurs through the combination of two vegetating components, are an obvious argument in favour of the hypothesis of florigen as a bicomponental system of plant flowering hormones.

3 Extraction of Flowering Substances and Testing Their Effects

Numerous studies on the extraction of flowering substances and the testing of their effects on different subjects have been conducted. Our experiments were conducted on the extraction of gibberellinlike substances from the leaves of vegetating and flowering plants of long-day tobacco, *Nicotiana sylvestris*, and short-day tobacco, *Nicotiana tabacum*, Maryland Mommoth variety. The gibberellinlike substances were extracted with acetone and tested for their effect in extracts on rosette plants of *Rudbeckia bicolor*, continuously subjected to short-day conditions.

The experiments showed that the stem growth and flowering of *Rudbeckia* occurred under the effect of gibberellinlike substance extracts, isolated from the leaves of both tobacco varieties, i.e. the flowering, long-day *N. sylvestris* and the vegetating, short-day Maryland Mammoth. In the second case, the flowering of *Rudbeckia* occurred under the effect of a combination of two vegetating components [8].

The scheme would be fully confirmed, if a combination of substances of two vegetating components was achieved in a reverse direction, i.e. of anthesins, extracted from the leaves of a rosette plant, vegetating under short days, and gibberellins in the stem buds of a short-day plant, vegetating under long days.

Many attempts were made to extract an anthesinlike substance, using flowering plants as sources in all cases; the extracts of the latter were tested on similar plants, but under long-day conditions or continuous light [14].

In experiments carried out together with the laboratory of the Institute of Organic Chemistry, for testing the effect of ethanol extracts from the leaves of flowering plants of short-day Mammoth tobacco, the seedlings of *Chenopodium rubrum* started flowering under continuous light, i.e. under non-inductive conditions [6], while the control seedlings, treated with an extract from Mammoth tobacco leaves, vegetating under long days, remained in the vegetative state.

In later experiments, the flowering of *Chenopodium rubrum* seedlings under continuous light was obtained under the effect of ethanol extracts from the leaves of long-day *N. sylvestris* tobacco plants, vegetating under short days [10].

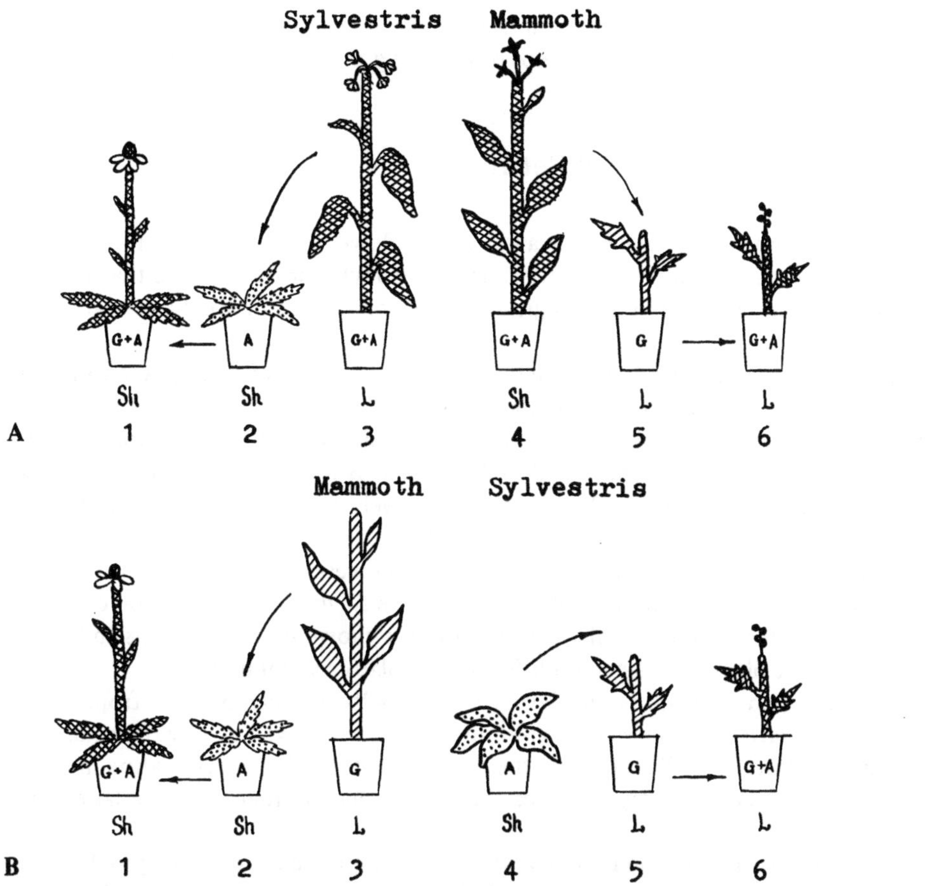

Fig. 1A,B. Effect of extracts from the leaves of short-day Mammoth tobacco and long-day *Sylvestris* tobacco on flowering of *Rudbeckia* and *Chenopodium rubrum* plants.
A *1* The flowering *Rudbeckia* under short days (Sh); *2* the vegetating *Rudbeckia* under short days (Sh); *3* the flowering *Sylvestris* under long days (L); *4* the flowering Mammoth under short days (Sh); *5* the vegetating *Chenopodium* under long days (L); *6* the flowering *Chenopodium* under long days (L).
B *1* The flowering *Rudbeckia* under short days (Sh); *2* the vegetating *Rudbeckia* under short days (Sh); *3* the vegetating Mammoth under long days (L); *4* the vegetating *Sylvestris* under short days (Sh); *5* the vegetating *Chenopodium* under long days (L); *6* the flowering *Chenopodium* under long days (L).
Abbreviations: *G* Gibberellins; *A* anthesins; *G + A* gibberellins + anthesins; *A + G* anthesins + gibberellins

The basic scheme of florigen was fully confirmed by a recent experiment: the extraction of anthesins from the leaves of *N. sylvestris* rosette tobacco plants, vegetating under short days, is possible, and when applying them to stem buds of short-day *Chenopodium rubrum*, vegetating under long days, and containing gibberellins, flowering is observed.

A series of experimental results on the extraction of flowering substances from tobacco plants and their effect on *Rudbeckia* and *Chenopodium rubrum* flowering is presented in the scheme, thus indicating the possibility of flower initiation in combination with two different groups of flowering hormones [4] (Fig. 1).

4 Flowering Stimulants and Retardants

Apart from gibberellins, which play a key role in flower induction in plants, there are cases of flower acceleration or initiation caused by phytohormones, other than florigen components, as well as by physiologically active compounds of non-hormonal nature [1].

However, the data on flower acceleration under the effect of these phytohormones do not offer essential grounds for more concrete concepts of their role as specific hormonal factors of flowering regulation. Meanwhile, they offer a basis for recognizing auxins, cytokinins, abscisic acid and ethylene as hormonal system components producing an effect on plant flowering cooperatively with their flowering hormones [3].

The regulation of flowering, as with any other process of plant vitality, is performed both by stimulating and inhibiting factors. The flower-inhibiting action of leaves, under conditions of unfavourable day length, was shown on many plants and it was found that the plants contained both flower stimulators and inhibitors.

Plant-flowering inhibitors are devoid of specificity and can be both trophic (nutrient) and active substances of non-hormonal nature, as well as phytohormones in certain concentrations. The effect of these compounds on flower retardation in short- and long-day species is varying, since the former lack anthesins and the latter, gibberellins.

In recent experiments on mutual grafting of tobacco plants, belonging to different photoperiodic groups, intercomponental effects, resulting in flower inhibition, were revealed [9, 13]. Hence, the formation of antiflorigen, a system of flower-inhibiting substances, was suggested [12].

5 Conclusions

Flower initiation in plants consists of two phases; induction and evocation.

Flower induction varies in biotypes with a different photoperiodism: the long- and short-day species are marked by mainly ecological by photoperiodic induction, and their transition to the mature flower state and the formation of a florigen system occur in different ways. The photoperiodically neutral species are characterized by age induction, and florigen system formation is related to age changes. Evocation initiation is manifested by the emergence of the bicomponental, complementary flowering hormone system in plant stem buds.

Thus, florigen is the final link in the chain of induction reactions and the initial link in the chain of flower evocation reactions, ending in floral rudiment formation.

While the pathways to florigen formation in stem buds are different at induction in short-day, long-day, and photoperiodically neutral species, the physiological and structural changes of apices, occurring at flower evocation, are similar in different biotypes [3].

References

1. Bernier G, Kinet J, Sachs RM (1981) The physiology of flowering, vol 11. Transition to reproductive growth. CRC, Boka Raton, Florida, 231 p
2. Chailakhyan MK (1968) In: Biochemistry and physiology of plant growth substances. Runge, Ottawa, p 1317
3. Chailakhyan MK (1982) Fiziol Rast 29:234
4. Chailakhyan MK, Lozhnikova VN (1985) Florigen hypothesis and its substantiation by extraction of substances inducing plant flowering. Fiziol Rast (in press)
5. Chailakhyan MK, Yanina LI (1973) Dokl Akad Nauk SSSR 208:749
6. Chailakhyan MK, Grigoryeva NY, Lozhnikova VN (1977) Dokl Akad Nauk SSSR 236:773
7. Chailakhyan MK, Khazhakyan KK, Agamyan LB (1976) Dokl Akad Nauk SSSR 230:1002
8. Chailakhyan MK, Khlopenkova LP, Lozhnikova VN (1964) Dokl Akad Nauk SSSR 38:45
9. Chailakhyan MK, Yanina LI, Lotova GN (1979) Dokl Akad Nauk SSSR 248:1513
10. Chailakhyan MK, Lozhnikova VN, Grigoryeva NY, Dudko ND (1984) Dokl Adad Nauk SSSR 279:1276
11. Lang A (1965) Encyclopedia plant physiology vol 15/1, p 1380
12. Lang A (1980) In: Skoog F (ed) Plant growth substances. Springer, Berlin, Heidelberg New York, p 310
13. Lang A, Chailakhyan MK, Frolova IA (1977) Proc Natl Acad Sci USA 74:2412
14. Lincoln RG, Gunningham A, Hammer KG (1961) Nature 202:559
15. Pol PA van de (1972) Medeo Landboiwhegesch, Wageningen, vol 72, p 1
16. Zeevaart JAD (1958) Medeo Landbouwhogesch, Wageningen, vol 58, p 1

Phytohormones in Seed and Fruit Development

C. BULARD and M.Th. LE PAGE-DEGIVRY [1]

1 Introduction

Research on hormone regulation of seeds and fruits has developed considerably over the last few years with, in the background, the aim of improving seed quality. Major projects have thus been launched, for example, concerning soybean in the USA. Several reviews have recently surveyed the evolution of hormone levels and their possible roles during maturation, in the case of gibberellins [22], cytokinins [40] or abscisic acid [5, 23]. After consideration of the possible mechanisms for regulation of hormone levels, attention will be focused on their role during fruit and seed development. In the majority of studies, conclusions are based on the simultaneous occurrence of peak values in hormone content and certain morphological and biochemical changes, or on the results of an application of growth regulators to intact plants. Karssen [21] has discussed the limitations of these experimental approaches. Nooden [28], in his presentation of an integration of soybean pod development and monocarpic senescence, has shown how difficult it is to interpret a study on a whole plant. However, recent anatomical, physiological, and biochemical research has furnished us with tools, which are indispensible for a better understanding of the levels at which different hormones act. Thus, new interest has recently been shown in storage proteins [19, 30]. The development of precise analytical techniques (immuno-assay, use of cDNA clones) has made it possible, in the case of a few species, to characterize their main storage proteins and to monitor the kinetics of their accumulation. Also, a new approach has been made to the mechanisms of transport of photosynthetic products [37]. It is obvious that leaves are the main source of photosynthates for the fruits. The developing fruits and seeds act as major sinks, and there is a close source/sink relationship in which growth regulators may play a regulatory rôle. We shall see how these new approaches are beginning to be used to determine more precisely the role played by hormones in embryogenesis and seed filling, while at the same time, recalling the potential and difficulties of more global studies.

1 Laboratoire Physiologie Végétale, Faculté des Sciences et des Techniques, 28, Avenue Valrose, 06034 Nice Cedex, France

Plant Growth Substances 1985
(ed. by M. Bopp)

2 Regulation of Hormone Levels

2.1 Hormone Metabolism

The different hormone levels show transient peaks, which seem to appear in a fixed sequence. Metabolism of hormones is often proposed as a mechanism for the regulation of their levels. Thus, for gibberellins (GA_S), the hydroxylation at the 3 β- or 13-position appears to increase their activity and this feature is often associated with the active phase of seed growth. As the seeds approach maturation, GA_S are inactived by 2 β-hydroxylation, conjugation, and catabolism. Thus, acidic GA_S are the major components of early immature seeds and GA-conjugates, the major GA components of mature seeds [35]. Furthermore, active forms are located in the embryos where they could form GA-glucosyl esters, more likely candidates for the function of storage components, while 2 β-hydroxylated inactive end products accumulate in the testa [4].

Similar changes between free and conjugated cytokinins were demonstrated by Lorenzi et al. [25]; their data suggested an evolution from biologically very active but less polar types, such as zeatin at early stages, to more polar substances with relatively lower biological activity, such as zeatin riboside and glucoside at intermediate and late stages. However, in developing wheat grains, Jameson et al. [20] showed that the different types evolved in parallel.

For abscisic acid (ABA), an inactivation by conjugation into ABA-GE and polar metabolites has also been envisaged. Nevertheless, even when the increase in the level of conjugates coincides with the decrease in ABA level, it concerns only a small part of the metabolized free-ABA. Embryos of *Phaseolus vulgaris* isolated and cultivated in vitro in the presence of [2-^{14}C]ABA, presented the same capacity for conjugate formation at all stages of their development; on the other hand, the proportion of oxydative metabolic products increased considerably at maturation (Barthe and Le Page-Degivry, unpublished). We have, however, shown that this increase in phaseic and dihydrophaseic acid levels could be attributed to a marked slowing of uptake rather than to a greater rate of oxydative catabolism.

2.2 ABA Translocation and Its Regulation

A foliar origin has often been suggested for ABA accumulation in fruits. When [2-^{14}C]-ABA was applied to the leaf, 24 h later in early podfill most of the radioactivity was found in the carpels and the seed coats [7, 29], while, in late podfill, most had migrated to the embryos. Since the embryo is symplastically isolated from the seed coat, ABA arriving via the phloem must first be deposited in the seed coat where the vascular bundles end, before moving apoplastically to the embryo. The evolution of the intracellular pH of the embryo and the seed coat explains the distribution of the radioactivity between the different parts of the pods. The accumulation takes place in the more alkaline compartment [7] and as in generally accepted, ABA is distributed according to pH gradient. Intracellular pH gradients were also implicated in the control of ABA movement between seed coat and embryo of cotton [18].

3 Embryo Development

3.1 The Alternative Between Precocious Germination and Embryogenesis

When embryos were isolated in the course of their embryogenesis, thus freed from the constraints existing in the seeds and the fruits, they could germinate, i.e. exhibit root growth. The youngest embryos showed a severely restricted development, corresponding to a "precocious germination". When treatments were applied in order to obtain continuation of embryogenesis, they also inhibited this precocious germination.

In our experiments with *Phaseolus vulgaris*, embryos induced to germinate quickly by a transfer from a liquid medium to a fresh agar one, presented the characteristics of a precocious germination, while embryos of the same age, germinating after a long lag phase, had a normal development. This strongly suggests that during the lag phase, embryogenesis can continue [31]. The study of the synthesis of the seed storage protein led Long et al. [24] to similar conclusions; the lag period before germination represents continued embryonic growth, which is controlled by factors intrinsic to the embryo.

However, Finkelstein and Crouch [16] showed that embryonic and germination specific processes could occur concurrently; the transition from embryogeny to germination was monitored both morphologically and biochemically using synthesis of storage protein. The capacity to develop as normal seedlings was acquired gradually during the maturation process.

3.2 Precocious Germination

Several studies have implicated ABA in the regulation of germination. In particular, the coincidence between the decrease of ABA concentration and the ability to germinate has been reported by several authors. But Black [5] noticed that elongation always begins when the embryos are about the same age. This may not necessarily be related to the amount of ABA that is present, but to other changes. By providing a nutritive supply to the embryos cultivated in vitro, we were able to observe the germination of the youngest embryos and to reveal the decrease in the ability to germinate of embryos isolated before the ABA maximum level occurred. This coincidence between increasing inhibition of germination and increasing ABA content for these young *Phaseolus* embryos reinforces the hypothesis of a regulatory role of ABA in the control of germination [31]. Moreover, during in vitro culture the free-ABA embryo content decreased quickly during the first days and reached low values a few days before germination occurred. Ackerson [3], with *Glycine max.* noted a similar coincidence. Providing embryos were 3 weeks of age or older, precocious germination could be induced by treating them in a way that caused a reduction in their ABA content. Washing the embryos or allowing them to dry slowly within or out of detached pods caused a gradual decline in ABA content. The extent of germination is correlated with the length of washing or drying treatments, which in turn affects the level of endogenous ABA.

3.3 Embryogenesis

The authors recognize generally three phases of embryo development: phase 1 is when cell division is rapid, but dry weight and average cell size are small; phase 2 is characterized by rapid cell expansion, especially in the cotyledons and a dramatic rise in both fresh and dry weight; phase 3 is when chlorophyll and fresh weight decline as the seed matures.

3.3.1 Regulation of Embryo Cell Division (Phase 1)

The general picture, which emerges from the literature, is a close correlation between endogenous levels of cytokinins and periods of active division [40]. Hybrid embryos *(Phaseolus lunatus x P. vulgaris)*, which ceased to divide at the 4-celled stage, can be stimulated to progress to the preheart stage by supplying cytokinin to the female parent via hydroponic culture [26]. In an anise cell culture, in which somatic embryo differentiation could be induced by removing 2,4 D, the maxima of the free cytokinin levels were correlated with the logarithmic growth of embryonic cells, but not with a specific stage of embryo formation [15]. Furthermore, the effect of exogenous cytokinins on embryogenesis was only promotive until the third day of culture, coincident with cell growth rather than differentiation [14].

3.3.2 Regulation of Storage Protein Synthesis (Phase 2)

From a developmental point of view, storage proteins appeared as a good marker of the mid-embryogenetic process. When non-zygotic embryos were induced from microspores in anther culture of *Brassica napus* [8], all embryolike structures were found to contain storage protein, whereas precultured anthers did not have detectable storage protein.

In situ, a rapid phase of dry weight increase and protein accumulation often coincided with a peak concentration of ABA [2]. When the embryos were cultured in vitro, if ABA was included in the medium, precocious germination was inhibited and embryos had significantly higher storage protein content than those on the basal medium [9, 24]. In some cases, the addition of ABA had no influence on the accumulation of the storage protein, but this lack of effect could be correlated with high osmolarities of the culture medium [10]. Indeed, Crouch and Sussex [9] showed that high osmotica inhibited precocious germination and favored storage protein synthesis; there was simply a lag period before storage protein synthesis rates equalled those obtained on ABA media.

Moreover, the effect of ABA during soybean embryogenesis [2] is both age-dependent and apparently tissue-type-dependent. Young, predominantly meristematic embryos responded to low concentrations of ABA by stimulation of growth and protein accumulation. When embryos became predominantly storage organs, ABA acted to inhibit growth without affecting protein accumulation. Dure et al. [12] had already demonstrated that ABA, when applied to very young embryos whose endogenous ABA levels are low, did not have the same effect on the different proteins: those, whose synthesis normally becomes apparent in late embryogenesis are seen to have been "induced" prematurely by incubation in ABA, while those whose mRNA is not present in mature cotyledons are no longer synthesized. Daily injections of ABA into the pedicels of pods of *Pisum sativum*, spanning the time of maximum protein synthesis [33], caused an increase in vicilin, but albumin content was not affected. These observations concur with those of Dure et al. [12] if one takes into account the fact that the rate of legumin synthesis is roughly constant throughout the protein deposition period, contrary to the synthesis of vicilin, which fluctuates [17].

Taken together, these results shows how essential it is to understand the kinetics of accumulation of the different protein fractions and the corresponding mRNA in order to situate exactly the level of synthesis at which the hormone regulation intervenes.

Other hormones have not been studied as fully as ABA. In vitro embryos of *Phaseolus vulgaris* showed no response to other plant hormones (IAA, kinetin or gibberellin acid). Similarly, daily application of GA_3 into the pedicels of pods of *Pisum sativum* caused no changes in total protein content or in the levels of individual protein fractions [33]. In carrot and anise cell lines, high levels of a GA_1-like substance and a reduced ability to metabolize GA_1 are correlated with the absence of embryo development, while lowered levels of a GA_1-like substance and a rapid metabolism of GA_1 into GA_8 and GA-conjugates are correlated with continued embryo development [27]. Exogenous application of GA_3 is known to reduce somatic embryogenesis in carrot cultures; GA_4 was found to have the same effect in anise cultures. Thus, a role, albeit negative, in somatic embryogenesis for a polar, biologically active GA would be implied.

4 Seed Filling

4.1 Control at the Source Level

Most crop plants are monocarpic and their foliar senescence occurs after fruit maturation [28, 32]. Thus, a modification in longevity of leaves by treatment with hormones, which are known to have a role in controlling leaf senescence, might have some influence on yield. Ray et al. [32] showed the possibility of increased grain-filling and thus yield of rice due to delayed foliar senescence by kinetin treatment and decreased grain-filling due to hastening of senescence by ABA.

Moreover, ABA inhibited sucrose uptake by discs from *Ricinus* cotyledons [41]. As sucrose uptake into discs of cotyledons takes place mainly in the veins and can be considered as phloem loading, the effect of ABA could be considered as an inhibition of phloem loading.

4.2 Control at the Sink Level

4.2.1 Correlative Evidence and Application of Growth Regulators

On the basis of the simultaneous occurrence of peak values in hormone content and the dry weight increase, all groups of hormones have been more or less implicated in the regulation of seed filling. But the correlation was often not good. In grapevines, sugar accumulation in the berries occurs in the period of ripening when auxins, cytokinins and gibberellins were no longer detected [13]. In the same way, the peak of cytokinin activity occurred several days before the period of high soluble sugar content in wheat [20]. Application of exogenous regulators (auxin GA_3, kinetin) on the fruit of *Pisum sativum* increased, to some extent, [^{14}C]-sucrose translocation [1]. But the high sink strength probably belongs to an indirect hormonal mechanism of assimilate transport.

The correlation between dry matter accumulation and ABA content appeared much more general. Furthermore, ABA injected into the fruit cane or into the maturing grain enhanced movement of [^{14}C]-labelled assimilate from leaves to the berry [13] or to the barley grain [38].

4.2.2 Regulation of Phloem Unloading

The involvement of ABA in phloem unloading [11] has been proposed by During and Alleweldt [13] and Tietz et al. [38]. The comparison between three soybean lines, differing in seed size, led Schussler et al. [34] to support the hypothesis that a high concentration of ABA in the seed coat might enhance the rapid efflux of sucrose necessary to support the unusually high seed growth observed in the large-seeded line. Tanner [36] proposed a model in which ABA decreases the ATPase proton extrusion from the sieve tube elements, thus enhancing the efflux of sucrose into the apoplast by H^+-sucrose cotransport. Recently, Van Bel and Patrick [39] demonstrated that ABA retarded the rate of fusicoccin-induced proton extrusion from the seed coats of *Phaseolus vulgaris*.

4.2.3 Regulation of Sucrose Uptake

ABA was shown to stimulate in vitro sucrose uptake by excised soybean embryos [34], suggesting that ABA is also involved in regulating sucrose accumulation in soybean cotyledons. Similar enhancement of transport sugar accumulation by ABA has been reported in growing apple fruits [6] and in another type of sink.

5 Conclusion

The survey presented here reviewed the main results obtained during the last few years in three fields: regulation of hormone levels, embryo development and source/sink relationships. Among the possible mechanisms for regulation of hormone levels, attention has been focused on hormone metabolism (conjugates and catabolites) and more recently, in the case of ABA, the variations of intracellular pH. With the aid of in vitro cultures, the development of the embryo can be directed either towards embryogenesis or towards germination. Monitoring of development, both morphologically and biochemically (using storage proteins as markers) opens up interesting prospects. The analysis of the mechanism concerning source/sink relationships provides the basis for a study of hormone regulation of seed filling; recent, substantial evidence supports the idea that ABA is involved in the control of phloem unloading in seed coats and sucrose uptake by the embryo.

References

1. Acchireddy NR, Kirkwood RC, Van Den Berg G (1984) Physiol Plant 61:130
2. Ackerson RC (1984a) J Exp Bot 35:403
3. Ackerson RC (1984b) J Exp Bot 35:414
4. Albone KS, Gaskin P, MacMillan JM, Sponsel VM (1984) Planta (Berl) 162:560
5. Black M (1983) In: Addicott FT (ed) Abscisic acid. Praeger, Eastbourne, p 331
6. Berüter J (1983) J Exp Bot 34:737
7. Boulon B, Prevost I, Barthe P, Le Page-Degivry MT (1985) Workshop on seed germination, Wageningen
8. Crouch ML (1982) Planta (Berl) 156:520
9. Crouch ML, Sussex IM (1981) Planta (Berl) 153:64
10. Davies DR, Bedford ID (1982) Plant Sci Lett 27:337
11. Delrot S, Bonnemain JL (1985) Physiol Veg 23:199
12. Dure LS, Galau GA, Greenway S (1980/81) Isr J Bot 29:293
13. Düring H, Alleweldt G (1980) Ber Dtsch Bot Ges 93:339
14. Ernst D, Oesterhelt D (1984) Planta (Berl) 161:246
15. Ernst D, Oesterhelt D, Schäfer W (1984) Planta (Berl) 161:240
16. Finkelstein RR, Crouch ML (1984) Planta IBerl) 162:125
17. Gatehouse JA, Evans IM, Brown D, Croy RRD, Boulter D (1982) Biochem J 207:629
18. Hendrix DL, Radin JW (1984) Plant Physiol (Bethesda) 117:211
19. Higgins TJV (1984) Annu Rev Plant Physiol 35:191
20. Jameson PE, MacWha JA, Wright GJ (1982) Z Pflanzenphysiol 106:27
21. Karssen CM (1982) In: Wareing PF (ed) Plant growth substances. Academic Press, London, p 623
22. Khan AA (1982) In: Khan AA (ed) The physiology and biochemistry of seed development, dormancy and germination. Elsevier Biomedical, Amsterdam, p 111
23. King RW (1982) In: Kahn AA (ed) The physiology and biochemistry of seed development, dormancy and germination. Elsevier Biomedical, Amsterdam, p 157
24. Long SR, Dale RMK, Sussex IM (1981) Planta (Berl) 153:405
25. Lorenzi R, Bennici A, Cionini PG, Alpi A, D'Amato F (1978) Planta (Berl) 143:59
26. Mok DWS, Mok MC, Rabakoarihanta A, Shii CT (1985) In: Bajaj YPS (ed) In vitro improvement of crops. Springer, Berlin Heidelberg New York
27. Noma M, Huber J, Ernst D, Pharis RP (1982) Planta (Berl) 155:369
28. Nooden LD (1984) Physiol Plant 62:273
29. Nooden LD, Obermeyer WR (1981) Biochem Physiol Planz 176:859
30. Pernollet JC (1985) Physiol Veg 23:45
31. Prevost L, Le Page-Degivry MT (1985) J Exp Bot 36:1457
32. Ray S, Mondal WA, Choudhuri MA (1983) Physiol Plant 59:343
33. Schroeder HE (1984) J Exp Bot 35:813
34. Schussler JR, Brenner ML, Brun WA (1984) Plant Physiol (Bethesda) 76:301
35. Sponsel VM (1980) In: Lenton JR (ed) Gibberellins. British Plant Growth Regul Group Monograph 5, p 49
36. Tanner RW (1980) Ber Dtsch Bot Ges 93:349
37. Thorne JH, Rainbird RM (1983) Plant Physiol (Bethesda) 72:268
38. Tietz A, Ludewig M, Dingkuhn M, Dörffling K (1981) Planta (Berl) 152:557
39. Van Bel AJE, Patrick JW (1985) Plant Cell Environ 8:1
40. Van Staden J, Davey JE, Brown NAC (1982) In: Khan AA (ed) The physiology and biochemistry of seed development, dormancy and germination. Elsevier Biomedical, Amsterdam, p 137
41. Vreugdenhil D (1983) Physiol Plant 57:463

A Revision of the Hormone Balance Theory of Seed Dormancy: Studies on Gibberellin and/or Abscisic Acid-Deficient Mutants of *Arabidopsis thaliana*

C.M. KARSSEN and E. LAÇKA [1]

1 Introduction

The basic framework of the hormone theory of seed dormancy originates from Luckwill [9], who suggested that the dormancy of apple seeds depends on the interaction between naturally occurring growth-inhibiting and growth-promoting substances. In later years, the interaction has often been described as a balance between simultaneously occurring promotive hormones, such as gibberellins (GAs) and cytokinins, and the inhibitory hormone, abscisic acid (ABA). It has been pointed out that the theory is mainly based on the effects of exogenous growth regulators on germination and dormancy, in which dormancy can be maintained, imposed and released by these chemicals singly or in combination [2]. Such effects are considered to reflect the action of naturally occurring hormones within seeds. Environmental factors, such as light and temperature conditions, are thought to operate on dormancy by causing changes in the balance between promotors and inhibitors.

Recent reviews unanimously judge the present experimental evidence insufficiently to justify the view that endogenous hormones control dormancy and germination, let alone that the control occurs in the form of a balance between promotors and inhibitors [2, 3, 13]. How is it that there is so little evidence? Have faulty techniques been applied or wrong hypotheses been formulated? Indeed, hormones often were determined by inadequate techniques. Improvement can be expected from a general adoption of recent advances of hormones assays, in particular, the use of immunoassays, which might also enable studies of hormone compartmentation and redistribution within seed tissues. It has been argued, however, that also with better assays the lack of correlation between dormancy and hormone levels might still exist, because the main regulatory principle is not based on changes in the levels of hormones, but in the sensitivity to hormones [11]. Such a hypothesis cannot be studied easily because endogenous hormone levels always interfere in the test for the sensitivity to applied growth regulators. Mutants, which lack the capacity to synthesize a hormone or a group of hormones, are unique tools to get around that dilemma. Moreover, they open possibilities to answer the question whether or not the presence of a hormone is required for a certain developmental process.

---

1 Department of Plant Physiology, Agricultural University, Arboretumlaan 4, 6703 BD Wageningen, The Netherlands

Plant Growth Substances 1985
(ed. by M. Bopp)

Recently, hormonal control of dormancy and germination was studied in ABA and GA-deficient mutants of *Arabidopsis thaliana* [8]. These studies unambiguously demonstrated the absolute requirement of ABA for the induction of dormancy during seed development [4]. They also showed that GA-deficient seeds never germinated without the addition of GA [5]. In this study, the mutants will be used to evaluate the validity of the hormone balance theory of seed dormancy.

2 Isolation and Characteristics of the Mutants

The isolation and genetic characterization of the GA- and ABA-deficient mutants of *A. thaliana* were described by Koornneef and Van der Veen [5] and Koornneef et al. [6], respectively. The latter mutants were isolated from the GA-deficient mutants via the induction of revertants (Table 1). The isolation started with the induction of non-germinating mutants that absolutely depended on the application of GA for both germination and seedling growth (gene symbol: *ga-1*). The mutant lacks endogenous GAs [14] and has a strongly reduced *ent*-kaurene synthesizing capacity in developing fruits and seeds [1]. Step 2 consisted of the mutagenic treatment of the *ga-1* mutant followed by selection for germinating seeds in the progeny. This step was based on

Table 1. Isolation of isogenic, recessive *Ga* and *Aba* mutants of *Arabidopsis thaliana*. (After [8])

Generation	Phenotype	Genotype
STAGE 1		
M_0	Wild type	*Ga-1/Ga-1, Aba/Aba*
↓ EMS[a]	↓	↓
M_1	Wild type	*Ga-1/ga-1, Aba/Aba*
↓ selfing	↓	↓
M_2	**GA mutant** *Germination:* GA-dependent *Plants:* dwarfs	*ga-1/ga-1/Aba/Aba*
STAGE 2		
M_0	GA mutant	*ga-1/ga-1, Aba/Aba*
↓ EMS[a]	↓	↓
M_1	GA mutant	*ga-1/ga-1, Aba/aba*
↓ selfing	↓	↓
M_2	**Revertant** *Germination:* GA-independent *Plants:* dwarfs	*ga-1/ga-1, aba/aba*
STAGE 3		
	revertant x wild type	*ga-1/ga-1, aba/aba* *Ga-1/Ga-1, Aba/Aba*
	↓	↓
F_1	wild type	*Ga-1/ga-1, Aba/aba*
↓ selfing	↓	↓
F_2	**ABA mutant** *Germination:* GA-independent *Plants:* normal height, wilting	*Ga-1/Ga-1, aba/aba*

[a] EMS = ethylmethanesulphonate; 10 mM applied for 24 h in darkness at 24 °C

the concept of a hormone balance. It was hypothesized that mutagenesis of the genes responsible for inhibitor production might restore the germination of GA-deficient seeds. Indeed, seeds were isolated that were reverted to the germination capacity of the wild type. Genetic analysis of these so-called revertants showed that the reversion was due to a mutation in a suppressor gene segregating independently from *ga-1*. As step 3 the second gene (symbol: *aba*) was isolated in a *Ga-1/Ga-1* background. The *aba* mutant had a tendency to wilt and the seeds were characterized by the absence of dormancy, much in contrast to the wild-type seeds that required a cold treatment to develop full germination capacity in water and light. It was shown, in particular during seed development, that the seeds of the *aba* mutant had very low ABA levels [4]. The isolation history suggests the existence of a hormone balance. This suggestion will be further analyzed in the present study.

For this study seeds of the different genotypes of *Arabidopsis thaliana* L. (Heynh.) were raised in the summer of 1983, according to techniques described by Koornneef et al. [6]. The isolation of unripe seeds and the performance of the germination test were described by Karssen et al. [4]. Germination was tested at 24 °C. White fluorescent tubes (Philips TL57) at an intensity of 8 W m^{-2} were used as light source.

3 Results

3.1 Induction of Dormancy

A previous study showed that part of the wild-type seeds germinate precociously when they are isolated from the fruits half-way development [4]. However, at maturity the seeds are fully dormant. In contrast, mature seeds of the *aba* mutant are not dormant, their germination capacity develops gradually from the 10th day after pollination onwards. It was concluded that endogenous ABA is responsible for the induction of dormancy in developing seeds of *A. thaliana*. The present experiments with seeds of wild type and *aba* mutant confirmed our previous conclusion (Fig. 1).

Seeds of the *ga-1* mutant did not germinate in water at any stage of development (Fig. 1). Application of 10 μM GA_{4+7} strongly stimulated precocious germination at days 14 and 18 after pollination, but failed to do so on day 22. Wild-type seeds behaved similarly. Therefore, it is concluded that in an *Aba/Aba* background, it is irrelevant for the induction of dormancy whether or not the seeds are able to synthesize GAs.

The partial germination of the seeds of the double mutant at maturity indicates that the absolute requirement of exogenous GAs, due to the GA deficiency, vanished in a number of seeds if they were also deficient for ABA. This means that seeds, which lack ABA during development, still need a certain level of GAs to germinate. Application of 10 μM GA_{4+7} fully satisfied that need, also in mature seeds. The clear distinction in the reaction of 10 μM GA_{4+7} in light between the seeds with the *Aba/Aba* genotype and the ones with the *aba/aba* genotype provides evidence that endogenous ABA plays a dominant role during the induction of dormancy, whereas GAs seem not to be involved.

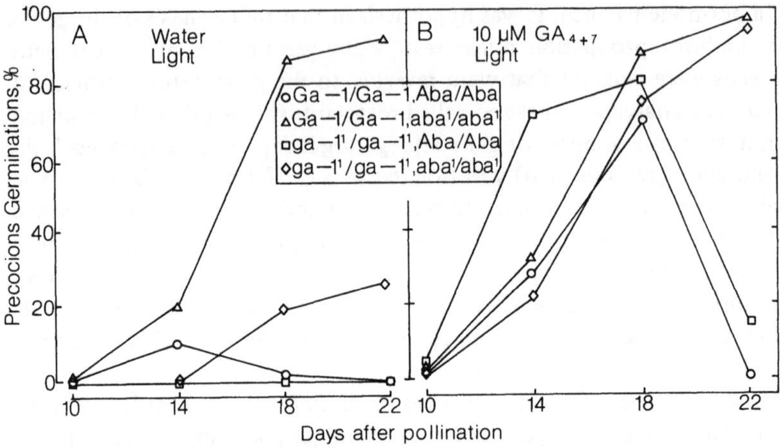

Fig. 1A,B. Precocious germination of seeds of four genotypes in water (A) or 10 μM GA_{4+7} (B) in continuous white light at 24 °C. Seeds were isolated from the siliquae at the indicated time and subsequently transferred to the germination conditions. Germination was counted after 14 days

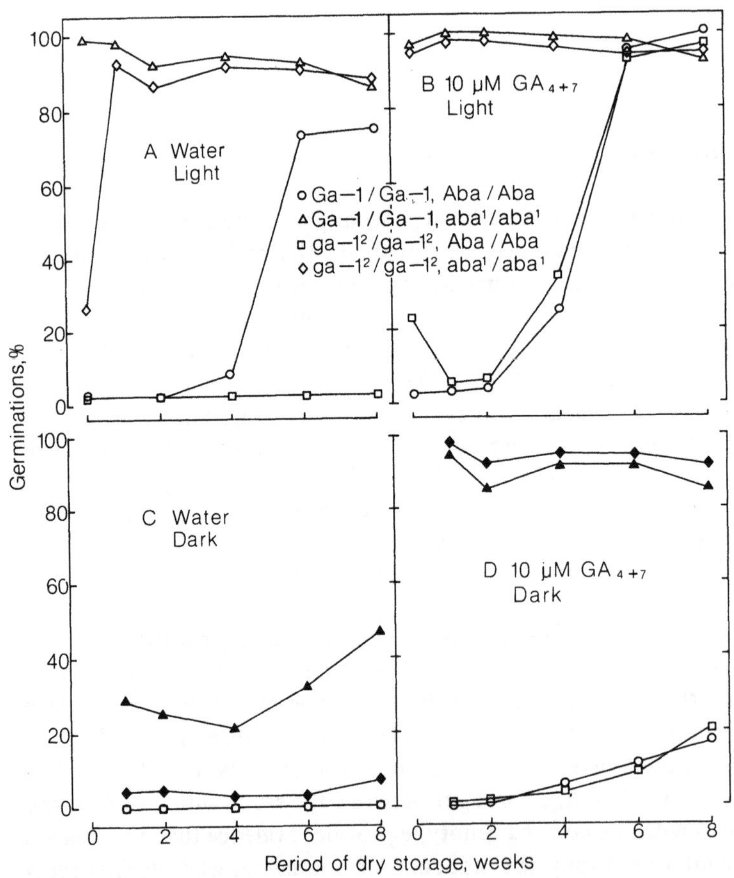

Fig. 2A–D. The effect of a period of dry storage at room temperature following directly upon the harvest of mature seeds of four genotypes on the germination in water (A,C) or 10 μM GA_{4+7} (B,D) in continuous white light (A,B) or darkness (C,D) at 24 °C

3.2 Release from Dormancy

Release from dormancy in seeds of the four different genotypes was studied during both dry storage (Fig. 2) and incubation at 2 °C as a chilling treatment (Fig. 3). The germination of wild-type seeds in light and water was stimulated during both treatments, although chilling acted faster, in days instead of weeks (Figs. 2A and 3A). Dark germination of wild-type seeds in water was not stimulated (Figs. 2C and 3C), seeds of *A. thaliana* showed a typical light-requiring germination pattern. Seeds of the *ga-1* mutant did not germinate in water, neither in light nor in darkness (Figs. 2A,C and 3A,C). Also after much prolonged treatments we never found an exception to that rule. Application of 10 μM GA_{4+7} immediately after harvest did not stimulate the germination of the GA-deficient seeds. However, both dormancy-releasing treatments increased the GA sensitivity of the *ga-1* seeds in light and chilling increased it also in darkness (Figs. 2B,D and 3B,D). The GA-sensitivity of the wild-type seeds rose in all situations in a similar way. These results demonstrate that the release from dormancy is independent of the capacity to synthesize GAs.

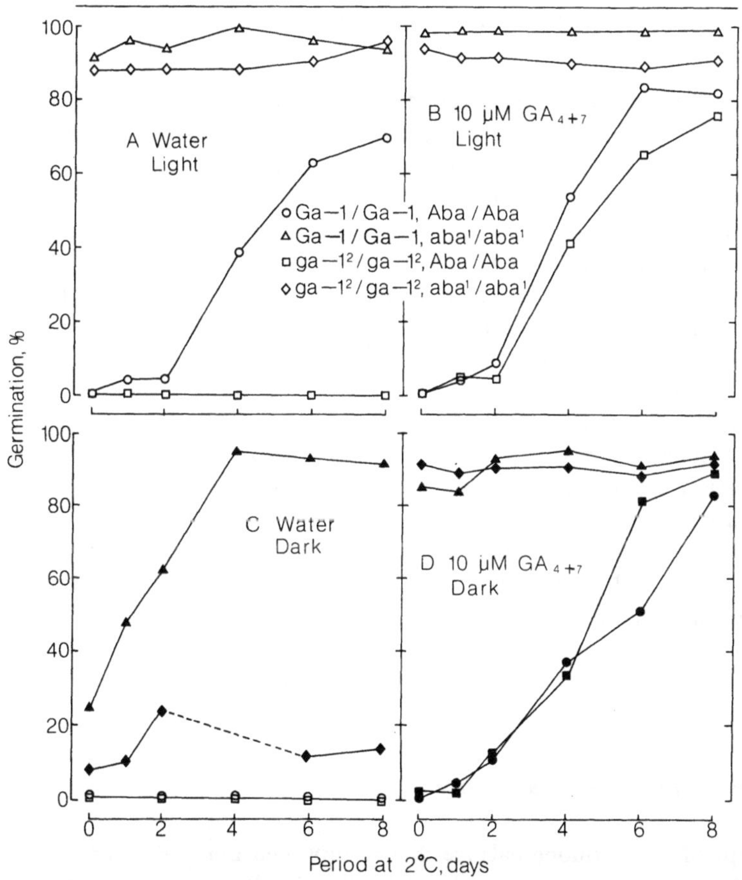

Fig. 3A–D. The effect of a period of incubation at 2 °C in water **(A,C)** or 10 μM GA_{4+7} **(B,D)** in darkness on the germination of seeds of four genotypes at 24 °C in the same media in continuous white light **(A,B)** or darkness **(C,D)**. The seeds had been stored at dry conditions during 2 weeks before the start of the experiment

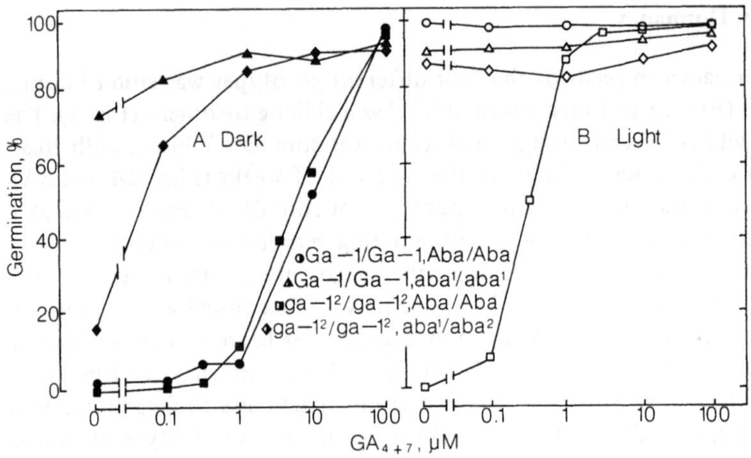

Fig. 4A,B. The effect of a concentration range of GA_{4+7} on the germination of seeds of four genotypes in continuous darkness (A) or white light (B) at 24 °C. Before the start of the incubation at 24 °C the seeds had been stored at dry conditions during 4 weeks and subsequently incubated in the different GA_{4+7} concentrations at 2 °C in darkness during 6 days

The germination of unripe seeds could not be tested in darkness because the isolation occurred in light. Germination tests of ripe seeds of the *aba* mutant in darkness revealed that for this reason our previous conclusion about the absence of dormancy in ABA-deficient seeds was slightly incorrect. Shortly after harvest, seeds of the *aba* mutant were obviously light-requiring, which is regarded as a form of dormancy (Figs. 2C and 3C). Particularly a chilling treatment was an effective means to rapidly render the seeds light-independent (Fig. 3C). Application of 10 μM GA_{4+7} replaced the light requirement directly after harvest and made dormancy breaking superfluous.

One week of dry storage was sufficient to induce full germination in light and water of the seeds of the double mutant (Fig. 2A). Apart from that, the double mutant only differed from the *aba* mutant in darkness (Figs. 2C and 3C). After 8 days of chilling, seeds of all four genotypes germinated in darkness in 10 μM GA_{4+7} (Fig. 3D). However, a germination test in darkness at a range of GA_{4+7} concentrations of 8-week-old seeds that had been chilled during 6 days revealed that the GA requirement of seeds with an *Aba/Aba* genotype was much higher than of seeds with an *aba/aba* genotype (Fig. 4A). In light, only the seeds of the *ga-1* mutant required application of GA_{4+7}, but at a lower level than in darkness (Fig. 4B).

4 Discussion

4.1 Revision of the Hormone Balance Theory

The classic concept of the hormone balance theory supposed that induction, maintenance and release of dormancy are regulated by the simultaneous action of promotive and inhibitory hormones, whereby the degree of dormancy depends on the ratio between the endogenous concentrations of the opposing groups of compounds.

Our studies with the ABA- and GA-deficient mutants of *Arabidopsis thaliana* no longer support such a concept.

Firstly, sufficient levels of ABA and GAs are probably never present simultaneously during any stage of the seed life. In wild-type seeds the major *ent*-kaurene synthesizing capacity was found during the first 7 days after pollination [1], whereas embryonal ABA – the only fraction relevant to the induction of dormancy – was present around day 16 [4]. It was also shown by these authors that the level of ABA, which remained in the dry, mature wild-type seeds, was insufficient to explain any inhibition of germination. Moreover, seeds of wild type and of the *aba* mutant were almost equally sensitive to applied ABA [6, 7]. Therefore, it is unlikely that the stimulative action of endogenous GAs on germination is directly opposed by endogenous ABA in wild-type seeds.

Secondly, the results of the present precocious-germination experiments clearly showed that the induction of seed dormancy during seed development was only controlled by the *Aba* gene and not by the *Ga-1* gene (Fig. 1). Also the release from ABA-induced dormancy during after-ripening treatments did not depend on the *Ga-1* gene (Figs. 2, 3). In both wild type and *ga-1* mutant the sensitivity of the seeds to 10 μM GA_{4+7} rose during dry storage and chilling in a similar way. It is concluded that ABA and GA do not directly interact during any stage of seed life. The action of the two hormones occurs separately in different developmental stages. ABA acted during seed development and GAs during germination. The only relationship between the two was that the GA requirement of the germination process dependend on the action of ABA during dormancy induction. Seeds with an *Aba/Aba* genotype required in darkness much higher levels of GA_{4+7} than *aba/aba* seeds (Fig. 4). The arguments mentioned above make it clear that the higher GA requirement was not due to higher ABA levels in the ripe seeds. It was caused by the much stronger physiological blockade of the germination process in the ABA-containing genotypes. Thus, high ABA caused deep dormancy, which required high levels of GA, whereas low ABA resulted in lightly dormant seeds, which required low GA levels to germinate (Fig. 5). Therefore, the original concept of a hormone balance has to be replaced by a kind of "remote control" of the GA requirement of germination by the ABA level of the developing seed through the intermediate dormancy condition.

4.2 Environmental Control

We have found no indication yet that environmental factors control the level of ABA during seed development. However, the release from dormancy and the germination

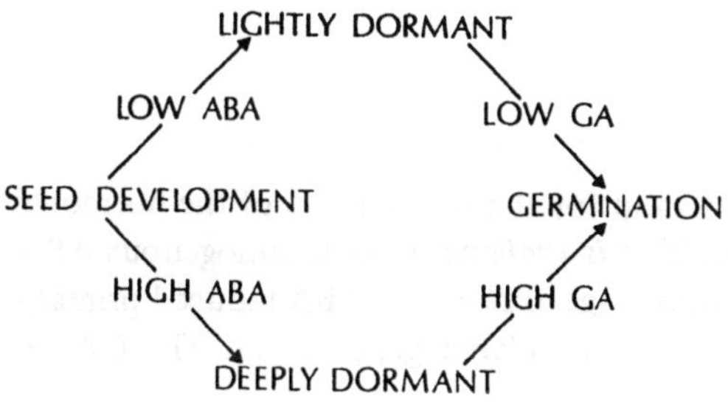

Fig. 5. Schematic presentation of the conclusions

process were clearly influenced by the environment. The use of GA-deficient mutants made it possible to conclude that dormancy-releasing treatments did not promote a rise in GA levels, as has been reported for several other species (see [13] for references), but increased the sensitivity to GAs. More detailed information on the changing GA sensitivity during dormancy release will be published in a separate report, which will also indicate that the induction of secondary dormancy is accompanied by a decrease of the GA sensitivity in the seeds of the *ga-1* mutant.

Light had a dual action on the germination of these seeds. The first light effect was dependent on GA biosynthesis. It made seeds of the wild type independent of a GA application, whereas seeds of the *ga-1* mutant remained GA-dependent (Figs. 2, 3, and 4). The second light effect became visible in the absence of GA biosynthesis. In the seeds of the *ga-1* mutant, light increased the sensitivity to GA_{4+7} (Fig. 4). Thus, light increased in the seeds of *A. thaliana* both the synthesis of GAs and the sensitivity to GA. This dual light effect was also found in the seeds of the related species *Sisymbrium officinale* (Hilhorst et al., in prep.). In those seeds GA biosynthesis depends on the simultaneous presence of light and nitrate, if nitrate was omitted from the medium, it was seen that light also increased the GA sensitivity.

Our study indicates that the environmental control of seed development and germination occurs both through changes in hormone levels and by changes in the sensitivity to hormones.

The low GA requirement of the seeds of the *aba* mutant could easily be replaced by light, probably via both light effects. Dormancy-releasing treatments made light even superfluous, probably because pre-exisiting phytochrome Pfr stimulated GA biosynthesis sufficiently in those seeds. It is not yet clear which factor stimulated the germination of the seeds of the double mutant in the absence of GA (Figs. 2 and 3). It might still be GA, because the GA deficiency is probably not absolute, enabling just enough GA biosynthesis to satisfy the low requirement of these seeds. Preliminary experiments indicate, however, that also the production of endogenous ethylene might play a role (S. Zagórski, pers. comm.). Application of ethylene bypassed the GA requirement of the *ga-1* mutant.

For the time being, our conclusions cannot be generalized to other species, in particular, not to seeds with embryo dormancy, which might have a different hormonal control. However, the lack of correlation between hormone levels and the degree of dormancy in many species makes it promising to consider a control via changing hormone sensitivity also in other species. Our revision of the concept of the hormone balance finds support in the strong doubts that have been expressed particularly about the role of ABA in the ripe seed [3, 10, 12].

5 Conclusions

1. The hormone-balance theory is not valid for the explanation of the hormonal regulation of dormancy in seeds of *Arabidopsis thaliana* because endogenous ABA and GA never acted simultaneously during any stage of seed life. ABA induced primary dormancy during seed development and GAs stimulated germination. The GA re-

quirement of germination depended on the degree of dormancy which, in turn, was determined by the level of ABA during development.

2. The release from dormancy during dry storage and chilling also occurred in the absence of GA biosynthesis. Both treatments increased the sensitivity of the seeds to applied GAs.

3. Light had a dual action. It stimulated both GA biosynthesis and the sensitivity to GA.

Acknowledgment. The authors are much indebted to Dr. M. Koornneef (Department of Genetics, Agricultural University, Wageningen) for the use of the mutants and for many stimulating discussions.

References

1. Barendse GWM, Koornneef M (1982) Arabid Inf Serv 19:25
2. Bewley JD, Black M (1982) Physiology and biochemistry of seeds in relation to germination, vol 2. Springer, Berlin Heidelberg New York
3. Black M (1980/81) Isr J Bot 29:181
4. Karssen CM, Brinkhorst-van der Swan DLC, Breekland AE, Koornneef M (1983) Planta 157: 158
5. Koornneef M, Van der Veen JH (1980) Theor Appl Genet 58:257
6. Koornneef M, Jorna ML, Brinkhorst-van der Swan DLC, Karssen CM (1982) Theor Appl Genet 61:385
7. Koornneef M, Reuling G, Karssen CM (1984) Physiol Plant 61:377
8. Koornneef M, Cone JW, Karssen CM, Kendrick RE, Van der Veen JH, Zeevaart JAD (1985) In: Freeling M (ed) Plant genetics. UCLA Symp Mol Cell Biol, new ser, vol 35. Liss, New York
9. Luckwill LC (1952) J Hortic Sci 27:53
10. Tilberg E (1983) Physiol Plant 58:243
11. Trewavas A (1981) Plant Cell Environ 4:203
12. Walton DC (1980/81) Isr J Bot 29:168
13. Wareing PF (1982) In: Khan AA (ed) The physiology and biochemistry of seed development, dormancy and germination. Elsevier Biomedical, Amsterdam
14. Zeevaart JAD (1983) In: 18th annual report MSU-DOE Plant research laboratory

Cytokinin Control of Monocarpic Senescence in Soybean

L.D. NOODÉN[1] and D.S. LETHAM[2]

1 Introduction

1.1 The Senescence Concept

Senescence is an internally controlled degeneration leading to death [4]. From the cell level to the whole plant level, most processes decline; however, this loss of vital functions and disassembly of the cell's internal structure is a highly ordered process [17, 19]. Indeed, senescence serves a variety of important functions in plant development [4]. While the concept of senescence seems clear, its precise biochemistry is not [19]. In any case, it should not be assumed that all downward adjustments of metabolism represent senescence.

1.2 Monocarpic Senescence

Many plants, including annuals, biennials, and even some perennials have a single reproductive phase (monocarpy), which is concluded by degeneration (monocarpic senescence) leading to death [13]. For many monocarpic plants, particularly soybeans and several other field crops, the degeneration preceding death is dramatic as well as rapid. Most prominent (and possibly most important) among these changes in soybeans is the yellowing and shedding of the leaves. These parameters make good, though not infallible, measures of senescence [15].

Since the monocarpic plant's assimilatory capacities decrease at just the time when they appear to be needed most to support fruit development, monocarpic senescence seems to be an important factor in determining the yields in several important crops including soybeans, but yield limitation is more than a single factor problem [14]. Senescence and pod development in the soybean are integrated or coordinated through a series of regulatory signals [14]. Because there are some significant physiological differences between soybean and other species [13], this report will focus on soybean.

1 Botany Department, University of Michigan, Ann Arbor, MI 48109-1048, USA
2 Department of Developmental Biology, Research School of Biological Sciences, Australian National University, Canberra, ACT 2601, Australia

Plant Growth Substances 1985
(ed. by M. Bopp)

1.3 Control of Monocarpic Senescence

Monocarpic senescence in soybean is regulated by the most advanced reproductive structures, which induce a rapid degenerative syndrome [13–15]. Removal of the pods before the end of fill can prevent this rapid death of the soybean plant [3, 7, 13, 14]. Since many processes and changes occur in the different parts of the plant, the total picture obviously must be very complex, but a striking feature of soybean maturation is the correlative control (exchange of regulatory signals) between different parts of the plant, particularly the influences exerted by the most advanced pods [13, 14]. Although the identity of these signals is unknown, at least some are probably hormonal in nature [14]. Understanding these regulatory signals seems essential to unravelling the biochemistry of the senescence process. In addition, major assumptions underlying the roles of hormonal regulatory signals in the whole plant (e.g. [22]) still have not been tested adequately, and some of these can be studied conveniently in the maturing soybean plant.

1.4 Two Major Phases in Monocarpic Senescence

In an effort to simplify and organize a complex phenomenon, it seems valid to view soybean monocarpic senescence in terms of a preparatory and a final phase [14, 15]. The preparatory phase seems to occur during early reproductive development (flowering through mid-podfill) and the final phase from mid-podfill until death; however, there may not be a sharp demarcation between these phases. Early in their development, the soybean pods cause a decline in many physiological functions, particularly root growth and assimilation, but they may also exert some promotive effects [13, 14, 20]. Sometime during late podfall, the plant passes a point of no return beyond which depodding no longer prevents rapid death [13].

2 The Preparatory Phase in Monocarpic Senescence of Soybean

2.1 Role of Cytokinin and Minerals from the Roots

Because the young root apices are generally more active than older ones in production of cytokinins and assimilation of mineral nutrients for supply to the foliage and pods, the cessation of root growth seems particularly critical [5, 13, 15, 21]. It has, however, been difficult to analyze the role of mineral and cytokinin flux from the roots in regulating foliar senescence and pod development. In general, it does appear that mineral uptake and cytokinin production [2, 13] decrease during podfill when root growth would also be declining [14, 20]. Correlations of this type are suggestive, but do not constitute proof of causality.

2.2 The Soybean Explant as a Tool for Studying Root Contributions

Our [12] recently developed explant system (a cutting with a leaf, one or more pods, and a subtending stem segment without roots) offers opportunities for more direct analysis of the root contribution to pod development and foliar maintenance; defined solutions can be substituted for the roots. Explants cut early in podfill complete pod development and foliar senescence in a manner which appears normal by a variety of parameters; however, explands cultured on water only show greatly accelerated pod development and leaf senescence. Either mineral nutrients (at a concentration roughly approximating the elemental composition of xylem sap) or cytokinin delay (normalize) pod development and foliar senescence, but the combination is particularly effective [12, 15]. Thus, it appears that a decrease in the flux of mineral nutrients and cytokinins up through the xylem is an important factor in senescence of the leaves; however, this is not in itself causal. Neither excised leaf disks nor depodded explants senesce (or at least not normally), and their life is greatly prolonged compared to their in situ counterparts even when cultured on water only [12, 15]. This indicates that the pods and not the decreased cytokinin and mineral flux are the principal cause of monocarpic senescence [15]. A decrease in root assimilates is nonetheless an important contributory factor in the early (preparatory) phase of whole plant senescence, for maintaining a high influx of minerals and especially cytokinin can override the influence of the pods. The ability of foliar applications of benzyladenine to delay senescence of soybean plants points in the same direction [8, 16].

3 Cytokinin Translocation and Metabolism

3.1 Partitioning of Cytokinin Flux from the Roots

Of all the known hormones, cytokinin is the most clearly implicated in senescence and the most extensively studied [5, 17]. Cytokinin seems to be *the* major endogenous senescence retardant. The developing roots seem to be the major source, and this cytokinin is pulled up into the shoot via the xylem as the leaves transpire [5, 21]. One mechanism proposed for the fruit induction of monocarpic senescence holds that the developing fruit monopolize the supply of nutrients and necessary hormones, including cytokinins, thereby creating a deficiency in the leaves and, hence, senescence [13]. The explant system can be used to test this idea directly.

Figure 1 shows the distribution of soluble [^{3}H] derived from [^{3}H]-zeatin riboside (ZR) supplied as a pulse (1 h) through the base of an explant at early podfill and 24 h afterwards. This soluble fraction would be expected to contain not only [^{3}H]-ZR, but a range of metabolites [6], so it only provides a crude upper limit on cytokinin distribution. Clearly, relatively little [^{3}H] occurs in the fruit (2.1% of the soluble [^{3}H] at the end of the pulse), and most of that is in the carpels with very little in the seeds, particularly the embryos. This is not the flux pattern which one would expect if the pods (or the "hungry" seeds) were competing with the leaves for cytokinin fluxing up the stem. Instead, this distribution of [^{3}H] roughly follows the partitioning of water through the xylem into the leaves, carpels, and seed coats [18]. The [^{3}H] distribution on a percentage basis is similar after 24 h; however, it is noteworthy that the total soluble [^{3}H] decreases in most tissues. This drop reflects metabolism of the [^{3}H]-labeled compounds and their incorporation into insoluble materials [18]. Similar results were obtained for [^{3}H]-zeatin (Z) supplied to explants at the same stage (early podfill) and [^{3}H]-ZR in explants taken earlier (prepodfill).

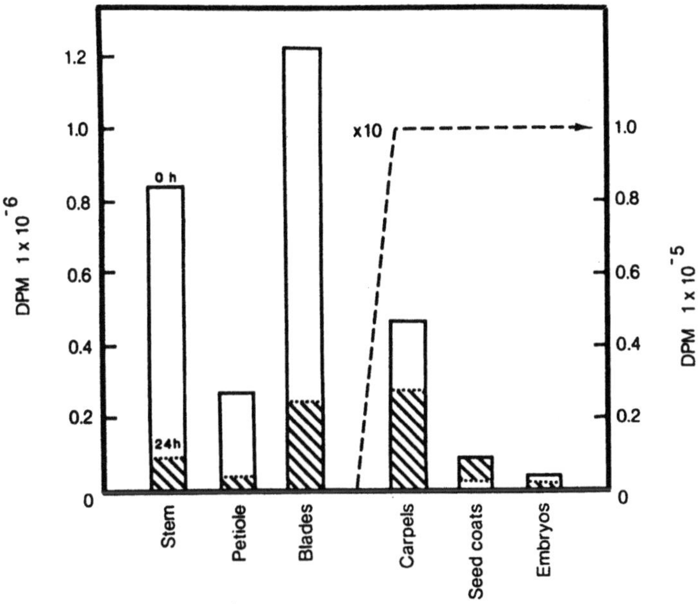

Fig. 1. Distribution of [^{3}H] from [^{3}H] (ring-labeled)-zeatin riboside (2.6 mCi) in H_2O supplied as a 1-h pulse via the xylem (through the cut base) of soybean explants cut in early podfill followed by a 24-h chase with H_2O. The soluble [^{3}H] was extracted with cold methanol/$CHCl_3$/conc. formic acid/H_2O (12:2:1:2, v/v) [18]. The open areas represent levels at the end of the pulse (0 h); the cross-hatched areas show levels at 24 h

3.2 Metabolism of the Xylem-Transported Cytokinin in Various Parts of an Explant

When the soluble [^{3}H] extracted from the leaves of explants fed [^{3}H]-ZR (ring-labeled) is fractionated by two-dimensional thin layer chromatography (TLC) and high pressure liquid chromatography (HPLC), only a small percentage of the label corresponds to the parent compound even at the end of the pulse (1 h), and by 24 h, this declines to almost zero (Fig. 2). At the end of the pulse, a wide range of metabolites are produced including dihydrozeatin riboside (DZR), but zeatin nucleotide (ZNT) is especially prominent. After 24 h, ZR, ZNT, and closely related metabolites have essentially disappeared. The proportion of 0-glucosides in the soluble [^{3}H], however, increases. Adenosine (AR) and adenine (A) are also prominent metabolites, and these apparently are incorporated into nucleic acids [18].

Rapid metabolism also occurs in other parts of the explant, but the pattern of metabolism in these parts differs at least quantitatively from that in leaves. This is illustrated by the differences in the major [^{3}H]-compounds (Table 1). The metabolites in the stem and petiole seem unlikely to be predominantly compounds in transit in the xylem. In the carpels, the ribose appears to be rapidly cleaved from ZR producing Z. Our studies also provide information on the relative stabilities of the cytokinin metabolites. ZNT is rapidly metabolized and may be intermediate in breakdown. The reduction of the side chain (dihydro-derivatives) does not completely stabilize the cytokinins, but the 0-glucosides seem to be a more stable cytokinin pool.

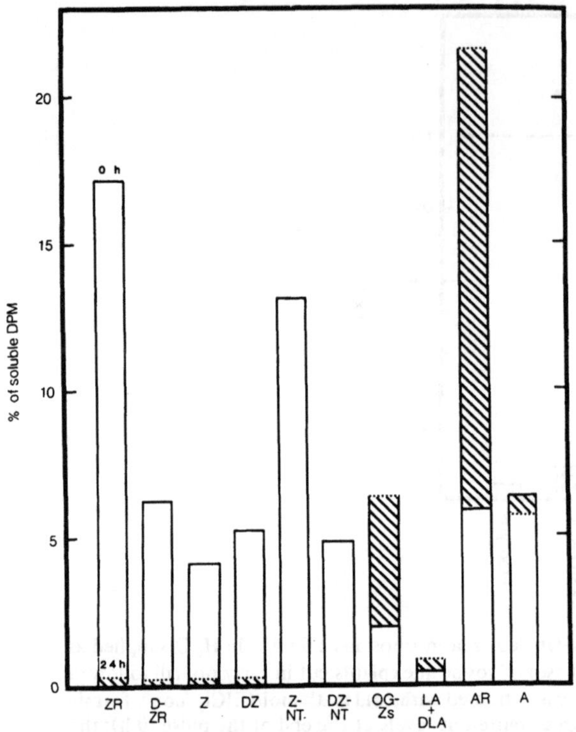

Fig. 2. Metabolites of [^{3}H]-zeatin riboside in leaf blades from soybean explants (early podfill) supplied as a 1-h pulse via the base. Same explants as in Fig. 1. The soluble [^{3}H] was extracted as in Fig. 1 and separated by two-dimensional TLC (silica) [18]. Solvent 1: *n*-butanol/conc. acetic acid/H_2O (12:3:5, v/v) and solvent 2:*n*-butanol/conc. NH_4OH/H_2O (6:1:2, v/v, upper phase). Eluants of the appropriate regions of the gel layers were then rechromatographed (TLC or HPLC) to establish the [^{3}H] due to each metabolite. TLC on paraffin-coated silica gel or HPLC on C_8 was used to resolve Z from DZ, ZR from DZR, etc. The metabolites shown are zeatin riboside (ZR), zeatin (Z), zeatin nucleotide (ZNT), lupinic acid (LA), and their dihydro-forms (DZR, etc.). OG-Zs represents the 0-glucosides, A adenine, and AR adenosine. The 7- and 9-glucopyranosyl derivatives of Z were not detected. Open bars represent levels at the end of the pulse, the cross-hatched bars 24 h

Table 1. Major [^{3}H]-compounds from various parts of soybean explants (early podfill) given [^{3}H]-zeatin riboside (ring-labeled) as a pulse via the base at early podfill[a]

Explant part	Major [^{3}H]-labeled compounds	
	At end of pulse (1 h)	After 24 h
Stem	Zeatin nucleotide	Adenosine
Petiole	Zeatin nucleotide	Adenosine
Blades	Zeatin riboside, zeatin nucleotide	Adenosine, adenine, and 0-glucosides
Carpels	Zeatin, dihydrozeatin, zeatin riboside, adenosine, and adenine	Adenosine and adenine

[a] Same explants and procedures as in Figs. 1 and 2

Table 2. [^{3}H]-compounds from the leaf blades of soybean explants supplied [^{3}H]-zeatin riboside via the base[a]

Developmental stage and treatment of the plants supplying explants		Total soluble DPM	Total soluble DPM (%)									
			ZR	DZR	Z	DZ	OGZR	OGDZR	OGZ	OGDZ	A	AR
A. Full extension	T = 1 h	94,100	1.08	0.82	0.78	0.28	0.05	0.03	0.46	0.03	4.5	4.3
Full extension	T = 24 h	138,400	0.51	0.23	0.26	0.06	0.16	0.21	0.56	0.05	14.4	21.0
B. Late podfill	T = 1 h	187,100	0.31	0.49	1.38	1.31	0.08	0.01	0.30	0.36	8.5	1.9
Late podfill	T = 24 h	86,800	0.98	0.13	0.67	0.04	0.00	0.21	1.78	0.81	12.2	6.6
C. Same as B, but depodded from full extension	T = 1 h	126,300	0.32	0.64	0.32	0.33	0.82	0.49	1.37	0.30	3.5	2.6
Same as B, but depodded from full extension	T = 24 h	153,760	0.02	0.15	0.57	0.26	0.25	0.14	1.06	1.05	9.6	4.2
D. Early leaf-yellowing	T = 1 h	249,680	5.46	1.03	2.82	1.20	0.44	0.82	3.38	0.88	4.3	4.1
Early leaf-yellowing	T = 24 h	54,510	0.51	0.86	0.20	0.51	1.29	0.62	9.58	1.72	5.1	3.7
E. Same as D, but depodded from full extension	T = 1 h	144,760	0.48	0.73	1.78	1.34	0.33	0.24	0.57	0.58	11.1	5.4
Same as D, but depodded from full extension	T = 24 h	84,350	0.56	0.38	0.52	0.28	1.10	0.38	0.75	1.15	9.7	4.3

[a] The explants were given [^{3}H]-ZR in water for 1 h, and then chased with H_2O for 1 h or 24 h. The soluble [^{3}H] was extracted as in Fig. 1 and passed over a cellulose phosphate column. The 0.5 M NH_4OH eluant was neutralizd with acetic acid, lyophilized and run on silica gel TLC in *n*-butanol/acetic acid/H_2O (450:113:18, v/v). A zone corresponding to A/AR was eluted and run on TLC with *n*-butanol/conc. NH_4OH/H_2O (6:1:2, v/v, upper phase). Zones containing the ZR/DZR/Z/DZ or the 0-glucosides were eluted and separated by HPLC on a C_8 column using a methanol gradient (in water, 1% acetic acid)

Returning to Fig. 1, there is some [^{3}H] in the seed coats and embryos. Two-dimensional TLC indicates that small amounts in seed coats correspond to ZR/DZR and ZMP (13% and 16%, respectively of the total extractable [^{3}H] at the end of the pulse), but this decreases to 0.3% and 8%, respectively, at 24 h [18]. The minute amount of cytokinin, which does enter the seed coat, is related to the tiny xylem connection running in from the stem. By contrast, the [^{3}H] in the embryo extract does not yield any activity peaks corresponding to ZR or other known cytokinins. If [^{3}H]-ZR does enter the embryo, which does not have direct vascular or symplastic connections with the seed coat, it must be quite a small amount, less than 0.0006% of the [^{3}H]-ZR applied originally. Since the cultured embryos do not seem to require exogenous cytokinin, it can be questioned whether any of the high levels of cytokinin in the embryos comes from the roots [8, 18]. While cytokinin does not seem to accumulate in the seed coat or to be transferred to the embryo as photosynthate, abscisic acid, and phosphate are, [^{3}H]-adenosine is not excluded, and it behaves more like photosynthetic products etc. [11, 18]. This restriction in the movement of cytokinin, no doubt, plays a role in its regulatory function.

3.3 Effects of Senescence and Depodding on the Metabolism of [^{3}H]-ZR in Soybean Explants

The spectrum of [^{3}H]-metabolites from [^{3}H]-ZR, appearing in the leaf blades of the explants represented in Table 2, is very similar to that in Fig. 2. These metabolites were emphasized, because previous studies (e.g., Fig. 2) showed them to be dominant forms. Aside from different purification methods, the main procedural difference in Table 2 is a 1-h chase following the pulse with [^{3}H]-ZR. ZNT and DZNT are not prevalent (in the runoff from cellulose phosphate) after the 1-h chase, which shows their highly transitory nature. In analyzing these data, it must be borne in mind that the level of [^{3}H] in any category shown is a steady-state level, which may be influenced by the rates of metabolism, by continued influx through the xylem and by other factors. Comparing A (full extension), B (late podfill), and D (early leaf yellowing) in Table 2, there seems to be only relatively minor quantitative differences in cytokinin metabolism. Similarly, depodding (C and E), which prevents rapid leaf yellowing and death of the plants, does not produce any striking alterations in cytokinin metabolism. Thus, it appears that leaf senescence is not related to any major changes in the rate or pattern of cytokinin metabolism in the shoot system.

A parallel set of studies on the metabolism of [^{3}H]-benzyladenine (BA) applied to leaves of explants taken at various stages shows that 9-alanyl-BA (identified by mass spectrometry) is the predominent BA metabolite and that the pattern of metabolism of BA does not change from mid-podfill through early leaf yellowing (Zhang, Letham, and Noodén, unpubl. data).

4 Relative Activity of the Optical Isomers of Dihydrozeatin

The explant system allows comparison of the activity of cytokinins supplied in the same way as they would be naturally, from the roots through the xylem, thereby

Table 3. Comparison of delayed leaf yellowing by cytokinins supplied in complete mineral nutrient solutions

Cytokinin		Delay in days to 50% of leaves ⩽ 1/2 yellow relative to minerals only
10^{-8} M	zeatin	5 ± 2.4
	(±) dihydrozeatin	7 ± 2.8
	(+) dihydrozeatin	15 ± 1.4
	(–) dihydrozeatin	3 ± 2.1
10^{-7} M	zeatin	12 ± 1.8
	(±) dihydrozeatin	> 17
	(+) dihydrozeatin	≫ 17
	(–) dihydrozeatin	17
Intact plants		7 ± 0.9

avoiding a number of complications [1]. DZ and DZR are important xylem sap constituents [2, 5, 18, 21], but only one of the optical isomers (–) of the side chain seems to occur naturally [9]. We find this racemic mixture of DZ or DZR isomers, made by reducing Z or ZR, respectively, to be very active in the soybean explant system, more active than the unsaturated forms [1]. The two optical isomers of DZ show quite different activity, the unnatural form (+) being more active than the natural form (–) (Table 3). Matsubara et al. [9] have reported a similar pattern of activity for these isomers in other bioassays.

5 Conclusions

During pod development, soybean plants undergo a physiological decline (monocarpic senescence), which leads to death. Root growth and assimilatory functions (e.g., mineral uptake and cytokinin production) decrease. The role of root assimilates in monocarpic senescence can be assessed through the use of explants, by substituting defined solutions for the roots. While reduced mineral and cytokinin flux from the roots do not by themselves cause monocarpic senescence, these decreased fluxes appear to be necessary (preparatory phase) for the senescence-inducing influence of the pods.

Supplying [^{3}H]-ZR or -Z (ring-labeled) to explants with young pods through the transpiration stream, much like cytokinin coming from the roots, indicates that (1) the pods do not compete with the leaves for the flux of [^{3}H]-cytokinin and (2) the cytokinin is not redistributed from the foliage to the pods. In fact, only very small amounts enter the seed coats, and it appears that very little or none of this [^{3}H]-cytokinin is transferred from the seed coats to the embryos. In contrast, [^{3}H]-adenosine, a major cytokinin metabolite, readily moves from the seed coat to the embryo. The earlier ideas about the flux of cytokinin into the seeds and fruit competition with the leaves for cytokinin (e.g., [22]) never did have a firm foundation in the literature [13], but direct evidence now indicates they are incorrect.

The [^{3}H]-cytokinin supplied through the xylem is metabolized very rapidly, and very little is left after 24 h. Following the pulse of [^{3}H]-ZR, [^{3}H]-ZNT is a dominant metabolite in the leaf blades, but substantial [^{3}H]-ZR is also present. Even 1 h after the pulse, very little [^{3}H]-ZR and no [^{3}H]-ZNT are left. After 24 h, [^{3}H]-AR (and -A) predominate in the soluble [^{3}H], but [^{3}H]-0-glucosides also occur. The metabolism of [^{3}H]-ZR does not seem to be substantially altered as senescence progresses or by depodding. The pattern of metabolism for [^{3}H]-ZR in the stem, petiole, and carpels differs from that in the blades. The rapid metabolism of ZR and Z is important, but the levels of Z and ZR in the leaves are probably determined more by the flux from the root system.

Studies on the optical isomers of DZ show that the unnatural form (+) is much more active than the natural (−) form, which indicates a need for using the purified isomers rather than racemic mixtures of the dihydroforms.

Acknowledgments. We gratefully acknowledge the support of NSF grants (PCM-7909434-01 and PCM-8302707 to LDN), a Fulbright Senior Research Scholarship (LDN) and Dr. J. Corse for supplying the isomers of dihydrozeatin. We wish to acknowledge the contribution of S. Singh, Zhang Ren, and Dr. Paula E. Jameson to the studies of cytokinin metabolism.

References

1. Garrison FR, Brinker AM, Noodén LD (1984) Plant Cell Physiol 25:213
2. Heindl JC, Carlson DR, Brun WA, Brenner ML (1982) Plant Physiol (Bethesda) 70:1619
3. Leopold AC, Niedergang-Kamien E, Janick J (1959) Plant Physiol (Bethesda) 34:570
4. Leopold AC (1961) Science 134:1727
5. Letham DS (1978) Cytokinins. In: Letham DS, Goodwin PB, Higgins TJV (eds) Phytohormones and related compounds – a comprehensive treatise, vol 1. Elsevier, Amsterdam
6. Letham DS, Palni LMS (1983) Annu Rev Plant Physiol 34:163
7. Lindoo SJ, Noodén LD (1976) Bot Gaz 137:218
8. Lindoo SJ, Noodén LD (1978) Plant Cell Physiol 19:997
9. Matsubara S, Shiojiri S, Fujii T, Ogawa N, Imamura K, Yamagishi K, Koshimizu K (1977) Phytochemistry (Oxf) 16:933
10. Neumann PM, Noodén LD (1983) J Plant Nutr 6:735
11. Neumann PM, Noodén LD (1984) Physiol Plant 60:166
12. Neumann PM, Tucker AT, Noodén LD (1983) Plant Physiol (Bethesda) 72:182
13. Noodén LD (1980) In: Thimann KV (ed) Senescence in plants. CRC, Boca Raton, Florida, p 219
14. Noodén LD (1984) Physiol Plant 62:273
15. Noodén LD (1985) In: Shibles R (ed) Proc world soybean res conf, vol 3. Westview, Boulder, Colerado
16. Noodén LD, Kahanak GM, Okatan Y (1979) Science 206:841
17. Noodén LD, Leopold AC (1978) In: Letham DS, Goodwin PB, Higgins TJ (eds) Phytohormones and related compounds – a comprehensive treatise, vol 2. Elsevier/North-Holland Biomedical, Amsterdam, p 329
18. Noodén LD, Letham DS (1984) J Plant Growth Regul 2:265
19. Noodén LD, Thompson JE (1985) In: Finch CE, Schneider EL (eds) Handbook of the biology of aging, 2nd ed. Van Nostrand Reinhold, New York, p 105
20. Shibles R, Anderson IC, Gibson AH (1975) In: Evans LT (ed) Crop physiology, some case Histories. Cambridge University Press, Cambridge
21. van Staden J, Davey JE (1979) Plant Cell Environ 2:93
22. Woolhouse HW (1982) In: Meudt WJ (ed) Strategies of plant reproduction. Littlefield-Adams, Totawa, New Jersey, p 201

Epigenetic and Genetic Factors Regulating the Cytokinin Requirement of Cultured Cells

F. MEINS, Jr. and C.E. HANSEN [1]

1 Introduction

It is believed that cell-division factors, such as the cytokinins, play an important role in regulating plant growth and development [17]. In a classical series of experiments, F. Skoog and his associates showed that the continuous proliferation of tobacco-pith tissues on an otherwise complete culture medium requires an endogenous source of both auxin and cytokinin [38]. Moreover, the relative abundance of auxin and cytokinin in the culture medium also determined the types of organogenesis of the tissues. When the concentrations of cytokinin are high relative to auxin, shoots form; when the concentrations of cytokinin are low relative to auxin, roots form; and at intermediate concentrations the tissues grow as unorganized callus.

Once established, root meristems cultured in isolation continue to form roots; whereas, shoot meristems continue to form shoots [36, 39, 41]. This shows that hormone treatment induces stable phenotypic changes that persist in the absence of the agent that originally induced the change [24]. These stable changes are of particular importance because they allow plant structures to "remember" their past and provide the basis for progressive determination.

Our work focuses on the relationship between cytokinin action and stable changes in development. As a model system we are studying variation in the cytokinin requirement of cultured, totipotent tobacco cells. This system has the particular advantage that cells exhibiting different heritable phenotypes can be assayed for their genetic integrity and developmental potential by regenerating plants from the cells and analyzing the plants in breeding experiments.

In this chapter, we review evidence that cell-heritable states of cytokinin requirement result from epigenetic changes, show that these states depend on specific genes, and provide evidence that the genes have an oncogenic function.

2 Epigenetic Changes in Cytokinin Requirement

Pith parenchyma tissue of Havana 425 tobacco sometimes loses its requirement for cytokinin for continuous growth in culture. This phenomenon is known as cytokinin habituation [20]. Cloning experiments show that the pith tissue is composed of two

1 Friedrich-Miescher-Institut, Postfach 2543, 4002 Basel, Switzerland

Plant Growth Substances 1985
(ed. by M. Bopp)

types of cells [30]: Approximately 70% of the cells exhibit the *inducible* phenotype. When pith explants are incubated on medium containing auxin and cytokinin at the standard temperature, 25 °C, or on auxin-containing medium at 35 °C, the inducible, cytokinin-requiring (C^-) cells rapidly shift to the cytokinin-autotrophic (C^+), i.e. habituated, state [29]. The remaining 30% of the cells, which exhibit the *non-inducible* phenotype, remain C^- under inductive conditions. Both the C^- and C^+ phenotypes persist when the cells are cloned, indicating that they are inherited by individual cells.

In principle, the cell-heritable shift of C^- cells to the C^+ state could result from either rare, random genetic mutations or from epigenetic (i.e. developmental) changes that do not permanently alter the genetic constitution of the cells. Epigenetic changes differ from classical mutations in several important ways [21]. First, the heritable cellular alteration is directed, i.e. it occurs at high rates, generally $> 10^{-3}$ per cell generation, under inductive conditions. Second, because the genome has not been permanently altered, the change is potentially reversible. Third, the range of phenotypes generated is limited by the genetic potential of the cell. Finally, by definition epigenetic changes are not transmitted meiotically.

Epigenetic changes can be distinguished from classical mutations by regenerating plants from cloned lines of cells exhibiting the phenotype and comparing the phenotypes of tissues cultured from the regenerates and from their sexual progeny. Epigenetic changes revert at some point during regeneration or meiosis, whereas genetic mutations should persist. In the case of cytokinin habituation, inducible pith cells rapidly shift to the C^+ phenotype under inductive conditions at rates $> 4 \times 10^{-3}$ per cell generation [29] and the capacity for habituation depends on both the physiological and developmental state of the cell [29, 31]. Plants can be regularly regenerated from C^+ cells and the pith tissue from these plants exhibits the C^- inducible phenotype typical of pith tissue from seed-grown plants [8]. Moreover, the rate of plant regeneration (and, hence, reversion) is high, $> 4 \times 10^{-3}$ per cell generation [25], indicating that habituation is a directed, reversible process. Thus, cytokinin habituation satisfies the major criteria for epigenetic changes. It appears that the C^+ and C^- states of pith are alternative, heritable, developmental states that are stable, but not permanent.

Evidence that a process similar to cytokinin habituation occurs during development has come from experiments in which the cytokinin requirement of tissues from different parts of the tobacco plant was compared [28]. Tissues cultured from the leaf lamina exhibit the C^- non-inducible phenotype; whereas tissues cultured from the cortex of the stem exhibit the C^+ phenotype. Both phenotypes persist when the two types of cells are cloned, indicating that the leaf and cortex-specific states of cytokinin requirement are inherited by individual cells.

The developmental potential of the two cell types was assayed by regenerating plants from cloned lines of the cells [22]. The results, summarized in Fig. 1, show that regardless of the clone used as a source of the regenerated plants, leaf-derived tissues exhibited the C^- phenotype and cortex-derived tissues exhibited the C^+ phenotype expected for the comparable tissues cultured from seed-grown plants. Evidently, the leaf and cortex states of cytokinin requirement are "reset" at some point during plant regeneration. Resetting is a directed process that occurs rapidly relative to classical mutations [25] and involves heritable changes that regularly arise in development. Therefore, the shift of cells between the leaf and cortex states, like cytokinin habituation of cultured pith cells, results from epigenetic changes.

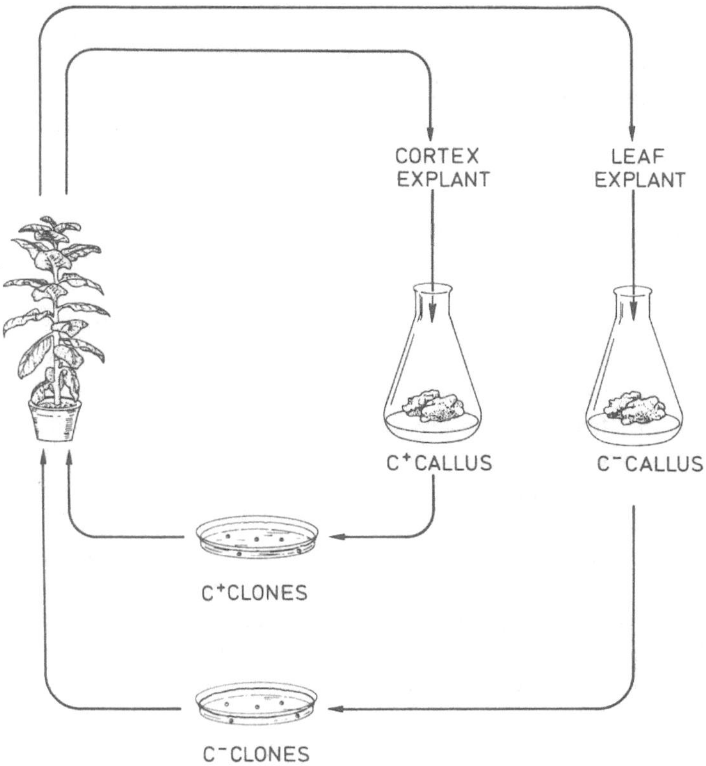

Fig. 1. The plant regeneration experiment demonstrating epigenetic changes in the resetting of the cortex and leaf cytokinin phenotypes [32]

3 A Positive-Feedback Model

The results summarized above pose the fundamental problem of how genetically equivalent, totipotent cells can inherit different phenotypes. We have proposed a positive-feedback hypothesis for cytokinin habituation that can account for both stability and reversibility of the habituated state [23]. This model specifies that cytokinins, or cell-division factors with similar properties, induce their own synthesis. Provided that some mechanism exists for degrading cytokinins, the biosynthetic pathway could exist in two states: an "off" state in which cytokinin is not produced; and an "on" state in which cytokinin is produced and reaches a steady-state concentration. If this biosynthetic system persists in dividing cells, then the different stages of cytokinin production can be transmitted to daughter cells. Thus, cells of the same genotype could express different phenotypes depending on their history.

According to the model, the system is poised to produce cytokinin, but can only do so when the cytokinin concentration reaches a critical threshold. We have found that cytokinins, such as kinetin, can induce cytokinin habituation of pith explants. The dose response shows, moreover, that this inductive effect exhibits a sharp threshold [29]. Once the putative-feedback loop has been established, breaking the loop momentarily by interrupting cytokinin production should induce C^+ cells to shift to the C^- state. Cloned lines of C^+ cells incubated at 16 °C or on culture medium containing lower than standard concentrations of auxin exhibit the C^- phenotype and appear to be blocked in cytokinin production [6, 9]. When these C^+ tissues are incubated under non-permissive conditions on concentrations of cytokinin below the threshold required for habituation and then shifted to permissive conditions, cells return to the C^- state. On the other hand, in control experiments, in which the tissues are incubated under non-permissive conditions with concentrations above the threshhold, the C^+ tissues continue to exhibit the C^+ phenotype, as predicted by the model [6, 23].

The evidence for this model at the physiological level of cytokinin biosynthesis and cytokinin content is conflicting. Studies with labelled precursors clearly show that C^+ tissues of tobacco can produce a variety of cytokinins [11, 14, 34]. What is still unclear is the relationship between the capacity for cytokinin autotrophic growth and cytokinin content. There are reports that cytokinins are present in C^+, but not C^- tissues [13, 14]; and others report that cytokinins are not detectable in either cell type [15, 42]. Interpretation of the cytokinin content data is further complicated by the observation that auxin added to the medium to support growth can also block cytokinin accumulation [15]. Thus, at present the physiological data do not argue strongly either for or against the working hypothesis. The critical prediction of the positive-feedback model is that cytokinins acting directly or indirectly should induce cytokinin production. Tests of this prediction are in progress.

4 Genetic Factors: The Habituated Leaf (Hl) Gene

Although positive-feedback regulation provides a plausible mechanism for cytokinin habituation, there is no evidence that it is involved in stabilizing different states of competence for habituation. We have been exploring the possibility that epigenetic changes in these states have a genetic basis.

The first hints that specific genes influence the cytokinin requirement of cultured tobacco cells came from experiments with cloned lines of C^+ cortex and C^- leaf cells from one specific plant [27]. When plants were regenerated from these lines and the leaf tissues assayed, it was found that the leaf tissue derived from cortex cells unexpectedly exhibited the C^+ phenotype typical of cortex cells. This *habituated leaf* (Hl) trait segregated when the regenerated plants were selfed. Breeding experiments were consistent with the hypothesis that Hl is inherited as a single, partially dominant, Mendelian factor.

Although there are no readily apparent differences in the morphology or development of homozygous Hl/Hl plants and wild-type plants, pith and leaf tissues of Hl/Hl plants consistently exhibit the C^+ phenotype in culture. Thus, the Hl allele appears to convert both inducible and non-inducible pith cells as well as non-inducible leaf cells to the constitutive C^+ phenotype characteristic of cortex cells. This suggests that epigenetic changes in habituation affect expression of the Hl gene or that this gene regulates resetting of epigenetic states of habituation during plant regeneration.

Direct evidence that the Hl trait can arise in culture was obtained by selecting for C^+ variants from cloned lines of non-inducible C^- pith and leaf cells [22, 26]. Since C^- cells cannot proliferate on cytokinin-free medium, we reasoned that C^+ variants could be enriched by transferring clones of non-inducible cells sequentially on media containing 10% of the cytokinin content used in the previous transfer. Using this procedure, C^+ sublines were obtained after only 2-3 subculture cycles.

In studies with lines derived from leaf cells, plants were regenerated from 19 subclones of the C^+ tissue and from 23 subclones of the original cloned line of C^- cells maintained on medium containing the standard cytokinin concentration. Leaf tissues cultured from plants regenerated from the C^- subclones behaved normally, i.e. they exhibited the C^- phenotype. In contrast, leaf tissues cultured from plants regenerated from the C^+ subclones exhibited the habituated-leaf phenotype. Breeding experiments carried out with the regenerated plants showed that this new habituated-leaf trait, designated Hl-2, was inherited as expected for a single, dominant, Mendelian factor [22]. Therefore, two types of cytokinin habituation can occur in culture: an epigenetic form, which is reset during plant regeneration, and a genetic form, which can be transmitted meiotically.

At present the relationship between these two forms of habituation is not known. Our working hypothesis is that conservative genetic changes at the Hl-2 locus might play a role in resetting of the cortex, pith, and leaf phenotypes. Indirect estimates of the rate of Hl-2 variation in culture, which give values of ca. 6×10^{-3} per cell generation [22], suggest that genetic variation could account for the high rate of epigenetic changes observed in phenotypic resetting.

5 Oncogenic Functions of the Hl Gene

There are striking formal similarities between habituation and tumor formation in crown gall disease [20]. The tumor phenotype, once established, persists when cells are cloned, indicating that crown-gall transformation, like habituation, results in alterations in cellular heredity. Conversion of normal cells to fully autonomous tumor cells involved a change from the hormone-requiring to the hormone-autotrophic state. Finally, certain tissues habituated for both auxin and cytokinin, like crown gall tissues, form tumors when grafted on the host plant.

In the case of crown-gall transformation, a portion of the tumor-inducing (ti) plasmid, the T-DNA, is integrated into the genome of the plant cell [5]. Neoplastic growth of the transformants requires expression of gene 4 of the T-DNA, which codes for a key enzyme in cytokinin biosynthesis, and genes 1 and 2 of the T-DNA, which code for two enzymes in auxin biosynthesis [1, 3, 37, 40].

The fact that the Ti-plasmid gene 4 is required for cytokinin-autotrophic growth of crown gall cells [33] suggested that the cellular gene, Hl, might have similar functions. To test this hypothesis, we transformed wild-type hl/hl and Hl/Hl tobacco cells with two nopaline-type pTiT37 plasmids: (1) *cyt*$^+$, the wild-type plasmid with a functional gene 4; and (2) *cyt*$^-$, a gene 4-defective derivative of *cyt*$^+$ obtained by inserting the Tn5 transposon into the BamHl fragment 14a in gene 4 [18]. Plants were inoculated with strains of *Agrobacterium tumefaciens* containing the *cyt*$^+$ and *cyt*$^-$ plasmids, and cloned lines isolated from tissues cultured from the inoculation sites by our standard methods [15].

Clones of Hl/Hl *cyt*⁻ and hl/hl *cyt*⁻ transformants were scored for the tumor phenotype using three commonly accepted criteria [33]: (1) capacity to grow on hormone-free medium; (2) inhibition of plant regeneration; and (3) tumorigenicity in grafting tests. The results obtained with these tests are summarized in Table 1. Wild-type hl/hl cells transformed with the virulent cyt^{+} plasmid can proliferate when cultured on hormone-free medium, whereas hl/hl cyt^{-} cells require cytokinin, but not auxin for proliferation in culture. Clones of Hl/Hl cyt^{-} transformants, like the hl/hl cyt^{+} transformants, did not require hormones for growth. The presence of a functional gene 4 blocks the regeneration of plants from transformed hl/hl cells [7]. We found that whereas hl/hl cyt^{-} cells could form plants, Hl/Hl cyt^{-} cells could not. Finally, tumor tissues form tumors when grafted onto the host plant. Clones of hl/hl cyt^{-} were only slightly tumorigenic; five grafts formed very small tumors, and only two grafts formed tumors comparable in size to hl/hl cyt^{+} clones. In contrast, the Hl/Hl cyt^{-} clones were highly tumorigenic; 95% of the grafts formed large tumors.

The findings summarized here show that three characteristics of autonomous crown gall cells – hormone autotrophy, inability to regenerate plants and tumorigenicity in grafting tests – are restored in combinations of Hl/Hl cells with the cyt^{-} plasmid. This provides direct evidence that the Hl gene has oncogenic functions similar to the *cyt* gene in the T-DNA. Braun [10] proposed that the neoplastic growth of plants as well as animals depends on the expression of cellular genes of the host concerned with production of growth-promoting substances. The existence of these oncogenic genes in plants is supported by the observations that transformation can occur in the abscence of infectious agents [4, 20]. The significance of our findings is that a specific gene, Hl, has been identified with this function.

Table 1. Comparison of the neoplastic properties of Hl/Hl and hl/hl cells transformed with the cyt^{-} Ti plasmid

Criterion for tumor phenotype	Number of cultures scored		Number of cultures exhibiting the tumor phenotype	
	hl/hl cyt^{-}	Hl/Hl cyt^{-}	hl/hl cyt^{-}	Hl/Hl cyt^{-}
Hormone-autotrophic growth:[a]	9	69	0	69
Inhibition of plant regeneration:[b]	15	35	0	35
Tumorigenicity in grafting tests:[c]	38	41	7[d]	39

[a] Based on growth when cultured under standard conditions [15] medium with and without kinetin and α-naphthalene acetic acid
[b] Tissues scored after incubation under the shoot-inducing conditions recommended by Binns [7]
[c] Tissues grafted on the side of Havana 425 tobacco plants as described by Meins [19]
[d] Five of the seven grafts formed small tumors

6 Conclusions

Studies of variation in the requirement of tobacco cells for cytokinins provide direct evidence that states of determination arising in ontogeny can be inherited by individual cells. These states, although stable for many cell generations, are not permanent. Cells determined under one set of conditions can nevertheless remain totipotent.

A positive-feedback model has been proposed, which can account for both the heritability and potential reversibility of cytokinin habituation. There is also evidence that auxin, ethylene, and gibberellins can induce their own synthesis in certain tissues [2, 12, 16, 35], suggesting that positive-feedback regulation may be a more general phenomenon. The fact that these hormones can also regulate morphogenesis suggests that these feedback loops could also play a causal role in plant development.

A specific gene has been identified that is involved in epigenetic changes in cytokinin requirement. There are hints that the Hl gene can undergo rapid genetic changes, suggesting that directed alterations in the Hl locus might provide a genetic mechanism for phenotypic resetting during plant regeneration. Although the functions of Hl in the normal plant are not known, complementation studies with *cyt*⁻ Ti plasmids show that Hl can serve an oncogenic function similar to gene 4 in the T-DNA.

References

1. Akiyoshi DE, Klee H, Amasino RM, Nester EW, Gordon MP (1984) Proc Natl Acad Sci USA 81:5994
2. Atzorn R, Weiler EW (1983) Planta (Berl) 159:289
3. Barry GF, Rogers SG, Fraley RT, Brand L (1984) Proc Natl Acad Sci USA 81:4776
4. Bayer MH (1982) In: Kahl G, Schell JS (eds) Molecular biology of plant tumors. Academic Press, New York, p 33
5. Bevan MW, Chilton M-D (1982) Annu Rev Genet 16:357
6. Binns AN (1979) Thesis, Princeton University, Princeton
7. Binns AN (1983) Planta (Berl) 158:272
8. Binns A, Meins F Jr (1973) Proc Natl Acad Sci USA 70:2660
9. Binns AN, Meins F Jr (1977) Planta (Berl) 145:365
10. Braun AC (1961) Harvey Lect 56:191
11. Burrows WJ, Fuell KJ (1981) In: Guern J, Peaud-Lenoel C (eds) Metabolism and molecular activities of cytokinins. Springer, Berlin Heidelberg New York, p 44
12. Cheng T-Y (1972) Plant Physiol (Bethesda) 50:723
13. Dyson WH, Hall RH (1972) Plant Physiol (Bethesda) 59:45
14. Einset JW, Skoog F (1973) Proc Natl Acad Sci USA 70:658
15. Hansen CE, Meins F Jr, Milani A (1985) Differentiation 29:1
16. Kende H, Baumgartner B (1974) Planta (Berl) 116:279
17. Letham DS, Goodwin PB, Higgins TJV (eds) (1978) Elsevier/North Holland, Amsterdam
18. Matzke AJM, Chilton M-D (1981) J Mol Appl Genet 1:39
19. Meins F Jr (1973) Differentiation 1:21
20. Meins F Jr (1982) In: Kahl G, Schell JS (eds) Molecular biology of plant tumors. Academic Press, New York, p 3
21. Meins F Jr (1983) Annu Rev Plant Physiol 34:327
22. Meins F Jr (1986) UCLA Symp Mol Cell Biol (in press)
23. Meins F Jr, Binns AN (1978) Soc Dev Biol Symp 36:185
24. Meins F Jr, Binns AN (1979) Bioscience 29:221

25. Meins F Jr, Binns AN (1982) Differentiation 23:10
26. Meins F Jr, Foster R (1985) Dev Biol 108:1
27. Meins F Jr, Foster R, Lutz JD (1983) Dev Genet 4:129
28. Msins F Jr, Lutz J (1979) Differentiation 15:1
29. Meins F Jr, Lutz J (1980) Planta (Berl) 149:402
30. Meins F Jr, Lutz J, Binns AN (1980) Differentiation 16:71
31. Meins F Jr, Lutz J, Foster R (1980) Planta (Berl) 150:264
32. Meins F Jr, Wenzler H (1986) Soc Exp Biol Symp (in press)
33. Nester EW, Gordon MP, Amasino RM, Yanofsky MF (1984) Annu Rev Plant Physiol 35:387
34. Nishinari N, Syono K (1980) Z Pflanzenphysiol 99:383
35. Nishinari N, Yamaki T (1976) Bot Mag Tokyo 89:73
36. Reinhard E (1954) Z Bot 42:353
37. Schröder G, Waffenschmidt S, Weiler EW, Schröder J (1984) Eur J Biochem 138:387
38. Skoog F, Miller CO (1957) Symp Soc Exp Biol 11:118
39. Smith RH, Murashige T (1970) Am J Bot 57:562
40. Thomashow LS, Reeves S, Thomashow MF (1984) Proc Natl Acad Sci USA 81:5071
41. Torrey JG (1954) Plant Physiol (Bethesda) 29:279
42. Yokota T, Takahashi N (1981) Int Assoc Plant Tissue Culture Newslett 34:2

Apical Dominance and Correlations by Hormones

J.R. HILLMAN [1]

1 Introduction

Correlative control systems provide most of the rationale and impetus for research on plant growth substances, elevating the research beyond natural product chemistry. The phenomenon of apical dominance is widely assumed to be a classical and exemplary example of hormonally integrated growth in plants, principally brought about by indole-3-acetic acid (IAA). This may be a correct assumption, but the available evidence in favour of the hypotheses based on the dogma is equivocal. Some of the major deficiencies encountered in devising working hypotheses to explain merely one aspect of apical dominance - the correlative inhibition of lateral buds by the apical region of the shoot – are the subject of this chapter. Several general reviews of correlative inhibition have been published [1, 3, 9, 16, 22–24].

2 Apical Dominance and Modular Growth

A modular growth strategy typifies higher plants: constructional units arise repeatedly from long-lived, vegetative, apical meristems. To some extent the modules are autonomous, but where each or most of the modules have lateral meristems, some form of positional signalling must occur, permitting the repair of damaged or replacement of lost modules, and the exploitation of the environment. Thus, potential sites of growth tend not to be activated, the suppression or "dormancy" providing a mechanism to ensure the availability of reserve modules. That the modules retain their totipotency in the vegetative condition, yet sustain constant phenotypic characters despite open-ended development, may point towards a relatively loose form of growth correlation. Hardwick [5] has reviewed elegantly these physiological consequences of modular growth in plants, and notes that there is cooperation and not competition between modules: the inclusive fitness of the genome appears to be paramount, and spatial organisation of the phenotype is advantageous in this respect.

1 Botany Department, The University of Glasgow, Glasgow G12 8QQ, Great Britain

Plant Growth Substances 1985
(ed. by M. Bopp)

3 Signalling Systems in Correlative Inhibition

A long-standing awareness of the range of morphogenetic signals in plants does not diminish the fact that most research has been devoted to the known plant growth substances and how they relate to environmental signals and regulate receptive, sensitive cells. In studies on correlative inhibition all of the plant growth substances, individually and in various combinations, have been reported to affect bud growth. Do such observations really indicate a correlative role for endogenous growth substances? Is positional signalling required? In an attempt to answer these questions, the case of the dwarf French bean *Phaseolus vulgaris* L. cv. Canadian Wonder may be considered (see also [9, 10]).

3.1 Phaseolus vulgaris

Lateral bud growth in plants up to about 30 days after sowing is suppressed by the presence of an intact shoot apical region. Photographic analyses of the buds in the axils of the primary leaves and first trifoliate leaf reveal that the suppression is not absolute and extension growth of 1-5 $\mu m\ h^{-1}$ can be detected.

Elimination of dominance can be achieved by various means (surgical removal, isolation or restriction of the apical region; modification of the environment or cultural system; application of synthetic and naturally occurring growth regulators; induction of the sexual reproductive condition; gravimorphic treatments, etc.) leading to enhanced growth of the lower lateral (usually axillary) buds. The source of bud inhibition appears to be the rapidly expanding trifoliate leaves, and removal or physical constriction of these leaves are equally effective bud-activation treatments. A root system does not seem to be essential for the maintenance of correlative inhibition.

When assessed as bud extension growth, release of suppressed laterals in *P. vulgaris* is not instantaneous. The lag period depends on the position of the bud and is noted to vary between 0.5 to 6 h. Cell extension in the bud internode accounts for all of the extension growth in the first 24 h after the removal of inhibition. When, however, the criteria of bud activation do not rely on bud extension measurements, then there is the distinct possibility that the signalling between the source of inhibition and the buds is exceptionally fast. In the nodes of *Bidens*, for example, an apparent accumulation of K^+ takes place within 5 min of decapitation [14]. Most studies indicate that the transmission of the signal requires living continuity and is presumably symplastic.

3.2 Plant Growth Substances as Correlative Agents

Evaluation of the role of plant growth substances in correlative phenomena such as lateral bud inhibition is difficult at the present state of knowledge. The experimental approaches adopted have been logical, based on observations of both exogenous and endogenous compounds [13].

3.2.1 Auxins

Most attention has been given to indole-3-acetic acid (IAA) and the significance of the observation based on the movement of radioactivity from radiolabelled IAA that there is a long-range continuous flux of "auxin" from the shoot apex to the root tip. Although in coleoptiles and roots of *Zea* there is substantial and rapid metabolism of the applied IAA [19–21], the polarly transported moiety appears to be pure IAA. Combined gas chromatography-mass spectrometry and bioassays show that young leaves of *P. vulgaris* are a source of auxin/IAA [15, 36]. [^{14}C] from both [1-^{14}C]-IAA and [2-^{14}C]-IAA move out of young leaves, whereas [^{14}C] from [U-^{14}C]-sucrose, [2-^{14}C]-ABA and 8-methylene-[^{14}C]-GA_3 do not (Table 1). The fact that the three latter radiochemicals move out of the young leaves is consistent with the observation from applications to older, mature primary leaves that there is a strong flow of assimilates and growth factors into growing trifoliate leaves, and indicates a degree of specificity in the translocation of [^{14}C] from [^{14}C]-IAA [34]. Perhaps, if the findings of Nonhebel et al. [21] are applicable to *P. vulgaris*, apically-derived IAA can reach lateral buds.

Further evidence supporting a fundamental role for endogenous IAA transport in correlative inhibition comes from the fact that inhibitors of exogenous IAA transport effectively release lateral buds on intact plants basipetal to the point of application, and this occurs in a wide range of species, including *P. vulgaris*. Two of these inhibitors, triiodobenzoic acid and morphactin, however, induce abscission of the young leaves and their effects on lateral bud growth could be indirect [35].

Researchers such as Snow [29], Hall and Hillman [4], and Morris [17] have questioned whether the path of IAA transport matches that of the path of the inhibitory influence from the shoot apex. Until quantitative analyses of the apically-derived IAA have been carried out, and its pathway elucidated, the role of the IAA remains controversial. Most recent reviews of correlative inhibition (see [9]) indicate that any role of apically-synthesized IAA must be indirect. Application of IAA to inhibited buds stimulates bud release (see Sect. 3.2.5) and measurements of endogenous IAA 24 h after decapitation of the shoot reveal a rise in levels in the released buds. IAA is not, therefore, associated solely with bud inhibition. Such observations go against, but do not disprove the pioneering proposals [25, 27, 31] in support of a "direct" theory of auxin.

Went [33] laid the basis of what is often referred to as the "nutrient-diversion" theory, in which endogenous or exogenous IAA creates a flow of growth factors to the point of auxin production or application, thereby depleting resources from the laterals. This is an attractive theory, and leads on to the concept of hormone-directed metabolite transport [23].

The classic experimental system of exogenous IAA or synthetic auxin replacing the apex is not strictly valid. Naylor [18], using *Tradescantia*, pointed out that naphthalene acetic acid (NAA) does not completely mimic the effects of the apex because it stimulates DNA synthesis in the region of the parent stem adjacent to the lateral buds, and there are also nuclear changes which take place in the presence of NAA and not in buds of plants with an intact shoot apex. Jacobs [12] and White et al. [36] pointed out the vast discrepancy between natural auxin production by shoot tips and

Table 1. Distribution of radioactivity, assessed by liquid scintillation spectrometry, in 14-day-old plants of *Phaseolus vulgaris* 24 h after agar-block donation of [^{14}C]-sucrose, [1-^{14}C]-IAA, [2-^{14}C]-IAA, [^{14}C]-ABA, and [^{14}C]-GA_3 to an abraded young trifoliate leaf[a]

Part A

$$\text{Recovery} = \frac{\text{Radioactivity in plant + donor block}}{\text{Radioactivity in unused donor block}} \times 100$$

$$\text{Uptake} = \frac{\text{Radioactivity in plant}}{\text{Radioactivity in plant + donor block}} \times 100$$

$$\text{Other figures} = \frac{\text{Radioactivity in plant segment}}{\text{Radioactivity in plant}} \times 100$$

Growth substance	[^{14}C]-sucrose	[1-^{14}C]-IAA	[2-^{14}C]-IAA	[^{14}C]-ABA	[^{14}C]-GA_3
No. of plants	15	5	8	13	5
Recovery	51.6 ± 6.2	85.6	39.6 ± 6.6	85.2 ± 5.1	40.6 ± 1.6
Uptake	93.8 ± 4.2	99.0	96.9 ± 0.5	98.1 ± 0.4	69.0 ± 3.6
Donor leaflet	98.7 ± 0.5	93.3	95.5 ± 0.9	97.2 ± 1.1	97.3 ± 1.1
Lateral leaflets	0.5 ± 0.2	0.6	0.0 ± 0.0	2.2 ± 0.9	2.6 ± 1.1
Petiole	0.5 ± 0.3	1.8	0.8 ± 0.4	0.1 ± 0.1	0.0 ± 0.0
DONOR LEAF	99.7 ± 0.1	95.7	96.4 ± 0.7	99.5 ± 0.3	99.9 ± 0.1
UPPER SHOOT	0.0 ± 0.0	0.8	0.2 ± 0.1	0.0 ± 0.0	0.0 ± 0.0
Primary midrib + peticle		0.0	0.1 ± 0.0		
Lower stem		3.0	1.8 ± 0.5		
Primary leaf axillary buds		0.2	0.0 ± 0.0		
LOWER SHOOT	0.2 ± 0.1	3.2	1.9 ± 0.5	0.3 ± 0.2	0.0 ± 0.0
ROOTS	0.1 ± 0.0	0.3	1.4 ± 0.3	0.2 ± 0.1	0.0 ± 0.1
Transported from donor leaf	0.3 ± 0.1	4.3	3.6 ± 0.7	0.5 ± 0.3	0.1 ± 0.1
Transported from donor area	23.8 ± 5.1	35.4	19.6 ± 3.7	33.8 ± 5.1	29.3 ± 2.5

Part B

Growth substance	[^{14}C]-Sucrose	[1-^{14}C]-IAA	[2-^{14}C]-IAA	[^{14}C]-ABA	[^{14}C]-GA_3
Concentration in donor block (μmol dm^{-3})	0.25	1.5	1.5	5	50
[^{14}C] in donor block (bq)	167.1	105.6	165.0	128.1	216.0
[^{14}C] recovered (Bq)	84.0	90.3	81.1	108.1	87.8
[^{14}C] transported from donor leaf (Bq)	0.6	3.7	3.0	0.6	0.0
Length of donor leaflet at application (mm)	40.4	46.4	40.4	39.7	37.2
Length of donor leaflet at harvest (mm)	50.3	52.6	49.1	48.5	56.4

[a] Radio thin-layer chromatography of extracts showed extensive metabolism of all of the applied compounds

the amounts of exogenous auxin required to sustain bud inhibition. Moreover, the region of IAA application becomes a metabolic sink affecting the importation of [^{14}C]-labelled photosynthetic assimilates into the buds [4].

3.2.2 *Cytokinins*

A wide range of exogenous cytokinins effectively release inhibited buds of many species and the release may be sustained by supplementary treatment with other growth substances (see [23]). Their role in correlative inhibition is unknown, however, for there are no thorough analyses relating to the control of bud growth. While their effectiveness is undisputed, the fact that compounds which show anti-cytokinin activity also promote bud growth causes problems of interpretation [26]. Nevertheless, cytokinins offer one of the most promising lines of research in the phenomenon. Their effects on buds should be carefully compared with changes brought about by other forms of dominance removal. Another important objective should be the characterisation of endogenous cytokinins and the sensitivity of regions of the bud to these compounds.

3.2.3 *Gibberellins*

Despite the tendency to dismiss gibberellins in the correlative control of bud growth (see [9, 22, 23], their dramatic effects on internodal extension of released buds may highlight an important role, perhaps in the immediate post-release phase.

3.2.4 *Abscisic Acid (ABA)*

ABA is not the correlative inhibitor originally postulated by Snow [9]: low doses actually promote bud growth in *P. coccineus* [7] and in *P. vulgaris* the decline in ABA levels in the bud occurs after bud release. Under stress conditions, ABA could achieve inhibitory levels.

3.2.5 *Ethylene (Ethene)*

Ethylene appears to have an indirect role in correlative inhibition in shoots of *P. vulgaris* [10, 37]. Those treatments which affect the production, release, availability and effects of ethylene, modify bud growth. Physical constriction of the apical shoot releases buds and is associated with enhancement of the internal ethylene content (concentration in gases recovered under reduced pressure) and rate of ethylene emanation (release) in the treated region. The effects of physical constriction on bud growth can be duplicated with ethylene or ethephon treatment, but in this instance the rate of ethylene emanation was not affected. Yeang and Hillman [37] suggested that the onset of axillary bud growth is associated with the availability of freely diffusible (intercellular, non-bound) ethylene in the apical region of the shoot.

In the region of the lateral buds of *P. vulgaris* ethylene may have a controlling role in growth because aminoethoxyvinyl glycine (AVG) and Ag^+ inhibit bud outgrowth in decapitated plants; neither ethylene nor ethephon have an effect on axillary bud growth of intact plants, so normal correlative inhibition is not attributable to a lack of free ethylene.

Following decapitation there is a general decline in ethylene levels and emanation in nodal and internodal regions, the decline is not restricted to the nodal region. Unlike pea [2] replacement of the apex with IAA did not affect the ethylene relations of the nodal region of *P. vulgaris* yet dramatically affected emanation in the treated zone of the stem. Thus, the inhibitory effects of IAA cannot be explained in terms of IAA-induced ethylene in the node. Additionally, although direct applications of IAA to inhibited buds induces bud-internodal elongation, this release can be eliminated with AVG applied at levels which do not affect the release of buds following shoot decapitation. Hence, IAA-induced growth might be anomalous as might be the kinetin-induced bud outgrowth in *Avena* stem segments [6]. Inexplicably, nodes contain high levels (5–8 mm^3 dm^{-3}) of ethane and exhibit high rates of ethane emanation (0.2–0.7 mm^3 kg^{-1} h^{-1}).

Whilst ethylene might be important in bud growth per se, its signalling role is dubious. Transport of ethylene precursors is largely an unexplored area and the movement of ethylene in the plant is poor [38]. Its relationship with ethane is problematical.

3.2.6 Other Compounds and Interactions

A vast range of structurally unrelated compounds affect lateral bud growth, some possibly by affecting the apical region (e.g. some chemical pruning agents). Fatty acids, steroidal oestrogens [8] and polyamines [28] are strong candidates for involvement.

Throughout the literature on the physiology of correlative inhibition are references to the effects of various combinations of growth substances [9, 22, 23, 32], but the observations have limited value due to lack of parallel studies on the respective endogenous compounds.

Potato (*Solanum tuberosum* L. cv. Arran Pilot) shoots respond rapidly to reorientation, ethylene, and IAA. Neither gravity nor ethylene treatments modify the basipetal and lateral redistribution of exogenous [^{14}C]- and [^{3}H]-IAA (as assessed by agar-block asymmetrical donors, soluble-compound microautoradiography and microdonation techniques) before or during the period of physiological response [10]. If these results hold true for other dicotyledonous species, then they raise doubts about the significance of IAA as a correlative agent in dicotyledonous shoots.

4 Conclusions

The correlative signal(s) in apical dominance phenomena and the characteristics of the exact path of transport, rate of movement and primary site of action are presently conjectural. Greater attention needs to be paid to the source of the signal. Equally meritorious is a systematic metabolic profile of the changes taking place in the bud during and following removal of dominance; proteins, carbohydrates, lipids, growth substances, inorganic ions, and water utilisation require detailed microanalysis, preferably in parallel with fine-structural investigations. Despite the failure to demonstrate their involvement as hormones or activators, following the definitions of Starling [30] and Huxley [11], the plant growth substances are likely to be involved at a secondary level in correlative inhibition; application of radioimmunoassay and immunocytochemistry techniques are likely to play an important part in revealing their time role. The entire study demands work at the genomic level, establishing those genes which are activated or repressed in the process of bud release. A first approach could be to monitor the protein population by two-dimensional electrophoresis, and to construct a cDNA library and screen for recombinants that represent mRNAs which change in abundance in response to the removal of dominance by the many treatments which activate quiescent buds.

References

1. Audus LJ (1959) Correlations. J Linn Soc Bot 56:177
2. Blake TJ, Reid DM, Rood SB (1983) Physiol Plant 59:481
3. Brown CL, McAlpine RG, Kormanik PP (1967) Am J Bot 54:153
4. Hall SM, Hillman JR (1975) Planta (Berl) 123:137
5. Hardwick RC (1986) Philos Trans Roy Soc Lond B Biol Sci (in press)
6. Harrison MA, Kaufman PB (1982) Plant Physiol (Bethesda) 70:811
7. Hartung W, Steigerwald F (1977) Planta (Berl) 134:295
8. Hewitt S, Hillman JR, Knights BA (1980) New Phytol 85:329
9. Hillman JR (1984) In: Wilkins MB (ed) Advanced plant physiology. Pitman, London, p 127
10. Hillman JR, Yeang HY, Fairhurst VJ (1985) In: Roberts JA, Tucker GA (eds) Ethylene in plant development. Butterworths, London, p 213
11. Huxley JS (1935) Biol Rev 10:427
12. Jacobs WP (1959) Proc 9th Int Bot Congr 2:178
13. Knox JP (1982) Ph.D. Thesis, University of Wales
14. Kramer D, Desbiez M-O, Garrec JP, Thellier M, Fourcy A, Bossy JP (1980) J Exp Bot 31:771
15. McDougall J, Hillman JR (1980) Z Pflanzenphysiol 97:367
16. McIntyre GI (1977) In: Jennings DH (ed) Integration of activity in the higher plant. Exp Biol Symp, vol 31. Cambridge University Press, Cambridge, p 251
17. Morris DA (1977) Planta (Berl) 136:91
18. Naylor JM (1958) Can J Bot 36:221
19. Nonhebel HM, Crozier A, Hillman JR (1983) Physiol Plant 57:129
20. Nonhebel HM, Hillman JR, Crozier A, Wilkins MB (1985a) Planta (Berl) 164:105
21. Nonhebel HM, Hillman JR, Crozier A, Wilkins MB (1985b) J Exp Bot 162:99
22. Phillips IDJ (1969) In: Wilkins MB (ed) The physiology of plant growth and development. McGraw-Hill, London, p 165
23. Phillips IDJ (1975) Annu Rev Plant Physiol 26:341
24. Rubinstein B, Nagao MA (1976) Bot Rev 42:83
25. Skoog F (1939) Am J Bot 26:702
26. Skoog F, Abdul Ghani AKB (1980) In: Guern J, Peaud-Lenoël C (eds) Metabolism and molecular activities of cytokinins. Springer, Berlin Heidelberg New York, p 140
27. Skoog F, Thimann KV (1934) Proc Natl Acad Sci USA 20:480
28. Smith TA (1985) Annu Rev Plant Physiol 36:117
29. Snow R (1940) New Phytol 39:177
30. Starling EH (1905) Lancet 2:339

31. Thimann KV (1937) Am J Bot 24:407
32. Thimann KV (1959) Proc 9th Int Bot Congr 2:396
33. Went FW (1936) Biol Zentralbl 56:449
34. White JC (1973) Ph.D. Thesis, University of Glasgow
35. White JC, Hillman JR (1972) Planta (Berl) 107:257
36. White JC, Medlow GC, Hillman JR, Wilkins MB (1975) J Exp Bot 26:419
37. Yeang HY, Hillman JR (1984) Physiol Plant 60:275
38. Zeroni M, Jerie PH, Hall MA (1977) Planta (Berl) 134:119

Hormonal Control of Assimilate Movement and Compartmentation

T.H. THOMAS [1]

1 Introduction

Correlated growth and development of plant organs is generally regarded as being controlled by a balance of endogenous hormones and/or responsiveness of the tissues to the hormones present. For many years, several approaches haven been followed to investigate the possibility that hormones may be involved in the production and movement of assimilates to developing organs. These approaches include the collation of correlative evidence, pharmacological studies, manipulation of endogenous hormones by environment or genotype and hormone replacement studies following sink excision [68]. All these approaches have yielded information tending to support the concept of hormonal control of assimilate partitioning, but the general conclusion seems to be that the major rate-limiting steps arise from or occur in the sink tissues themselves. This idea is supported in that sink tissues, particularly fruits and seeds, are often the richest sources of plant hormones as demonstrated by the use of modern hormone analytical procedures [38]. However, detailed knowledge of the distribution of hormones in various tissues and their fluxes within the plant are still required before the exact role of hormones in assimilate synthesis and partitioning can be elucidated. Since the topic of assimilate control has received considerable attention recently this brief review will attempt to highlight some of the current evidence in favour of hormonal control, particularly in relation to sink activity.

2 Correlative Evidence for Hormonal Control

Recent reviews [38, 52] have identified the potential role for certain plant hormones in developing reproductive sinks. Generally, cytokinins and possibly gibberellins and auxins all seem to be involved in initiating sink activity and determining "priority demand" in different reproductive and vegetation sinks, e.g. tomato [41], wheat [72], apple [38], and vegetable storage roots [43]. Since growth rates of fruit often correlate with gibberellin concentration in the seeds [38] and high GA/IAA ratios favour the early growth of parthenocarpic fruit [81, 82] it would appear that gibberellins may

1 National Vegetable Research Station, Wellesbourne, Warwick CV35 9EF, Great Britain
New aadress: Long Ashton Research Station, Bristol BS18 9AF, Great Britain

Plant Growth Substances 1985
(ed. by M. Bopp)

be involved in the early establishment of sink strength. In relation to sink growth per se, no clear role has been identified for plant hormones, as explained for tomato fruit development [41], other than perhaps for ABA. Although roles have been suggested for the control of assimilate import into developing fruit, detailed studies of the changes in hormone levels have failed to relate them clearly to changes in fruit growth.

However, evidence is accumulating that ABA may be involved in the control of sugar import into seeds, fruits and vegetative storage organs following the establishment of sink activity. Thus, ABA application to ears of wheat [13] and barley [87] stimulated assimilate transfer from the flag leaf to the ear. Growth rate and endogenous ABA concentrations have been shown to be correlated in wheat, grape, apple, pea, and bean [6]. ABA has also been shown to enhance sucrose uptake into storage sinks such as soybean seeds [6], grape berries [16], maize kernels [34], and sugar beet tap root tissue [77]. Extensive studies on soybean have demonstrated that ABA concentrations in embryos of three different genetic lines correlated well with seed growth rates in vivo and with [^{14}C] sucrose uptake rates in vitro. The highest ABA concentration occurred in seed coats and cotyledons as compared with embryonic tissue, indicating possible ABA control of sucrose unloading into the seed coat apoplast. It was suggested that ABA present in cotyledons may also enhance sucrose uptake [6]. In further studies it was demonstrated that water-stressing increased both the endogenous ABA levels in vivo and sucrose uptake in vitro into soybean embryos, supporting the hypothesis that ABA stimulates sucrose movement in filling sinks.

3 Hormones and Source Activity

The processes occurring during the production and movement of assimilate in and from source leaves involve sucrose biosynthesis, partitioning of sucrose in photosynthetic cells, alteration of intra- and intercellular distribution of this assimilate, control of localised release of sucrose into the apoplast and the control of the active loading system at the sieve-element plasma membrane. In addition to inorganic phosphate and sucrose, plant hormones have been suggested as regulatory substances in the control of source photosynthesis by sinks [40]. Hormone movement across the chloroplast envelope and the boundary between the source and primary sinks together with their possible effects on transport processes, are all potentially important for assimilate production [52].

There is evidence that both GA and IAA biosynthesis occur in plastids [39, 78], though not exclusively so. Both IAA and ABA appear to be synthesised in the cytoplasm, ABA being metabolised and accumulated in chloroplast stroma in the light [35]. It has been suggested that ABA may couple stomatal conductance and photosynthetic carbon metabolism in leaves [9] though the results of Cornic and Miginiac [8] indicate lack of a direct effect of ABA on the photosynthetic reaction mechanism. The fact that restriction of ABA export from leaves increases stomatal resistance and reduces photosynthesis [79] suggests that non-removal of source leaf ABA by phloem may induce a negative feed-back control of source assimilate supply to sinks. On the other hand, the contribution that import of cytokinins and nutrient ions in the transpiration stream may have on chloroplast integrity and stomatal conductance cannot be discounted [52].

Regulation of sucrose biosynthesis in leaves is thought to occur by modulation of the supply of substrates derived from metabolites transported across the chloroplast envelope. The principal site for this regulation seems to be by cytoplasmic fructose -1.6 bisphosphatase and associated control pathways [71], but sucrose phosphate synthase (SPS) may also be important by controlling the balance between sucrose and starch biosynthesis [44]. This idea is supported by recent results [70], which demonstrate a rapid increase in SPS activity in *Lolium temulentum* leaves on illumination. This enzyme is assumed to play an important role in export since a temporal correlation between the appearance of its activity and the induction of exporting capacity in developing leaves has been demonstrated previously [30, 80]. Other enzymes involves in the regulation of sucrose concentrations in leaves are sucrose synthase (SS) and invertase.

There is little direct evidence for direct hormonal control of the activity of enzymes involved in the regulation of sucrose production or metabolism in source tissues [88]. Recent evidence has been presented [10] indicating that ABA, GA, and IAA may regulate the activities of SPS, SS, and invertase in *Phaseolus coccineus* source leaves and may also modify phloem loading of sucrose and leaf starch metabolism. Such effects are consistent with the observation that as individual leaves change from net importers to net exporters changes in hormonal complement may occur (see [10]). In particular, auxin concentrations have been demonstrated to reach their maxima during early stages of leaf expansion in tobacco [92], *Ricinus communis* L [2] and *Phaseolus vulgaris* [46] declining during and after leaf expansion ceases. Since the highest auxin concentrations occur around the time of peak acid invertase activity in bean leaves [60] and in view of the stimulatory effects of applied auxins on bean internode acid invertases [63], it has been suggested that acid invertases in expanding leaves of *P. vulgaris* may be hormonally regulated. In contrast to these observed correlations between hormones and leaf function, Blechschmidt (see [18]) has demonstrated that artificial conversion of mature leaves of C_4-plants to sinks by omitting light and CO_2, does not affect either auxin or ABA levels, indicating no effect of these hormones on the direction of translocation.

There is considerable evidence that following its formation in the mesophyll cytoplasm, sucrose is compartmented within the mesophyll cell itself and among the various cells situated within the source leaf. A rapid turnover of sucrose within the cytoplasm, indicative of a transport pool leading towards the phloem loading sites has been demonstrated [23, 24]. In addition, there is evidence for a dynamic vacuolar pool, which has been suggested to be a temporary site of storage and which may contribute sucrose for export in response to changes in the rate of sucrose export from the leaf [86].

The biochemical mechanisms and possible hormonal control involved in the sucrose compartmentation process have not been elucidated. It is clear from the data of Farrar and Farrar [21], using barley plants, that the rate of sucrose synthesis in leaves does not drive the rate of translocation. Since the leaf cell vacuoles store about half the sucrose produced, it seems likely that the vacuole may be a stronger competitor than the phloem for cytoplasmic sucrose. Net transport across the tonoplast only falls when sucrose concentrations in the vacuole are high, resulting in translocation of sucrose. It would seem that sugar storage in the leaf has priority over translocation, indicating that unloading in the sinks may be a more likely candidate as a control mechanism for regulating assimilate transport and, hence, for hormonal control.

4 Hormones and Phloem Loading

The movement of assimilates to the region of the sieve-element companion-cell complex of the phloem is generally thought to be via the symplastic continuity between the photosynthetic cells in the leaves.

Alternatively, sucrose could follow an intercellular apoplastic route involving the crossing of the cell membrane and cell walls. Most evidence favours a symplastic route, which may not involve any direct energy-dependent step. Such a route has been proposed for wheat, C_4 plants and legumes [19, 20, 22, 25, 26, 48].

The major step in phloem loading is generally considered to involve the transfer of sucrose from the vascular boundary into the sieve elements. The bulk of evidence indicates that movement of sucrose into the sieve element is by an active mechanism allowing loading against a concentration gradient from the source leaf apoplast. However, in some plants, particularly cucurbits, an entirely symplastic pathway seems possible [54], though somewhat controversial [32]. The way in which sugar moves into the apoplast prior to loading is not at all clear, though the calculated efflux rates for sucrose from the mesophyll in *Beta vulgaris* implies the operation of an active transfer mechanism rather than a passive efflux [15]. Since monovalent cations seem to be involved in this efflux process [15, 36, 69], a sucrose cation symport mechanism for sucrose efflux, which may be associated with the phloem loading process [28, 45], is indicated. This step could be a candidate for hormonal involvement.

The mechanism of phloem loading is fairly well resolved and has been reviewed recently [3]. The loading step is assumed to involve proton cotransport of sugars driven by a proton extrusion pump [42]. This active pump is probably a vectorial proton-translocating plasma membrane ATPase, and this has been demonstrated in the sieve element in *Ricinus* [4] and a number of other species [7] by cytochemical techniques. Thus the driving force for loading is the gradient of electro-chemical potential for protons, which is maintained by a proton-extruding ATPase located at the plasma membrane of the sieve elements and companion cells of the phloem tissue, throughout the whole length of the pathway. The question of whether phloem loading is hormonally controlled has also been reviewed [3].

In *Ricinus* [55, 56] both fusicoccin (FC) and IAA promoted, whereas ABA inhibited, sucrose loading into the phloem, though in contrast to FC, there was no effect of IAA on the pH of the perfusing solutions. Sucrose uptake by leaf discs from *Phaseolus vulgaris* was also stimulated by FC or IAA, again more actively by FC [23]. In more recent studies using the *Ricinus* perfused petiole system [3], both sucrose and potassium concentrations in the exudate of treated plants were increased by the cytokinins BA and kinetin and by a number of synthetic auxins, particularly NAA. GA_3 has been shown to increase the rate of phloem loading in *Vicia faba* plants trimmed to all but one source and one young sink leaf [1].

Plant growth regulators have been reported to stimulate proton extrusion in other systems and so it was suggested that their action merely reflects a stimulated proton pumping. The possibility that external acidification could account for increased proton extrusion has been rejected on the basis of pH and kinetic studies in *Phaseolus vulgaris, Beta vulgaris*, and *Vicia faba* [12, 29, 83]. An alternative possibility is that the sucrose carrier itself may be hormonally controlled [73].

In the proton-efflux pump model, lateral leakage from the pathway is thought to be prevented by the presence and activity of this pumping mechanism. Lenton [52] has raised the interesting question as to whether this pump is present in regions of sucrose unloading and if so, is it specifically inhibited. Such an inhibitory role could be played by ABA, which inhibits sucrose uptake in castor bean cotyledons [89] and petioles [56]. It has been suggested that ABA may stimulate unloading by increasing passive efflux or inhibiting reloading.

It is fairly certain that potassium plays an important role in the loading process by increasing sucrose availability in the apoplast and stimulating the proton-efflux pump. Baker [3] suggests that there may not be a close hormonal control of the loading process and that unloading of sucrose in the sink tissue may be more important.

There are some grounds to suppose that sink demand may be transmitted to source tissues via a reduction in sieve-element turgor, since proton extrusion in leaf tissues can be enhanced by high external concentrations [11, 74]. In this respect, Lang [49] has proposed that turgor-pressure differences in the phloem depend on a gradient both of sucrose concentration and potassium ions in the sieve tube. K^+ exchange would thus occur between the sieve tubes and their surroundings throughout their length and K^+ would be recycled in the xylem stream. The K^+ gradient would thus arise from radial exchange of K^+ with the xylem. A K^+ gradient of this type was demonstrated in willow stems [33] and more recently by *Manihot exculenta* and *Ricinus communis* [90]. On the basis of these results, Wareing and Adkin [91] suggest that since the phloem transport process is likely to be active, it is easier to understand how IAA may operate throughout the transport pathway, especially in view of the known effects of IAA on proton excretion and K^+ exchange.

5 Hormones in Sucrose Translocation and Unloading

Although it is generally accepted that phloem loading generates the driving force for the mass flow of assimilates from source to sink by the generation of a water potential gradient and water influx into the phloem [31], sucrose transport is influenced by the steepness of concentration gradients in the sieve tubes linking sources and sinks. The sites of sucrose unloading from the phloem may be remote from the consumer tissues as in developing seeds or may be in close association as in storage parenchyma. Because of the wide range of sinks occurring in plants, several unloading mechanisms are possible depending on the nature of the sink tissue [27]. A symplastic route involving little expenditure of energy may be common to developing organs and some meristematic tissues, though an active transport mechanism may be required at the sink cell tonoplast to maintain the necessary adjustment of sucrose concentration to maintain flow. Conversely, an apoplastic route has been clearly identified in storage parenchyma and developing seeds, which require transport across both the plasmalemma of the sieve tube and the membranes of the receiving sink cell. In some sink tissues such as sugar cane, sucrose is inverted during unloading, while in others hexoses may accumulate as storage materials. In such tissues metabolic conversion by invertase is an important component and may be one of the steps at which hormonal control is exerted [58].

The role of hormones in controlling assimilate unloading from the phloem and accumulation in the sink tissues is not clear. There are several steps in the sugar transport system at which direct or indirect control by hormones could occur. These include the control of active transport processes located at the phloem and/or sink plasma membranes where sucrose is loaded intact or the control of a wall-located invertase before transport of hexose across the sink plasma membrane in systems where sucrose hydrolysis is a prerequisite for sugar uptake. Where sucrose is loaded intact across the apoplast, sucrose concentration gradients may be maintained by the hydrolysis of sucrose by a cytoplasmic invertase or by active loading into the cell vacuole, these steps also presenting points of possible hormonal control.

Hormonal control of the transfer of assimilate across the phloem-boundary membranes into the sink free space has been suggested by Patrick and Wareing [68]. Some support for this proposal is provided by the demonstration that C unloading to the apoplast in *Phaseolus vulgaris* is specifically stimulated by kinetin and GA_3 [37]. In addition, Patrick and Wareing [68] demonstrated that sugar levels in the stem apoplast were elevated by kinetin, so enhancing uptake by the storage cells in the stem. Patrick [65] thus proposes that sink-mediated hormone regulation of phloem unloading may be a mechanism to drive turgor-dependent carbon flow, whilst maintaining apoplast C at levels for optimal sink uptake.

However, the maintenance of apoplastic sugar levels depends on the integration of phloem unloading and sink uptake [66, 67]. Such integration could be exerted by extracellular invertase [17] such that both the steepness of the sucrose concentration gradient and the levels of hexoses available for accumulation could be controlled. Enzyme activity could be mediated by an outward-directed proton pumping ATPase, which regulates apoplast pH [17] and creates conditions favourable for proton co-transport of the hexoses [47]. In the absence of extracellular hydrolysis, the ATPase could provide energy to drive a sucrose-proton antiport mechanism for phloem unloading and proton co-transport for sink uptake [53]. In either case, the vectorial ATPase could be a candidate for sink-produced hormonal control.

6 Hormones and Sink Activity

6.1 Effects on Membrane Transport

As already indicated, an apoplastic route following phloem unloading has been clearly identified in several maturing sink systems. As the consumer cells possess the capacity for active accumulation and rapid utilization of the unloaded solutes, efflux from the phloem could occur entirely in response to the resulting concentration gradient.

In sugar cane, sugars are accumulated by parenchyma cells at a rate dependent on the action of cell wall invertase to hydrolyse sucrose. The resulting hexoses are converted to sucrose phosphate by sucrose phosphate synthetase in the cytoplasm. Vacuolar storage of sucrose is dependent on the activity of a UDP-glucose-dependent group translocator, which utilises a highly co-ordinated multi-enzymic reaction to synthesise and simultaneously transport sucrose phosphate across the tonoplast using UDP-glucose as the sole precursor [84]. This group translocator consists of UDPG pyropyrophosphorylase, phosphoglucose isomerase, phosphoglucomutase, sucrose-P synthase, and sucrose phosphatase [85]. Hydrolysis of sucrose phosphate within the vacuole is thought to maintain a steep concentration gradient of sucrose across the tonoplast. Accumulation of sucrose into sugar beet vacuoles does not involve hydrolysis and resynthesis of sucrose [29]. The most favoured mechanism for transport of sucrose into the vacuole is by a proton-sucrose antiport dependent on the activity of an ATP-driven proton pump [14, 50, 76]. However, a group translocation mechanism has also been demonstrated in sugar beet [51] suggesting that the tonoplast is not simply involved in vectorial transport of solutes, but also in sugar interconversion [5].

Whether hormones are involved in the control of either of these mechanisms is not proven. As mentioned previously, ABA has been demonstrated to stimulate the accumulation of sucrose in a range of tissues and there is often a correlation between dry matter accumulation and endogenous ABA levels in sink regions. In sugar beet root tissues, ABA stimulation of sucrose uptake was inhibited by CCCP, indicating an effect of this hormone on membrane-controlled translocation [77]. However, the exact nature of this effect, whether on the sucrose carrier, membrane potential, pH or physical characteristics of the membrane was not clear. In contrast, IAA inhibited the uptake of sucrose. Although auxins promote sucrose uptake in leaf tissue, it was pointed out that this apparent contradiction could be explained by the evidence of a proton/sucrose co-transport mechanism in shoot tissues [3] as opposed to the proton/sucrose antiport system in sugar beet tissue [76].

There is some evidence that GA may inhibit phloem unloading and sucrose accumulation in sink tissues. In carrots, application of GA_3 to either shoots or roots alters the ratio of shoot to storage root growth in favour of the shoot [57]. Incubation of carrot root discs in GA_3 solutions reduces the active uptake of [^{14}C]-sucrose into the tissues, resulting in reduced sucrose accumulation (T.H. Thomas, unpubl. data). This suggests that GA effects on shoot/storage root ratio are not exclusively through the establishment of a strong shoot "sink", but also operate by direct inhibition of the storage root sink, possibly at the tonoplast. GA_3 was also shown to reduce acidification and decrease proton flux across membrane vesicles prepared from *Vica faba* stem sections [1].

6.2 Invertase and Sink Activity

In many sink tissues there is a close correlation between invertase (EC 3.2.1.26) activity and rates of carbohydrate import and/or tissue growth. The controversial possibility that hormonal regulation of assimilate flow may be mediated by this enzyme has been reviewed recently [59, 62]. Within sink tissues invertases may be located in the cell wall so that apoplastic transfer can occur via hydrolysis, as in sugar cane stems

[47], in the cytoplasm to maintain a downhill sucrose concentration gradient as in carrot tap roots [75], or in the vacuole itself. In sinks, which store hexose and/or polysaccharides, cells which are growing by expansion or those which have a high demand for hexose for respiratory or synthetic metabolism, sucrose hydrolysis will be important to maintain the concentrations necessary to drive phloem transport. In such cases it seems more than likely that invertase activity will play a central role in the regulation of sucrose import. Hormonal regulation of invertase activity or synthesis may be one of the ways in which utilisation and consequent transport of carbohydrate substrates is controlled in growing sinks.

Changes in invertase levels have been shown in a wide range of growing and storage sinks following application of plant growth substances to intact plants or isolated tissues (see [59]). In leaves [60] and stems [63] of *Phaseolus vulgaris* L, maximum invertase activity and rates of cell expansion coincide with those of IAA activity. Application of exogenous IAA substitutes for the shoot apical bud both in inducing invertase activity and in attracting assimilate movement towards the apex. However, such close correlations between invertase activity and growth have led to the suggestion that invertase changes may be a result of rather than a cause of cell growth. Contradictory evidence has been provided by Morris and Arthur [61], who showed that in bean internode segments the stimulation of acid invertase activity by IAA occurred both in the presence and absence of growth inhibitory Ca^{2+} ions and mannitol. Although GA_3 had little effect on invertase activity or bean internode elongation when applied alone, it greatly enhanced the effect of IAA on both parameters [64].

There is evidence to suggest that GA itself may affect assimilate partitioning between root and shoot systems by its ability to increase invertase activity in the shoot sink, the growth of which is limited by the availability of hexose. In *Phaseolus vulgaris* [64], GA_3 stimulated growth of stem internodes, reduced root growth and increased upward translocation of sucrose to the elongating stem region at the expense of transport to the hypocotyl and root system. These changes were correlated with increases in acid invertase in the stem internodes and decreases in invertase in the hypocotyl and root. Similar effects of GA application were obtained with carrot [57] where it was shown that the establishment of a strong shoot sink was correlated with reduced growth, sugar accumulation and invertase activity in the storage root. Thus, it would appear that by stimulating invertase synthesis in shoots and reducing it in roots, gibberellins may act to establish a more favourable sucrose gradient between sink and source leaves.

7 Conclusions

Evidence from correlative and pharmacological studies tend to support the contention that plant hormones are likely to be involved in the control of assimilate production and movement from source to sink tissues. It would seem that most transfer processes involved in assimilate transport would be suitable for hormonal control, particularly those involving energy-dependent movement across membrane or enzymatic conversion of sugars. Although significant progress has been made towards understanding the regulation of carbon allocation between starch and sucrose in source leaves, evidence for the involvement of hormones in these processes is tenuous. Since availability of assimilate from source tissues is rarely a yield-determining factor, there also may be no need for close hormonal control of the sucrose-loading process. It would seem from studies on membrane-transport mechanisms and carbohydrate-conversion enzymes that unloading of sucrose in sink tissues is the most likely candidate for hormonal control, though hormonal communication between sources and competing sinks cannot be disregarded.

References

1. Aloni B, Wyse R (1985) Plant Physiol (Bethesda) 77:685
2. Allen JRP, Baker DA (1980) Planta (Berl) 148:69
3. Baker DA (1985) Plant Growth Regul Group Monograph 12:163
4. Browning AJ, Hall JL, Baker DA (1980) Protoplasma 104:55
5. Brown SC, Coombe BG (1984) Physiol Veg 22:231
6. Brun WA, Brenner ML, Schussler J (1985) Br Plant Growth Regul Group Monograph 12:257
7. Cronshaw J (1981) Ann Rev Plant Physiol 32:465
8. Cornic CG, Miginiac E (1983) Plant Physiol (Bethesda) 73:529
9. Cowan IR, Raven JA, Hartung W, Farquhar GD (1982) Aust J Plant Physiol 9:489
10. Daie J (1984) Proc 11th Annu Meet Plant Growth Regul Soc Am 32
11. Delrot S, Bonneman JL (1979) Physiol Veg 17:247
12. Delrot S, Bonneman JL (1981) Plant Physiol (Bethesda) 67:560
13. Dewdney SJ, McWha JA (1979) Z Pflanzenphysiol 92:183
14. Doll S, Rodier F, Willenbrink J (1979) Planta (Berl) 144:407
15. Doman DC, Geiger DR (1979) Plant Physiol (Bethesda) 64:528
16. Düring H, Alleweldt G (1980) Ber Dtsch Bot Ges 93:339
17. Eschrich W (1980) Ber Dtsch Bot Ges 93:363
18. Eschrich W (1984) Ber Dtsch Bot Ges 97:5
19. Evert RF, Eschrich W, Heyser W (1977) Planta (Berl) 136:77
20. Evert RF, Eschrich W, Heyser W (1978) Planta (Berl) 138:279
21. Farrar SC, Farrar JF (1985) Br Plant Growth Regul Group Monograph 12:67
22. Fisher DB (1967) Bot Gaz 128:215
23. Fisher DB, Outlaw WH Jr (1979) Plant Physiol (Bethesda) 64:481
24. Fondy BR, Geiger DR (1980) Plant Physiol (Bethesda) 66:945
25. Francheschi VR, Giaquinta RT (1983) Planta (Berl) 157:411
26. Francheschi VR, Giaquinta RT (1983) Planta (Berl) 157:422
27. Geiger DR, Fondy BR (1982) What's New Plant Physiol 11:25
28. Geiger DR, Giaquinta RT (1982) In: Govindjee J (ed) Photosynthesis: CO_2 assimilation and plant productivity. Academic Press, New York, p 345
29. Giaquinta RT (1977) Plant Physiol (Bethesda) 60:339
30. Giaquinta RT (1978) Plant Physiol (Bethesda) 61:380
31. Giaquinta RT (1980) Ber Dtsch Bot Ges 93:187
32. Giaquinta RT (1983) In: The biochemistry of plants. Academic Press, New York, p 371
33. Grange RJ, Peel AJ (1978) Planta (Berl) 138:15
34. Griffith SM, Jones RJ, Brenner ML (1985) Plant Physiol (Bethesda) 77:Suppl. 687
35. Hartung W, Gimmler H, Heilmann B (1982) In: Wareing PF (ed) Plant growth substances. Academic Press, London, p 325
36. Hawker JS, Marschner H, Downton WJS (1974) Aust J Plant Physiol 1:491
37. Hayes PM, Patrick JW (1985) Planta (Berl) 166:371
38. Hedden P, Hoad GV (1985) Plant Growth Regul Group Monograph 12:211
39. Hedden P, MacMillan J, Phinney BO (1978) Annu Rev Plant Physiol 29:142
40. Herold A (1980) New Phytol 86:131
41. Ho LC (1984) Plant Growth Regul 2:277

42. Ho LC, Baker DA (1982) Physiol Plant 56:225
43. Hosford DJ, Lenton JR, Milford GFJ, Pocock TO, Elliott MC (1984) Plant Growth Regul 2:371
44. Huber SC (1983) Plant Physiol (Bethesda) 71:818
45. Huber SC, Moreland DE (1981) Plant Physiol (Bethesda) 67:163
46. Humphries EC, Wheeler AW (1964) In: Regulateurs naturels de la croissance vegetale. Paris, p 505
47. Komor E, Thom M, Maretzki A (1981) Planta (Berl) 153:181
48. Kuo J, O'Brian TP, Canny MJ (1974) Planta (Berl) 121:97
49. Lang A (1983) Plant Cell Environ 6:683
50. Leigh RA (1984) Plant Growth Regul 2:339
51. Leigh RA, Thom M, Maretzki A (1985) Plant Physiol (Bethesda) (in press)
52. Lenton JR (1984) Plant Growth Regul 2:267
53. Lichtner FT, Spanswick RM (1981) Plant Physiol (Bethesda) 67:869
54. Madore M, Webb JA (1982) Can J Bot 60:126
55. Malek F, Baker DA (1977) Planta (Berl) 135:297
56. Malek F, Baker DA (1978) Plant Sci Lett 11:233
57. McKee JMT, Thomas TH, Hole CC (1984) Plant Growth Regul 2:359
58. Morris DA (1982) In: Wareing PF (ed) Plant growth substances 1982. Academic Press, London, p 659
59. Morris DA (1983) News Bull Br Plant Growth Regul Group 6:23
60. Morris DA, Arthur ED (1984) J Exp Bot 35:1369
61. Morris DA, Arthur ED (1984b) Phytochemistry (Oxf) 23:2163
62. Morris DA, Arthur ED (1984c) Plant Growth Regul 2:327
63. Morris DA, Arthur ED (1985) J Exp Bot 36:623
64. Morris DA, Arthur ED (1985) Physiol Plant 65:257
65. Patrick JW (1984) In: Pearson CJ (ed) Control of crop productivity. Academic Press, Sydney, p 59
66. Patrick JW, McDonald R (1980) Aust J Plant Physiol 7:671
67. Patrick JW, Turvey PM (1981) Ann Bot 47:611
68. Patrick JW, Wareing PF (1981) Br Plant Growth Regul Group Monogr 6:65
69. Peel AJ, Rogers S (1982) Planta (Berl) 154:94
70. Pollock CJ, Housley TL (1985) Ann Bot 55:593
71. Priess J (1984) Trends Biochem Sci 9:24
72. Rademacher W, Graebe JE (1984) Ber Dtsch Bot Ges 97:167
73. Reinhold L, Kaplan A (1984) Annu Rev Plant Physiol 35:45
74. Reinhold L, Seiden A, Volokita M (1984) Plant Physiol (Bethesda) 75:846
75. Ricardo CPP, apRees T (1970) Phytochemistry (Oxf) 9:239
76. Saftner RA, Wyse RE (1980) Plant Physiol (Bethesda) 66:884
77. Saftner RA, Wyse RE (1984) Plant Physiol (Bethesda) 74:951
78. Sandberg G, Jensen E, Crozier A (1982) Planta (Berl) 156:541
79. Setter TL, Brun WA, Brenner ML (1980) Plant Physiol (Bethesda) 65:1111
80. Silvius JE, Kremer DF, Lee DR (1978) Plant Physiol (Bethesda) 62:54
81. Sjut V, Bangerth F (1983) Plant Growth Regul 1:243
82. Sjut V, Bangerth F (1984) Plant Growth Regul 2:49
83. Sturgis JN, Rubery PH (1982) Plant Sci Lett 24:319
84. Thom M, Maretzki A (1985) Plant Physiol (Bethesda) 77:Suppl. 688
85. Thom M, Maretzki A (1985) Proc Natl Acad Sci USA 82:4697
86. Thorne JH, Giaquinta RT (1984) In: Lewis DH (ed) Storage carbohydrates in vascular plants. Cambridge University Press, London, p 75
87. Tietz A, Ludewig M, Dingkuhn M, Dorffling K (1981) Planta (Berl) 152:557
88. Treharne KJ (1978) Br Plant Growth Regul Group Monograph 1:153
89. Vreugdenhil D (1983) Physiol Plant 57:463
90. Vreugdenhil D (1985) Planta (Berl) 163:238
91. Wareing PF, Adkin MP (1984) Plant Growth Regul 2:287
92. Wightman F (1977) In: Pilet PE (ed) Plant growth regulation. Springer, Berlin Heidelberg New York, p 77

V. Hormones: Applications

Aims and Possibilities of Applied Plant Bioregulation*

J. JUNG[1]

1 Introduction

Whenever a substance with growth-regulating properties is discovered, the question of its practical use arises. This applies both to the five phytohormone groups so far known and to a constantly increasing number of exogenous substances. A new group of biological active substances has established itself in the meantime under the collective term "plant bioregulators" = PBRs – substances that are increasingly gaining acceptance in agriculture and horticulture[2] [4, 7, 8].

The principal purpose of using these substances is to optimize plant production by modifying growth, development and stress behavior and the quantitative and qualitative yield of crop plants. A survey will be given here of the objectives, status and prospects of the plant bioregulators that are being used.

2 Aims of Applied Plant Bioregulation

The modification of crop plants by using appropriate substances is the principal aim of applied plant bioregulation. There are, however, other aspects and starting points in this field. In a general survey the present objectives and focal points of interest may be outlined as follows.

1. Evaluation and utilization of the results of basic research into hormone physiology and the adaptation of the relevant methods of investigation for characterizing and quantifying processes that are important for the cultivation of crop plants and their yields.
2. Exploitation of the possibilities mentioned under (1) as guidelines and tools for plant breeding.
3. Use of substances with bioregulatory action for effecting certain modifications to crop plants.

A list of starting points for desirable modifications and useful effects that can be derived from these is given:

---

* The following contributions (pp. 363–409) are part of workshop *Applied Aspects*

1 BASF Agricultural Research Center, D-6703 Limburgerhof, FRG

2 In the following chapters the term "plant growth regulators" (PGR) is mainly used

Plant Growth Substances 1985
(ed. by M. Bopp)

Objectives of modification	Possible results
Sequence and duration of development steps during vegetative and reproductive growth (germination, tillering, shooting, flowering, fruit development)	Enhanced establishment and biomass production Increased or improved formation of yield-determining organs
Anatomical and morphological structure	Enhanced "harvest index" Improved standing ability, improved light interception, improved shoot/root ratio
Behavior towards drought, extreme temperature and diseases	Enhanced resistance to environmental stress and infections
Photosynthesis and metabolism	Increased photoproductivity, enhanced content of yield- and quality-determining ingredients
Senescence, ripening, abscission and dormancy	Control of ripening processes, promotion of mechanical harvest, control of germination

4. Use in vegetative reproduction and in tissue and cell culture.
5. Application-relevant exploitation of lead substances for the synthesis of new compounds with a homologous or modified spectrum of action ("biorational design").

The realization of these objectives is dependent on various conditions – in particular a wide palette of substances which have greatly differentiating actions and are thus able to achieve as many of the effects that are aimed at as possible. This point will be dealt with in more detail in the following section.

3 Available Compounds and Possible Uses

The range of available compounds can be subdivided into substances with hormone-like or hormone-interfering actions and into those with a different kind of mechanism of action or one that has not yet been discovered. The extent to which the phytohormone groups known so far can be used for plant bioregulation in practice has been investigated intensively for many years and with notable success. In connection with the further development of the range of products, interest is concentrated both on substances with a homologous action and on those with modified and opposite effects. Particularly, this last type of substance deserves special interest on account of its ability to antagonize phytohormone action.

Auxins. There have not recently been any special new developments in *auxins* and their antagonists. If the borderline area of the hormone-type herbicides is disregarded, auxins still have only limited practical application. Even substances with an antagonistic mode of action have not been able to establish themselves on a major scale.

Cytokinins. In contrast, *cytokinins* have experienced a remarkable development. On the one hand, it has been possible to supplement the classical purin-cytokinin type and, on the other hand, the range of substances has been extended by additional cytokinin-active urea derivatives [3, 5]. The defoliation action of phenylthiadiazolylurea is worth noting.

Gibberellins. The list of *gibberellins* has been extended particularly effectively in recent years. However, no substantially, additional applications have resulted. The successful application of the "mirror-image concept" in modifying phytohormones must be emphasized all the more. In recent years, namely, the most intensive development within the range of bioregulators has been in GA antagonists [2]. Starting with the classical type of onium compounds, new, highly effective compounds have been found in the group of pyrimidines, norbornenodiazetines and triazoles, which are kaurene oxidase inhibitors [9].

Onium compounds have found very broad application in cereals in Europe as agents for improving standing ability. In the Federal Republic of Germany and some other European countries the major areas with wheat and winter barley are treated with substances of this type, also in combination with ethylene generators. In cotton the onium compound N,N-dimethylpiperidinium chloride (mepiquat chloride) is also used as a growth regulator on a large scale; there are numerous established uses in the cultivation of ornamentals as well. Possibilities are opening up in rice for the application of norbornenodiazetine derivatives and triazoles. Moreover, the evaluation of these new classes of substances is still in full progress.

Abscisins. In the case of *abscisins* there are now indications of the palette being extended as well [1]. Testing abscisic acid itself for relevance to application has so far had to be restricted mainly to laboratory and model trials because of the limited availability of sample material. It cannot, therefore, be adequately assessed to what extent the interesting effects of this substance on water status, transport and storage processes and senescence may be exploited in practice.

Ethylene. Ethylene generators have found versatile application in practice through ethephon. Induction of flowering in pineapples, determination of ripening in tomatoes, fruit abscission in cherries, acceleration of ripening in tobacco, stem stabilization in barley and increasing latex production in hevea are well-known examples of their use. However, the endogenous substance ACC (l-aminocyclopropane-1-carboxylic acid) has not yet been able to find an application.

There is also special interest in the further development of inhibitors of ethylene biosynthesis [6, 10], the possibility of reducing flowering and fruit abortion processes being investigated recently to an increasing extent.

Beside the groups of substances with a hormone-related principle of action mentioned up to now, the heterogenous group of substances with a non-hormonal mechanism of action cannot be dealt with here. It should merely be pointed out that increasing interest is being devoted to substances for pollen suppression (gametocides) as aids especially in the hybrid breeding of self-fertilizing plant species.

Summing up at the end of this only incomplete survey of the range of bioregulatory active substances, it may be said that they already represent a strongly differentiated spectrum of action. In some fields of crop production, they have also become important aids. They have not, however, followed the steep development trend that was predicted for them in various quarters.

4 Prospective Assessments of Applied Plant Bioregulation

The following factors in particular will be of importance for the further development of applied bioregulation.

1. *The Particular Research and Development Concepts.* Experience suggests that the most promising route is to influence individual parameters in the development behavior and yield formation. It will probably be very difficult to find a regulator that increases yield directly (anabolic).
2. *Quantity and Efficiency of Available Substances.* Whereas up to now by far the majority of synthetic active substances have been found via the empirical approach, in future the biorational design will probably gain in importance. This will particularly be the case if hormone physiology and bioregulator research can be suitably extended on the basis of intensifying structure activity relationships (QSAR), identifying phytohormone receptors, etc. Greater knowledge of the relations between effects in chiral compounds will probably not be the least significant factor in opening up new aspects. The establishment of "mirror-image" groups of active substances may also be regarded as an interesting aspect (see gibberellins and GA-antagonists). However, in discovering and evaluating active substances, much will continue to depend on the biological selection methods and the screening systems.
3. *Coordination with Possibilities that can be Realized in Plant Breeding.* In principle, attempts should be made to use plant bioregulators to achieve what breeding cannot or can no longer achieve (for example, because the plant is already being cultivated) or cannot achieve quickly or economically enough. When coordinated with plant breeding in this way, phytohormonal bioregulation can provide crop production with additional possibilities.
4. *Projection of the Bioregulatory Substances into the Cropping Concept.* When considering the projection of bioregulatory active substances into the cropping system, a differentiation must be made between highly developed and extensive production techniques. Bioregulators should be given chances – even if different ones – of being used at both levels. However, the fact should not be overlooked that the possibilities of application will be infinitely greater in highly developed crop production.

5. *Research and Development Expenditure.* The level of these costs entails minimum requirements being made on the area of application and on the farmer's expenditure. This can mean, for example, that only a plant species with a high production value and a high level of intensity where there is a correspondingly large cultivation area justifies the producer's investment risk and the farmer's expenditure.

5 Conclusions

In summary, it can be said that the conditions, on which the further development of applied plant bioregulation will depend, are not only numerous, but also very varied. However, the principle of this possibility of regulation and the techniques derived from it are probably of such interest for crop production that increasing acceptance may be expected.

Acknowledgments. My thanks are due to Mr. B.E. Byrt for the translation of the manuscript.

References

1. Grossmann K, Jung J (1984) J Agron Crop Sci 153:14
2. Jung J (1985) In: Hedin P, American Chem Soc Symp Ser 276. Washington, DC, p 95
3. Matsubara S (1980) Phytochemistry (Oxf) 19:2239 (review)
4. McLaren JS (ed) (1982) Chemical manipulation of crop growth and development. Butterworth, London
5. Mok MC, Mok DWS, Armstrong DJ, Shudo K, Isogai Y, Okamoto T (1982) Phytochemistry (Oxf) 21:1509
6. Morgan P (1980) Bot Gaz 141 (4):337
7. Nickell LG (ed) (1983) Plant growth regulating chemicals. CRC Boca Raton, Florida
8. Ory RL, Rittig FR (eds) (1984) Bioregulators – chemistry and uses. ACS Symp Ser 257, Washington DC
9. Rademacher W, Jung J, Graebe JE, Schwenen L (1984) In: Monograph No 11. Br Plant Growth Regul Group, p 1
10. Yang SF, Hoffman NE (1984) Annu Rev Plant Physiol 35:155

Growth Regulation of Arable Crops

K.J. TREHARNE, R.D. CHILD, H. ANDERSON, and G.V. HOAD[1]

1 Introduction

Plant growth regulators (PGRs) are increasingly used as part of the farmer's strategic chemical management of cereals and other arable crops. Over the past decade yields per hectare of wheat and barley have virtually doubled as improved varieties, intensified use of fertilizers and agrochemical sprays have been adopted. But economic and environmental pressures on farming practices are influencing the approach in chemical use toward cost effectiveness and minimum risk to non-target organisms; the objectives of agricultural research are likewise changing in emphasis from increasing production levels to better quality and improving efficiency of crop development and yield formation. The scope for chemical growth control is obvious in this context, but the sensible use of PRGs needs to be developed on a scientific basis rather than "spray and pray".

The targets for fungicide, pesticide and herbicide use are reasonably identifiable as the "biological" constraints to crop production. Growth and yield potential are genetically determined, but the underlying, interacting, physiological processes, which are adversely affected, or limited, by environmental factors, are less defined and difficult to prioritise either as targets for PGR modulation, or for genetic improvement either by conventional methods or new "bio-engineering technology".

2 Physiological and Agronomic Targets for PGR Use in Cereals

Chemical growth retardants have been in use for over 30 years and have been particularly successful in improving harvestable "agronomic" yield when applied to wheat and barley cultivars with long or weak straw, especially when soil fertility is high. Until recently the major PGR in use was Chlormequat (2-chloroethyl-trimethyl-ammonium chloride, CCC), which retards stem elongation through a mode of action which is considered to be antagonism of the endogenous gibberellins. The response to application of this compound is a shortening of the crop and a reduction in stem lodging. There are, however, many claims of cereal yield increases using CCC and/or

1 Long Ashton Research Station, Department of Agricultural Sciences, University of Bristol, Long Ashton, Bristol BS18 9AF, Great Britain

Plant Growth Substances 1985
(ed. by M. Bopp)

other retardants in situations where no crop lodging occurred (e.g. [12]). A large body of field-trial information on yield response in wheat and barley to applied PGRs [18] shows considerable variability between and within species. Many field studies have considered formulations of the chemical, timing and dose, where little quantitative information concerning effectiveness of uptake, penetration, transport and metabolism of the "active ingredient" is gathered, against which to judge morphological, anatomical or physiological response, or lack thereof. These data are very necessary if we are to understand what may be achieved by use of PGRs but is obviously difficult in variable field conditions where temperature and moisture affect markedly, spray chemical behaviour at the time of application.

PGR response is also dependent upon the physiological "status" of the plant as affected by its environmental history and conditions after application. Many studies have established the rapid uptake of CCC by roots from soil drenches and its subsequent translocation through the cereal plant [8, 17]. In a definitive study in our laboratories [2], factors affecting leaf uptake of CCC in wheat and barley were compared under controlled environments and field conditions. Using a quantitative microsprayer, [6] with [^{14}C]-CCC, retention of spray droplets on initial impaction was greater in wheat than in barley leaves, irrespective of growth stage or site of application. Retention by the leaf was a declining function of leaf age but increased with surfactant concentration and with decrease in droplet size. Uptake of chemical was greater in wheat than in barley and increased with temperature and with a lowered irradiance, coupled with increase in moisture status. These factors influencing uptake were related to quantity and composition of leaf epicuticular waxes in the controlled environment (C.E.) cultured plants. In contrast to these C.E. grown plants, field-grown crops had extensive cuticular damage and chemical spread more extensively over the leaf surface. A most important result was that CCC translocation occurred readily only in immature tissues, but the conclusion was reached that biological effectiveness of PGR sprays applied to crop canopies at late stages of development is most likely to be affected by droplet size and surfactant concentration rather than physiological consideration. This has important implications to studies of formulations and new PGR compounds.

Although it may be naive to consider that any single compound, like the GA-antagonist retardants, can be used at different stages of crop development to modulate growth, which ultimately improves realisation of the yield potential, there is some evidence that this is the case. Target processes range from encouragement of root growth, control of tillering through suppression of apical dominance, prevention of spikelet and embryo abortion to improved partitioning of assimilates to developing grain (Table 1).

A similar summary of phenological events may be compiled for other arable crops. Differences between and within species in relative efficiency of physiological attributes is quite evident and with gramineaceous crops evaluation of problems, and the ability chemically to modulate events, is made difficult by their growth compensating ability.

In previous studies [5] with wheat genotypes varying in straw height, CCC treatments were not, unexpectedly, more effective with long-stemmed types. With young seedlings, an increased proportion of dry matter was partitioned to roots; CCC

Table 1. Potential use of PGRs during wheat development

Zadoks growth stage	Developmental phase	Physiological/agronomic targets
0	Germination	Seedling establishment; emergence control – frost protection
10– 12	Seedlings growth, tiller initiation	Improved rooting, synchrony of tiller development
13– 14	Tiller growth, spike induction	Uniformity of tiller production and growth control of early canopy architecture
15– 30	Spike development, stem extension	Control of fertile tiller survival, synchrony of spikelet development, less abortion
40–100	Grain development	Control of canopy, leaf activity, optimal assimilate partitioning, uniformity of grain fill

reduced total dry weight increment compared with untreated plants. Similar effects were obtained using the newer triazolyl retardants, such as paclobutrazol, which are known to inhibit GA biosynthesis [10]. The short-strawed wheat cultivars with Norin 10 dwarfing Rht_1 and Rht_2 genes are relatively non-responsive to CCC applied at a number of different growth stages during tiller production and stem extension [5], with no effect on yield. Other work, where yield increases were obtained, has demonstrated increases in ear and grain number [13] or attributed improvement in canopy structure to CCC treatment [3].

Retardant applications to winter barley at the early stem extension stage result in an increase in the number of fertile stems [15], but do not necessarily result in yield improvement, since compensation between grain number and size occurs [9, 14]. In certain seasons early CCC treatment improved fertile tiller survival and increased yield. Treatment of winter barley (cv. Sonja) at Zadoks growth stage 30 (and later) with CCC or growth regulators containing ethephon (Table 2), show some potential for yield improvement.

Ethephon, which on application releases ethylene, is an effective stem-shortener [11]; its use has concentrated on reducing "brackling" and "necking" in barley. Whilst the GA antagonists inhibit elongation and result in thicker stems with enhanced development of sclerenchyma and vascular bundles, the action of ethylene on stem growth is less clearly defined. Yield increases (Table 2) were related to degree of growth reduction, but were particularly related to shortening of the upper internodes by the ethephon-containing chemicals. With spring barley cv. Triumph similar effects were observed in 1984. No yield response was obtained with CCC treatments at GS 10 or GS 32, but Cerone applied at GS 49 significantly increased grain yield. Whether assimilates are made more available for grain growth rather than development of stem tissues following ethephon application is now being tested.

Table 2. Yield response to timing of PGR application in winter barley cv. Sonja (GS – Zadoks Growth Stage)

PGR treatment		Grain yield (t ha^{-1})	Yield difference (t ha^{-1})
CCC	GS 13	8.21	–0.16
CCC	GS 23	8.64	+0.27
CCC	GS 30	8.99	+0.62
CCC	GS 32	9.35	+0.98
CCC	GS 49	8.81	+0.44
Terpal	GS 37	9.00	+0.63
Cerone	GS 49	9.32	+0.95
Unsprayed "control"		8.37	0

The most effective treatments in our experiments are at later growth stages and imply emphasis should be placed on use of these PGRs to improve grain filling and thus grain size and quality. In a comparative study of sites, spring barley cv. Maris Mink, showed greater grain yield in longer and cooler grain-filling conditions [7] of Scotland than at LARS. Ethephon responses may thus have potential for grain size improvement at warmer latitudes with a relatively shorter grain-fill period.

Grain legumes such as *Vicia faba* (field bean) with indeterminate growth habits are susceptible to lodging. Response to chlormequat is slight, whereas daminozide produces rapid, marked effects on stem extension. To test whether pod development was enhanced through greater availability of assimilates, as a result of inhibiting stem growth, bean plants cv. Maris Blaze were daminozide treated when the fifth node carried expanded leaves. Treated plants were significantly shorter than untreated controls, when harvested 2 weeks later, but no difference was observed in total leaf area or plant dry weight. Within a further 2 weeks the dwarfing effect had disappeared. Treatments at later stages, e.g. 15th node with expanded leaves, resulted in internode shortening, which was sustained to plant maturity. Although total pod yield was unaffected by daminozide application, distribution of pods on the plant was modified.

Effects of retardants on pod set in oilseed rape are dependent upon the degree to which growth reduction is sustained during this period [4]. Both daminozide and ethephon reduce extension growth and act as anti-lodging agents. Daminozide reduced seed yield when applied at early stem elongation but increased pod number from application at early flowering, significantly reducing terminal raceme and branch lengths. Although total seed yield was unaffected, a different pod distribution and canopy architecture resulted. Ethephon treatments at mid flowering in oilseed rape reduced set of the late-opening flowers and gave a larger concentration of pods at the base of the branches. Both treatments have potential for increasing harvestable yield by manipulating the distribution of pods within the canopy. Thus, significant losses in harvestable yield can occur in many crops, in addition to those associated with crop lodging.

In cereals (oats) grain loss can occur with delayed harvest, but losses are more appreciable in pod-bearing crops such as rape, evening primrose, borage and grain legumes. Major targets for PGR use toward better agronomic yield are improving synchrony of flowering and seed development which, because of the maturity range in the indeterminate crops, result in yield loss through pod shatter, seed shedding or a proportion of immature seed. Better synchrony of yield formation would theoretically improve efficiency of utilisation of assimilates and nutrients into realisable yield. Our experiments with ethephon applied at 70% pod set in rape prevented further set and despite a reduction of some 30% in pod number compared with untreated plants, resulted in a small increase in seed yield through larger seed weight. Precision of timing requires investigation in relation to optimising the compensation in partitioning which occurred.

Similar responses to ethephon are evident with field bean. Both in winter and spring cultivars internode extension was reduced and unopened and unpollinated flowers continued to abscind for 3 weeks after application. Pollinated flowers continued to develop and in some plants a terminal flower formed instead of the normal vegetative apex.

3 PGR Modification of Growth to Avoid Environmental Stress

Plant breeding has successfully combined characters to improve straw strength, harvest index and winter hardiness of wheat so the need for chemical control is less evident than in other cereal crops where less progress has been made. In winter oats, as an example, the relationship between yield and biomass remains unbroken; selection for reduction in straw length, increased tolerance to frost or water stress impose yield penalties [16]. An effective PGR treatment, e.g. to improve winter hardiness of oats, would relieve pressure to select for this character at the expense of other yield-related attributes. Similarly, improved tolerance to winter damage

Table 3. Effect of seed dressing with tetcyclacis on winter damage in oats

	Dead plant tissue (%)			
	Bulwark		Peniarth	
Sowing depth (cm)	Control	Treated [a]	Control	Treated [a]
2	53	17	37	13
4	54	15	28	14
6	56	14	26	12
8	38	14	24	13

[a] Tetcyclacis 0.1 g kg^{-1} seed/1.8 l water, seed soaked 16 h

could extend the environmental range for production of many crops or enable high quality spring cultivars to be autumn sown with yield advantage.

The triazole compounds, such as triadimenol, primarily developed as fungicides, when used as seed dressings reduce lower stem internode length of wheat and barley cultivars [1]. Winter oats, seed-treated with tetcyclacis, were sown at different depths in October 1984. Reduced shoot extension following treatment, with lower crown depth and reduced above-ground leaf area resulted in significantly less winter damage to foliage and the apex (Table 3). The PGR treatment had a greater advantage in the more susceptible cultivar Bulwark. Sowing depths had an effect on frost damage in untreated plants, but PGR treatment protected the plants to a similar extent, irrespective of planting depth.

4 Conclusions

Growth retardants can play a role in crop management to overcome lodging and especially in arable crops not yet advantaged through dwarfing genes. Newer classes of retardant, notably the triazoles, produce more marked responses, but the inherent persistency of their effects may be a disadvantage. Promise is evident in modifying growth to "avoid" environmental stress and in changing crop architecture. In pod crops, use of ethylene-releasing chemicals provides opportunities for improving uniformity of yield formation with economic yield advantages. Understanding the mechanisms controlling growth and yield formation is required to place the use of PGRs on a scientific basis.

Acknowledgments. We thank the Agricultural and Food Research Council for funding this work as part of the LARS Growth Substance Research Centre; we acknowledge co-sponsorship from BASF, Bayer U.K., Cyanamid and Union Carbide. The excellent scientific assistance of Jill Adam and Hazel Atkins is gratefully acknowledged.

References

1. Anderson HM, Huband NDS, Murphy PJ, Child RD (1985) Proc Crop Protection Conf – Weeds (in press)
2. Baker EA, Hunt GM (1985) Ann Appl Biol 106:579
3. Bruinsma J (1982) In: McLaren JS (ed) Chemical manipulation of crop growth and development. Butterworths, London, p 1
4. Child RD (1984) Aspects Appl Biol 6–127
5. Child RD, Adam JS, Atkins HA, Arnold GM (1984) In: Marketable yield of cereals. NAC, Stoneleigh, p 85
6. Coggins SJ, Baker EA (1983) Ann Appl Biol 102:149
7. Cottrell J, Easton R, Dale JE, Wadsworth AC, Adam JS, Child RD, Hoad GV (1985) Ann Appl Biol 106:365
8. Dekhuijzen H, Vonk CR (1974) Pestic Biochem Physiol 4:346
9. Green CF, Dawkins TCK, McDonald HG (1985) J Exp Bot 36(168):1126
10. Hedden P (1985) J Plant Growth Regul 4:111
11. Hill DM, Joice R, Squires NRW (1982) In: McLaren JS (ed) Chemical manipulation of growth and development. Butterworths, London, p 391

12. Humphries EC (1968) Field Crop Abstracts 21:91
13. Humphries EC, Welbank PJ, Witts KJ (1965) Ann Appl Biol 56:351
14. Jordan VWL, Stinchcombe GR (1985) Appl Biol 108 (in press)
15. Koranteng GO, Matthews S (1982) In: McLaren JS (ed) Chemical manipulation of growth and development. Butterworths, London, p 343
16. Lawes DA, Valentine J, Jones JE (1982) Ann Rep Welsh Plant Breeding Station for 1982, p 198
17. Tolbert E (1960) Plant Physiol (Bethesda) 35:380
18. Woolley EW (1982) In: Hawkins AE, Jeffcoat B (eds) Opportunities for manipulation of cereal productivity. BPGRG Monograph No 7, p 44

Ethylene as an Indicator and Regulator in the Development of Field Crops

P.W. MORGAN[1]

1 Introduction

Ethylene was used to "degreen" citrus in 1924 [1, 19], but it did not become a major plant growth regulator until the discovery of ethephon during the 1960s [6]. New, major trends are occurring in the use of ethylene as a developmental regulator.

2 Ethylene as a Regulator of Plant Development

2.1 Agricultural Uses

Ethylene became an important regulator because: (1) it naturally regulates a wide array of developmental processes throughout the plant life cycle [1]; (2) exogenous ethylene will modify many of the processes ethylene naturally regulates [1]; (3) diagnostic tests will identify ethylene's participation in specific developmental processes [1, 20]; (4) the ethylene physiology of plants can be manipulated in many ways [17, 20]; (5) ethylene is active at low concentrations [1]; (6) ethylene is not an environmental hazard [20].

2.2 "Liquid" Ethylene

The discovery of ethephon [6] led to the very rapid commercial use of ethylene substrates in agriculture [7]. Ethephon is probably the most widely used plant growth regulator (Table 1).

The natural precursor of ethylene, 1-amino-1-cyclopropane carboxylic acid [2, 14], is available as another "liquid" ethylene. When more is known about localized, tissue-specific actions of ethylene, the potential value of ACC can be assessed.

1 Department of Soil and Crop Sciences, Texas Agricultural Experiment Station and Texas A&M University, College Station, TX 77843, USA

Plant Growth Substances 1985
(ed. by M. Bopp)

Table 1. Current commercial uses of Ethephon as Ethrel, Florel, Cerone, and Prep registered in the USA [a, c]

Effect	Crops
Vegetative stage	
Regulates seedling growth, reduces lodging[b], reduces height	Tobacco, wheat, barley, daffodils, Hyacinths, geraniums
Promotes lateral branching	Azaleas
Reproductive stage	
Modifies sex expression	Hybrid cucumbers/squash
Fruit set	Pickling cucumbers
Delays flowering, hardens buds	Cherries
Induces flowering	Bromeliads, pineapple
Harvest stage	
Fruit harvest aid	Garden tomatoes, apples, blackberries, Cantaloupes, cherries, cranberries, Figs, filberts, table/raisin grapes, peppers, fresh market/processed tomatoes, walnuts, lemons, tangerines
Remove fruit, cure leaves	Crab apples, carob, olive, tobacco
Defoliate	Roses, tallhedge, apple nursery stock
Fruit dehiscence	Cotton

[a] Source: Union Carbide 1985 Chemical Guide Specimen Labels
[b] Registration in progress for lodging resistance in corn
[c] International uses include latex flow in rubber, harvest aid in coffee, pineapple and sugar cane, prevents flowering and promotes growth in sugar cane, induces ripening in pineapple

2.3 Trends in Uses of Ethephon

The commercial uses of ethephon have tended to be on horticultural crops with limited acreage [19], especially in the USA. Internationally, ethephon has been used on wheat, barley, sugar cane, and pineapples, i.e., "field" crops, but still these markets involve rather limited acreages. During the last 3 years the use of ethephon on field crops in the USA has increased so that it represents the major volume of sales and use on horticultural crops represents less than half of the total (D.L. Page, market manager, Union Carbide Corp.). This trend resulted from use on three crops: wheat, barley, and cotton. Ethephon use on field crops is expected to continue to increase (D.L. Page). An additional use being developed is to increase stem stiffness and reduce lodging in sweet corn, popcorn, and irrigated field corn.

2.4 Use of "Prep" on Cotton

The use of ethephon on cotton was made possible by studies on that plants's ethylene physiology. Young fruits and flower buds are more sensitive to ethylene- and ethephon-induced abscission than leaves [10, 16]. Ethylene naturally regulates cotton flower bud and young fruit abscission as well as dehiscence of mature fruit [11–13]. Ethephon and ethylene promote dehiscence, but not abscission of mature fruit [12, 16]. Since several cotton insects enter diapause in response to photoperiod [4], the possibility of reducing the number of overwintering insects by eliminating food (flower buds and young fruit) during the critical period was recognized [3]. These facts led to a dual goal for preharvest manipulation of cotton: to promote fruit dehiscence and to remove late season fruit [18]. Testing revealed that ethephon applied after fiber maturity: (1) promotes rapid, uniform fruit dehiscence and earlier harvest; (2) allows once-over harvesting; (3) increases picker efficiency and yield; (4) removes flowers and young fruits; and (5) predisposes leaves to defoliation [5, 8, 15]. The practice has been adopted by growers, and it has become the largest single use of ethephon.

It should be pointed out that the use of ethephon to hasten cotton fruit dehiscence and to reduce lodging in small grains and corn are not practices which increase basic productivity; however, they may prevent losses. What these practices do is subject the plant's development to managerial control. They are risk management tools. As expressed in 1973, one goal for growth regulator use in cotton was to give the grower "... management tools that will allow him to manipulate his crops to fit seasonal climatic variations, pests, and other variables ..." [8].

3 Ethylene as an Indicator of Plant Development

3.1 Hormonal Balance and Development

Although development is assumed to be controlled by the balance of the hormone complement interacting with genetics and environment, under certain circumstances one substance can be the dominant hormone. For ethylene this role is easily recognized because it is usually associated with an elevated rate of production of the gas and with the sensitivity of the process to exogenous ethylene [1]. Since such developmental processes represent opportunities to regulate plant behaviour via ethylene, their identification serve practical and scientific purposes.

3.2 Stress

The capacity of stress to promote ethylene synthesis is well-known [1, 25]. Thus, the production of unusually high levels of ethylene may be an indicator of stress. The diagnostic value of ethylene production as a means to detect stress may be limited by practical considerations, but there may be circumstances where this relationship could be used in screening systems. If ethylene is involved in the plant response, susceptible lines may be distinguished from resistant lines by exposing them to ethylene or ethephon [24].

The participation of ethylene in plant responses to stress, especially biological stress, can lead to an understanding of the mechanism of damage. The sequence of events may be attack by biological agent, induction of stress-ethylene production, physiological action by the ethylene, and development of symptoms. Stress ethylene appears to be the cause of insect-induced abscission of cotton flower buds that we are studying [22], but the induction of ethylene production could be due to mechanical injury or something secreted during the feeding process, such as a hormone, ACC, a microbe, an enzyme, or an elicitor [22, 25]. Our recent data (unpublished) indicate that microbes are involved; the microorganism may be producing the ethylene or inducing the plant to produce it [25].

3.3 The Indicator Role in Flowering

Apparently, flowering is a developmental event in which ethylene acts as a natural regulator and also as a means to respond to stress. Aside from initiation of flowering in bromeliads [1], increased ethylene production occurs in connection with opening, pollination, petal fading, and abscission [1, 21, 23]. These fluctuations in ethylene production appear to occur within a balance of the total hormone complement that allows "normal" development. After pollination in cotton, the daily peaks of ethylene production decrease on fruit which continue development. However, in plants exhibiting high fruit abscission these daily peaks remain high or even increase prior to abscission [13]. Conditions, which promote young fruit abscission, also promote ethylene production [9]. Thus, much evidence implicates ethylene as a fruit-shedding regulator in cotton. Therefore, it is necessary to distinguish between ethylene that regulates normal developmental events and ethylene that initiates fruit abscission in response to stress. It may be possible and desirable to suppress stress-induced abscission with inhibitors of ethylene synthesis or action [17–20]. This opportunity may also occur in crops, such as seed legumes, where stress ethylene causes fruit abscission [24].

The suppression of ethylene action has not been employed extensively in the field; most treatments are adapted for use in postharvest situations. Effective inhibitors of ethylene action or synthesis are needed. AVG, AOA, Ag^+, and Co^+ are available [20], but they have not proven to be commercially useful. New inhibitors might open up new ways to regulate plant development.

4 Conclusions

The use of ethylene as an agricultural plant growth regulator is increasing. Most applications of ethylene actually involve ethephon rather than the gas itself. The proportion of ethephon used on horticultural crops is declining, while the proportion used for field crops is increasing. On the world market, the primary field-crop uses of ethephon have been to reduce lodging in small grains and to "ripen" sugar cane. In the USA, relatively recent uses of ethephon on field crops include promoting cotton fruit opening and reducing lodging in small grains. A use of ethephon under development is to reduce lodging in corn.

Ethylene production is now well recognized to be an indicator of certain developmental events and of stress. Ethylene is the means by which some biological stresses, including insect and disease organisms, produce some of their symptoms. It may not always be practical to monitor the occurrence of stress by measuring ethylene production; however, inhibitors of ethylene synthesis or action may be useful in reducing damage. In addition, ethephon can be used to test for stress "resistance".

References

1. Abeles FB (1973) Academic Press, New York
2. Adams DO, Yang SF (1979) Proc Natl Acad Sci USA 79:170
3. Adkisson PL (1962) J Econ Entomol 55:949
4. Adkisson PL (1966) Science 54:234
5. Cathey GW, Luckett K (1980) Proc Beltwide Cotton Prod Res Conf (Nat Cotton Council, Memphis, Tenn 38112, USA), p 35
6. Cooke AR, Randall DI (1968) Nature 218:974
7. deWilde RC (1971) Hortic Sci 6:364
8. Dunster KW, Fosse RA, Lavoy JD, Jarinko PA (1980) Proc Beltwide Cotton Prod Res Conf, p 35
9. Guinn G (1982) Plant Physiol (Bethesda) 69:349 (and references cited therein)
10. Hall WC, Truchelut GB, Leinweber CL, Herrero FA (1957) Physiol Plant 10:306
11. Lipe JA, Morgan PW (1972a) Plant Physiol (Bethesda) 50:759
12. Lipe JA, Morgan PW (1972b) Plant Physiol (Bethesda) 50:765
13. Lipe JA, Morgan PW (1973) Plant Physiol (Bethesda) 51:949
14. Lursen K, Naumann K, Schroder R (1979) Z Pflanzenphysiol 92:285
15. Metzer RB (1984) Proc Beltwide Cotton Prod Res Conf, p 73
16. Morgan PW (1969) Plant Physiol (Bethesda) 44:337
17. Morgan PW (1973a) Acta Hortic (The Hague), 34:41
18. Morgan PW (1973b) Proc Beltwide Cotton Prod Res Conf, p 95 (also see 1977, Adv Chem Ser No 159, p 42, Am Chem Soc)
19. Morgan PW (1982) Agric Forestry Bull (University Alberta, Canada) 5:29
20. Morgan PW (1984) In: Fuchs Y, Chalutz E (eds) Ethylene: biochemical, physiological and applied aspects. Nijhoff, The Hague, p 231
21. Morgan PW, Durham JI, Lipe JA (1974) In: Sumiki Y (ed) Plant growth substances 1973. Hirokawa, Tokyo, p 1062
22. Morgan PW, Powell RD, Grisham MP, Sterling WL, Cooper L (1983) Proc Beltwide Cotton Prod Res Conf, p 54
23. Suttle JC, Kende H (1978) Plant Physiol (Bethesda) 62:267
24. Wien HC (1979) Plant Physiol (Bethesda) Suppl 63:92
25. Yang SF, Pratt HK (1978) In: Kahl G (ed) Biochemistry of wounded plant tissues. Gruyter, Berlin, p 595

Effects of Endogenous and Exogenous Plant Growth Substances on Development and Yield of Soybeans

M.L. BRENNER[1], W.A. BRUN[2], J. SCHUSSLER[1], and N. CHEIKH[1]

1 Introduction

Development of commercially useful plant growth regulators (PGR) for soybeans has proven to be a major challenge. It seems that the regulation of soybean yield is highly self-compensating. Our program therefore, was first to understand the endogenous regulation of yield components. We hope information from this work will provide a basis for biorational selection processes for PGRs. Endogenous plant growth substances (PGS) have been implicated in the regulation of reproductive development of soybeans [1–3, 5, 7, 8, 14–18]. This report describes the long-distance transport of PGSs throughout soybean plants and suggests roles for PGS in the movement of photoassimilates to developing seeds.

2 Movement of PGSs in Soybeans

Developing soybean seeds contain substantial quantities of both ABA and IAA, especially during their mid-pod filling stage [7, 18]. It appears that developing soybean seeds are both a sink and source of PGR during their development. Previous work in our group established that depodding the entire soybean plant results in accumulation of ABA in leaves [16, 17]. In addition, substantially more ABA was recovered in EDTA-enhanced phloem exudate from leaves transported basipetally than that which moved acropetally [8].

We have recently completed a series of experiments to further characterize how ABA is transported throughout soybean plants. In these experiments, we simultaneously injected trace levels of [^{3}H]±ABA (5 ng) and [^{14}C]sucrose (451 ng) in 10 μl of ethanol into the petiole of fully expanded trifoliolate leaves. Entire plants were sampled at several time intervals, subdivided into parts, and dried. The tissue was pulverized, combusted in a tissue oxidizer, and radioactivity measured by liquid scintillation counting. In previous work, we had shown that ABA, injected in this manner, moves to the leaf blade via the transpiration stream and is then exported

1 Department of Horticultural Science and Landscape Architecture, University of Minnesota, St. Paul, MN 55108, USA

2 Department of Agronomy and Plant Genetics, University of Minnesota, St. Paul, MN 55108, USA

Plant Growth Substances 1985
(ed. by M. Bopp)

from the leaf via the phloem [17]. In another experiment, we have established that approximately 84% of the exported ABA is transported in the phloem or xylem as ABA, with small amounts of PA and DPA [4]. The current work, using both vegetative and reproductive plants, we find [^{3}H]ABA to be present in all parts of the plant within 15 min (data not shown). The tracers were injected into the petiole of the most recently fully expanded trifoliolate leaf of vegetative plants and the fourth fully expanded leaf of reproductive plants.

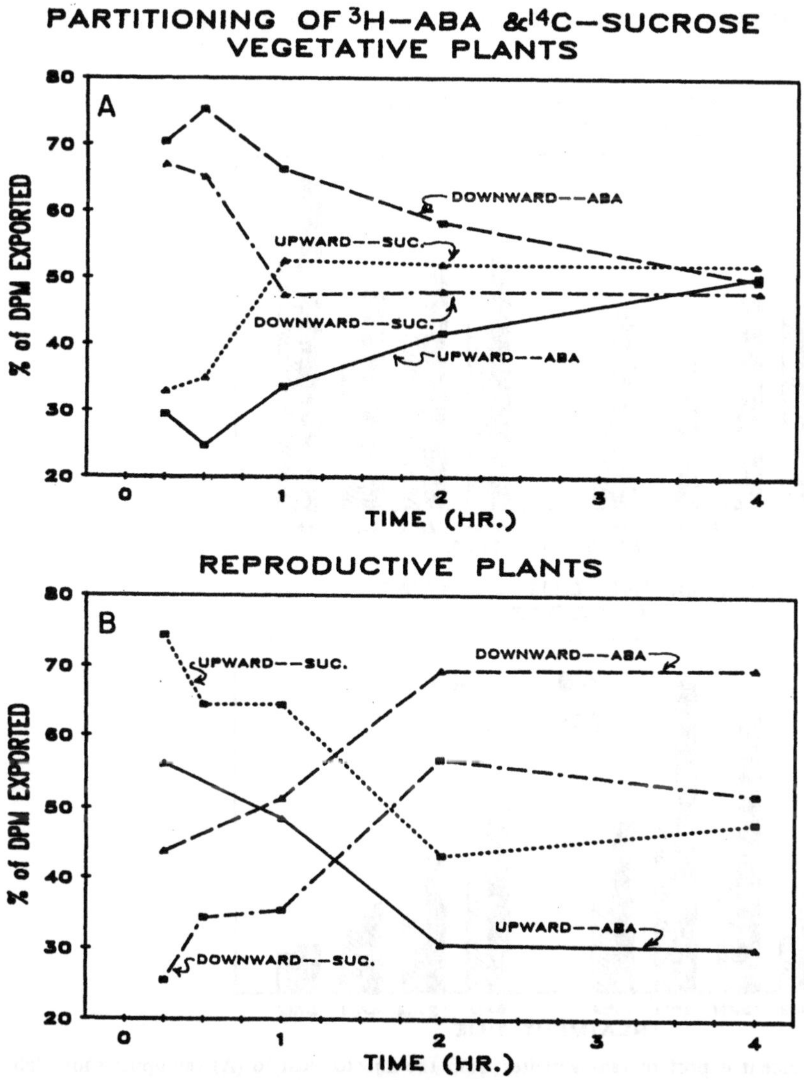

Fig. 1A, B. Percent distribution of radioactivities above and below the node of the injected leaf as a function of the time after the simultaneous injection of [^{3}H]ABA and [^{14}C]sucrose into petiole of (A) vegetative and (B) reproductive soybean plants. Means of five replicates of plants grown in a controlled environment

In the vegetative plants, initially most of the exported labeled material was found below the injected leaf. By 4 h both tracers were found equally distributed above and below the injected leaf (Fig. 1A). In reproductive plants as much as 37% of the [^{14}C] and 15% of the [^{3}H] exported was found in the pods subtended by the injected leaf 4 h after injection. The partitioning pattern of that which moved to other parts of the plant, however, was quite different from that seen in the vegetative plants. Following the injection, the [^{14}C]-labeled material first moved upward, but then was found equally above and below the injection point. In contrast, the [^{3}H] was initially equally distributed after the injection; but after 2 h, the [^{3}H], was found primarily below the injection point (Fig. 1B), especially in the lower shoot tissue (data not shown).

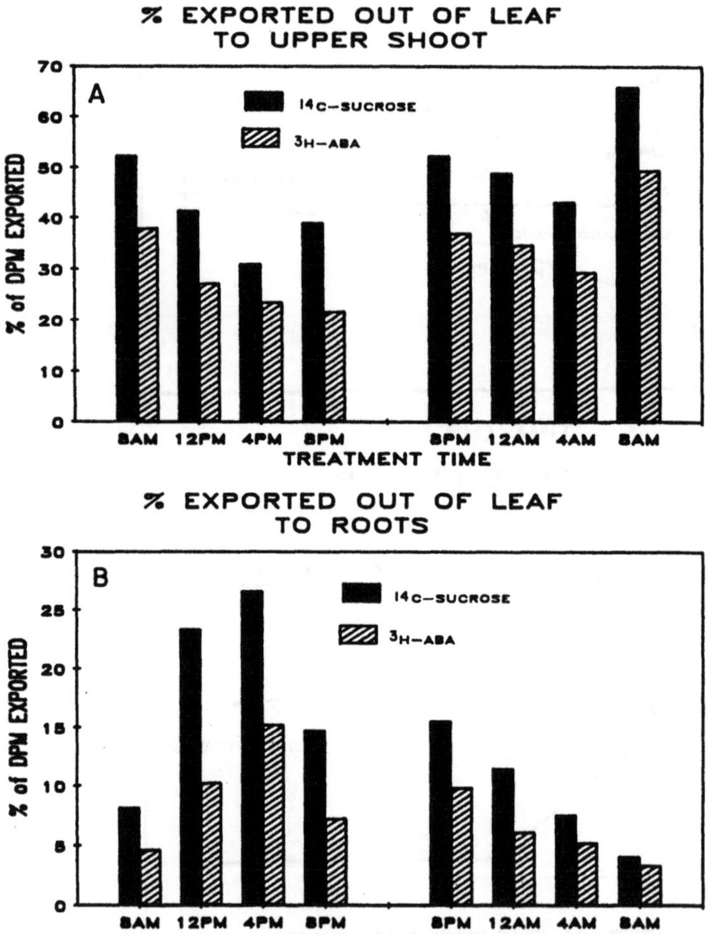

Fig. 2A, B. Percent export of radioactivites from the injected leaf to (**A**) the upper shoot (all tissue above the node of the injected leaf) and to (**B**) the root as a function of the time after the simultaneous injection of [^{3}H]ABA and [^{14}C]sucrose into the petiole of the fifth leaf of vegetative plants. Injections were made every 4 h and the plants were harvested 4 later. The experiment was conducted in two 12 h segments separated by a 36 hr interval. The vegetative plants were maintained in a growth chamber with a 14 h (7:30 am to 9:30 pm)

Adding an additional complication is the observation that the partitioning pattern for both tracers appears to be dependent on the time of day at which the injection is made (Fig. 2). During the day, the relative amounts of both label going to the upper shoot (Fig. 2A) first decreased, then leveled off, while that going to the root (Fig. 2B) first increased and then dropped. During the night, the amounts of label exported to the upper shoot showed no clear trends, while that going to the root clearly decreased with time. It is not known if the observed diurnal pattern is due to differences in the endogenous pools of ABA and sucrose, or due to differences in the actual export of the substances. We have observed that ABA concentration does exhibit a diurnal pattern (C. Lecoq and M. Brenner, unpubl.).

3 ABA as a Promotor of the Accumulation of Photoassimilates by Seeds

There are many reports of a positive relationship between endogenous or exogenous ABA and the accumulation of photoassimilates by seed [4]. When three soybean plant introduction lines that varied in seed growth rates and final seed size were studied, we found a highly significant positive correlation (r^2 = 0.60) between ABA content and in vitro sucrose uptake by excised cotyledons [18]. Correspondingly, sucrose uptake is a good indicator of seed growth rates.

Though we [18] and Saftner and Wyse [14] have been able to stimulate the in vitro accumulation of sucrose as a result of ABA application, we have only been able to do this with growth-chamber grown plants. Extensive tests on field-grown material have shown no effect of the applied ABA. It is likely that field-grown material is subjected to greater periods of drought stress and may, therefore, have higher endogenous ABA levels.

Under field conditions, even if soil moisture is apparently adequate, soybean plants may be exposed to intermittent water stress. That is, mid-afternoon periods of water stress are interspersed with late night and early morning periods of stress recovery [19]. Boyer et al. [2] monitored mid-afternoon leaf water potentials of eight soybean cultivars growing in the fertile soils of central Illinois for 3 years during which soil water availability was high. They found that despite the favorable soil-moisture conditions, leaf-water potentials at mid-afternoon frequently fell below-11 bars, which they defined as water stress.

It is well-documented that with decreasing leaf-water potentials, ABA content of leaves rapidly increases and stomata close [12]. When leaf turgor is restored, the ABA within the leaf gradually decreases due to its catabolism and/or export from the leaf [20, 21]. ABA is very mobile within the plant as described above. Zeevaart and Boyer [21] have shown that in *Xanthium.* ABA is translocated from its site of synthesis in mature source leaves down to the root and back up to the young leaves, where it accumulates.

With this background, we tested the hypothesis that exposing the parent soybean plant to periods of stress and stress recovery during rapid seed filling would elevate the concentration of ABA in the seeds. As shown in Table 1, the stress cycling caused a 31% and 37% increase in the ABA concentrations of seed coats and developing

Table 1. Effect of three water-stress cycles on endogenous ABA content and in vitro sucrose uptake by soybean embryos

	Nonstressed	Stressed [a]
Testa ABA (ng mg^{-1} fresh wt.)	4.8 ± 0.43	6.3 ± 0.31
Cotyledon ABA (ng mg^{-1} fresh wt.)	5.1 ± 0.42	6.9 ± 0.43
Sucrose uptake (μmol/100 mg dry wt. x h)	0.8 ± 0.02	1.2 ± 0.03

[a] Soybean plants were subjected to three periods of drought by withholding water until slight wilting occurred, followed by 24 h of recovery after rewatering. Experimental methods for ABA content and sucrose uptake were the same as previously reported [18] using soybean Pl 416.845

cotyledons, respectively. These responses were associated with a 50% increase in the rate of in vitro sucrose uptake by the isolated cotyledons.

In a second experiment, a single stress cycle (water withheld for 48 h) was followed by a 24 h recorvery period (well watered), after which the developing cotyledons were excised and pre-incubated for 1 h in either 150 or 10 mM sucrose. Following this incubation period, the cotyledons were transferred to 10 mM [^{14}C]-sucrose for 3 h to determine their rate of in vitro sucrose uptake. The uptake was stimulated, in this case 30%, by the stress cycle. There was no difference between

Table 2. The effect of exogenous treatments on sucrose-phosphate-synthase and fructose-1,6-bisphosphatase of soybean leaves [a]

Treatment	SPS	F-1,6-BPase
	(μmol of Product/dm^2 x h)	
Experiment I: Vegetative plants – V3		
Control	167 ± 33	165 ± 19
10^{-6} M GA_3	282 ± 100	137 ± 19
Experiment II: Vegetative plants – V3		
Control	148 ± 32	120 ± 13
10^{-5} M $GA_{4/7}$	193 ± 7	237 ± 27
Experiment III: Reproductive plants – R4		
Control	80 ± 8	480 ± 11
Stressed-cycled	124 ± 8	318 ± 37

[a] GAs or water were applied for 24 h prior to sampling via a capillary wick. The wick system was developed by I. Wellerstein (pers. comm. [4]). It consisted of embroidery thread drawn through plastic capillary tubing and was threaded through leaf petiole of Clay soybean at stage V3. Enzyme assays were as follows: SPS [10], F-1,6,BPase [10, 11, 13]. Tissue for enzyme analysis was collected 2.5 h after the illumination period commenced in the growth chamber

the 150 and 10 mM sucrose pre-incubation treatments, indicating that the stimulation of in vitro sucrose uptake was not caused by differential apoplastic pools.

From these two experiments, we conclude that following recovery from a mild water stress, in vitro sucrose uptake rates by isolated soybean cotyledons is stimulated, and that this stimulation is associated with elevated ABA concentrations, but is not related to decreased sucrose concentrations in the cotyledons.

4 The Effect of Applied PGRs on Enzymes Involved with Phloem Loading

Kerr et al. [10] have reviewed that sucrose-phosphate-synthase (SPS) and fructose-1,6-bisphosphotase (F-1,6-BPase) are rate-limiting enzymes controlling the level of sucrose in the leaf available for phloem loading. Our preliminary data indicates that application of low concentrations of gibberellins stimulates both of these enzymes (Table 2). In addition, subjecting the soybean plants to two stress cycles resulted in increasing SPS activity, while inhibiting F-1,6-BPase.

5 Conclusions

ABA movement about the plant is dynamic. It has previously been shown that ABA will rapidly move from mature leaves to all other parts of vegetative plants and that it generally accumulates in sink tissues [6, 9, 21]. Our data are consistent with these observations, but we now know that the pattern of ABA distribution changes diurnally and with the stage of plant development.

The accumulation of ABA in sinks, such as filling soybeans seeds, appears to function as a promoter of assimilate import. Increased mobilization of ABA to seeds in plants recovering from stress may be a mechanism to accelerate recovery from the stress.

Acknowledgments. The technical assistance by M. Clement is greatly appreciated. Support for this research came from the Minnesota Agricultural Experiment Station, USDA 83-CRSR-2-2275, Minnesota Soybean Research an Promotion Council, and Monsanto Agricultural Products Co.

References

1. Ackerson RC (1984) J Exp Bot 35(152):403
2. Boyer JS, Johnson RR, Saupe SG (1980) Agric J 72:981
3. Brenner ML, Hein MB, Schussler J, Daie J, Brun WA (1982) In: Wareing PF (ed) Plant growth substance 1982. Academic Press, New York, p 343
4. Brun WA, Brenner ML, Schussler J (1985) Br Plant Growth Regul Group Monograph 12:257
5. Eeuwens CJ, Schwabe PB (1975) J Exp Bot 26:1
6. Everat-Bourbouloux A (1982) Physiol Plant 54:431

7. Hein MB, Brenner ML, Brun WA (1984) Plant Physiol (Bethesda) 76:951
8. Hein MB, Brenner ML, Brun WA (1984) Plant Physiol (Bethesda) 76:955
9. Hocking TJ, Hillman JR, Wilkins MB (1972) Nature New Biol 235:124
10. Kerr PS, Huber SC, Israel DW (1984) Plant Physiol (Bethesda) 75:483
11. Pontremoli S, Melloni E (1975) Methods Enzymol 42:354
12. Raschke K (1982) In: Wareing PF (ed) Plant growth substance 1982, Academic Press, New York, p 581
13. Rathbun WA, Betlach MV (1969) Anal Biochem 28:436
14. Saftner RA, Wyse RE (1984) Plant Physiol (Bethesda) 74:951
15. Setter TL, Brun WA, Brenner ML (1980) Plant Physiol (Bethesda) 65:884
16. Setter TL, Brun WA, Brenner ML (1980) Plant Physiol (Bethesda) 65:1111
17. Setter TL, Brun WA, Brenner ML (1981) Plant Physiol (Bethesda) 67:774
18. Schussler JR, Brenner ML, Brun WA (1984) Plant Physiol (Bethesda) 76:301
19. Slatyer RO (1967) Academic Press, New York
20. Zeevaart JAD (1983) Plant Physiol (Bethesda) 71:477
21. Zeevaart JAD, Boyer GL (1984) Plant Physiol (Bethesda) 74:934

Natural and Synthetic Plant Growth Substances in Fruit Growing

F. BANGERTH[1]

1 Introduction

Fruit production is a laborious and expensive business. Many technological steps and a great number of physiological processes are involved in fruit production, beginning with a nursery grown tree and ending with the storage of fruit in a sophisticated storage room. Many of these steps and processes can be influenced by plant growth regulators (PGR) in order to curtail costs of production, to increase yield, or to improve fruit quality. Thus, the potential use of PGRs in fruit production is greater than in any other crop and it is impossible in a short review to cover in detail even the most important of them. The potential uses for PGRs in fruit production are, therefore, summarized in Table 1, which also includes the mentioning of more specific reviews or at least publications with extented reference lists. Hopefully, this will allow the interested reader an easy access to current literature and more selective information.

Despite this multitude of potential uses of PGRs in fruit production, large-scale applications are still restricted to a few areas. Some of the principal limitations for a more widespread use of PGRs are:

1. Great variability and inconsistency of effects.
2. Dependency on prevailing weather conditions and physiological stage of the plant.
3. Numerous and often negative side effects.
4. Species- and even cultivar-specific response.

These potential, undesirable effects of PGRs make recommendations for their use and predictions of their effects often unreliable. Most of these disturbances are to be expected, because PGRs usually interfere with a very dynamic, endogenous hormone balance and thus affect a multitude of physiological processes, in addition to the one intended to modify. Whether it will be possible to overcome most of these problems in the future will depend mainly on our progress in understanding the regulation of specific metabolic and developmental processes in plants and how hormones and PGRs interfere with them.

1 Institut für Obst-, Gemüse- und Weinbau, Universität Hohenheim, D-7000 Stuttgart 70, FRG

Plant Growth Substances 1985
(ed. by M. Bopp)

Table 1. Potential uses of plant growth regulators in fruit production

Purpose of application		Substances used [a]	Physiological reaction involved	Review articles or articles with extendend literature citations
1	In the nursery			
1.1	Seed stratification	GA_S, CK_S	Overcoming dormancy	[22]
1.2	Rooting of cuttings	IBA, NAA, etc.	Initiation of adventitious roots	[15] [18] [1]
1.3	Propagation by tissue culture	Auxins, cytokinins	Cell division, tissue differentiation	[29] [11]
1.4	Increasing lateral branching and branching angle	Auxin transport inhibitors, BA	Weakening, but not prevention, of apical dominance	[26] [7]
1.5	Stimulating leaf shedding in atumn	Ethylene generators, various PGRs	Stimulating leaf senescence and abscission	[21] [27]
2	In the orchard			
2.1	Enhancing bud break	GA_S, CK, DNOC, mineral oil	Dormancy release	[8] [9]
2.2	Delaying bud break	GA_3, ethephon	Delaying blooming after dormancy release	[30]
2.3	Enhancing early bearing	Alar, ethephon	Shortening secondary juvenility	[19] [23]
2.4	Reducing tree height to ease pruning and harvesting	Various growth retardants	Reducing vegetative growth	[3] [24] [23]
2.5	Reducing regrowth after severe pruning	NAA in special paints	Enhancing apical dominance	[4] [10]
2.6	Affecting yield			
2.6.1	Preventing alternate bearing	Alar, Ethrel, GA_3	Reducing or stimulating flower induction	[5] [16]
2.6.2	Stimulating fruit set	GA_S, auxins, CK_S, AVG, Alar	Parthenocarpic fruit development, preventing excessive fruit drop	[12] [25]
2.6.3	Reducing fruit set	NAA, Carbaryl, ethylene generators, Figaron	Abscission of young fruitlets	[6, 31] [25]
2.6.4	Reducing preharvest drop	NAA, 2,4,5 TP AVG	Preventing abscission	[14, 24]
2.7	Mechanical harvest aids	Ethylene generators, substances enhancing C_2H_4 production	Stimulating abscission	[17] [32] [13]

Table 1 (cont.)

Purpose of application		Substances used [a]	Physiological reaction involved	Review articles or articles with extendend literature citations
2.8	Improving fruit quality	Various PGRs	Changing fruit constituents	[2]
2.9	Affecting ripening			
2.9.1	Enhancing fruit ripening	Ethylene generators, auxins	Induction of autocatalytic ethylene production	[2]
2.9.2	Delaying fruit ripening	AVG, GA_S, CK_S, Alar	Inhibition of C_2H_4 biosynthesis	[2]
3	Improving storage life	AVG, O_2, CO_2, Alar	Inhibition of C_2H_4 synthesis and action, retardation of senescence	[20] [28]

[a] *Abbreviations:* AVG: Aminoethoxyvinylglycine; BA: Benzyladenine; C_2H_4: Ethylene: CK_S: Cytokinins; DNOC: Dinitroorthocresol; GA_S: Gibberellins; IBA: Indolylbutyric acid; NAA: Naphthaleneacetic acid; PGR: plant growth regulators 2,4,5 TP: 2,4,5-Trichlorophenoxypropionic acid

2 Conclusion

PGRs have been used for a variety of purposes in fruit production for a long time. In fact, some of the earliest practical applications of these substances (e.g. rooting of cuttings, preventing pre-harvest drop) took place in fruit production. Despite this long-term use of PGRs serious problems still exist. These problems will most likely remain until we are able to better understand the homeostatic control mechanisms responsible for regulating hormone levels and how these hormones interfere with developmental processes, such as fruit set, growth, drop, ripening, etc. Only then can we develop new strategies in the search for new and probably more specific PGRs.

References

1. Al Barazi Z, Schwabe WW (1984) J Hortic SCI 59: 453
2. Bangerth F (1983) In: Lieberman M (ed) Post-harvest Physiology and crop preserveration, Plenum, New York, p 331
3. Bartolomew DP, Criley RA (1983) In: Nickel LG (ed) Plant growth regulating chemicals, vol 2. CRC, Boca Ration, Florida, p 1
4. Brana BA, Jackson JE (1982) J Hortic Sci 57:17 and 31
5. Buban T, Faust M (1981) Hortic Rev 4:174

6. Edgerton LJ (1973) Acta Hortic (The Hague) 34:333
7. Elfing DC (1984) J Am Soc Hortic Sci 109:476
8. Erez A, Lavee S (1974) Proc 19th Int Hortic Congr Warszawa 3:69
9. Erez A, Zur A (1981) Sci Hortic (Amst) 14:47
10. Freund H, Bünemann G (1982) Erwerbsobstbau 24:56
11. George EF, Sherrinton PD (1984) Plant propagation by tissue culture. Exegetics, Eversly, England, p 448
12. Goldwin GK (1984) Acta Hortic (The Hague) 149:161
13. Goren R (1983) Acta Hortic (The Hague) 137:15
14. Halder-Doll H, Bangerth F (1985) Gartenbauwiss 50 (in press)
15. Hartmann HT, Kester DE (1975) Plant propagation principles and practices. Prentice Hall, New Yersey
16. Hoad GV (1984) Acta Hortic (The Hague) 149:13
17. Holm RE, Wilson WC (1977) J Am Soc Hortic Sci 102:576
18. Howard BH, Bassuk NL, Kim YK (1981) Acta Hortic (The Hague) 120:199
19. Karnatz A, Simon P (1980) Aktuelle Literaturinformation aus dem Obstbau Nr. 100, Berlin (IBSN 3 7983 04092)
20. Knee M (1985) In: Roberts JA, Tucker GA (ed) Ethylene and plant development. Butterworth, London, p 297
21. Larsen FE, Lowell GD (1977) Hortic Sci 20:45
22. Lewak S (1981) Acta Hortic (The Hague) 120:149
23. Looney NE (1983) In: Nickel LG (ed) Plant growth regulating chemicals. CRC, Boca Ration, Florida, p 1
24. Monselise SP (1979) Sci Hortic (Amst) 11:151
25. Monselise SP (1983) Proc 21st Int Hortic Congr, Hamburg, vol, p 468
26. Quinlan JD (1981) Acta Hortic (The Hague) 120:95
27. Reid MS (1985) Hort Sci 20:45
28. Sherman M (1985) Hort Sci 20:57
29. Styer DJ, Chin CK (1983) Hortic Rev 5:221
30. Webster AD (1984) J Hortic Sci 59:377
31. Williams MW (1979) Hortic Rev 1:270
32. Zahara M, Johnson SS (1979) Hort Sci 14:578

Recent Advances in the Use of Growth Substances in Ornamental Horticulture

A.H. HALEVY[1]

1 Introduction

The use of plant growth regulators (PGR) for the manipulation of growth and development in ornamentals is more common than in most other commercial crops. The number and variety of practical uses to which PGR can be put, is also probably greater in ornamentals than in food and plantation crops relative to the very small percentage of the total agricultural and horticultural growing area. There are several reasons for this: (1) the number and variety of cultivated ornamental species are probably greater than the total of all other cultivated plants. (2) The cropping of cut flowers and potted plants often has to be accurately timed for specific occasions, such as pointsettias for Christmas and *Lilium longiflorum* for Easter. (3) The shape of the plants contributes to their ornamental value, i.e., size and color of flowers and foliage, length and strength of stems, branching pattern, etc. These traits can be manipulated by PGR. (4) Flowers and flowering potted-plants are very vulnerable and are often short-lived. Flowers are often harvested and shipped in the bud stage and are intended to continue to develop even after being purchased by the customer. PGR are used both to promote flower development and opening and to extend their longevity. (5) There are two economical factors which justify the extensive use of PGR in ornamentals. First, since ornamentals are luxury commodities, their relatively high prices can cover the cost of certain treatments, such as those with PGR that are too expensive to use for food crops. Second, it is probably easier to secure official clearance for the use of new PGR in ornamentals than in edible crops.

The following review of recent developments in the use of PGR in ornamentals, is confirmed mainly to treatments that are already commercially available. Furthermore, only those applications of PGR, which are unique to specific ornamental plants, are described here. Of the five known endogenous plant hormones, abscisic acid does not appear to be used in ornamentals, and there are only a few recently developed uses of auxins. This review will, therefore, concentrate on ethylene and ethylene antagonists, gibberellin (GA) and GA-antagonists (growth retardants), and cytokinins.

1 The Hebrew University of Jerusalem, Department of Ornamental Horticulture, Rehovot 76100, Israel

Plant Growth Substances 1985
(ed. by M. Bopp)

2 Ethylene and Ethylene Inhibitors

More progress has been made in understanding the physiology, biochemistry [64] and practical uses of ethylene than of other PGR (compare Kende et al., this vol.).

Until recently it was possible to apply ethylene to plants only by treating them with ethylene gas or with ethylene-releasing liquid preparations, such as (2-chloroethyl) phosphonic acid (ethephon) (see Morgan, this vol.). It is now also possible to treat plants with ACC, which in most tissues becomes endogenously converted almost immediately to ethylene [5].

Several ethylene antagonist are now known, which act as inhibitors of ethylene biosynthesis or action [64]. The most effective inhibitors of ACC synthase are rhizobitoxin and its analogs such as aminoethyoxyvinylglycine (AVG). However, these chemicals are costly and not readily available. An inexpensive alternative is aminooxyacetic acid (AOA), which also inhibits ACC formation in plant tissues. The oxidation of ACC to ethylene is inhibited by Co^{2+} ions, as well as by several uncouplers and free-radical scavengers, such as propyl gallate. The most effective inhibitor of ethylene action, both endogenous and exogenous, is silver. Since silver ions are almost immobile in plant tissues, the introduction of the mobile anionic silver thiosulfate complex (STS) represents a very important practical development [62].

One of the most important uses of ethylene is in the promotion of flower formation in ornamental bromeliads, such as *Aechmea, Ananas, Bromelia, Billbergia, Guzmania, Qusnelia,* and *Varisea.* Ethephon is the agent normally used, either as a foliar spray or poured onto the center of plants of appropriate size and maturity stage. Similar treatments using ACC (20–50 mg l^{-1}) have recently been introduced, with very good results [8]. The control of flowering in bromeliads also includes the prevention of flower formation when not desired. Single treatment with AVG (10 mg l^{-1}) or AOA (0.2 μM) can prevent flowering for several months, and subsequent treatment with ACC will then induce flower formation [33]. The ability of the tissue to convert ACC to ethylene serves as an indicator for the maturity of the plants, i.e., their ability to respond to the flower-promoting effect of ethylene (M. De Proft, pers. comm.).

Some plants flower abundantly after fire. This phenomenon known to botanists for many years, was recently attributed mainly to the effect of ethylene [12]. Japanese workers have used smoke to promote the flowering of iris bulbs and to break the dormancy of *Freesia* corms. The most effective constituent in the smoke was found to be ethylene [25, 31, 60]. Ethylene also appears to be the active compound in the smoke treatment applied to *Narcissus tazetta* cv. Grand Soleil d'Or in Southern England [24, 58, 59].

Iris bulbs of flowering size sometimes fail to form flowers when forced, and instead produce vegetative plants with three leaves. Flower formation in iris is greatly enhanced by spraying the fields with ethephon prior to bulb harvesting [10, 26]. Similar results were obtained when freshly harvested iris bulbs cv. Ideal were treated with ethylene (5 μl l^{-1}, 24 h) [59] or smoke [25], or were dipped in ethephon [56]. These treatments may also make it possible to use smaller bulbs for winter forcing and increase planting density [59]. In the production of bulbs flowering is undesirable. Storage of iris bulbs in an atmosphere containing 5% CO_2 considerably reduced flower formation, apparently by antagonizing the ambient and endogenously formed ethylene (W. J. De Munk, pers. comm.).

When tulip plants are grown for bulb production, flower formation is undesirable since the flowers compete with the new bulbs for the allocation of metabolites. The exposure of tulip bulbs to ethylene (0.5 $\mu l\ l^{-1}$, 3 days) causes total flower blasting and promotes bulb development [26].

Ethephon has also been shown to break the dormancy of *Gladiolus* [19] and *Liatris* [27] corms. Ethephon has also been used to ocercome the inhibition of buds at the basal region of rosa plants, thus promoting the production of vigorous renewal canes [66].

Some bulbous plants, such as hyacinths and daffodils do not respond to the common growth retardants. In such cases ethephon (1000–2000 mg l^{-1}) may be used as a soil drench in the production of potted-plants [9].

In many plants ethylene is directly involved in the enhancement of flower senescence or abscission [17] and in the promotion of flower bud abortion or blasting [26]. In these cases ethylene antagonists, and especially STS, may promote flower development to full anthesis, delay flower senescence, and prevent abscission. Postharvest, short-term treatment (a few minutes to a few hours) with STS to the base of the cut flowers is now commonly applied to carnations, since it doubles or triples the life of the cut flower and enables their long-term storage [14, 62]. The preshipment spraying of flowering potted-plants and the postharvest base treatment of cut flowers with STS is very effective in preventing or reducing flower bud and petal drop in many plants, e.g., sweet pea, snapdragon, lily, lupine, *Delphinium, Pelargonium, Schlumbergera, Hibiscus, Calceolaria, Bougainvillea,* and *Streptocarpus* [1, 4, 13, 37, 62], as well as in preventing or reducing leaf drop in several foliage plants [2, 43].

Treatment of hybrid lily bulbs with STS prior to planting or spraying the plants with STS in the greenhouse greatly reduces flower but abortion [55, 61].

Other important effects of STS include the reduction of geotropic stem bending in some flowers (such as *Kniphofia*) and the enhancement of flower bud openings in cut flowers of iris [9] and *Freesia* [52]. In several cultivars of *Begonia Cheimantha* grown as potted-plants, the flowers fail to develop normally, do not open or abscise, when grown at high temperatures in summer. STS spray (2 mM) enabled normal development and opening of flowers [34].

3 Gibberellins

Gibberellins (GA) are known to promote flower formation and development in many long-day and cold-requiring plants as well as in several coniferous trees [3]. Surprisingly, GA promote flower formatiln also in some short-day plants, such as the ornamentals *Zinnia elegans* [44] and *Impatiens balsamia* [39]

In spite of the fact that GA has been shown to induce flower formation in many plants, they do not appear to be used *commercially* to induce flowering under strictly non-inductive conditions. Under marginal or non-optimal environmental conditions, however, GA are sometimes used to promote flower production.

In field-grown Killian daisies (*Chrysanthemum x superbum,* a long-day plant) in Southern California, night-break light is provided to facilitate winter production. GA (2 sprayings of 100 ppm, 3 weeks apart) to plants at the rosette stage advanced flowering under these conditions, but GA is not able to act as a substitute for the light treatment.

To obtain winter production in Israel of *Gypsophila paniculata,* a long-day plant, lighting is provided for 5–6 h at night. However, the treatment is effective only when night temperatures are above 12°C. GA promotes flowering of lit plants when night temeratures are low, but even massive and repeated applications of GA cannot induce flowering in short days [47, 48]. Thus, in *Gysophila* GA substitute only for the high night temperature, which is required as a background for the photoperiodic induction.

Flowering of the annual statice (*Limonium sinuatum*) is induced by vernalization and promoted by long days. GA promotes flower production in statice plants grown at moderate temperatures and 12 h day [46].

Injection, vacuum infiltration, or soaking of tulip bulbs with GA (250–500 mg l^{-1}, 24 h) advanced flowering of greenhouse-forced plants by 1–2 weeks and enhanced flower size. The treatment is used with cooled bulbs as a supplement to the normal cold treatment, but not as a replacement for it [20, 21]. Treatments with GA as well as with cytokinins also reduce the susceptibility of flowers to floral bud blasting, a common occurrence in winter forcing of several tulip cultivars [10, 26].

An important recent finding is the promotion of flower production by GA in *Cordyline* (Agavaceae) [11] and in plants of the Araceae, which are photoperiodically neutral and do not respond to cold. A single foliar spray with GA (250 mg l^{-1}) induced profuse flowering 10 to 14 weeks after treatment in several ornamental aroids, such as *Aglaonema, Dieffenbachia, Spathiphyllum,* and *Caladium.* The GA-induced flowers were fully fertile and produced many seeds, but in a few *Spathiphyllum* cultivars some floral deformities were observed [22, 23].

GA may also promote flower development and even opening, as in the case of *Exacum* [29] and statice [50]. "Bullhead" is a severe floral malformation, which occurs under certain growing conditions, in several rose cultivars. In "Baccara" roses it is prevalent at low temperature and is characterized by an increased number of short petals and the occasinal appearance of secondary distorted florets in the center of the flower [30]. Injection of GA (5 μg) into the receptacle of young flowers prevented the development of this malformation [68].

The most common effect of GA, namely, stem elongation is used in practice in several flower crops. Low concentrations of GA are normally used to enhance stem and penduncle elongation of flowers, such as *Chysanthemum* [28], *Cyclamen* [63] and *Anemone.* In *Liatris* GA is also applied to tubers prior to planting in order to promote bud sprouting and growth [65].

Substituted phthalimides can replace GA treatments for the elongation of peduncles in *Cyclamen* [42] and spray-chrysanthemum [32].

In most flowers the shortest-lived organs are the petals, and the shelf-life of the flowering stem is determined by petal longevity. In some plants, however, the green foliage leaves on the flowering stems senesce even before the flowers do. In such cases it is important to treat the foliage in order to delay their deterioration. This is usually done with cytokinins. However, in *Lilium* and statice (*Limonium sinuatum*), GA is more effective than cytokinins in retarding chlorophyll destruction in cut flowering stems held in the dark, as under normal shipping conditions [51].

4 Growth Retardants

The most widely used PGR in ornamentals are the growth retarding chemicals, which are utilized commercially in order to produce compact, sturdy potted and bedding plants, to enhance the green color of the foliage, to strengthen flower stems, and to promote resistance of plants to environmental stresses [7, 40]. Of the many retardants available, only three: chlormequat chloride (CCC, Cycocel), daminozide (B-nine, Alar), and ancymidol – are widely used. Chlorphonium (phosfon) has a limited use only as soil applications.

GA inhibit flower initiation in several woody plants [3]. Growth retardants, which inhibit GA biosynthesis or action, stimulate flower formation in these plants. Growth retardants are widely used to enhance flowering of azaleas [53, 54], and occasionally also of other plants such as *Syringa* [57] and *Camellia* [54].

In *Gladiolus* [18] and roses [36] chlormequat enhanced flower production by reducing the incidence of flower abortion.

One of the main disadvantages of the available growth retardants is the restriction of their activity to certain plants. Their activity is especially low in woody plants, geophytes ("bulbs"), and foliage plants. Several new growth retardants have been introduced in recent years, but only one – paclobutrazol (PP-333, Bonzi) – seems to have a significant advantage over those mentioned above [15]. It has a wider range of activity than the other growth retardants. It can be applied as a foliar spray, as a soil drench, and by injection into woody plants [49]. It is not phytotoxic even at high concentrations and its active concentrations are lower than those of most other growth retardants. In addition to controlling stem extension, paclobutrazol also stimulates branching and greatly enhances flowering in some woody ornamentals, such as *Ixora* [70].

5 Cytokinins

Cytokinins have been shown to variously promote or inhibit flower formation in a number of plants [3], but are not used to enhance flowering directly in flower crops. However, they effectively release lateral buds from correlative inhibition, thus promoting bud breaking and increasing the number of lateral shoots. Cytokinins are, therefore, used to promote the production of cuttings in mother stocks of some plants. In self-inductive plants, such as roses and carnations, flower production is releated to the number of lateral shoots produced. Cytokinins, such as benzyladenine (BA), may therefore indreictly, promote flower production [38, 67].

Indeed, an increase in the flower production of roses was reported following the direct application of BA to selected lateral buds that otherwise did not sprout [41, 69]. BA applied to the base of rose plants promotes the production of vigorous bottom breaks, which are essential for continous renewal of roses [6].

In several cut flowers, such as statice, *Alstroemeria*, and some spray-chrysanthemum cultivars, yellowing and senescence of the green foliage occur before petal senescence. A pre- or postharvest spraying with cytokinins may be used in such cases to delay foliage deterioration [17].

A preshipment spraying with cytokinins is used to prevent flower bud drop in certain potted rose cultivars [16].

6 Auxins

The only practical use of auxins in the control of flowering crops is in the reduction of flower bud abscission in a few cut flowers and flowering potted-plants, such as *Bougainvillea* and *Chamaeleucium uncinatum* [17]. The usual compound used is naphthalene acetic acid (NAA), applied in concentrations of 20 to 50 mg l^{-1}. In cut *Agapanthus* flowers the most effective treatment against floret bud drop is a combination of stem base treatment with STS (4 mM, 3 h) and spraying with NAA (30 ppm) [35].

7 Conclusions

The most important recent developments in the use of growth regulators in ornamentals involve the use of ethylene or ethylene inhibitors. Instead of ethephon, ACC may be applied to promote flowering in bromeliads. Iris and narcissus may be smoked or treated with ethylene to advance flowering and reduce flower blasting.

Silver thiosulfate, an inhibitor of ethylene action, is now widely used to extend the longevity of several cut flowers and to delay flower abscission in cut flowers and flowering potted-plants. GA is used to promote flower formation, flower stem elongation, and flower opening in several ornamental plants. Yellowing of foliage is normally controlled by cytokinin treatment, as in *Chrysanthemum*, *Alstromeria*, and statice. In statice and lily GA is more active in delaying foliage senescence. Cytokinins are also used to enhance the growth of inhibited axillary buds, thereby increasing the production of cuttings in mother plants and flower production in roses and carnations.

Paclobutrazol is an important new growth retardant. It is active in a wide range of plants, including several bulbous and woody plants, which did not respond to other retardants.

The sensitivity of plant tissues to PGR varies with the type of plant and organ and their stage of development. A major obstacle to understanding the mode of action of PGR and extending their practical use is the lack of information about factors determining such sensitivity.

References

1. Agnew NH, Albrecht ML, Kimmis RK (1985) Hortic Sci 20:118
2. Auer CA, McDonnell DB (1984) Hortic Sci 19:517
3. Bernier G, Kinet JM, Sachs R (1981) The physiology of flowering, vol 1 and 2. CRC Boca Raton, Florida
4. Cameron AC, Raid MS (1983) Sci Hortic (Amst) 19:373
5. Cameron AC, Fenton CAL, Yu Y, Adams DO, Yang SF (1979) Hortic Sci 14:178
6. Carpenter WJ (1975) Hortic Sci 10:605
7. Cathey HM (1964) Annu Rev Plant Physiol 15:271
8. De Greef JA, van Dijck R, De Proft M, Mekers O (1983) Acta Hortic (The Hague) 137:211
9. De Hertogh AH (1985) Holland bulb forcer's guide, 3rd Edn. Int Flower Bulb Center, Hillegorn
10. De Munk WJ, Gijzenberg J (1977) Sci Hortic (Amst) 7:255
11. Fisher JB (1985) In: Halevy AH (ed) Handbook of flowering, vol 1. CRC, Boca Raton, Florida, p 394
12. Gill AM, Groves RH, Noble TR (eds) (1981) Fire and the Australian Biota. Aust Acad Sci Canberra, p 243
13. Goh CJ, Halevy AH, Engel R, Kofranek AM (1985) Sci Hortic (Amst) 26:57
14. Goszcynska D, Rudnicki R (1983) Acta Hortic (The Hague) 141:203
15. Goulston GH, Shearing SJ (1985) Acata Hortic (The Hague) 167:339
16. Halvey AH, Kofranek AM (1976) J Am Soc Hortic Sci 101:658
17. Halevy AH, Mayak S (1981) Hortic Rev 3:59
18. Halevy AH, Shilo R (1970) Physiol Plant 23:820
19. Halevy AH, Shilo R, Simchon S (1970) J Hortic Sci 45:427
20. Hanks GR (1984) Sci Hortic (Amst) 23:379
21. Hanks GR, Rees AR (1983) Sci Hortic (Amst) 21:159
22. Henny RJ (1985) In: Halevy AH (ed) Handbook of flowering, vol 2, CRC, Boca Raton, Florida, p 443
23. Henny RJ (1985) In: Halevy AH (ed) Handbook of flowering, vol 4. CRC, Boca Raton, Florida, p 382
24. Imanishi H (1983) Sci Hortic (Amst) 21:173
25. Imanishi H, Fortanier EJ (1982) Bull Univ Osaka Perf Ser B 34:1
26. Kamerbeek GA, de Munk WJ (1976) Sci Hortic (Amst) 4:101
27. Keren-Paz V, Borochov A (1984) In: Fuchs Y, Chalutz E (eds) Ethylene: biochemical physiological and applied aspects. Nijhoff, The Hague, p 163
28. Kofranek AM (1980) In: Larson RA (ed) Introduction of floriculture. Academic Press, New York, p 3
29. Kofranek AM, Halevy AH, Mor Y, Kubota J (1984) Flor Rev 174(4506):33
30. Lindenbaum S, Ginzburg C, Halevy AH (1975) Ann Bot 39:219
31. Masuda M, Asahira T (1980) Sci Hortic (Amst) 13:85
32. McDaniel GL (1984) Hortic Sci 19:434
33. Mekers O, De Proft M, Jacobs L (1983) Acta Hortic (The Hague) 137:217
34. Moe R, Smith-Ericksen A (1986) Acta Hortic (The Hague) (in press)
35. Mor Y, Halevy AH, Kofranek AM (1984) J Am Soc Hortic Sci 109:494
36. Mor Y, Mishaeli G, Zieslin N, Halevy AH (1977) Ann Rept for 1975/77. Dept Orn Hortic Hebrew Univ Rehovot, p 24
37. Mor Y, Raid MS, Kofranek AM (1984) J Am Soc Hortic Sci 109:866
38. Mor Y, Spiegelstein H, Halevy AH (1981) Physiol Plant 52:197
39. Nanda KK, Krishnamoorthy HN, Krishan L, Anuradha TA (1967) Planta (Berl) 76:367
40. Nickell LG (1982) Plant growth regulators – agricultural uses. Springer, Berlin Heidelberg New York
41. Ohkawa K (1984) Sci Hortic (Amst) 24:379
42. Parups EV (1979) Hortic Sci 14:279
43. Reid MS (1985) Hortic Sci 20:45

44. Sawhney S, Sawhney N (1976) Planta (Berl) 131:207
45. Schipper JA (1982) Neth J Agric Sci 29:173
46. Shillo R (1985) In: Halevy AH (ed) Handbook of flowering, vol 3. CRC, Boca Raton, Florida, p 292
47. Shillo R, Halevy AH (1982) Sci Hortic (Amst) 16:385
48. Shlomo E, Shillo R, Halevy AH (1985) Sci Hortic (Amst) 26:69
49. Sterrett JP (1985) J Am Soc Hortic Sci 110:4
50. Steinitz B, Cohen A (1982) Hortic Sci 17:903
51. Steinitz B, Cohen A, Leshem B (1980) Z Pflanzenphysiol 100:343
52. Systema W (1986) Acta Hortic (The Hague) (in press)
53. Stuart NW (1961) Science, 134:50
54. Stuart NW (1964) Proc 16th Int Hortic Congr 5:58
55. Swart A (1980) Acta Hortic (The Hague) 113:45
56. Swart A, Schipper JA (1982) Hortic Sci 17:905
57. Systema W (1969) Acta Hortic (The Hague) 14:205
58. Tompsett AA (1980) Acta Hortic (The Hague) 109:57
59. Tompsett AA (1985) Grower Feb 21:83
60. Uyemura S, Imanishi H (1983) Sci Hortic (Amst) 20:91
61. Van Meeteren U, De Proft M (1982) Physiol Plant 56:236
62. Veen H (1983) Sci Hortic (Amst) 20:211
63. Widmer RA (1980) In: Larson RA (ed) Introduction to floriculture. Academic Press, New York, p 373
64. Yang SF, Hoffman NE (1984) Annu Rev Plant Physiol 35:155
65. Zieslin N (1985) In: Halevy AH (ed) Handbook of flowering, vol 3. CRC, Boca Raton, Florida, p 287
66. Zieslin N, Halevy AH, Mor Y, Bachrach A, Sapir I (1972) Hortic Sci 7:75
67. Zieslin N, Halevy AH (1976) Z Pflanzenphysiol 77:160
68. Zieslin N, Madori G, Halevy AH (1979) J Exp Bot 30:15
69. Zieslin N, Mor Y, Khayat E, Levy M (1985) Acta Hortic (The Hague) 167:433
70. Zouk H (1986) MS Thesis, The Hebrew University, Rehovot

Interrelations of Plant Growth Substances, Mineral Nutrition and Crop Yield

H. BERINGER[1]

1 Introduction

Synthetic growth regulators are occasionally applied in intensive farming. Most of those used for cereals interfere with gibberellic acid (GA) metabolism, i.e. they reduce shoot length and the risk of yield losses due to lodging. This, however, requires correct timing and dosage and reveals the importance of the plant's developmental stage for the response to the synthetic growth regulators applied [19].

Interrelations between mineral nutrition of plants and growth regulating substances have been repeatedly observed [4, 9, 11, 16]. For this reason only some illustrative examples will be given in this chapter. They concern the role of the root in mineral nutrition and plant water relationships with particular regard to the development of the storage tissue, which represents the economic yield. In addition, these examples were chosen to emphasize the need for a greater consideration of biophysical aspects in the hormonal regulation of growth.

2 Effect of Synthetic Growth Regulators on Root Growth

In several screening experiments synthetic growth regulators stimulated root growth or influenced root morphology [12]. A recently released inhibitor of GA synthesis, Tetcyclacis ®, has been found to reduce shoot length and to increase root weight of rice seedlings. Thus, transplantation stress was decreased, but final yield remaind unaffected [17]. Sprays of chlorocholine chloride (CCC), also an antagonist of GA, stimulated root growth and root density of wheat, whereas nutrient uptake and grain yield were not higher than in the controls [2, 3].

These results are in line with the finding that "root growth is at least adequate to impose no limitations on productivity when the rooting medium is uniformly favourable" [18]. In the field, however, root development may be mechanically impeded as well as be affected by localized availability of nutrients and water or by inoculations by soil microorganisms. Such effects are likely to change the balance of endogenous phytohormones [4, 9]. Considering that synthetic growth regulators

1 Landwirtschaftliche Forschungsanstalt Büntehof, Bünteweg 8, D-3000 Hannover 71, FRG

Plant Growth Substances 1985
(ed. by M. Bopp)

may increase the root-shoot ratio as indicated above, they likely affect the root more as a source of endogenous hormones than in its nutrient absorbing capacity [4, 11].

3 The Root as a Site of Nutrient and Growth Regulator Interaction

3.1 Initiation and Growth of the Potato Tuber

The number of tubers per plant is an important component of potato yield. Tuber initiation depends on temperature, photoperiod and mineral nutrition. Repeated short-term discontinuation of N supply in the nutrient solution leads each time to the initiation of a new tuber, thus producing chain like tubers at the stolon tip [6, 15]. The receptor for this signal seems to reside in the root, where N stress reduces cytokinin (CYT) activity in stem exudate and shoot. As a consequence, the shoot slightly senesces and shows a decline in GA, but an increase in ABA concentration. Thus, more abscisic acid ABA, relative to GA, is translocated to the stolon, which initiates the new tuber [6, 15]. This hormonal approach is supported by the finding that reduced tuber initiation under long-day or high-temperature conditions is accompanied by higher GA concentrations [10]. Similarly, high soil temperature effects can be mitigated by exogenous GA applications [6]. Once the tuber is initiated, the rate of tuber growth is also affected by hormonal changes. It is positively correlated with the indoleacetic acid (IAA) concentration in the tuber, but negatively correlated with ABA, and as soon as the ABA:IAA ratio exceeds 0.7, the tuber growth rate becomes very low [8].

In crops with vegetative storage tissues, like potato and sugar beet yield seems to be more limited by the source than by the sink. A more recent hypothesis even suggests that the distribution of assimilates in the turgor-regulated phloem transport is decisive and that in this respect the lateral exchange of K^+ between sieve tubes and xylem is of tremendous importance [7]. This gives tise to the question to which extent interactions between the nutrient status and growth substances are also affected by osmotic relationships. These influence membrane integrity and thus the recognition of environmental or hormonal stimuli.

The growth rate of a cell can be expressed by the following formula:

$$\frac{dV}{dt \cdot Vo} = \frac{Eg \cdot C}{Eg + C} \cdot (\Delta\psi_w + \Delta\psi_p).$$

This means that the volume increase (dV) of a cell (Vo) in time (dt) depends on the cell wall extensibility (Eg), the hydraulic conductivity (C), the water potential gradient between the cell and its surrounding ($\Delta\psi_w$) and the turgor pressure, which exceeds the threshold turgor needed for cell extension ($\Delta\psi_p$). In view of these biophysical parameters, the yield differences between the two potato cultivars reported in Table 1 may be due to K-related differences in $\Delta\psi_p$ of tubers and to varietal differences in C. Admittedly this is a speculation, which can even be extended by assuming that tubers with higher water content may be physiologically

Table 1. Effect of K supply on yield of two potato cultivars of comparable maturity (Forster, pers. comm.)

	cv	K_1	K_1	K_3
Tuber yield	A	581	728	697
g fresh wt./plant	B	730	861	872
Tuber-starch	A	116	126	111
g/plant	B	120	140	124
Tuber-moisture	A	2.65	3.15	3.48
g/g dry wt.	B	3.24	3.46	3.98

younger and may, therefore, contain a higher IAA:ABA ratio [8]. Evidence for varietal differences in the hydraulic conductivity is provided by studies on cotton, N and P deficiency were found to limit growth not by reduced $\Delta\psi_p$ of leaves, but by decreased C. In addition, the CYT/ABA ratio was decreased, which is indicative for accelerated leaf senescence [13, 14].

3.2 Effect of Nutrient and Water Supply on Hormonal Control of Seed Yield

In developing seeds of cereals the maxima of the individual phytohormones generally occur in the sequence: CYT, followed by GA and IAA and finally by ABA. They can be modified by the quantitiy and distribution of fertilizer (primarily N) and by

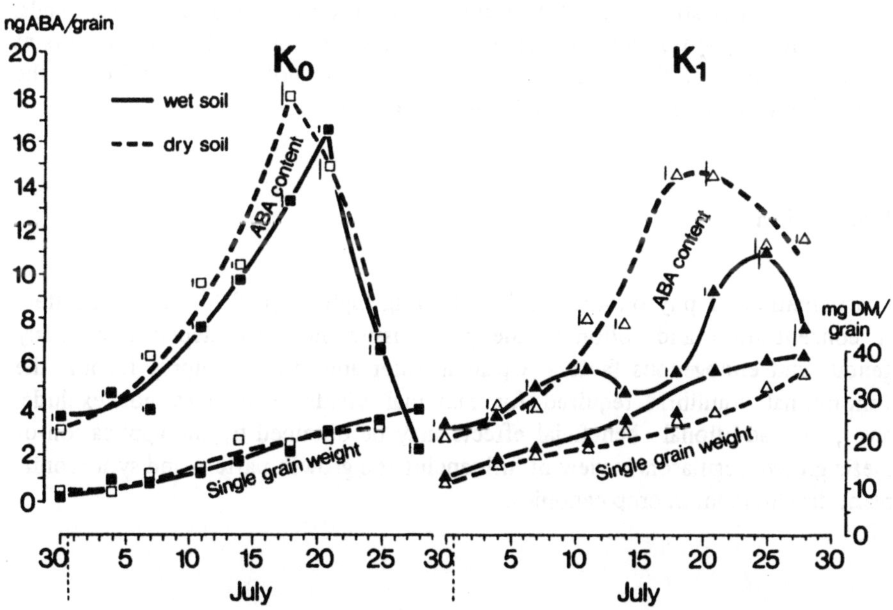

Fig. 1. Influence of 1-week water-stress periods on ABA content and weight of barley grains [5]

Table 2. Effect of water shortage on grain parameters of two wheat cultivars (Schacherer and Rademacher, pers. commun.)

CV	Treatment	IAA[a] ppb fresh grain	ABA[a] ppb fresh grain	Cells[b] 10^{-3} grain	Dry wt.[c] mg/grain
A	Control	10	50	269	46
	Stress	284	341	299[d]	36[d]
B	Control	56	35	246	30
	Stress	202	808	267	30

a–c Measurements were made 8 (a) and 20 days after anthesis (b) and at maturity (c)
d Significant at $P = 0.05$

the water supply [11]. The experiment in Fig. 1 was designed by considering the participation of K^+ and ABA in osmoregulation. It shows that single grain weight of K_1 plants in well-watered soil is higher and that the peak of ABA is smaller and delayed as compared to K_0. When water ist withheld for 1 week before the dates indicated on the abscissa, ABA increase starts earlier, associated with a decline in grain size. This applies especially to K_1 plants because of their greater leaf area and faster depletion of soil moisture.

Water stress induces many and very complex physiological changes (Table 2). When water supply of two wheat cultivars differing genetically in grain size was reduced only during the first 10 days after anthesis, a remarkable increase in ABA as well as in IAA was noticed. Contrary to the expectation of a lower rate of cell division under water stress and ABA accumulation, the number of endosperm cells, i.e. the storage capacity/grain, was raised. Nevertheless, the single-grain weights at maturity indicate that this higher storage capacity was not utilized, although the plants were rewatered after the stress like in the control.

4 Conclusion

The participation of phytohormones in achieving high crop yields is uncontested. Their concentration and action in the plant tissue, however, depends on many exogenous and endogenous factors. Optimal water and nutrient supply favours the phytohormonal equilibria required for maximal yields. This does not exclude, however, that additional beneficial effects may be obtained by the application of synthetic growth regulators in view of influencing the growth pattern and synchronizing plant development in crop canopies.

References

1. Bradford KJ, Hsiao TC (1982) In: Physiological plant ecology II. Springer, Berlin Heidelberg New York, p 264
2. Bragg PL, Rubino P, Henderson FKG, Fielding WJ, Cannell RQ (1984) J Agric Sci 103:257
3. Cooke DT, Hoad GV, Child RD (1983) In: Jackson MB, Stead AD (eds) Growth regulators in root development. Br Plant Growth Regul Group Monograph 10
4. Feldman LJ (1984) Annu Rev Plant Physiol 35:223
5. Haeder HE, Beringer H (1981) J Sci Food Agric 32:552
6. Krauss A (1981) Potato Res 24:435
7. Lang A (1983) Plant Cell Environ 6:683
8. Marschner H, Sattelmacher B, Bangerth F (1984) Physiol Plant 60:16
9. Matthysse AG, Scott TK (1984) In: Scott TK (ed) Encyclopedia of plant physiology new ser, vol 10. Hormonal regulation of development II. Springer, Berlin Heidelberg New York, p 219
10. Menzel CM (1983) Ann Bot 52:65
11. Michael G, Beringer H (1980) Proc 15th Colloq Int Potash Inst Berne, p 85
12. Rademacher W, Jung J, Hildebrandt E, Graebe JE (1983) In: Bangerth F (ed) Regulation des Phytohormongehaltes und seine Beeinflussung durch synthethische Wachstumregulatoren. Ulmer, Stuttgart, p 132
13. Radin JW, Boyer JS (1982) Plant Physiol (Bethesda) 69:771
14. Radin JW, Eidenbock MP (1984) Plant Physiol (Bethesda) 75:372
15. Sattelmacher B, Marschner H (1979) Potato Res 22:49
16. Schilling G (1985) In: Podlesak W, Grün M (eds) Mikronährstoff-Forschung, Akad Landwirtschaftswissenschaften der DDR, Jena, p 61
17. Schott PE, Knittel H, Klapproth H (1984) In: Ory R, Rittig F (eds) Bioregulators: chemistry and uses Symposium series. ACS, p 45
18. Scott RR (1981) Proc 16th Colloq Int Potash Inst Berne, p 23
19. Trewavas AJ (1982) Physiol Plant 55:60

Examples of the Use of Natural and Synthetic Plant Growth Substances in Cereal Breeding

S.A. QUARRIE and M.D. GALE[1]

1 Introduction

As the limits to cereal yield improvement through redistribution of dry matter are approached, a greater awareness has developed of the need to extend selection criteria beyond visible morphological and developmental characteristics. There is a wealth of literature on the possible roles of endogenous growth substances in many aspects of plant growth and development, and genetic variation in plant hormone systems has been reported in many species (review [5]). However, in only a few instances has this variation been studied specifically as a means of breeding improved crop varieties.

Responses to endogenous hormones may be modified in various ways, among which are control of hormone synthesis and control of the response system. The two examples described here illustrate these two alternatives as applied to wheat: the effect on height and yield of major genes, which restrict the plant's sensitivity to gibberellin (GA) and the effect of variation in endogenous abscisic acid (ABA) concentration on water use and yield.

The results of previous experiments, using varietal comparisons to study the pleiotropic effects of both GA-insensitive dwarfing genes and variation in ABA accumulation, have been confounded by variation in the genetic background of the varieties tested [2, 3]. In the experiments described here, plants were either isogenic, as for the *Rht* genes or random genetic lines from a single cross selected for ABA differences.

2 The Effects of GA Insensitivity on Height and Yield in Winter Wheat

The Norin 10 genes, *Rht1* and *Rht2*, were originally sought by breeders as a source of stiff straw to circumvent lodging under high fertility conditions. However, the worldwide success of these genes has probably been due to their positive effects on yield even in the absence of lodging.

GA is characteristically a potent stimulator of elongation growth, but wheat plants containing *Rht1* or *Rht2* are characterised by insensitivity to applied GA.

1 Plant Breeding Institute, Maris Lane, Trumpington, Cambridge, CB2 2LQ, Great Britain

Plant Growth Substances 1985
(ed. by M. Bopp)

In addition, levels of endogenous GA are about ten times higher than normal (J.R. Lenton, pers. comm.) and the primary gene products are thought to cause a lesion in the active turnover of the hormone (review [7]).

A seedling GA-response test [4], which allows the genes to be unequivocally identified in segregating families, has been the key to much of our present knowledge. Effectively, GA-insensitivity has transformed plant height, a complex quantitative genetic trait, into a simply classified, qualitative character.

The seedlings test is being used by breeders to recognise dwarfs in early generations, providing considerable savings in time and resources. The test has also facilitated the genetic characterisation and chromosomal mapping of *Rht1* on chromosome 4A and *Rht2* on 4D. Most importantly, the test has allowed the production of precise genetic stocks, including isogenic series of *Rht1* and *Rht2*. Trials with these genotypes have provided clear evidence of the direct effects of the genes upon yield and are beginning to give answers to important questions concerning interactions with agricultural and edaphic factors.

Yield. The results of trials carried out in 1984 are summarised in Table 1. It is clear that the effect of the *Rht* genes is to increase yield by more than 20%. This is achieved by increased numbers of slightly smaller grains within the spikelets.

Drought. The relative performance of semi-dwarfs and talls under water-stressed conditions is a particularly important component of overall adaptability of *Rht* genotypes. Experiments conducted under mobile plot shelters with the same genotypes discussed above (Table 1) have shown that the yield reductions obtained with both pre- and post-anthesis droughts are similar for both *rht* and *Rht* lines (Fig. 1). Water-extraction rates were also very similar for *rht* and *Rht* lines [8], indicating

Table 1. Yield and related characters in tall and semi-dwarf isogenic lines

	Tall	P [a]	Semi-dwarf	sed
Height (cm)	125	***	110	0.8
Grain yield (kg m^{-2})	0.94	***	1.20	0.018
Biomass (kg m^{-2})	2.68		2.77	0.096
Grain No. ($m^{-2} \times 10^3$)	16.7	***	22.9	0.41
1000 grain wt. (g)	56.2	***	52.4	0.68
Tiller No. (m^{-2})	488		486	18.6
Spike yield (g)	2.25	**	2.76	0.239
Spike grain No.	37.0	***	48.5	2.23

Note: Data from Gale and Salter (unpubl.). Plot data from 4 x 1.16 m drills meaned over Huntsman, Bersée and Widgeon isogenic series. Semi-dwarf data meaned over *Rht1* and *Rht2* lines. Individual spike data from sampled ears within plots

[a] Significance of differences between *rht* and *Rht* means; *, **, *** = less than 5%, 1% and 0.1%, respectively

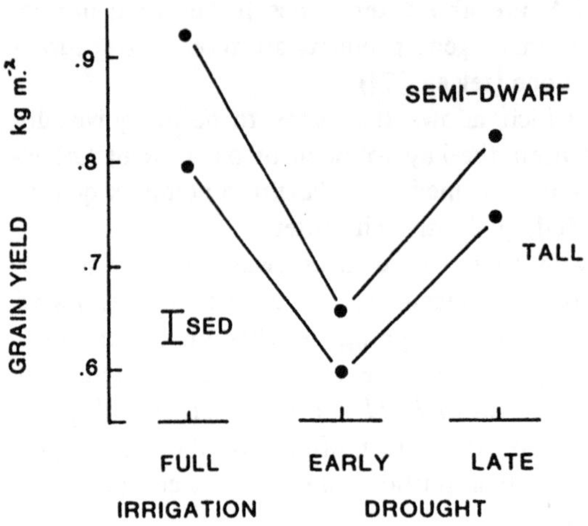

Fig. 1. Grain yields of *Rht* and *rht* lines grown with different water availabilities

similar rooting patterns. This lack of interaction between dwarfing gene and drought treatment shows that Norin 10 semi-dwarfs should be widely adapted.

CCC. Both *Rht* genes and chlormequat appear to reduce the physiological availability of endogenous GA [1]. If this is so, then it may be expected that many of the agronomic benefits of CCC application to wheat could be restricted in semi-dwarf varieties.

Experiments to test this hypothesis, employing isogenic lines of winter wheats, indicate that this is indeed so. With CCC treatment the proportionate reduction in height and beneficial effects on yield components were much less in the *Rht* lines. The results for the most effective CCC treatment in the experiment are shown in Table 2. This may indicate that the cost effectiveness of CCC application is less with dwarf varieties. The question as to whether the growth regulator significantly improves straw strenght in semi-dwarfs is still open because all the plants in these experiments were supported to avoid lodging.

Table 2. The effect of CCC on height and ear yield components in tall and semi-dwarf isogenic lines

	Tall			Semi-dwarf			
	Control	CCC	Difference (%)	Control	CCC	Difference (%)	LSD (5%)
Height (cm)	113	102	– 10% ***	88	86	– 2%	2.2
Spike yield (g)	2.47	2.71	+ 10% *	2.46	2.47	0	0.17
Grain No. per spike	44	51	+ 16% ***	54	55	+ 2%	4.1
Grain wt. (mg)	55	54	– 2%	46	45	– 2%	1.4

Note: Data derived from Gale and Youssefian [6]. Plots were 1x1.16 m, at 147 plants m^{-2}. CCC was applied at 3.1 l ha^{-1} at Zadocs' growth stage 3.3. Results are means of Huntsman and Widgeon isogenic series meaned over *Rht1* and *Rht2* lines. Significance of differences as in Table 1

The precise mechanism of GA-insensitive dwarfism is still not clearly understood. However, it is likely that the gene products interfere with the normal active turnover of GA in vegetative tissues, possibly by direct effects on a receptor mechanism. Whatever their mode of action, the genes do not affect plant developmental events (S. Youssefian, pers. comm.) and reduce the size only of vegetative plant parts.

In any event, further investigation to identify the gene product, as a potential PGR, and to isolate and clone these genes is of paramount importance. It is possible that the wheat GA-insensitive genes could be of great value when introduced into other species in which the economic yield is obtained from reproductive structures.

3 The Effects of Variation in ABA Accumulation on Water Use and Yield of Spring Wheat

Endogenous ABA levels are very sensitive to plant water status and increase as plants become stressed. ABA is known to be a potent anti-transpirant and is also an effective inhibitor of leaf and shoot growth in wheat [11]. Consequently, endogenous ABA is potentially important in controlling water loss in the short-term through the stomata and in the long-term by modifying leaf areas. We were, therefore, interested in testing the effect on water-use efficiency and yield of genetic variation in the drought-induced accumulation of ABA.

Random lines from a cross between the breeding line TW 269/9 and Highbury were selected on the basis of ABA accumulation in a detached leaf test [10]. At the F_6 generation, 20 high- and 20 low-ABA lines, showing stable differences of ca. 80% in ABA accumulation, were tested under the mobile shelters with fully irrigated, pre- and post-anthesis drought treatments as above.

Table 3. Grain yield and water use of high- and low-ABA lines derived from a cross between TW269/9 and Highbury grown with different water availabilities

		Grain yield ($t\ ha^{-1}$)	Difference[a] (%)	Water use (mm)	Water use efficiency for grain yield[a] ($t\ ha^{-1}\ 100\ mm^{-1}$)
Full irrigation	High	7.53		320	2.35
			+ 5.5*[a]		
	Low	7.14		322	2.22
Early drought	High	4.92		216	2.28
			− 2.2	*	
	Low	5.03		224	2.25
Late drought	High	6.39		216	2.96
			+ 6.1*		***
	Low	6.02		220	2.74

[a] Significance of differences between high and low lines as in Table 1

Table 4. Grain yields of four high- and four low-ABA lines derived from a cross between TW269/9 and Highbury at two sites

	Gran yield ($t\ ha^{-1}$)		
	High ABA	Low ABA	Difference[a] (%)
Dry site	8.22	7.08	+ 16.1*
Damp site	9.30	8.47	+ 9.8**

[a] Significance of differences between high and low lines as in Table 1

High-ABA lines used slightly less water than low-ABA lines in each of the treatments and had higher water-use efficiencies (Table 3), and this was associated with significantly higher yields (5–6%) for high-ABA lines in the fully irrigated and post-anthesis drought treatments. This was due primarily to larger grains (13–15%) in the high-ABA lines in these treatments. The more efficient water-use of high-ABA lines was associated with slightly lower conductances (6–7%), confirming a previous finding [11], and smaller leaf area indices [9]. Both of these effects of selecting for high ABA accumulation were consistent with the effects of applied ABA on stomatal conductance and leaf growth in wheat [11].

Four high- and four low-ABA lines from the same cross were tested on a separate occasion in drilled plots at a wet and a dry site. Again the high-ABA lines outyielded the low-ABA (Table 4) with the yield advantage being more pronounced at the dry site where overall ABA production was expected to have been higher.

Although it is possible that the water use and yield differences between high- and low-ABA lines were the results of inadvertent co-selection in an early generation for an independently controlled character, such as leaf size, these results are consistent with the expected pleiotropic effects of increased endogenous ABA levels viz. lower stomatal conductance and a higher water-use efficiency, leading to a higher yield when water availability is limited. It remains to be seen whether such yield advantage of high-ABA levels would be present consistently in different genetic background and with plants grown in different environments and water regimes.

4 Conclusions

The experiments described here illustrate the value of working with genetically defined stocks so that confounding effects of background variation are either eliminated or taken into account. GA-insensitivity in wheat in the form of the *Rht1* and *Rht2* dwarfing genes is shown not only to reduce height, but also to confer a consistent and significant yield advantage in a range of genetic background and with different water regimes. High stress-induced ABA accumulation in wheat also appears to be beneficial for improved water-use efficiency and, in a number of environments, increasing yield.

References

1. Dicks JW (1979) Joint SCI/BPGRG Monograph 4:1
2. Fischer JE (1973) Plant Sci 53:7
3. Durley RC, Kannangara T, Seetharama N, Simpson GM (1983) Can J Plant Sci 63:131
4. Gale MD, Gregory RS (1977) Euphytica 26:733
5. Gale MD, Hanson PR (1982) In: McLaren JS (ed) Chemical manipulation of crop growth and development. Butterworth, London, p 425
6. Gale MD, Youssefian S (1983) Proc 6th Int Wheat Genet Symp, Kyoto 1983:271
7. Gale MD, Youssefian S (1985) In: Russell GE (ed) Progress in Plant Breading, vol 1. Butterworth, London, p 1
8. Innes P, Blackwell RD (1985) Plant Breeding Inst Annu Rep 1984, p 107
9. Innes P, Blackwell RD, Quarrie SA (1984) J Agric Sci 102:341
10. Quarrie SA (1981) Plant Cell Environ 4:147
11. Quarrie SA (1982) In: Wareing PF (ed) Plant growth substances 1982. Academic Press, London, p 609

Subject Index

Plant Cell Reports

ISSN 0721-7714 Title No. 299

Managing Editors for European Countries:
K. Hahlbrock, Cologne

For Countries outside Europe:
O. L. Gamborg, Belmont

Plant Cell Reports is a journal specifically devoted to the rapid publication of short communications by investigators in all fields of plant cell research. Scientists working in cell biology, agronomy, botany, plant physiology, phytopathology, plant biochemistry, molecular biology, plant genetics, and horticulture will find the very forefront of plant cell research contained in its articles. Quickly becoming one of the leading journals in its field, **Plant Cell Reports** gives the reader the opportunity to become informed of new developments as they happen.

Fields of interest: All aspects of plant cell and plant cell culture research, e. g., physiology, cytology, biochemistry, molecular biology, genetics, phytopathology, and morphogenesis including plant regeneration from protoplasts, cells, tissues, and organs.

Abstracted/Indexed in: Biosis, CAS, Current Contents.

Springer-Verlag
Berlin Heidelberg
New York Tokyo

Springer

MIX
Papier aus verantwortungsvollen Quellen
Paper from responsible sources
FSC® C105338

If you have any concerns about our products,
you can contact us on
ProductSafety@springernature.com

In case Publisher is established outside the EU,
the EU authorized representative is:
Springer Nature Customer Service Center GmbH
Europaplatz 3, 69115 Heidelberg, Germany

Printed by Libri Plureos GmbH
in Hamburg, Germany